Energy Materials

Energy Materials: A Circular Economy Approach emphasizes the engineering scalability of a circular economy approach to development and use of energy materials. It focuses on waste minimization and its valorization, recycling and reuse, and emerging sustainable materials and technologies. It offers a view of the eco-friendly energy materials and state-of-the-art technologies required for production of these materials in the process industry and manufacturing sectors.

- Covers fundamentals, concepts, and current initiatives within the circular economy
- Outlines technologies and materials with specific applications for energy systems and societal benefits
- Focuses on detailed aspects of process scale-up, kinetics, and application of circular economy in waste utilization and valorization
- Discusses technologies, processing methods, and production of materials related to fuel cells, carbon capture, catalysis, functional materials, nanotechnology, biofuels, solar energy, and fine chemicals
- Details topics related to synthesis and application of energy materials, their recycle, reuse, and life cycle.

This book is aimed at researchers and professional engineers and scientists working in chemical, materials, energy, and environmental engineering, as well as materials chemistry.

Emerging Materials and Technologies

Series Editor: Boris I. Kharissov

The *Emerging Materials and Technologies* series is devoted to highlighting publications centered on emerging advanced materials and novel technologies. Attention is paid to those newly discovered or applied materials with potential to solve pressing societal problems and improve quality of life, corresponding to environmental protection, medicine, communications, energy, transportation, advanced manufacturing, and related areas.

The series takes into account that, under present strong demands for energy, material, and cost savings, as well as heavy contamination problems and worldwide pandemic conditions, the area of emerging materials and related scalable technologies is a highly interdisciplinary field, with the need for researchers, professionals, and academics across the spectrum of engineering and technological disciplines. The main objective of this book series is to attract more attention to these materials and technologies and invite conversation among the international R&D community.

Shape Memory Polymer Composites
Characterization and Modeling
Nilesh Tiwari and Kanif M. Markad

Impedance Spectroscopy and its Application in Biological Detection
Edited by Geeta Bhatt, Manoj Bhatt and Shantanu Bhattacharya

Nanofillers for Sustainable Applications
Edited by N.M. Nurazzi, E. Bayraktar, M.N.F. Norrrahim, H.A. Aisyah, N. Abdullah, and M.R.M. Asyraf

Chemistry of Dehydrogenation Reactions and its Applications
Edited by Syed Shahabuddin, Rama Gaur and Nandini Mukherjee

Biosorbents
Diversity, Bioprocessing, and Applications
Edited by Pramod Kumar Mahish, Dakeshwar Kumar Verma and Shailesh Kumar Jadhav

Principles and Applications of Nanotherapeutics
Imalka Munaweera and Piumika Yapa

Energy Materials
A Circular Economy Approach
Edited by Surinder Singh, Suresh Sundaramurthy, Alex Ibhadon, Faisal Khan, Sushil Kumar Kansal, and S.K. Mehta

For more information about this series, please visit: www.routledge.com/Emerging-Materials-and-Technologies/book-series/CRCEMT

Energy Materials

A Circular Economy Approach

Edited by Surinder Singh, Suresh Sundaramurthy,
Alex Ibhadon, Faisal Khan, Sushil Kumar Kansal,
and S.K. Mehta

CRC Press
Taylor & Francis Group
Boca Raton London New York

CRC Press is an imprint of the
Taylor & Francis Group, an **informa** business

Designed cover image: Shutterstock

MATLAB® and Simulink® are trademarks of The MathWorks, Inc. and are used with permission. The MathWorks does not warrant the accuracy of the text or exercises in this book. This book's use or discussion of MATLAB® or Simulink® software or related products does not constitute endorsement or sponsorship by The MathWorks of a particular pedagogical approach or particular use of the MATLAB® and Simulink® software.

First edition published 2024
by CRC Press
2385 NW Executive Center Drive, Suite 320, Boca Raton FL 33431

and by CRC Press
4 Park Square, Milton Park, Abingdon, Oxon, OX14 4RN

CRC Press is an imprint of Taylor & Francis Group, LLC

© 2024 selection and editorial matter, Surinder Singh, Suresh Sundaramurthy, Alex Ibhadon, Faisal Khan, Sushil Kumar Kansal, and S.K. Mehta; individual chapters, the contributors

Library of Congress Cataloging-in-Publication Data
Names: Singh, Surinder, (Chemical engineer), editor. | Suresh, Sundaramurthy, editor. |
    Ibhadon, Alex, editor. | Khan, Faisal, (Chemical engineer), editor. | Kansal, Sushil Kumar, editor. |
    Mehta, S. K., editor.
Title: Energy materials : a circular economy approach / edited by Surinder Singh, Suresh Sundaramurthy,
    Alex Ibhadon, Faisal Khan, Sushil Kumar Kansal, and S.K. Mehta.
Other titles: Energy materials (CRC Press : 2024)
Description: First edition. | Boca Raton, FL : CRC Press, 2024. | Series: Emerging materials and
    technologies | Includes bibliographical references and index.
Identifiers: LCCN 2023045575 (print) | LCCN 2023045576 (ebook) | ISBN 9781032217260 (hbk) |
    ISBN 9781032217277 (pbk) | ISBN 9781003269779 (ebk)
Subjects: LCSH: Energy storage—Materials. | Electric batteries—Materials. | Green chemistry. |
    Waste minimization—Economic aspects. | Circular economy.
Classification: LCC TK2910 .E53 2024 (print) | LCC TK2910 (ebook) | DDC 660.028/6—
    dc23/eng/20231208
LC record available at https://lccn.loc.gov/2023045575
LC ebook record available at https://lccn.loc.gov/2023045576

ISBN: 978-1-032-21726-0 (hbk)
ISBN: 978-1-032-21727-7 (pbk)
ISBN: 978-1-003-26977-9 (ebk)

DOI: 10.1201/9781003269779

Typeset in Times
by Apex CoVantage, LLC

Contents

*Amandeep Kaur, Sandeep Singh, Niraj Bala, Surinder Singh,
and Sushil Kumar Kansal*

Manjeet Singh, Kulbir Singh, Surinder Singh,
Arashdeep Singh, and Harjot Gill

Navneet Kaur Bhullar and Ashok Prabhakar

Editor Biographies

Surinder Singh (PU, Chandigarh, India) Dr. Surinder Singh is working as Faculty Member at Dr. S. S. Bhatnagar University Institute of Chemical Engineering and Technology, Panjab University, Chandigarh, since 2013 and has been involved in teaching and research since 2001. He earned his PhD from University School of Chemical Technology, GGSIPU University Delhi, and master's in chemical engineering from Panjab University, Chandigarh. He did his B. Tech in chemical engineering in the year 2000. He also possesses a master's of science in information technology (2005) and postgraduate diploma in ecology and environment (2002). He has guided about 30 postgraduate candidates and currently is guiding two PhD candidates. He has a Scopus h-index of 21 and i-10 index of 39 (Google Scholar) with 75 national and international publications in reputed journals and book chapters. He has five international utility patents/patents to his credit and has filed two Indian patents. He has presented his research papers in the United States (2009), Canada (2011), and Singapore (2016). He received the Certificate of Merit award for his research paper during WCECS-2009 in the United States. He has delivered around 20 invited talks related to the themes of environment, chemical, and food technology in reputed institutes.

He has organized two international conferences and one International GIAN Workshop with collaborating faculty from Memorial University, Newfoundland, Canada, and many national conferences. He is a reviewer of many reputed international journals including *Journal of Colloid and Interface Science* (JCIS); *Food Bioscience; Nanotechnology for Environmental Engg, Foods, Separations, processes.* He is presently secretary of Catalysis Society of India (CSI), Chandigarh Chapter. He is a life member of various professional bodies. He has organized many technical workshops, symposia, conferences, industry lecture series, and start-up workshops. He has active collaborations with industry and government research laboratories. His main areas of research include separation technology, chemical engineering, environmental engineering, plant bioactivities and nutraceuticals, food technology, nanotechnology and heterogeneous catalysis, renewable energy technologies, and modeling and simulation. He is also actively involved in technology development and transferand in areas such as edible films for enhancing shelf life of foods, sustainable waste water treatment, and tapping of renewable energy using microgrids.

Suresh Sundaramurthy (NIT Bhopal, India) Prof. Suresh Sundaramurthy is a Faculty and Former Head of Department, Chemical Engineering, Maulana Azad-National Institute of Technology (MA-NIT) Bhopal, India. He received a PhD in chemical engineering from Indian Institute of Technology Roorkee, India. Prof. Suresh was the recipient of a Post-Doctoral Fellow award by IUSSTF, Govt. of India. At that time, he was associated at the City University of New York

(CUNY), United States. He was also the recipient of Visiting Faculty and Researcher awarded by Govt. of India. During this award, he has deputed to Asian Institute of Technology (AIT) Thailand and International Centre for Materials Science (JNCASR), Bengaluru, India. He has also worked at Pondicherry University and Indian Institute of Technology Kanpur, India.

He has supervised nine PhD students, 55 Mtech/MSc/MDS students, 250 Btech/ project students, and several students along with co-supervisors. He has more than 120 research journal publications in the areas of environmental biotechnology, new sorbents and catalysts, reactor design, waste-to-energy, nanoengineered materials, biofuels and intermediates from CO_2 and biomass decarbonization, energy transition, technical textiles, gas separation, air pollution abatement, sensing of toxic gases, process safety, hazard and disaster management, and more than 66 presentations/proceedings, several of which are co-authored. He has undertaken 16 R&D and consultancy projects from ten agencies. In addition to this credential, he has published and been granted two Indian patents, four are under processing, and one is technology synchronized.

He has conducted 14 seminar/workshops, ten faculty development programs, 11 national and international conferences in collaboration with National Mission on Education through ICT (MoE, Govt. of India) and IIT Bombay, SCHEMCON-2021 (IIChE, Kolkata), and GIAN course (IIT Kharagpur), WeenTech, UK, SusTanCon, United States. He has guest edited special issues published by Springer and Elsevier (ScienceDirect) including *Environmental Science and Pollution Research* and *Carbon Capture Science and Technology*. In recognition of his research contributions, he has received a number of awards and honors including National Innovation-Rashtrapati Bhavan, office of the President of India, Young Scientist Award; Visiting Research Fellowship; the Prof. R.C. Singh Medal; and IEI Young Engineers Award. He has been invited for special lectures at national and international platforms in the United States, Singapore, UAE, Thailand, and Malaysia. He has also been nominated as an expert member in organizations such as National Clean Air Programme of MoEF&CC, GoI, M.P. Pollution Control Board, MP Animal & Poultry Development Corporation, and Ujjain Engg. College.

Alex Ibhadon (Hull, UK) Alex Ibhadon, PhD Cchem FRSC MIM SFHEA, Cert Ed., MSPE Professor of Decarbonization, Sustainability and Reactor Engineering.

Dr. Ibhadon graduated with a PhD in physical polymer chemistry from the University of Birmingham, United Kingdom, before a lectureship appointment in chemistry at the University of Hull in 1996. While at the Centre for Environmental and Marine Sciences, he was Senior Lecturer in Environmental Chemistry for over ten years, teaching courses in environmental monitoring and analysis, pollution and toxicology, and environmental and soil chemistry. He is a Fellow of the Royal Society of Chemistry (Cchem), Senior Fellow of the UK Higher Education Academy, and Member of the Institute of Materials. He was the Director of the School of Environmental Sciences (2015–2016), Programme Director of Chemical

Engineering (2017–2022), University of Hull; and a Visiting Professor in Catalysis at Panjab University, India.

He has received more than £7.08 million in research funding (22 grant awards) from the UK Royal Society, UK Engineering and Physical Sciences Research Council (EPSRC), Newton, EU, NetZero Innovation and the Commonwealth and is a co-founder of a spin-out company between the University of Hull and the University of Warwick. He has published over 76 articles in leading journals in the area of catalysis, materials and chemical synthesis and has made over 94 national and international conference presentations. He is a reviewer for several organizations including the EU, EPSRC, UKRI, Netherlands, Estonia, Russia, and Israel, and several high-impact journals such as *Applied Catalysis B, Energy and Environmental Science, ChemCatChem, Plasma Science and Technology,* and *American Chemical Society.*

He is the Editor of *Frontiers in Heterogeneous Catalysis, Catalyst and Synthesis,* among others. He is an advisory board member of the EPSRC Centre for Energy Systems Integration (CESI/HI-ACT) and a consultant to many chemical and energy industries with contributions in hydrogen and ammonia generation as well as in low-carbon materials (LCM) for applications in construction and energy storage.

He coordinated the EU project on Microwave, Acoustic and Plasma Synthesis (2013–2016) that was described by Retell Publications as a "research and innovation success story." An extensive array of strategic academic and industrial partnerships in the UK, EU, United States, and Asia have enabled him to make significant international reach and impact in energy generation as well as contributing to research projects enabling some countries in the global South to be more resilient in the face of global challenges relating to water and sanitation, antimicrobial technology, environmental monitoring, renewable and sustainable energy access.

Faisal Khan (Texas A&M, United States) Dr. Faisal Khan is Mike O'Connor II Chair and Interim Department Head & Professor, Department of Chemical Engineering at Texas A&M University, United States. He is also the Director of Mary Kay O'Connor Process Safety Center and the Director of Ocean Energy Safety Institute. His research interests include safety and risk engineering, asset integrity assessment management, process modeling and system design, and environment system modelling and management.

He is a former Professor and Canada Research Chair (Tier I) of Offshore Safety and Risk Engineering at Memorial University of Newfoundland, Canada. He founded the Centre for Risk Integrity and Safety and Engineering, which has over 100 research members. His research interests include offshore safety, drilling safety, extreme event modelling, asset integrity, and risk engineering. He is the recipient of many national and international awards. He is actively involved with multinational oil and gas industries on the issue of safety, risk, and asset integrity. In 2006, he spent eight months as a risk and integrity expert with Lloyd's Register (UK), a risk management organization. He also served as Safety and Risk Advisor to the Government of Newfoundland, Canada. He continues to be a subject matter expert for many organizations. From 2008 to 2010, he visited Qatar University and Qatargas LNG Company as Process

Safety and Risk Management Research Chair. From 2012 to 2014, he served as a Visiting Professor of Offshore and Marine Engineering at Australian Maritime College (AMC), University of Tasmania, Australia, where he led the development of an offshore safety and risk engineering group and an initiative of global engagement with many international institutions. He is a Fellow of the Canada Academy of Engineers. He has mentored 86 PhDs and 87 master students. He has co-authored over 800 peer-reviewed journal publications and has an h-index of 99. He is the recipient of many international awards. He is Editor-in-Chief of the *Journal of Process Safety & Environmental Protection and Safety in Harsh Environments.*

He has found traditional safety and risk assessment techniques do not adequately capture the hazards and risk factors faced by complex process designs and operations in the variable, dynamic conditions experienced in today's system. His team is establishing a lab-scale plant to test and evaluate dynamic risk assessment models and to verify which models provide a reliable indication of process safety performance. Their goals are to apply the knowledge gained to real-life industrial applications, such as detection systems; to educate and train highly qualified personnel to help advance the mission of sustainable, safe, and green development of natural resources and to improve risk management.

Sushil Kumar Kansal (PU, Chandigarh, India) Sushil Kumar Kansal is a Professor of Chemical Engg. at Panjab University, Chandigarh. Presently he is also Dean of International Students at PU, Chandigarh. He has more than 23 years of teaching as well as research experience in Chemical Engg./Environmental Engg. He has many fellowships and awards in his credit. He is a regular reviewer of about 30 national and international journals. He has 166 publications, 143 of which are in international journals and 23 are book chapters. He has an h-index of 49, total citations of more than 8100 (Google Scholar), and is involved in a number of research projects funded by many agencies including UGC, AICTE, DST, and MHRD, TEQIP. He has been awarded with Smt. Prem Lata Jain Best Researcher Award in 2017 and 2019 by Smt. Prem Lata Jain and Prof. D.V.S. Jain Research Foundation, Panjab University, Chandigarh. He has also been awarded with "The Think of Ecology" Award by Hiyoshi Corporation, Japan, for the year 2019. His name has appeared in the list of Top 2% World's Scientists in 2020 and 2021).

He is a Fellow of Royal Society of Chemistry and Institution of Engineers and Fellow of Panjab University Senate.

S.K. Mehta (University of Ladhak, India) Prof. S.K. Mehta, Fellow Royal Society of Chemistry (FRSC) is the Vice Chancellor of University of Ladakh, UT, and Professor at the Department of Chemistry. He is the Ex-Chairman, Chemistry and Ex-Director SAIF/CIL/UCIM at Panjab University, Chandigarh. He is among the world's top 2% scientists according to a study by Stanford University in 2021. He did exceptional work as Coordinator, Chandigarh Region

Innovation Knowledge Cluster (CRIKC), Local Coordinator MHRD initiative GIAN and Coordinator UGC CAS at Panjab University, Chandigarh.

He is credited with two patents and more than 450 publications in international journals of repute with h-index of 64, citation index of 13,242, i10-index of 272, and is an author of about 50 books/chapters. He is recipient of renowned DAAD and JSPS fellowships, a Bronze medal from Chemical Research Society of India (CRSI), an Authors award by Royal Society of Chemistry (UK), the Haryana Vigyan Ratna award, and the Prof. W.U. Malik Memorial Award of Indian Council of Chemists (ICC) and STE award for his outstanding contribution in research.

He has been a visiting scientist to UK, Germany, Japan, United States, and France and has guided 15 post-doctoral, 48 PhD, and 50 master's students and handled 20 research projects. He has completed or is completing 5 DST, 5 CSIR, 6 UGC, Indo-German, Indo-UK and Indo-Japan Research/Exchange projects and workshops. He was also Coordinator, Chandigarh Region Innovation Knowledge Cluster (CRIKC), Local coordinator MHRD initiative GIAN and Coordinator UGC CAS at Pu Chandigarh. Prof. Mehta has more than 450 publications in international journals of repute with h-index of 64, i10–272, Citations 13,242 and is an author of about 50 books/chapters. He has been nominated as member of several DST, CSIR and UGC national committees. He is recipient of renowned DAAD and JSPS fellowships several times, Bronze medal from *Chemical Research Society of India (CRSI)*, authors award by Royal Society of Chemistry (UK), Haryana Vigyan Ratna award and Prof. W.U. Malik Memorial Award of Indian Council of Chemists (ICC) for the year 2015 for his outstanding contribution in research.

He has made noteworthy contributions in the field of chemistry which range from indulging in the basics to applicability for societal needs. He is actively engaged in deciphering the dynamics of colloidal chemistry and nanochemistry. Initially, he immersed himself in the field of soft assemblies such as metallosurfactants, emulsions, and inclusion complexes. He made valuable contribution in understanding the intricate framework of these complex systems along with thermodynamic/physical aspects. He was successful in concocting an anti-tuberculosis drug rifampicin in non-amphiphile aggregate with enhanced solubility, stability, and bioavailability.

Contributors

Ali, Hadi
Department of Chemistry
University of Ladakh Kargil
Ladakh, India

Anand, Neeru
University School of Chemical
    Technology
Guru Gobind Singh Indraprastha
    University
Delhi, India

Alex Ibhadon
School of Engineering,
University of Hull,
Hull, United Kingdom.

Bala, Niraj
National Institute of Technical Teachers
    Training and Research
Chandigarh, Punjab, India

Bansal, Himanshi
Energy Research Centre
Panjab University
Chandigarh, India

Bhullar, Navneet Kaur
Department of Chemical
    Engineering
Chandigarh University
Mohali, Punjab, India

Chakinala, Anand G.
Department of Chemical Engineering
Manipal University
Jaipur, Rajasthan, India

Chakinala, Nandana
Department of Chemical Engineering
Manipal University
Jaipur, Rajasthan, India

Dehiya, Brijnandan Singh
Department of Chemical Engineering
    Deen Bandhu Chotu Ram
University of Science and Technology
Murthal, Sonipat, Haryana, India

Gaba, Rekha
Department of Chemistry
DAV University
Jalandhar, Punjab, India

Gill, Harjot
Department of Electrical Engineering
Chandigarh University
Mohali, Punjab, India

Jasra, Raksh Vir
Reliance Technology Group
Reliance Industries Limited
Gujarat, India

Kansal, Sushil Kumar
Dr. S. S. Bhatnagar University Institute of
    Chemical Engineering and Technology
Panjab University
Chandigarh, India

Kataria, Ramesh
Department of Chemistry
Panjab University
Chandigarh, India

Kaur, Amandeep
National Institute of Technical Teachers
    Training and Research
Chandigarh, Punjab, India

Kaur, Harjit
Dr. S. S. Bhatnagar University Institute of
    Chemical Engineering and Technology
Panjab University
Chandigarh, India

Kim, Byungki
Future Convergence Engineering Korea
University of Technology and Education
Cheonan, Chungnam, Republic of Korea

Kumar, Naveen
Department of Chemistry and Centre of
    Advanced Studies in Chemistry
Panjab University
Chandigarh, India

Kumar, Sahil
Department of Chemistry
Punjabi University
Patiala, Punjab, India

Mehta, Neena
Department of Biochemistry
Rayat and Bahra University
Mohali, Punjab, India

Mehta, S.K.
Department of Chemistry and Centre of
    Advanced Studies in Chemistry
Panjab University
Chandigarh, India

Mitra, Chanchal Kumar
Former Professor, Department of
    Biochemistry
University of Hyderabad
Hyderabad, Telangana, India

Prabhakar, Ashok
Department of Chemical Engineering
Chandigarh University
Mohali, Punjab, India

Prasad, Majeti Narasimha Vara
Honorary Emeritus Professor, School of
    Life Sciences
University of Hyderabad
Hyderabad, Telangana, India

Sasikumar, Ragu
School of Mechatronics Engineering
    Korea

University of Technology and Education
Cheonan, Chungnam, Republic of Korea

Sharma, Surendra Kumar
University School of Chemical Technology
Guru Gobind Singh Indraprastha
    University
Delhi, India

Singh, Arashdeep
Department of Electrical Engineering
Chandigarh University
Mohali, Punjab, India

Singh, Kulbir
Department of Electronics and
    Communication Engineering
Thapar Institute of Engineering &
    Technology
Patiala, Punjab, India

Singh, Mangat
Faculty of Education and Methodology
Jayoti Vidyapeeth Women's University
Jaipur, Rajasthan, India

Singh, Manjeet
Department of Electrical Engineering
Chandigarh University
Mohali, Punjab, India

Singh, Sandeep
Department of Mechanical Engineering
Punjabi University
Patiala, Punjab, India

Singh, Surinder
Dr. S. S. Bhatnagar University Institute of
    Chemical Engineering and Technology
Panjab University
Chandigarh, India

Singhal, Nidhi
Dr. S. S. Bhatnagar University Institute of
    Chemical Engineering and Technology
Panjab University
Chandigarh, India

Sundaramurthy, Suresh
Department of Chemical Engineering
Maulana Azad National Institute of
    Engineering and Technology
Bhopal, Madhya Pradesh, India

Surolia, Praveen K.
Department of Chemistry
Manipal University
Jaipur, Rajasthan, India

Tyagi, Uplabdhi
University School of Chemical
    Technology
Guru Gobind Singh Indraprastha
    University
Delhi, India

Preface

In an era defined by environmental concerns, resource scarcity, and the pressing need for sustainable solutions, the world finds itself at a crossroads. The intersection of energy production, material science, and circular economy principles has emerged as a beacon of hope, offering a path towards a more sustainable and regenerative future. This book, *Energy Materials: A Circular Economy Approach*, is a testament to the pressing need for transformation in our energy systems and materials utilization.

It is an exploration of how circular economy principles can revolutionize the way we source, produce, utilize, and dispose of energy materials. Our journey begins with a profound understanding of the challenges that have driven us to re-evaluate our energy and materials paradigms. Climate change, resource depletion, and mounting waste have compelled us to question the linear "take-make-dispose" model that has characterized our economic systems for so long. The circular economy emerges as a compelling alternative, emphasizing the sustainability and longevity of materials and energy resources.

In these pages, you will embark on a voyage through the multifaceted landscape of energy materials. We delve into the realm of supercapacitor materials, renewable energy sources, exploring how materials innovation can unlock the full potential of solar, wind, and battery technologies. We scrutinize the life cycles of these materials, unraveling the complexities of their production, utilization, and end-of-life management. In our exploration, we engage with the pioneers and visionaries who are driving the transition towards circular energy materials. We spotlight the innovative technologies, policies, and strategies that are reshaping industries and economies.

From recycling and repurposing to sustainable sourcing and eco-design, the circular economy principles showcased in this book provide a blueprint for a more sustainable and resilient energy future. However, this journey is not without its challenges. We confront the barriers and limitations that impede the widespread adoption of circular energy materials. We examine the economic, social, and regulatory factors that influence decision-making and explore how we can overcome these obstacles to create a more sustainable world.

As we move forward, we must recognize that the transition to a circular economy for energy materials is not just a technological endeavor but a societal transformation. It requires collaboration, innovation, and a shared commitment to the well-being of our planet and future generations. The chapters that follow offer a comprehensive and insightful exploration of the diverse facets of energy materials in a circular economy context. They represent the collective efforts of experts, researchers, and practitioners who are dedicated to shaping a sustainable energy future. We invite you to embark on this transformative journey through the pages of *Energy Materials: A Circular Economy Approach*.

May the knowledge contained herein inspire and empower you to contribute to the transition towards a more sustainable, circular, and prosperous world. Together, we can redefine the energy materials landscape and illuminate a path towards a brighter and more sustainable future for all.

On behalf of all Editors,

Prof. S.K. Mehta
Vice Chancellor,
University of Ladhak, Leh,
UT of Ladhak, India

Foreword

Boundless scientific ingenuity and entrepreneurial spirit of humankind has provided path breaking solutions to many of the societal problems related to energy, materials, health, and food post industrial revolution. However, it is also feared that the rapid development and growth coupled with increased global population could push the Earth system outside its stable state of 10,000 years. In fact, climate change due to greenhouse emissions resulting from unprecedented energy and material consumption has made sustainability a real issue today. This has driven researchers to look for sustainable alternative energy solutions and materials. As a result, the world is witnessing a great global energy shift from fossil fuels to renewables along with heightened sustainable awareness leading to circular materials. Scientific literature is brimming with alternative energy options like hydrogen as an energy carrier, biomass, battery, solar, and wind. A major thrust is also seen in recyclability of materials, especially polymers, besides production from renewable biomass.

In this context, the present edited volume, *Energy Materials: A Circular Economy Approach*, is an encouraging collective endeavour to present research efforts and directions to achieve harmonious human progress with nature. This book is a pleasing effort in collaboration of minds on the subject of energy, materials, and circularity with an objective to illuminate a path forward.

The chapters in this volume encompass a diverse spectrum of topics ranging from biomass to the potential of supercapacitors, from the transformative power of hydrogen energy to the intricate webs of microgrids. The exploration of energy storage by harnessing nanohybrids and nanomaterials, lingo-cellulosic materials to valuable chemicals, the imperative of plastic recycling, and the reimagining of solar photovoltaic panels and recycling within the chemical industry waste all coalesce to present a comprehensive mosaic of sustainable energy practices.

Each chapter contributes a thread to the overarching tapestry of circularity. The insights into electrochemical sensors, batteries, and low-cost energy storage devices underscore the invaluable role of innovation in shaping the energy landscape of tomorrow. As we grapple with challenges on both local and global scales, the wisdom shared within these pages serves as a compass guiding us towards a regenerative, cyclical approach to materials and energy.

I am profoundly impressed by the dedication and expertise demonstrated by the editors and contributors of this volume. Their collaborative spirit and intellectual rigor is reflected in the analysis and the clarity of vision present in each chapter.

I must say that the book is more than a compilation of research; it provides a trigger for future discussion to develop a path forward towards achieving sustainability in energy and materials through recycle economy. As we confront the urgency of our sustainability challenges, research efforts and approaches discussed in the book *Energy Materials: A Circular Economy Approach* also renew our confidence in the power of scientific knowledge and human collaboration to overcome these challenges. I am confident that this volume will inspire researchers, practitioners, policymakers, and students alike to engage with renewed vigour in shaping a future that is sustainable, just, and abundant for all.

Dr. Raksh Vir Jasra, FNA, FNAE, FICS, FGSA
Senior Vice President, R&D Centre, Vadodara
Reliance Industries Limited, India

Acknowledgments

The creation of this book, *Energy Materials: A Circular Economy Approach*, has been a collaborative effort that wouldn't have been possible without the dedication, expertise, and support of numerous individuals and institutions. We extend our heartfelt gratitude to those who have contributed to this endeavor.

First and foremost, we would like to express our appreciation to the authors and contributors whose insights, research, and expertise have enriched the content of this book. Your commitment to advancing the knowledge in the field of energy materials and circular economy principles is commendable. We would like to thank the reviewers and experts who generously shared their time and provided invaluable feedback, ensuring the quality and rigor of the content presented in this book.

Our gratitude also extends to the editorial and production teams who worked tirelessly behind the scenes to bring this book to fruition. Your insights and discussions have been a source of inspiration and motivation. Furthermore, we extend our thanks to the academic institutions, research organizations, and funding agencies that have supported the research and projects contributing to the content of this book.

Our heartfelt and special thanks are due to Himanshi Bansal for handling this book project along with the editors and giving her round-the-clock support while she put forward her hardest efforts and valuable time. Without her continuous and critical efforts, this book could have not been presented to our readers on time.

Last but not least, we would like to express our gratitude to our families and loved ones for their unwavering support and understanding during the process of creating this book. Your patience and encouragement have been a constant source of strength.

In conclusion, the creation of *Energy Materials: A Circular Economy Approach* has been a collaborative effort that reflects the collective commitment to addressing the challenges of our time. We hope that this book serves as a valuable resource for researchers, policymakers, practitioners, students, institutions, and all those who share our vision of a more sustainable and circular energy future.

With sincere appreciation,
On behalf of all Editors,

Dr. Surinder Singh
Panjab University, Chandigarh
August 31, 2023

1 Circular Economy for Energy Materials

State-of-the-Art Initiatives and Regulatory Issues

Harjit Kaur* and Nidhi Singhal

1.1 INTRODUCTION TO CIRCULAR ECONOMY AND ENERGY MATERIALS

The circular economy (CE) delves to find methods to ensure coexistence of economy and environment in tandem with each other in contrast to linear economy that operates without any inbuilt system to create the balance. Foremost objective of circular economy is to decouple the environment from economic growth [1]. Circular economy draws lineage from the first law of thermodynamics and argues that it is not possible to destroy the resources used for production and consumption of products and those resources end up in the environment in the form of waste. Contrary to it, the linear economy emanating from price theory neglects the material point of view. Therefore, it is essential correct this dynamics between material and value viewpoints to establish a correct relationship between the two and arranging the two in such a way that both the views are given their due importance [2]. The primary challenge of circular economy is to deliver products with minimum energy [3] and cope with critical challenges of scarce resources, environmental impact and generating economic profits altogether. It implies that transformation to circular economy is a highly complex process considering the integration of resources and energy, economics, product design and modeling, process development and products, service and distribution of materials or products, and data handling and management. Circular economy is about building economic processes built upon principles of "spiral loop system" with aim of utilization of products, contrary to their disposal using R-principles [4]. The main drivers of circular economy are to take advantage of the recycling the waste/materials and manage sustainable growth without hampering the environment using low-carbon development strategies. The core principle of CE is to assess valorization of materials utilizing a closed-loop strategy minimizing waste and sustaining economic gains at the same time [5].

The circular economy paradigm has been receiving a significant amount of attention from policy inceptors like European Union and business sustenance organizations such as Ellen MacArthur Foundation in the discussions related to industrial

---

* harjitkaur.uicet@gmail.com

DOI: 10.1201/9781003269779-1

1

development as a means to overcome environmental challenges and advance sustainable development [6]. The growing concerns about climate change and need to shift to low carbon energy materials and systems is being advocated strongly both in developed and developing countries [7]. The worldwide energy scenario is facing a severely uncertain future dependent on the options available in the years to come with colossal outcomes affecting climate and current civilization. The worldwide share in clean technologies supported by renewable power based and electric vehicles transportation was estimated to $1.1 trillion in 2022 that was more than by 31% than in the year 2021 [8].

Energy materials are defined as those that can store large amounts of chemical energy that can be released later on. The energy materials have applications in energy conversion or transmission, reducing power consumption, increasing the efficiency of existing gadgets. There are wide ranges of energy materials available such as common fuels like gasoline and diesel used to power automobiles to highly explosive materials like gunpowder and dynamite. In other words, energy materials include variety of substances that are used for the production conversion, storage and transmission of energy. The circular economy for energy materials focuses on improving the efficiency of materials and energy usage to minimize waste, pollution and natural resource consumption. There is a huge potential to improve the sustainability and reducing harmful effects of energy production and consumption by adopting circular economy principles in development of energy materials. The areas under energy materials where circular economy principles can be applied include efficient manufacturing of energy materials that will exist longer and have been designed for reform, reuse and recycling on retirement to make certain that resources and supply chains for switching to clean energy are safe and sustainable [9]. This approach involves designing energy materials with longer life span and incorporating principles of reuse and recycling into their production and disposal. Circular economy for energy materials spans across numerous disciplines such as engineering, materials science, renewable energy, waste management, supply chain management, modeling and simulation and policy development. It directly affects various industries such as batteries, building materials, renewable energy technologies, energy storage systems, electric vehicles, polymers, composites etc. Despite the growing interest in the circular economy as a sustainable development strategy for energy materials, it is important to recognize that the concept is still evolving and further research and development is necessary to fully understand its potential and effectively implement it in practice.

This is an introductory chapter about fundamentals of circular economy and waste minimization using materials recycling and saving money and environment with sustainability, how it is applicable to industries in general and to chemical industry and energy materials/energy providing processes and energy generation and storage. The chapter also covers the regulatory issues involved in implementing circular economy concepts. The objective of the chapter is to make the readers understand circular economy, its application to industries and energy industry, environment and sustainability and how sustainability can be achieved using circular economy (as there is no direct relation between these two terms) and the regulatory concerns in India and globe concerning circular economy.

1.1.1 Definition, Principles and Implementation Strategies

The idea of CE bears its roots in varying scientific disciplines and semi-scientific concepts. CE concepts originate from point of departure from established research fields like ecological economics, industrial ecology, industrial symbiosis, cleaner production, product service systems, eco-efficiency, cradle-to-cradle design, bio-mimicry, resilience of social-ecological systems, the performance economy, natural capitalism, zero emissions, and many more [5,10,11]. The concept of materials recy-cling has been around since the beginning of the industrial revolution. However, the concept of CE is different from traditional recycling in the sense that it lays stress on product, inventory and material reuse, remanufacturing, refurbishment, repair, cascading and upgrading. It is a sustainable approach and caters to business devel-opment that supports the potential renewable energy resources and their recycling like solar, geothermal, wind, tidal, biomass and waste-derived energy capitalization involving supply chain and product value using cradle-to-cradle framework. A fully developed CE approach will encourage high-value materials cycles instead of recy-cling low value feedstocks as in case of linear recycling. The current concept of CE developed by [6] is presented in Figure 1.1. It indicates that inner circles of circular economy demand lesser resources and energy and are more economical. Therefore, these inner circles should be extended as long as possible [6].

Further concept of CE has a multi-disciplinary perspective beginning with lin-ear economy during the industrial revolution characterized by overexploitation of resources. This evolution was interrupted during 1960s with notable interest in

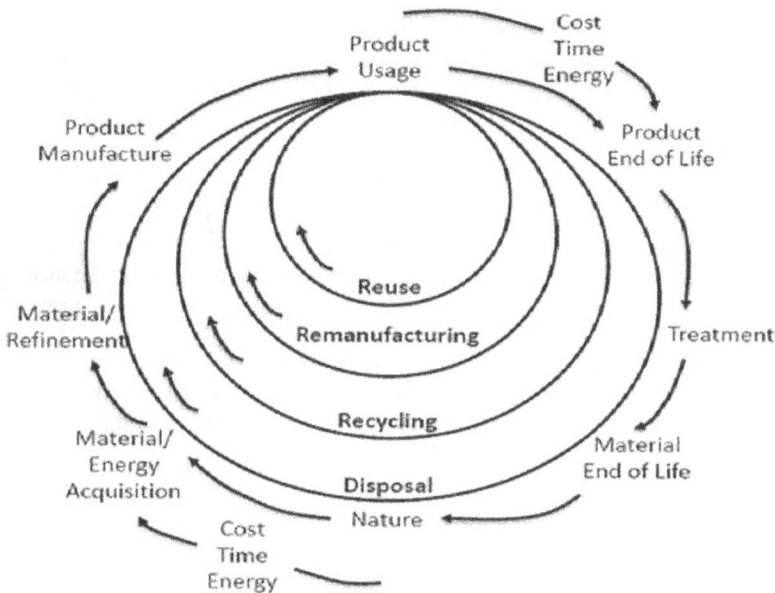

FIGURE 1.1   Current concept of circular economy.

Source: Copyright 2018. Reproduced with permission from Elsevier [6].

preserving the environment by ecologists and by economists positing earth as a cyclical ecological ecosystem. This was followed by the emergence of framework of "green economy" which sustained human growth and social equity minimizing the threats to environment and ecological paucities significantly. In the 1990s, circular economy emerged as a policy and model that aimed to promote economic expansion in a way that is sustainable and respects nature [11]. It was viewed as a regenerative system and a multi-level concept—micro, meso, macro and significance of sustainable business strategies [12]. Though there is varied deviation in the definitions of CE posed by different researchers, the majority of scholars agree that CE bears important role in accomplishing the UN Sustainable Development Goals. The four pertinent components to establish the notion of circular economy are recirculation of resources and energy, demand reduction and valorization of waste, multifarious approach, significance as a means to attain sustainable gains. and proximity with social innovation.

1.1.1.1 Defining Circular Economy

Pearce and Turner formally utilized the term "circular economy" for the first time in 1990 and several attempts have been made since then to define the concept ranging from resource-oriented definitions and interpretations, emphasis on closed loop material flows, reduction in use of fresh raw materials and diminishing the damaging impacts on the environment [13]. The commonly found keywords in the definitions of circular economy are economic growth, promotion of renewable energy, restoration and replenishment of resources [14]. The definitions of circular economy given in the literature may be classified into resource-oriented definitions and those that look beyond the notion of managing materials and include other dimensions. The resource-oriented dimensions lay emphasis on the requirement to build closed loops of material and energy flow and decreased use of fresh or virgin raw materials and focus on reducing the negative environment effects. The second category of definitions focuses on other dimensions like sustainability and social equity besides managing material resources. The most commonly referred to definition of CE given by the Ellen MacArthur Foundation is

> an industrial system that is restorative or regenerative by intention and design such that it replaces the "end-of-life" concept with restoration, shift towards use of renewable energy, elimination of use toxic chemicals that impair reuse and aims to eliminate wastage through superior design of materials, products, systems and business models. In other words, the circular economy is one where the value of products, materials, and resources is maintained in the economy as long as possible, and the generation of waste is minimized.
>
> [15]

A study investigated 114 definitions of CE and found that in most of the definitions the foremost objective of CE was economic success and safeguarding of the environment [11]. It is to be noted that most of the definitions ignore impact of circular economy on social equity and future generations and also explicit linkages to sustainable development are not highlighted. However, in the definition given by [12],

CE is acknowledged as a closed-loop strategy for recycling and a semantic approach advocating paradigm change leading to industrial metamorphosis or transfiguration in a broad sense. A variety of definitions have been proposed by different authors to describe the notion of CE and putting up it as a broad framework which could be expressed differently by different researchers. From the perspective of energy materials, circular economy definition, within the set of previous definitions, can be understood as optimizing the energy inputs, utilization of renewable energy and material recycling of energy materials [16]. To summarize, the key ideas that emerge from definitions of circular economy are closed loop flow of materials, efficient use of material resources and energy throughout various stages of production and consumption with the goal to achieve sustainable development. Implementing the circular economy approach requires holistic resource management and collaboration between manufacturers, consumers, and other societal agents.

1.1.2 Principles of Circular Economy

CE happens to be alternative economic strategy that converts "goods at the end of life cycle" into utilizable materials for others by circular flows in industrial networks and reducing the wastage of feedstocks. The CE business models may be classified into two groups. The first kind are those businesses that foster repair, remanufacture, upgrades and retrofits. The second category of businesses turns obsolete goods into novel feedstocks employing recycling of materials [17]. Historically, the 3R principles of *reduce, reuse of materials and recycle* [1,16] formed the foundation of circular economy with the aim of optimizing production through reduction in utility of natural materials, environment contaminants, emissions and wastes. The concept of lean manufacturing was based on 1R (reduce) that further led to the development of green manufacturing in 1990s that was based on the 3R principles. Later on, a green manufacturing framework emphasized wider, innovative 6R methodology—"reduce, reuse, recycle, recover, redesign, and remanufacture"—for products over *"multiple life-cycles"* [18]. Several R-frameworks proposed by scholars go beyond 3R or 4R frameworks such as the 6Rs, 9Rs [12] or even 10Rs (as mentioned in Table 1.1) [19]. The common feature of all these frameworks is that they share a hierarchy with the first R to be viewed as priority over subsequent Rs [12].

> **Refuse:** This concept is applicable to any consumption article and emphasizes the option to buy a smaller amount or use a lesser amount of something with the aim of preventing waste generation. It is generally applied to ensure rejection of packaging/covering material waste and shoppers/carry bags in case of consumers. From the producers' perspective, refuse is associated with "concept and design life cycle." The commodity designers can ensure refusing of particularly harmful materials, designs, manufacturing processes to evade waste [19,20].
>
> **Reduce:** The objective of this principle is to minimize the consumption of key raw materials, energy and reduce generation of waste through improvement in production efficiency, also called as eco-efficiency [21], and consumption processes such as introduction of better technologies, more lightweight

TABLE 1.1

Value Retention Options under Circular Economy

	Principle "R"	Key Activity of Consumer/ Manufacturer	Key Activity Market Actor
Efficient product utility and processing	1 – Refuse	Refrain from buying; product usage is to be stopped or utilizing radically upgraded product	See 2nd life cycle redesign
	2 – Reduce	Reduction in use of products; sharing products, less similar products accumulation, saving resources	See 2nd life cycle redesign
	3 – Reuse/resell	Give to others and sell the product for reuse	Resell, give away to others, inspect, clean, reuse
Extend life span of products and their parts	4 – Repair	The intended product after some repair can be utilized	Repairing or inserting malfunctioned parts
	5 – Refurbish	Remaking the product with minor issues to altogether working product without flaws	Restoration of failed components or modules, product life same as new
	6 – Remanufacture	Servicing of the product under contract or dispose (i.e. utilizing parts of abandoned product to make new product having intended function)	Return for service and remaking of product. Parts of product can be utilized afresh in new product.
	7 – Rethink	Think before buying, refuse to buy products which can't be recycled	Refuse, ensure recycling, prevent discarding
	8 – Recycle	Utilize the product for a number of times after recycling, material in use without degrading its quality	Recycle, quality check, down-cycle to other products
Useful application of materials	9 – Recover (energy)	Material back to the production cycle, recovery of useful materials/energy and its utilization, recover wastage/scrap and do its recycling	Material and energy recovery and useful waste/old product recycle, closed loop flow of materials
	10 – Repurpose	Buy and use secondary material, waste/old product can be reused in new form if repurposed	Reinvent, make use of component/product to a new purpose as down-cycle or up-cycle

Source: Adapted by author.

products, simple packaging, simpler lifestyle etc. It is about achieving efficiency by utilizing lesser materials/feedstocks per unit of product, replacing harmful materials with safer ones per unit of product. Zero-emissions framework seeks to maximize value of goods by minimizing or zero environment load [1]. The focus of this R-principle is on refusing, rethinking and redesigning including extending the life of products, minimization, reduction, prevention of resource use and preserving of natural capital [12,19,20].

Reuse/resell: EU (2008) refers to reuse as a process by which products or components that are not waste may be used again for the same purpose for which they were originally designed [1]. The discussion on the *reuse* principle is about reusing (excluding waste), repairing and refurbishing the products, thereby closing the loop [11]. The reuse of products can be classified into two types based on change in ownership. Relocated (gifted or discarded) or resold products are those in which there is a change of ownership and there is a need to align diffusion and efficiency in these marketplaces. Product-service systems (PSS) belong to the category of hired/shared/refunded deposit products in which there is no change in the ownership [20].

Repair: The common purpose of repair is to prolong the time for usage of the product and is described as "bringing back to working order" or "replacing broken parts." Repair activities can be undertaken by various parties, regardless of whether there is a change in ownership or not [19,20].

Refurbish: The term *refurbish* is employed to indicate the process of reconditioning a product where multiple components are either replaced or repaired without altering the overall structure of a complex multi-component item [19]. It is about restoring an old product and bringing it up to date by modernizing the function or product or upgrading [20]. This concept is used in context of airplanes, trains, mining etc. It is a medium long loop concerned with business activities with indirect links to the consumer [19]. It involves only replacement and not disassembly of components and is therefore called light manufacturing that brings back the product to satisfactory working or cosmetic conditions [20].

Remanufacture: It is also a medium long loop, also called second-life production that applies to the situations where the complete product constitution takes place. It involves disassembling, checking, cleaning and repairing or replacing (wherever necessary) in an industrial process [19]. It is also referred to as rebuilt/remold/rewound with the aim of reversing or postponing obsolescence and making product of same quality as a new one [20].

Repurpose: It is not a widely used concept and is largely used either in industrial design or by artists. A material gets a distinct new life cycle by adapting it to some other function rather than just discarding [18]. It is also referred to as contextualizing by using the discarded products or their parts in the formation of a new product with different functions or reusing a product for an altogether new purpose. Repurpose is different from other Rs because the original article gets a different identity [20].

Recycle: Recycling encompasses a range of processes aimed at transforming waste materials into new products or substances that can be utilized for their original purpose or other purposes. It also involves the reprocessing of organic material, although it does not involve energy recovery or transformation into materials suitable for use as fuels or in backfilling operations [1,12]. Recycling a long loop denotes traditional waste management activities that are considered to be least desirable [19].

Recover (energy): European Union Waste Framework Directive introduced the recover principle in 2008. Recover is used to describe "Collecting used products at the end-of-life and then disassembly, sorting and cleaning for use" (p. 257). The word *recover* is often used in connection to collection of recyclable products and materials. It is used in relation to the "energy recovery" from waste streams. Generally, it denotes capturing energy embodied in waste, linking it to incineration in combination with producing energy [19].

Re-mine: It is the most ignored option in operationalizing circular economy that is concerned with recovery of materials that have been disposed in a landfill [19].

1.1.3 Circular Economy Processes or Strategies

The description of circular economy processes provides an understanding about the practical adoption and implementation of circular economy through these processes or strategies by countries and companies [13].

Reducing use of primary resources: The key processes covered under this category to implement the concept of circular economy are recycling, efficient use of resources and using renewable energy resources [13,22]. Another such concept, extended producer's responsibility (EPR), was first proposed by Germany under the *reuse* principle in legislation on packaging in 1992 and later on by Korea and European Union. EPR is based on the "polluter pays" principle with the aim of enhancing circularity of products and materials by promoting reuse and recycling. From a producer's perspective, it is suggested that the costs associated with disposal and recovery should be shifted as an incentive towards reusing, recycling, or properly disposing of waste materials. It also proposes that the products that cannot be reused, recycled or disposed of should neither be produced by producers nor be consumed by consumers. It lays emphasis on shared responsibility among all stakeholders to achieve ambitious results in terms of waste collected for reuse or recycle [1].

Maintaining the highest value of materials and products: This category comprises remanufacturing, refurbishment, reuse of products and components and product-life extension [13]. This maximizes the durability of products by employing strategies such as reuse, repair, refurbishment, remanufacturing, and repurposing. These methods ensure that final products and their components remain in circulation within the economy for extended periods while preserving or increasing their value. The success of these approaches is contingent upon factors such as market acceptance, efficient reverse logistics systems, and financial viability for stakeholders, and implementation through various business models [20].

Changing utilization patterns: It focuses on three processes, which are product as service, sharing models and shift in consumption patterns [13]. The product-service system is an innovative business model that challenges traditional approaches to selling and ownership of physical products. It focuses on delivering value to customers through the provision of comprehensive bundles of products and services, without requiring customers to take ownership of the physical product itself [23]. Designing smarter product use and manufacture through use of refuse/rethink/reduce principles can lead to transition to circular economy before production takes place [20].

1.1.4 EVOLUTION AND GLOBAL ADOPTION OF CIRCULAR ECONOMY CONCEPTS

The circular economy has progressed in diverse ways under the influence of varied cultural, social and political systems. The concept was introduced in Germany in early 1990s as part of the environmental policy with the aim of attending to the problems related with raw material and natural resource use for sustained economic expansion [5]. Germany pioneered the adoption of circular economy with enactment of closed substance Cycle Waste Management Act in 1994, and Japan initiated a waste management plan to establish a recycling-based society in a comprehensive and systematic manner [24]. Chinese circular economy promotion laws view circular economy as generic concept used to represent the reduction, reuse and recycling processes and activities conducted during production, circulation and consumption of products and services [1]. China promoted the circular economy through an "econ-industrial park" model in late 1990s and introduced the concept of "harmonious society" in the mid-2000s stressing recycling waste after consumption and developing waste-based closed loops in an organization or among different producer and consumer groups [5]. The broader goals in countries like Europe, Japan, the United States, Korea and Vietnam focus on achieving the synergistic effects of circular economy with national polices and strategies to prevent landfills, resource procurement, reducing greenhouse gas emissions and managing hazardous waste [1]. In the cases of United Kingdom, Denmark, Switzerland and Portugal, circular economy principles were applied primarily for management of waste and some models were implemented in these countries that apply circular use or reuse of materials [5]. It is evident form some parts of Japan and Korea that circular economy–related initiatives seek to enhance the responsibility of consumers for material consumption and disposal. In North America and Europe, business corporations employ circular economy principles and concepts to enhance reduction, reutilization and recycling programs and to conduct product life cycle studies [5]. The literature on circular economy has identified that implementation of the concept in two directions. The first is systemic economy-wide implementation at micro, meso and macro levels, as done by China. The Netherlands has ambitious designs to implement circular economy and make the Netherlands "a circular hotspot." The Dutch government launched the Green Deal Initiative in 2013 and Realization of Acceleration of Circular Economy (RACE) in 2014 with work packages under RACE ranging from design and knowledge sharing to demonstration projects and community development. Another aspect of the circular economy that can be implemented on a systemic level is within the production sector. UNIDO/UNEP programs for sustainable production and consumption, National Cleaner Production Centers, and UNEP's life cycle initiative have launched transnational policies adopting circular economy principles. Additional measures to promote the adoption of the circular economy involve economic incentives and a commitment to implementing eco-design for energy-consuming devices. Another approach is targeted action in specific sectors such as plastics, critical raw materials, industrial waste, and mining waste. This includes proposed legislation for effective waste management practices and extended producer responsibility programs. The prioritized implementation of circular economy principles is in the product categories like electric and electronic equipment, textiles, furniture, packaging and tires [21].

1.1.5 CIRCULAR ECONOMY IN VARIOUS INDUSTRIES

Electrical and electronic equipment (EEE) waste is among the major waste categories that need to be managed. EEE industry has a high material consumption including metals such as iron, silver, copper, gold, manganese, chromium, zinc and several rare earth metals, and the rate of extraction of these materials is much higher compared to rate of their generation by the nature. Therefore, EEE waste is considered to be one of the richest sources for supply of secondary raw materials and has a major role to play in securing resources and environmental sustainability. It is a major challenge to collect EEE waste and makes it imperative to shift towards a circular economy. Regulations and policy measures have an important role to play in this sector through employing the concept of extended producer responsibility to make producers responsible for adoption of circularity principles and increasing resource efficiency. Offering incentives for eco-design, modular product designs to reduce rate of obsolescence, improved recycling and recovery of secondary resources can attract producers to adopt circularity, making them more cost-effective and competitive [25].

The world lithium-ion battery market is expected to multiply fivefold in the coming decade. These batteries are expected to play a critical role in creation of equitable clean-energy economy and mitigating the climate change impacts in the United States, according to National Blue Print offered for 2021–2030. It demands breakthrough scientific challenges for new materials and development of manufacturing base to meet the growing demand for electric vehicles and electrical grid storage markets [26].

The thermal performance of building requires replacement of insulation to optimize the thermal efficiency of building. Applying circular economy principles to keep materials at the highest possible value in case of buildings implies retention and retrofit instead of demolition as long as possible. In order to maintain highest value through retrofitting by remanufacturing the window in situ is preferred over recycling. The retention of materials leads to reduced greenhouse emissions from the generation and transportation of materials. The new wall designs should allow for easy replacement of insulation [27].

The transition to a circular economy has become an increasingly important global issue. Among firms pursuing circular economy principles and models of business, a distinction can be made between two clusters: long established and often large companies that are transitioning towards circularity, and start-ups that have circularity in their DNA since inception and are carriers of radical innovations, particularly in terms of business models. According to a recent book, the latter cluster of firms is called "born circular" enterprises [28]. The born circulars or circular-born start-ups have a unique advantage as they are able to build infrastructure and customer base from scratch, while incumbents have a continuous struggle with the legacy business models that are mostly not in sync with the principles of circular economy. Such start-ups have emerged as one of the most significant factors driving socioeconomic growth and benefit from the investment focused towards circular principles. Start-ups have the advantage of being able to adopt circularity more easily due to their smaller size and flexibility in adapting their business models. In addition, start-ups often bring radical innovations that disrupt traditional linear value chains, changing entire industry landscapes. Start-ups focused on circular economy aim to promote sustainable practices in the production and consumption of goods and services. They prioritize recyclability and resource efficiency throughout their product design, development and production processes. This way they reduce waste generation while creating new economic opportunities. Circular economy opportunities can be found across sectors in the global economy such as energy, agriculture, and manufacturing.

There are numerous examples of innovation in circularity across industries and in different phases of production such as product redesign to ensure recyclability, the use of renewable energy and biomaterials in production processes, and novel business models that prioritize sharing or leading innovative approaches to reduce waste generation.

1.2 THE ENERGY SECTOR AND ITS ENVIRONMENTAL IMPACT

In today's global world, sustainable development, zero emissions, and climate change have become pressing concerns. When choosing an energy storage technology, it is crucial to consider its environmental impact. Energy storage technologies can be classified into mechanical energy storage, electrochemical energy storage, electrical energy storage, chemical energy storage and thermal energy storage. Chemical energy storage technologies such as hydrogen fuel cells have a negligible impact on environment as the only byproducts are water and small amount of carbon dioxide [29]. The energy sector comprises transport, heat and power generation, residential and commercial buildings and industrial production and accounts for 76% of the total carbon emissions globally. Energy consumption is inherent in all human activities, and the environmental problems associated with this sector have reached their peak.

1.2.1 OVERVIEW OF THE ENERGY SECTOR

By 2023, it is expected that $1.7 trillion will be invested in clean energy initiatives including renewable energy, nuclear grids, storage, low-emission fuels, improvements in energy efficiency and end-use renewability and electrification. The rise in clean energy investments, especially in import-dependent economies, have been triggered by many factors such as better economies due to high and volatile prices of fossil fuels, policy support, alignment of climate and energy objectives. Additionally, the countries are trying to have strong footing in the clean energy economy through different ways while developing their industrial strategy [30].

1.2.2 ENVIRONMENTAL CHALLENGES AND SUSTAINABILITY ISSUES

There is an urgent call to shift to more sustainable socio-technical systems to address environmental problems, unmet societal expectations, and economic challenges. The conception of circular economy is gaining traction on the agendas of policymakers to address sustainability issues. The term *sustainability* refers to the balanced and systematic amalgamation of economic, ecological and social performance. Circularity in business models and supply chains is a prerequisite for sustainable manufacturing that further influences the economic and environmental performance of the developed and emerging nations. This makes principles of circularity important elements to achieve sustainable growth. To pursue long-term sustainability, mere adoption of service-based systems is not sufficient to achieve sustainability. There is a need to change other conditions like lifestyle and adoption of closed loop systems. There are several positive connotations between sustainability and circularity but the negative relationships cannot be overlooked. It is essential to balance the cost of circular systems in order to overcome any negative value generated by the system. The technical impracticality of creating a closed circle in combination with increased demand for products or energy required to recycle resources cannot be overlooked. This energy and its brunt may be higher for many materials than the overall environmental outcome of obtaining the materials from traditional sources like mining [31]. However, in the earlier definitions and interpretations of circularity and circular economy, sustainability and social dimensions are ignored. The only focus was on reducing the deterioration of environment.

1.2.3 ROLE OF CIRCULAR ECONOMY IN ADDRESSING ENERGY SECTOR CHALLENGES

The circular economy perspective in wind industry can be adopted to address the challenge of using blades and composite material for differing applications. In this case the end-of-life value of blades adds value through extended producer responsibility and application of R-principles at different stages of production and consumption [32]. It is a Herculean task to embrace and implement circularity concepts in energy and utilities sector. The businesses are finding it challenging to fully participate in the circular economy business model with rising consumer demands across a variety of industries in addition to lack of necessary expertise to bring about the change. The energy and utilities sector has an important role to play in the addressing the environmental need to adopt circular economy through advanced energy management for smart city initiatives, sustainable urban living by smart integration of electric vehicles, smart and sustainable homes and smart planning for optimal placement of wind turbines and solar plants. Providing clean electricity by bringing about a fundamental architectural change within the energy and utilities business is a major challenge facing this sector [33].

1.3 IMPORTANCE OF CIRCULAR ECONOMY IN THE ENERGY SECTOR

Many aspects of circularity address the challenges posed by the current climate crisis including rationalization of operations, building localized supply chains, and resilient systems. Circular economy principles have a very significant role to play in developing the strategies to address the problems raised by climate change and to alleviate the possibility of future crisis. Circularity in production can be brought in through maximum use of renewable resources and minimization of leakages across the value chain. The circular economy can address the circularity issues related to consumption in energy sector by reducing leakages by circulating the products and materials at their highest utility by sharing, reusing, repairing, remanufacturing and recycling. The value leakages at the end of product life cycle can be controlling the discarding of products and materials after use [34].

1.3.1 RESOURCE SCARCITY AND ENERGY SECURITY

The advancement of energy systems is determined by two major issues needing redressal—energy supply security and climate change. The associated problems with disrupted energy supply are energy and material shortages, too much dependence of imports, and waste generation. The two ways to manage these problems are via development of a low-carbon, sustainable, competitive and resourceful economy based on circular economy principles and closing the loop approach. Integrated waste management system closes the energy loop by converting the energy from waste to take the waste management and recovery chain forward [35]. The common denominator in almost all the definitions given by studies is to work towards saving the continually depleting natural resources and their overexploitation, reducing associated carbon emissions and curbing the ever-expanding mountains of waste and resultant pollution. The logical purpose of circular economy is to work towards reducing the resource exploitation and preventing waste generation [36].

1.3.2 REDUCING ENVIRONMENTAL FOOTPRINT AND EMISSIONS

The term *carbon footprint* is well understood as the "lifecycle carbon equivalent emissions and effects related to a product or service" (p. 1). Estimation of relative magnitude and growth trends of carbon footprint are important matters to investigate for any sector due to rising concern for global warming. There is a strong relationship between carbon footprint, energy consumption and energy supply. Therefore, it's important to understand relationship in the study of energy consumption [37]. A progress towards circular economy can significantly improve the environmental quality due to reduction in carbon dioxide emissions. This can be achieved through adoption of concepts of ecodesign, effective materials management and reuse [37]. Although shift to an electrical energy–based society is viewed as a major ecological improvement, the fact cannot be ignored that the electrical energy industry is a major contributor to carbon dioxide emissions, and renewable energies can still adversely affect the environment [39].

1.3.3 Transition to a Sustainable and Resilient Energy System

The objectives of both circular economy and energy transition are concerned with restoring the environment, economic growth and reaching a sustainable future. Like circular economy that aims to optimize the utilization of resources and eliminating waste, energy transition is concerned with shifting to sustainable energy system comprising varied and low carbon resources. The low-carbon energy transition is different from traditional energy transition in the sense that it is driven by climate change rather than by technological innovation and new resource acceptance [7]. A resilient system is one that has the capability to survive with change and grow with it with the aim of guarantying an adequate and unswerving flow of fundamental ecosystem services meeting the needs [40]. The circular economy needs a fundamental shift instead of incremental twisting of the current system. The transition to circular economy needs to happen at three levels that are interpreted as three levels of circular economy system: the macro, the meso, and the micro system. The macro systems perspective emphasizes the need to adjust industrial composition and constitution of the whole economy. The focus of the meso perspective is on development of eco-industrial parks as systems. This level is also called regional level. The micro systems perspective usually takes into account the products, individual ventures and requirements to increase the circularity in businesses as well as among consumers [11]. Energy transition is a crucial issue to achieve sustainability [40]. Transitioning to a circular economy can be facilitated through setting targets for reducing waste, closing production loops, efficient use of resources, and maximizing retention of economic value of materials and products [20]. A sustainable system is one that considers a wide range of environmental, social and economic factors associated with a process or a product and evaluates the manner of interactions between them. A complete and ample understanding of the impact on the overall system and of any potential risk shifting is required before an action, policy or product is considered to be sustainable. Therefore, from the sustainability viewpoint individual components cannot be optimized without optimizing the whole system based on an understanding of the relationship between a complete system and its environment. This gives rise to a large-scale balance and avoids shifting and unintended consequences. A circular economy offers an economic system with a cyclical low model that is regenerative by design and builds resilience against future disruptions like extreme weather events or effects of changing climate. Sustainability mediated by innovation can enable circular economy, and circular economy can be a stepping stone towards aligning the three dimensions of sustainability. A successful transition to sustainable circular economy requires acknowledgement of relationship and contribution of circular economy to sustainability and long-standing waste management principles [42]. In short, circular economy is a prerequisite or necessary condition for sustainability [43]. In the transition phase, the aim of circular economy is to achieve overall societal change. It involves a reduction in use of finite business resources like metals, recycling of metals/minerals across production and consumption cycles, production of bio-based, biodegradable materials that return to the environment at the end of useful life and makes it possible to achieve a fully renewable-based economic system at feasible economic, social and environmental costs. The major obstacles in transition to circular economy are political-regulatory, cultural, financial, economic, operational, techni-cal, informative etc. Nevertheless, some drivers that push the transition are business potential and economic benefits, sharing of information on waste streams and promo-tion of dialogue between the stakeholders throughout the value chain [42]. Transition to circular economy demands implementation of circular economy business models that convert waste to energy and enables it to meet rising energy consumption stan-dards. Circular economy business models contribute positively to society and envi-ronment through decrease in the carbon footprint, waste creation and dependence on virgin materials [45].

1.4 CIRCULAR ECONOMY APPROACHES IN THE ENERGY SECTOR

The energy sector covers human activities related to production and consumption of energy that are majorly accounted to combustion of fossil fuels such as oil, coal, and gas leading to release of pollutants to the atmosphere. The major approaches in energy sector include improving and optimizing the energy material efficiency, opting for circularity in renewable energy applications, improved energy storage devices and adopting circular economy concepts in fossil fuel industries.

1.4.1 ENERGY MATERIAL EFFICIENCY AND OPTIMIZATION

Circular economy is often portrayed as an efficient new paradigm with the ability to minimize material and energy demand and hence reducing the harmful environ-mental impacts, while enhancing the economic output. The claims about the intrin-sic efficiency of circular economy accompany the calls for shifting to a circular economy. Based on the notions of welfare maximization and Pareto optimality, eco-nomic efficiency refers to an economic state in which it is not possible to improve the condition of one part without imposing cost on others (i.e., pareto equilibrium). However, the neo-classical approach to efficiency is not employed in circular econ-omy. The term *eco-efficiency* implies production of goods and services from fewer resources into a production process with creation of minimum waste and pollution. It is concerned with creating more value through increase in resource productivity and reduction in resource intensity, thus offering a competitive advantage for the busi-nesses. However, there are associated discrepancies in the conceptualization of eco-efficiency that need to be addressed for successful transition to a circular economy. Otherwise, the transition towards circular economy runs the same risk of repeating failures of linear economy and prioritization of over-production over creating a sus-tainable ecosystem [46].

1.4.2 RENEWABLE ENERGY AND CIRCULAR SYSTEMS

The increasing global energy demand due to world population growth, changing lifestyles and depleting fossil fuel resources is resulting in greater dependence on the renewable energy resources [47]. The renewable energy sources reported in literature are solar energy or photovoltaic energy including solar photovoltaic-grid-connected, solar photovoltaic isolated and thermal solar energy, wind energy, geothermal energy, biomass energy and hydropower energy. Solar and wind energy are cleanest forms of renewable energies but they are weather dependent [48]. However, one problem associated with renewable energy sources is unsteady electricity generation. Therefore, surplus energy needs to be stored to compensate for the fluctuations in the energy supply. Lithium-ion batteries (LIB) are an established and promising candidate for the energy storage in future due to its high energy density, high specific energy and good recharge capability [47]. It is also important to understand the effects of climate change on the production of renewable energy to attain sustainable future. The negative environmental effects of renewable energy resources can be mitigated through careful choice and utilization of these resources [48].

1.4.3 ENERGY STORAGE AND CIRCULAR SOLUTIONS

Energy storage has a substantial role to play with the extensive development and growth of renewable energy sources and in integration of renewable energy resources. This has primarily gained attention in the entire supply chain due to enhanced electric power quality, dependability and better grid stability [29]. LIB are being promoted as promising candidates for energy storage in future. These have developed into state-of-the art power sources for moveable electronic devices and in stationary or electric vehicle applications. Nevertheless, recycling and reutilization of their components demands great improvements [47]. The negative emission technologies (NETs) such as direct air capture and carbon capture and storage (CCS) play a very critical role in a very ambitious climate scenario aspiring to achieve international targets of restricting global temperature rise of 1.5 to 2 degree Celsius. Even more controversial technologies such as solar geoengineering may need to be adopted as long as fossil fuel use and greenhouse gas emissions continue to soar [8].

1.4.4 CIRCULAR ECONOMY IN FOSSIL FUEL INDUSTRIES

The availability of fossil fuels at cheap rates added to the idea of unlimited economic development, while resulting in an ecological crisis. Burning of fossil fuels generates huge quantity of carbon dioxide which lead to global warming and the current natural absorption of carbon dioxide is insufficient to manage the levels of carbon dioxide being produced through human activities [44]. Fossil fuels such as coal, oil, and natural gas meet 80% of the global energy needs but they are an unsustainable energy sources. Circular economy is regarded to be more resilient and sustainable approach to address resource conservation, waste reduction, and environment pollution related to use of fossil fuels. The processes in energy industry and fossil fuels are combination of linear and circular processes throughout the entire material life cycle phases as shown in Figure 1.2. The goal of this industry is to curtail the linear processes and replace them with circular processes. The energy consumption and wastage can be controlled or reduced by using machines that are more efficient. The different returning options to return used materials to natural environment that can be used in fossil fuel industry are carbon capture and storage, improving natural resiliency, improving energy efficiency and reducing emissions. The adoption of returning processes slows down the environmental deterioration caused by fossil fuels [49].

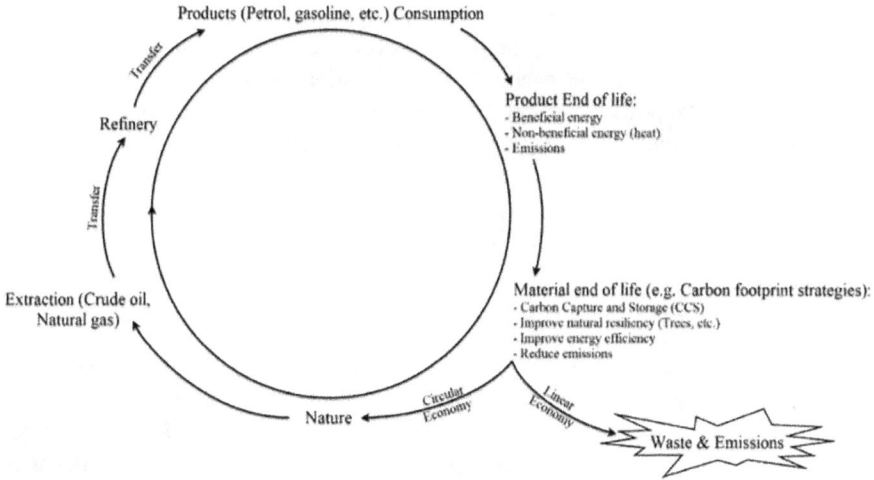

FIGURE 1.2   Circular economy (CE) model for fossil fuel material life cycle.

Source: Adapted from [49].

1.5   CIRCULAR ECONOMY AND THE ENERGY TRANSITION

Environment and attainment of economic growth promoting a sustainable future are primary concerns of both circular economy and energy transition. Circular economy is about optimizing the resource usage through elimination of waste during the product/service life cycle. Energy transition is concerned with shifting to sustainable energy system comprising varied and low-carbon options. It is about structural transforming the energy system in context of collecting, converting, transporting, consuming and managing the energy resources. The focus of energy transition studies have shifted from changes in energy usage to future prospects of energy generation components and consumption. The contemporary energy transition, also called "low-carbon" or sustainable energy transition is driven by global challenges like "climate change" instead of technical innovation and adoption of new resources. The primary approach to achieve the objectives of Paris Convention is by orienting the national policies towards reduction of carbon dioxide (a greenhouse gas) and developing renewable energy and improving energy efficiency through innovations like electric vehicles, smart meters, carbon pricing etc. [7]. The world is largely continuing its long history of adding to, rather than transitioning away from the older energy sources despite multiple pledges from governments and major corporations around the world to minimize greenhouse gas emissions. Although emissions and fossil fuel consumption remain at or near their all-time high globally, some regions, particularly Europe, appear to have entered a true energy transition. The fossil fuel sources are being displaced at a large scale by cleaner technologies [8]. Renewable energy resources are presented as ultimate panacea to decrease emission of environment polluting gases. However, their production is not entirely climate favorable. Therefore, society needs to adopt circular energy practices to achieve efficient energy transition. There is a need to cut down the existing energy intensity levels as a policy tool against controlling the present level of carbon dioxide emissions subject to adoption of economies encouraging exploration and use of renewable energy resources [50]. The policies need to be coherent and should target supply as well as demand side. Especially, the demand side policies need to be designed in such a way that they increase the attractiveness of circular materials for consumers with special emphasis on older, less qualified and wealthy consumers. The fiscal policy needs to provide incentives to increase circular economy [51].

1.5.1 ROLE OF CIRCULAR ECONOMY IN THE SHIFT TO CLEAN ENERGY

Efficient use of resources is an important circular economy process that results in use of less primary resources and is connected with the idea of clean production. The clean production promotes achievement of efficiency in material and energy processes through careful use and replacement of hazardous materials or materials with short life cycles. It is about improvements in both industrial production processes as well as products. Improvements in industrial processes implies conserving and reducing raw materials inputs, reducing energy consumption, reducing waste, avoiding toxic substances, processing and reducing toxic emissions and waste. Product improvements include reduction in environment, health and safety related harmful effects throughout the product life cycle [13]. The world carbon dioxide emissions surpassed the 2019 peaks by 1% illustrating the massive level of the global energy system and the challenge of shifting it not just toward clean sources but also away from polluting the causes [8]. The roadmap for shifting to clean energy through implementation of circular economy strategies is given by [16] in their scenario-based study conducted in a town of China. This transition can be achieved through cascading use of industrial excess heat to form symbiosis between factories and to meet the heat demand in buildings and through electrification of transport sector. The reuse of batteries for a second life as energy storage device is also a way to achieve the transition. The outcomes of circular economy are at least 7% better than a policy than simply focusing on energy efficiency improvements [16]. The critical barriers in use of LIB are heterogeneity of cell design and battery chemistries need to be addressed in developing recycling processes. For recycling to be economically viable and environmentally friendly in circular economy, life cycle assessment schemes should be used during experimental studies. Reverse logistics and collaboration among stakeholders can play an important role in managing efficient return of materials to manufacturer and practical application of extended producer responsibility in the LIB value chain. The role of electric vehicle manufacturers is very vital in battery value chain in establishing recycling plants and providing financial and R&D support to academics for new battery designs using renewable materials [52].

1.5.2 CIRCULAR ECONOMY AND DECENTRALIZED ENERGY SYSTEMS

The emerging research is stressing on the need to shift from a centralized energy production based on a large-scale production unit to a more distributed system based on a small-scale flexible production units [41].

1.5.3 Circular Economy and Circular Cities

The model of eco-towns or circular cities evolved in the 1980s in the United States within the "urban ecology mission." The aim was to revamp the cities that are compatible with environment preservation goals of zero emission, managing landfills, and revitalizing the local industry. Other examples of eco-cities are seen in Europe (Germany, Sweden, UK) as well as China (Beijing, Shanghai, Tianjin, Dalian). These cities were developed with the vision to evolve circular economy through efficient employment of resources, municipal waste management and treatment [1]. The concept of circular economy is interwoven with various other concepts like industrial symbiosis that predates idea of eco-cities evolving rapidly in Japan, Singapore and elsewhere [5]. Therefore, cities have a very critical role to play in transition to a circular economy as they represent around two-thirds of global energy demand, generate 50% of the solid waste and are accountable for 70% of greenhouse gas discharges. Cities hold the key competences on important sectors for the circular economy like waste management, water, urban planning and mobility. Unlocking the potential of the circular economy in cities needs coordination across people, policies and places: a "3P framework." Circular economy demands a cultural shift towards diverse production and utilization pathways, new business and governance models. A holistic and systemic approach is required to cut across sector policies and functional approach that goes much farther the administrative boundaries cities such that it closes narrows and slows the loops at precise scales [53].

1.6 BARRIERS AND CHALLENGES IN IMPLEMENTING CIRCULAR ECONOMY IN THE ENERGY SECTOR

The circular economy model provides the desired integration between sustainability and business development needed for disruptive changes and radical innovations. The transition to circular economy experiences barriers such as financial, structural, operational, attitudinal and technological [4] policy and regulatory, financial and economic, managerial, performance indicators, customers and social [54]. The implementation of energy projects in order to create a sustainable development demand shift from centralized systems to decentralized systems. Nevertheless, this shift requires addressing triple technical, legal and financial challenges to create a sustainable development. However, many technological solutions based on sustainable energy systems exist in the field of energy like electricity generation from renewable and decentralized technologies, electric vehicles for mobility or distribution, but these solutions are often too complex to be implemented on the territory. This is so because, unlike traditional energy companies, these projects focus not only on energy supply but also on providing a variety of services by being closer to the consumer [41].

1.6.1 Technological and Infrastructural Challenges

Product-service system (PSS) design models have the potential to play a pertinent role by supporting companies in designing sustainable PSS for different purposes.

Nevertheless, the major limitation to the development of sustainable PSS is lack of suitable activities, techniques or tools to steer practitioners in designing PSS with a sustainable approach [23]. The adoption of circular business models is interrupted by deficient technologies and technical skills, knowledge to transform current operations and information. Insufficient technical skills hinder identification assessment and implementation of more advanced technical options. The identification of business opportunities becomes difficult due to absence of databases for sharing information about waste [55]. Extending the product life span through circular economy strategies poses challenges in innovation and demands adjustments in revenue models and socio-economic patterns. Expanding the life span of products may slow down the development of new and more environmentally friendly products or may phase out new products or higher standards for safety, energy efficiency etc. [20]. For instance, using LIB as energy storage devices will require direct recycling approaches, that is, separating metals before actual recycling, to have a lesser environmental footprint. However, this transition will be highly dependent on the stabilization of battery chemistries and amplified ability to recover electrolytes [47]. The solar energy and wind energy are clean forms of energy but they have huge land requirements for their installation. Hydroelectric energy is considered most destructive for the environment and must be implemented cautiously. Similar conditions exist for solar photovoltaic energy [48].

1.6.2 POLICY AND REGULATORY BARRIERS

According to a study conducted to understand the regulatory barriers in implementing the circular economy principles in European Union, the nature of various policy and regulatory barriers can be summarized into six types. The most commonly occurring barriers may be classified as lack of definitions and gaps in legislation, unclear definitions of targets in definition, definition of hard numerical limits in legislations, incomplete implementation of legislation, different and conflicting national implementations of legislation, and conflicting legislations due to conflicts in values being represented. In most of the situations, the key determinant of legislative considerations is the context and application of circulated material. This requires more information to understand the scale of challenges and inefficiencies caused by the legislation. At present, one-size-fits-all legislation for whole sectors is applicable in absence of well-informed locally tailored decisions [15]. The most commonly observed regulatory barriers in a study conducted in European Union are lack of definitions and standards, lack of government enforcement and cooperation, and lack of harmonization in legislation [56].

1.6.3 FINANCIAL AND ECONOMIC IMPLICATIONS

The transition to sustainable energy systems faces great difficulty due to lack of current modern economy to take into account the environmental and social problems resulting from short-term and profitability thinking. It is too complex to finance a decentralized renewable energy infrastructures (e.g., in Germany) because public authorities lack capital and private investors are generally reluctant to invest due to high transaction costs and risk-return-concerns [41]. The economic uncertainty in defining and measuring the long-term benefits of circular economy due to lack of tools and methods is among the principal causes for ineffective growth of circular economy. The institutional and structural barriers such as unfavorable industrial policies, lack of expertise of political decision-makers, complications associated with laws and regulations harm the circular economy businesses [55]. Long-term economic, social and environmental stability is not the priority of the financial sector that stills gives more weight to short-term monetary growth. Governments are reluctant to take bold steps required to achieve ambitious targets set by circular economy out of fear that companies might depart to other locations known to be tax havens and have less stringent social and environmental standards causing a loss of income from taxes and increased unemployment affecting political stature and admiration of the government [36]. In case of oil and gas industry, a change to circular business model demands substantial amount of process modifications, planning, time and investments on the part of the company and availability of funds. High upfront costs and long payback periods hinder the adoption process [57].

1.6.4 STAKEHOLDER ENGAGEMENT AND COLLABORATION CHALLENGES

Organizational and supply chain challenges hamper the establishment of appropriate systems meeting circular economy requirements. In circular economy businesses, generally multiple stakeholders are involved. Absence of sufficient connections and network support and apt partners, probably due to puny environmental awareness of the stakeholders, creates barriers in adopting circular strategies. Absence of collaboration reduces the amount of available resource and hinders building to appropriate supply chains. The organizational challenges related to circular economy implementation include absence of skills and capabilities, operational incompatibility with existing systems, lack of systems thinking, incompatible existing business culture, long-term thinking mindset and communication inabilities; they present barriers to implementation of circular economy [55]. Implementation of circular economy principles is a complex process requiring parallel and consecutive adjustments from diverse stakeholders. For instance, availability of cheap material curbs the need for recycling resulting in an unexploited technical capability to recover materials in absence of resource recovery targets [36].

1.7 CONCLUSION

Circular economy is an economic model that is a departure from the traditional linear take-make-dispose economic model. This model is built around renting out, recycling, refurbishing, reusing and sharing of existing materials, products for as long as possible. It proposes to extend the life cycle of products and their components, thereby reducing the associated waste as well as requirement for fresh or virgin materials. This model is primary driven by R-principles to take advantage of material recycling, product life extension, minimizing the consumption to create a balance between the economic growth and environmental health. The growing concern for climate change and necessity to shift to low-carbon options is forcing the governments, corporations and societies to move towards renewable sources of energy and materials. In this context, energy materials have a very important role to play to meet the ever-increasing consumer demand for energy and to develop sustainable energy systems. The principles and strategies of circular economy can help in selecting appropriate energy management systems for generation, transportation, storage and consumption of renewable energies in such a way that is able to address the challenges of reducing emissions, carbon or environmental footprints, and smooth energy transition to sustainable and resilient energy systems. Adoption of circular economy for energy systems comes with technological and infrastructural challenges and policy and regulatory barriers that have numerous financial and economic implications. Policymakers, researchers and scientists must join hands to address the complex challenges jointly to achieve a successful future.

REFERENCES

[1] Ghisellini, P.; Cialani, C.; Ulgiati, S., A review on circular economy: The expected transition to a balanced interplay of environmental and economic systems, *Journal of Cleaner Production*, 114, 11, 2016.

[2] Heshmati, A. *A Review of the Circular Economy and its Implementation*. IZA Discussion Paper No. 9611, v. Available at SSRN: https://ssrn.com/abstract=2713032, 1–63, December 2015.

[3] Allwood, J. M., Squaring the circular economy: The role of recycling within a hierarchy of material management strategies, in *Handbook of Recycling*. Cambridge: Elsevier, 2014. Cap. 30. http://doi.org/10.1016/B978-0-12-396459-5.00030-1.

[4] Ritzen, S.; Sandstrom, G. O., *Barriers to the Circular Economy – integration of perspectives and domains*. The 9th CIRP IPSS Conference: Circular Perspectives on Product/Service-Systems. [S.l.]: Procedia CIRP 64-https://doi.org/10.1016/j.procir.2017.03.005. 2017. 7–12.

[5] Winans, K.; Kendall, A.; Deng, H., The history and current applications of the circular economy concept, *Renewable and Sustainable Energy Reviews*, 68, 825, 2017.

[6] Korhonen, J., et al., Circular economy as an essentially contested concept, *Journal of Cleaner Production*, 175, 544, 2018.

[7] Chen, W.-M.; Kim, H., Circular economy and energy transition, *Energy & Environment*, 30, 586, 2019.

[8] Raimi, D. et al. *Global Energy Outlook 2023: Sowing the Seeds of an Energy Transistion*. Washington, DC. 2023; https://www.rff.org/publications/reports/global-energy-outlook-2023/.

[9] Uceda-Rodríguez, M., et al., Comparative life cycle assessment of lightweight aggregates made from waste—applying the circular economy, *Applied Sciences*, 12, 1, 2022.

[10] Pauliuk, S., Critical appraisal of the circular economy standard BS 8001:2017 and a dashboard of quantitative system indicators for its implementation in organizations, *Resources, Conservation & Recycling*, 129, 81, 2018.

[11] Prieto-Sandoval, V.; Jaca, C.; Ormazabal, M., Towards a consensus on the circular economy, *Journal of Cleaner Production*, 179, 605, 2018.

[12] Kirchherr, J.; Reike, D.; Hekkert, M., Conceptualizing the circular economy: An analysis of 114 definitions, *Resources, Conservation & Recycling*, 127, 221, 2017.

[13] Rizos, V.; Tuokko, K.; Behrens, A. *The Circular Economy -A review of definitions, processes and impacts*. CEPS Energy Climate House. Brussels. 2017. (No 2017/08, April 2017). https://circulareconomy.europa.eu/platform/sites/default/files/rr2017-08_circulareconomy_0.pdf

[14] Prendeville, D. S. et al. *Circular Economy: Is it enough?* Ecodesign Centre (EDC). Cardiff, UK, pp. 1–18. 2014. https://www.researchgate.net/publication/301779162_Circular_Economy_Is_it_Enough.

[15] Barneveld, V. J. et al. *Regulatory barriers for the Circular Economy-Lessons from ten case studies*. Technopolis group. Amsterdam, The Netherlands, pp. 1–174. 2016. https://ec.europa.eu/newsroom/growth/items/455063

[16] Su, C.; Urban, F., Circular economy for clean energy transitions: A new opportunity under the COVID-19 pandemic, *Applied Energy*, 289, 1, 2021.

[17] Stahel, W. R., Circular economy, *Nature*, 531, 435, 2016.

[18] Jawahir, I. S.; Bradley, R. *Technological Elements of Circular Economy and the Principles of 6R-Based Closed-loop Material Flow in Sustainable Manufacturing*. *13th* Global Conference

on Sustainable Manufacturing – Decoupling Growth from Resource Use. Kentucky: https://www.sciencedirect.com/science/article/pii/S2212827116000822. 2016. pp. 103–108.

[19] Reike, D.; Vermeulen, W. J. V.; Witjes, S., The circular economy: New or refurbished as CE 3.0?—exploring controversies in the conceptualization of the circular economy through a focus on history and resource value retention options, *Resources, Conservation & Recycling*, 135, 246, 2018.

[20] Morseletto, P., Targets for a circular economy, *Resources, Conservation & Recycling*, 153, 1, 2020.

[21] Kalmykovaa, Y.; Sadagopan, M.; Rosado, L., Circular economy—From review of theories and practices to development of implementation tools, *Resources, Conservation & Recycling*, 135, 190, 2018.

[22] Tse, T.; Esposito, M.; Soufani, K., How businesses can support a circular economy. *Harvard Business Review*, 1, 2016.

[23] Pieroni, M. D. P. et al. *PSS design process models: are they sustainability-oriented?* The 9th CIRP IPSS Conference: Circular Perspectives on Product/Service-Systems. Kongens Lyngby, Denmark: Elsevier Procedia CIRP 64-https://doi.org/10.1016/j.procir.2017.03.040. 2017. pp. 67–72.

[24] BERG, A. et al. *Circular Economy for Sustainable Development.* Finnish Environment Institute. Helsinki, Finland, https://helda.helsinki.fi/server/api/core/bitstreams/75f6f473-a071-4340-b096-dbb2da1d13b5/content, pp. 1–24. 2018.

[25] Chatterjee, S. *Circular Economy in Electronics and Electrical Sector.* New Delhi, pp. 1–48. 2021. https://www.meity.gov.in/writereaddata/files/Circular_Economy_EEE-MeitY-May2021-ver7.pdf

[26] DOE-U.S. *National Blueprint for Lithium Batteries 2021–2030.* Washington, pp. 1–24. 2021. https://www.energy.gov/sites/default/files/2021-06/FCAB%20National%20Blueprint%20Lithium%20Batteries%200621_0.pdf.

[27] Tingley, D. D., Embed circular economy thinking into building retrofit, *Communications Engineering*, 1, 2022.

[28] Zucchella, A.; Urban, S., The circular enterprise, *Symphonya- Emerging Issues in Management*, 1, 62, 2020.

[29] Kumar, A., et al., An overview of energy storage and its importance in Indian renewable energy sector-Part I—Technologies and comparison, *Journal of Energy Storage*, 13, 10, 2017.

[30] IEA. *World Energy Investment 2023.* International Energy Agency. Paris, pp. 1–181. 2023. https://www.iea.org/reports/world-energy-investment-2023.

[31] Geissdoerfer, M., et al., The circular economy—A new sustainability paradigm? *Journal of Cleaner Production*, 143, 757, 2017.

[32] Jensen, J. P.; Skelton, K., Wind turbine blade recycling: Experiences, challenges and possibilities in a circular economy, *Renewable and Sustainable Energy Reviews*, 1, 2021.

[33] IBM, *The Future of Energy Business in a Circular Economy*, New York: IBM, 2022.

[34] PWC. *Taking on tomorrow- The rise of circularity in energy, utilities and resources.* PwC. Netherlands, pp. 1–23. 2020. https://www.pwc.com/gx/en/energy-utilities-mining/assets/pwc-the-rise-of-circularity-report.pdf.

[35] Tomić, T.; Schneider, D. R., The role of energy from waste in circular economy and closing the loop Concept-Energy analysis approach, *Renewable and Sustainable Energy Reviews*, 98, 268, 2018.

[36] Velenturf, A. P. M.; Purnell, P., Principles for a sustainable circular economy, *Sustainable Production and Consumption*, 1437, 2021.

[37] Malmodin, J.; Lunden, D., The energy and carbon footprint of the global ICT. *Sustainability*, 1, 2018.

[38] Hailemariam, A.; Erdiaw-Kwasie, M. O., Towards a circular economy: Implications for emission reduction and environmental sustainability, *Business Strategy and the Environment*, 1951, 2022.

[39] Farghali, M., et al., Social, environmental, and economic consequences of integrating renewable energies in the electricity sector: A review, *Environmental Chemistry Letters*, 21, 1381, 2023.

[40] Suarez-Eiroa, B.; Fernandez, E.; Mendez, G. Integration of the circular economy paradigm under the just and safe operating space narrative: Twelve operational principles based on circularity, sustainability and resilience. *Journal of Cleaner Production*, v. 322, pp. 1–13, 2021.

[41] Tyl, B.; Lizarralde, I. *The citizen funding: an alternative to finance renewable energy projects*. The 9th CIRP IPSS Conference: Circular Perspectives on Product/Service-Systems. France: Procedia CIRP 64-https://www.sciencedirect.com/science/article/pii/S2212827117301749. 2017. pp. 199–204.

[42] Meidl, R. A., Disentangling circular economy, sustainability, and waste management principles, *Rice University's Baker Institute for Public Policy*, pp. 1–6, 2021.

[43] Arrudaa, E. H. et al. Circular economy: A brief literature review (2015–2020). *Sustainable Operations and Computers*, pp. 79–86, 2021.

[44] Langen, S. K. V. et al. Promoting circular economy transition: A study about perceptions and awareness by different stakeholders groups. Journal of Cleaner Production, 316, 128–166, 2021.

[45] Foroozanfar, M. H.; Imanipour, N.; Sajadi, S. M. Integrating circular economy strategies and business models: a systematic literature review. *Journal of Entrepreneurship in Emerging Economies*, v. 14, n. 5, pp. 678–700, 2022.

[46] Bimpizas-Pinis, M. et al. Is efficiency enough for circular economy? *Resources, Conservation & Recycling*, v. 167, pp. 1–2, 2021.

[47] Neumann, J. et al. Recycling of Lithium-Ion Batteries—Current State of the Art, Circular Economy and Next Generation Recycling. *Advanced Energy Materials*, v. 12, pp. 1–26, 2022.

[48] Osman, A. I. et al. Cost, environmental impact, and resilience of renewable energy under a changing climate: a review. *Environmental Chemistry Letters*, v. 21, pp. 741–764, 2023.

[49] Hasheminasab, H. et al. A circular economy model for fossil fuel sustainable decisions based on MADM techniques. *Economic Research*, v. 35, n. 1, pp. 564–582, 2022.

[50] Ishaq, M. et al. From Fossil Energy to Renewable Energy: Why is Circular Economy Needed in the Energy Transition? *Frontiers in Environmental Science*, pp. 1–12, 2022.

[51] Neves, S. A., Marques, A. C., Drivers and barriers in the transition from a linear economy to a circular economy. *Journal of Cleaner Production*, 341, 1–13, 2022.

[52] Islam, M. T.; Iyer-Raniga, U. Lithium-Ion Battery Recycling in the Circular Economy: A Review. *Recycling*, v. 33, n. 7, pp. 1–40, 2022.

[53] OECD. *Towards a more resource-efficient and circular economy: The role of G20*. Italy. 2021.

[54] Galvão, G. D. A. et al. *Circular Economy: Overview of Barriers*. 10th CIRP Conference on Industrial Product-Service Systems. Linköping, Sweden: Elsevier Procedia-https://doi.org/10.1016/j.procir.2018.04.011. 2018. pp. 79–85.

[55] Tura, N. et al. Unlocking circular business: A framework of barriers and drivers. *Journal of Cleaner Production*, v. 212, pp. 90–98, 2019.

[56] Stumpf, L.; Schoggl, J.-P.; BAUMGARTNER, R. J. Climbing up the circularity ladder? – A mixed-methods analysis of circular economy in business practice. *Journal of Cleaner Production*, v. 316, pp. 1–16, 2021. https://doi.org/10.1016/j.jclepro.2021.128158.

[57] Sharma, M. et al. Overcoming barriers to circular economy implementation in the oil & gas industry: Environmental and social implications. *Journal of Cleaner Production*, v. 391, pp. 1–19, 2023. https://doi.org/10.1016/j.jclepro.2023.136133

2 Circular Economy
An Industrial Perspective for Sustainable Future

Amandeep Kaur, Sandeep Singh, Niraj Bala*,
Surinder Singh, and Sushil Kumar Kansal

2.1 CONCEPT OF CIRCULAR ECONOMY

The term *circular economy* (CE) was first coined in 1989 by Pearce and Turner, while the dawn of CE can be traced back to the era when it was suggested to implement a cyclic ecological system instead of the inefficient linear one [1]. The concept of CE has received a lot of attention since it was introduced to the global economy, and the regions which appear to be immensely fascinated and committed towards CE are China (2002) and Europe (2015) [2]. Earlier, it has been demonstrated that most of the CE case studies deal with the deployment of CE in China (Figure 2.1) because of its substantial environmental, economic, social and health concerns imposed by its quick and persistent socio-economic growth trend [3]. The circular economy is viewed as a unique economic model that is intended to result in better economic progress and a more united community.

Since the creation of this notion, numerous researchers over the past conceptualized CE in different ways. A CE is a restorative and regenerative industrial system. It strives to eliminate waste and other toxic chemicals from the environment, shifts the focus towards utilization of renewable energy sources, and replaces the end-of-life cycle approach with restoration [4]. This implies reuse of materials, quality products with more superior designs, and systems through innovative business models [5]. However, early articulation of this notion comes from the principle of industrial ecology, where each manufacturing process employs virgin raw materials to generate new products. In this, waste production and resource (material and energy) consumption were both reduced to a minimum [6,7]. The concept of CE clearly advocates the capacity to extend the productive life of resources as a way to create added value and to lessen value destruction [8]. This demonstrates how CE emphasizes resource life extension tactics by adhering to a number of imperatives, including reuse, reduction, remanufacture, waste to energy, product longevity approaches, and cascading of different substances [9].

In today's scenario, the linear production and consumption model of consumer society assumes falsely that natural resources are infinite, which has led to unprecedented levels of pollution and resource depletion. The current trends, such as

* nirajbala@nitttrchd.ac.in

DOI: 10.1201/9781003269779-2

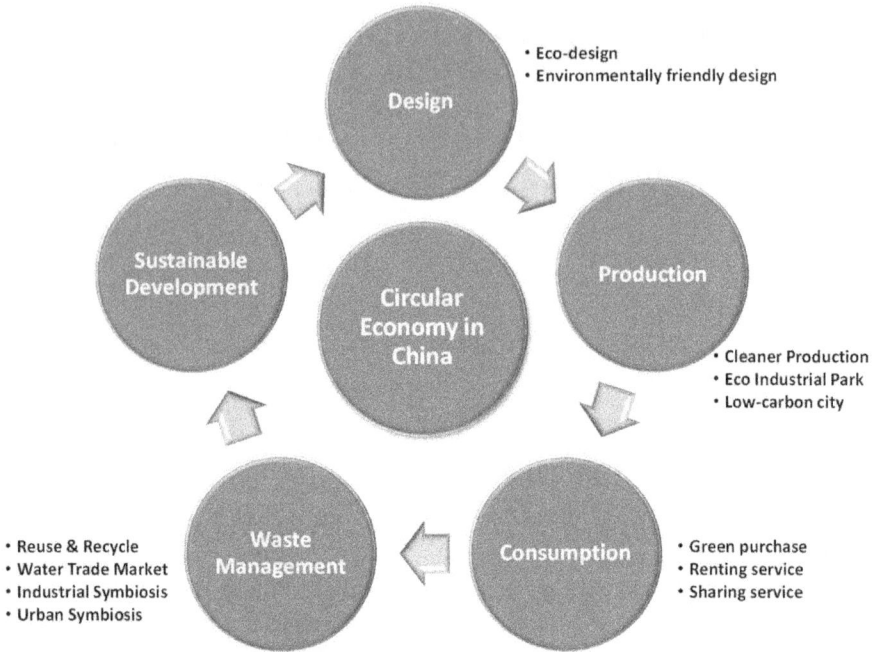

FIGURE 2.1   Circular economy development in China [3].

Source: Copyright 2010. Reproduced with permission from MDPI.

increasing consumption, new generation of consumers, urbanization and employment, tightening legislation and technology leaps, accelerate the transition to a circular economy [10]. In contrast to traditional approach of linear economy (LE), CE
works the central theme "matters matter too," that helps to extend approaches in
terms of efficiency of resources usage which in turn helps to enhance the current
equilibrium between the environment, the economy, and society [11]. As per the
Brundtland Report [12], environment, economy and society are the three universally
adopted major dimensions for sustainable development (SD). The report defines sustainability as "meeting the needs of the present without depriving the needs of the
future generations" [4,13]. In line with the aspect of sustainable development, the
concept of CE becoming more prevalent may result in more environmental benefits
as well as economic and social growth [14].

While at present, the industrial strategies focus more on environmentally related
aspects along with economic prosperity, but consumption as a part of social issues
needs to be thought of attentively and then interlinked. Another major challenge is
how to make industries more circular, because there is ongoing continuous use of
materials and there are physical boundaries of circular flow [4,13]. By concentrating
more on the development of business models under the conceptualization of CE as
a healthier approach to the environment and responsible consumer-led society leading to a better economy by sustainable usage of the pool of existing resources, these

challenges can be eliminated at the industrial level [10]. As per the literature, the main key features of CE are as follows:

- The CE deploys strategies and principles in business to achieve profitability and the UN Sustainable Development Goals (SDGs). The term *waste* is going to be obsolete very soon.
- To focus more on value added products and services, industries will collaborate with their eco partners to create, capture, and deliver sustainable products.
- The industries will focus more on increasing the life span of products and improving resource efficiency and resilience.
- The industries/organizations will work more closely with the customers to create new income resources [15].

2.2 DIFFERENCE BETWEEN LINEAR ECONOMY AND CIRCULAR ECONOMY

The linear economy (LE) in industries following the approach of take-make-dispose scheme concerns taking environmental, demographic and economic components into account. This open system places maximum emphasis on mass consumption and products, whereby raw materials are irrationally used and not fully converted to final products. This traditional LE approaches surpasses the natural resources that cause irreversible loss in terms of its capacity to sustain the biosphere as shown in Figure 2.2. This system not only causes the environmental burden, but also a social damage by exploiting workers to supplement the market needs. LE, the blueprint of existing economy, heavily relies on the use of resources, expensive human labour, supportive system of regulatory authorities, and a system in which producers are not as encouraged to take into account the external costs of their operations.

The LE is very successful in creating the material wealth, but fails in the millennium due to incremental prices/cost of material/resources. Another reason for its failure may be the demographic evolution of mankind in terms of shifting of population from industrialized nations to emerging markets. At present this LE approach is growing to supply constraints unprecedented [6,18,19], coupled with a fast-growing middle class of consumers unable to pay high prices of the products released. The severe consequences of LE endanger "economic sustainability" and the "stability of the ecological system," both of which are crucial to long-term survival [19]. In the past several years, there has been an increasing global interest in the new idea and development paradigm of CE with the goal of providing a suitable substitute to the prevalent economic growth model, as such "receive, produce, and dispose" [20].

Resource Extraction ▸ Production ▸ Distribution ▸ Consumption ▸ Waste

FIGURE 2.2   Linear economy model [16].

Source: Copyright 2010. Reproduced with permission from MDPI.

The CE works on the notion of spiral-loop (or closed-loop) self-replenishing economic construct, which eventually developed from the idea of "performance" economy. The core of the performance economy was to redefine the subject of production, sales, and maintenance, sharing based businesses models which focuses on market performance rather than goods, as shown in Figure 2.3.

Afterwards, the concept of performance economy was incorporated in successful cradle-to-cradle approach, which considers all material involved in industrial and commercial processes to be nutrients and majorly divided into two categories as technical and biological materials [17,18]. The technical materials are described "as material synthetic or mineral by nature and has the potential to remain in a closed loop system of manufacturing, recovery and reuse. Concurrently, it maintains its highest value through multiple product life cycles." On the other hand, biological materials are "non-toxic and can undergo composting while technical materials can be fed back into the cycle using recycle, remanufacture etc." [21]. In order to deal with industrial and commercial issues and compare operational efficiency to solutions found in nature, the economic system should mimic (as in learn from and imitate) the methods of nature since it is considered as another component and feature of the circular economy.

The circular economy also benefits from the interdisciplinary scientific method of industrial ecology, which focuses on energy and material flow of industrial systems [7]. This considers both natural and social environments at local, regional and global level to create closed loop processes that minimize waste. Further, the idea of "natural capitalism" that seeks to establish a common economic framework also takes into account the requirements of both the capital and the environment. The

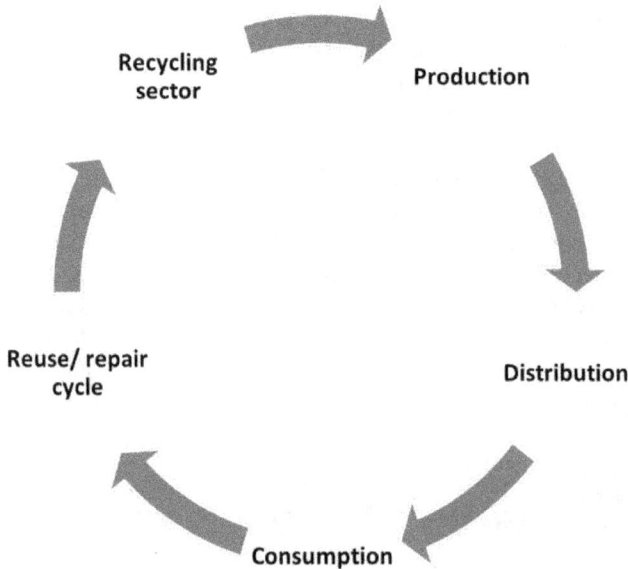

FIGURE 2.3   Circular economy model [16].

Source: Copyright 2010. Reproduced with permission from MDPI.

crux of natural capitalism is to redefine the producer-user contract as a "service and flow" model rather than a "sell" paradigm, increase resource efficiency to extend the availability of natural resources, and create closed-loop production cycles to reuse non-degradable materials [22]. Due to these conceptual evolvements over the time, CE business models have become more prevalent.

Over the time, CE concept also extends its approach in terms of waste usage, beyond internal closed-loop production cycles, in which waste of one industry should be regarded as a potential input to different sectors, whose arrangement is often dubbed as cascading [1].

In CE, the closed loop consists of two supply chains, a forward and a reverse chain. In a reverse chain, a recovered product re-enters the forward chain. Possibilities open up, for instance for business that provide solution and services along with the reverse cycle. This principle can be well adopted for the economic and environmental dimension of sustainability, but to work on social dimension of it, industries needs to bring change in customer behaviour too. At the same time, modification of consumer's behaviour, is also the prerequisite of economic dimension and perception of quality. For this, at industrial level, R-frameworks are being adopted. This system perspective can be considered as the main driving force to attain CE. The CE in industries following the approach take-make Rs imperatives [23,25–27].

The most used R imperative in literature related to CE are 3Rs, 4Rs, and 9 Rs, which have evolved to 10 Rs now. These R imperatives are shown in Figure 2.4.

In nutshell, CE business models are focused on services which are eco effective and reduces the ecological impact of environment inclusive of reduction in social and environmental cost to get the same output, whereas LE focuses on products which are eco efficient, that reduces the ecological impact of environment to get the same output.

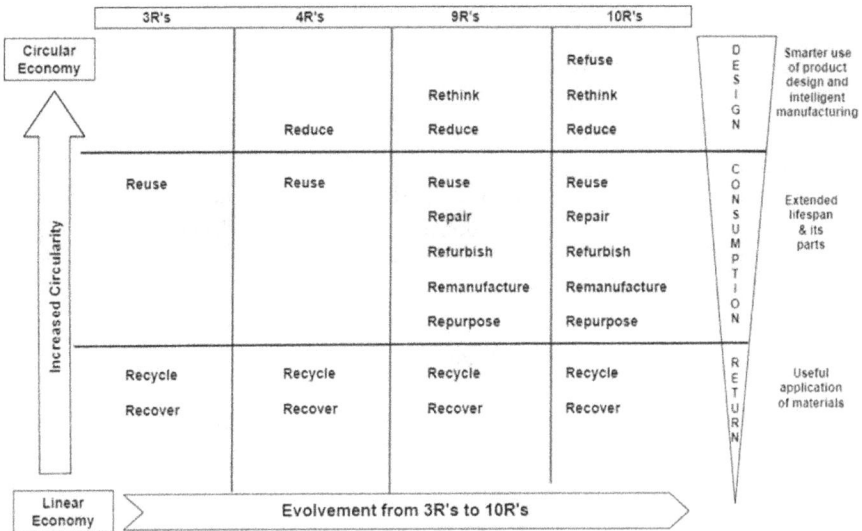

	3R's	4R's	9R's	10R's		
Circular Economy				Refuse	D E S I G N	Smarter use of product design and intelligent manufacturing
			Rethink	Rethink		
		Reduce	Reduce	Reduce		
	Reuse	Reuse	Reuse	Reuse	C O N S U M P T I O N	Extended lifespan & its parts
			Repair	Repair		
			Refurbish	Refurbish		
			Remanufacture	Remanufacture		
			Repurpose	Repurpose		
	Recycle	Recycle	Recycle	Recycle	R E T U R N	Useful application of materials
	Recover	Recover	Recover	Recover		
Linear Economy	Evolvement from 3R's to 10R's					

Increased Circularity

FIGURE 2.4 Evolvement of R imperatives to increased circularity [11,25–27].

Source: Copyright 2010. Reproduced with permission from Elsevier.

2.3   RELATIONSHIP BETWEEN CIRCULAR ECONOMY
AND SUSTAINABLE DEVELOPMENT

Sustainability as per the Brundtland Report is defined as "meeting the demands of the present without diminishing the ability of future generations to fulfil their requirements." It has three distinct dimensions that are generally referred to as social, economic, and environmental. It spurred further conceptual development of sustainability. The SDGs of the 2030 agenda for sustainable development, which were accepted by countries during a historic UN summit in September 2015, went into effect in 2016 [12]. These additional goals apply to all nations that are expected to meet the SDGs during the next 15 years. The new goals are unusual in that they call on all countries, wealthy and poor, to take action to create prosperity while safeguarding the environment. They acknowledge that eradicating poverty requires methods that promote economic growth and meet a variety of social needs such as education, health, social protection, and employment opportunities, as well as addressing climate change and environmental protection. They consist of 17 goals and 169 targets. They will shape the policy of the 193 countries that have agreed to them during the next 15 years. These objectives have the potential to revolutionize our planet and provide a future in which both people and the environment flourish. These goals are a significant step towards fair and ecologically sustainable economic development. They recognize that we all rely on the planet's natural resources for our social and economic well-being, including resources such as clean water, arable land, abundant fish, and wood, as well as ecosystem services such as pollination, nutrient cycling, erosion prevention, and resilience to climate change. The introduction of CE could have positive effects on the environment in addition to economic growth [13]. Holistic or sustainable growth of any industry refers to its economic benefits besides environmental and social value that could be captured from the commercial ventures. The transition from linear to circular economy is supposed to result in attainment of the SDGs, especially 6, 7, 8, 12, 13, and 15, which are:

> **Goal 6: Clean water and sanitation:** "Ensure universal access to and sustainable management of water and sanitation."
>
> **Goal 7: Affordable and clean energy:** "Ensure access to affordable, reliable, sustainable and modern energy for all."
>
> **Goal 8: Decent work and economic growth:** "Promote inclusive, long-term economic growth, full and productive employment, and decent work for all."
>
> **Goal 12: Responsible consumption and production:** "Ensure long-term consumption and production trends."
>
> **Goal 13: Climate action:** "Take immediate action to prevent climate change and its consequences by limiting emissions and encouraging renewable energy growth."
>
> **Goal 15: Life on land:** "Protect, promote and restore sustainable use of terrestrial ecosystems, sustainably manage forests, battle desertification, and halt and reverse land degradation and biodiversity loss."

The achievement of SDG 12 will also be the result of production and consumption habits that must primarily be sustainable. Sustainable progress necessitates an equal and concurrent evaluation of the socio-economic, ecological, intellectual, and social components of a studied economy, industry or particular manufacturing operation, as well as the interplay between all these components [27]. Circular economy helps to harmonize all of the elements because of its fundamental reasoning, which is primarily anchored in ecological and governmental policies in addition to business concerns [28]. CE encourages suitable and eco-friendly sound resource utilization in order to execute a sustainable economy, which is described by an innovative business framework and creative workforce opportunities, along with enhanced wellbeing and visible effects on equity within as well as in different eras in aspects of both commodity use and availability [29]. In this regard, the major objective of CE under the sustainable development framework is to centralize the economic growth from the utilization of finite resources. The waste generation and emission needs to be minimized within the ecosystem adsorption boundaries limits [30]. Dependent on perceptions and the kind of trade-off that needs to be determined, CE generally supports sustainability. The following categories can be used to generalize the link between CE and SD:

1. **Conditional relation:** To achieve the sustainable development objective, CE is a requirement. If CE promotes transformation for sustainable development, a strong relationship exists. It is occasionally regarded as a condition, but sustainability is not always the end consequence.
2. **Beneficiary relation:** The relations may be structured and unstructured in terms of sustainability, without referring to conditionality or alternative approach. It can be one among several solutions to foster sustainability. Another can be producing a certain level of sustainability with other ideas being more and or less sustainable but aids in progression.
   CE advocates for an economic system that dissociates environmental pressure from economic growth through circular production, where waste becomes a resource [31].
3. **Trade-off/cost benefits:** CE implication requires costs that in turn result in profits which are sustainable. Having cost and benefits in regard to sustainability can also lead to negative outcomes. The CE places a greater emphasis on eco-efficiency than on eco effectiveness, which significantly tightens its scope for achieving sustainability.
4. **Selective relation:** Focusing only on some aspects of sustainability selectively. Since different CE models' economic profitability and value proposition depend on recalculating actions that take place after the use phase, the uncertainties in these models are particularly significant [9,34]. CE, for example, has usually been interpreted as a strategy of more suitable waste management. One such restricted perspective may cause CE to collapse, as some reprocessing, reusable, or recuperation choices may be inappropriate in one frame of reference while being relevant in another. The transition possibilities of CE that rely on bioengineering and green chemistry may be more expensive and significant because they prioritize protection over treatment, as opposed to the acknowledged existing technologies. The transition possibilities of CE relying on green chemistry and

bioengineering which may probably be costlier and impactful than the traditional technologies acknowledged, which highlights protection over treatment. Overall, the task ahead in terms of preventative and renewable environmental growth is not one of "more of the same," requiring further deployment of "green" technologies [32]. However, it necessitates a wider and far more exhaustive glance at the layout of radically substitute strategies, across the whole life cycle of any system. It also acts as the interplay among the system and the surroundings in which it is integrated, so that rejuvenation is not only energy retrieval, but rather an advancement of the whole living and economical system comparison to prior organization economy and commodity management. CE has the capacity to identify and apply profoundly new patterns, assisting society in achieving enhanced stability and well-being with minimal or no material, resource, or ecological expenditure. Figure 2.5 shows the intellectual relationships among circular economy (CE) and sustainable development (SD) that may be classified as favourable/unfavourable and completely linked/distributed unequally [24].

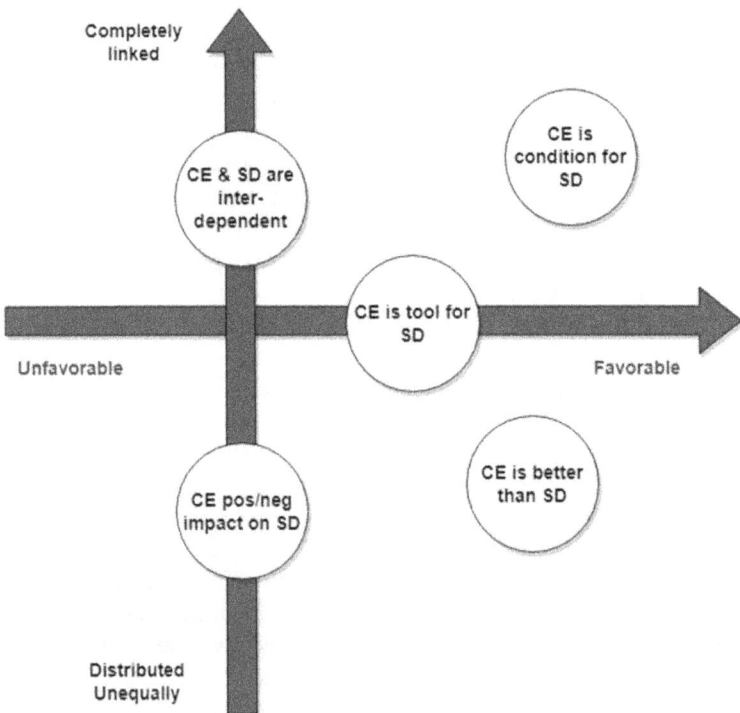

FIGURE 2.5 Relationships among circular economy (CE) and sustainable development (SD) [33].

Source: Copyright 2010. Reproduced with permission from Elsevier.

Finally, it may be emphasized that sustainable trends in line with CE not only entail novel ideas but also inventive actors. In fact, due to the intricacy of the green development concept, its execution is frequently aided by creative engineers and intermediates who offer ideas and concepts aimed at bringing about necessary radical changes in behaviours, regulations, and judgment tools [34].

In order to execute CE, socially desirable and efficient approaches must be included in order to advance economic development and environmental conservation in line with the sustainable development framework. A circular business model can be defined as the rationale of how an organization creates, delivers, and captures value with and within closed material loops. The idea is that a circular business model does not need to close material loops by itself within its internal system boundaries, but can also be part of a system of business models that together close a material loop in order to be regarded as circular [35]. The balance between profit and cost must also be taken into consideration while closing the loop under CE. Circular business model innovations are by nature networked: they require collaboration, communication, and coordination within complex networks of interdependent but independent actors/stakeholders. The challenge of re-designing business ecosystems is to find the "win-win-win" situations for all. Consumers contribute to a CE by using things longer, repairing them or replacing them in a loop for longer by using an R' imperative. There must be a conversation with customers about the use of R strategies in order to create a collaborative product development culture. Since considering the social dimension of SD which is an essential pillar, there is a need to explicitly adapt existing practices as cultural changes. The CE-based economy ecosystem, on the other hand, can benefit from the coordination of various organizations and industries where material replacements can be encouraged and constrain the capitalization of each other's market for profitability.

At large CE business models are operating at three distinct levels:

1. **Micro level:** These are the global trends and drivers which includes initiatives taken at cooperate level such as eco-design of manufacturing plants, waste minimization, cleaner production, and environmental management systems. The most significant activity to do is the cleaner production at industrial level, i.e. at product, company, and consumer levels [25,28,39].
2. **Meso level:** The initiatives taken at interfirm level, where springing of many eco-industrial parks have been developed in order to capitalize on the trading of industrial byproducts such as heat energy, wastewater and manufacturing wastes, etc. This emphasizes the establishment of integrated material, water and energy management system at industrial part level by supporting green supply chain management and reverse logistics [25,28,39].
3. **Macro level:** Thinking and planning for CE at social level. The consideration of CE at macro level approach where societal and stakeholder interests are considered as incentives for CE, in which CE could meet up with SDGs [36,37,39]. This level concerns both with production and consumption. From a production point of view, the circular economy concept encourages the establishment of regional eco-industrial networks, and seeks to create a circular society by optimizing material use eco-efficiency.

"Scavenger" companies, which perform waste recovery, reuse, repair and remanufacturing functions, and "decomposer" companies, which enable recycling by breaking down complex wastes into reusable organic, metal, plastic and other components [33], are being promoted by local governments. Preferential industrial recruitment and financial policies (such as low rents for land and low-interest loans) are being drafted in order to facilitate the operations of such companies. From a consumption viewpoint, the circular economy concept encourages the creation of a conservation-oriented society, seeking to reduce both total consumption and waste production. Both individuals and governments are encouraged to reduce the impacts of consumption, aiming to guide consumers away from wasteful forms of consumption in favour of energy preservation and environmental protection in their daily life. For example, urban residents are now given the choice of having agricultural products in the supermarket that have not been sprayed with pesticides. Some industrial products, like recycled paper, have been labelled as "green products," despite the fact that its production results in non-favourable environmental products [5].

This conception of CE was well adopted in China; a closed loop of material flow is implied and was considered as a new model to achieve sustainable development [38,39]. As per the report of the European Commission, 2014, where a product ends its life cycle it is supposed to be kept in the economic system as a resource which can be productively used again and again, thus creating a further additional value.

2.4 PRINCIPLES FOR A SUSTAINABLE CIRCULAR ECONOMY

It is crucial that the circular economy supports sustainable development given the increasing need to address sustainability issues. This section of the chapter examines significant sustainable development literature to identify principles and ideas that will serve as a framework for the critical assessment of existing principles and ideas for a sustainable circular economy. The three universally adopted dimensions as per sustainable development framework demonstrate the interdependencies between them. The Brundtland Report perpetuated the belief that economic growth could go hand in hand with the preservation or indeed the improvement of environmental resources, a belief that has found wide resonance in the circular economy. Economic growth was considered critical for reducing poverty, as was environmental quality. The risk was, however, that economic growth was associated with environmental degradation and it is here that circular economy professes to offer solutions. In this part of the chapter, an attempt has been made to map the sustainable development principles with CE. The various principles in this regard are explained in the following sections.

2.4.1 CONCURRENT EFFORTS TO SUSTAIN THE ECONOMY TO RAISE LIVING AMENITIES UNDER ENVIRONMENTAL CONSTRAINTS

There is an agreement that sustainability necessitates continuous progress in ecological, societal, and economical results. Figure 2.6 shows the perspectives on their

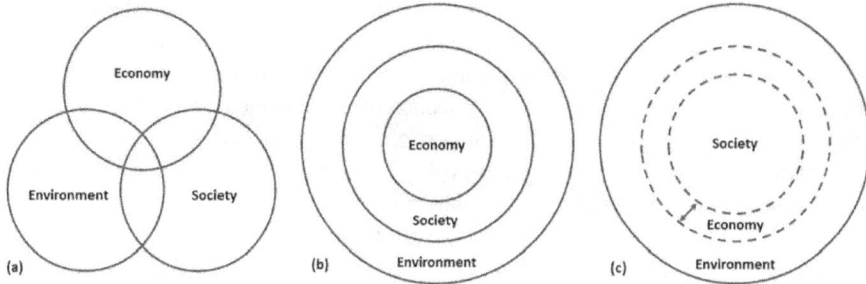

FIGURE 2.6 Viewpoints on sustainable development have progressed from the triple bottom sequence (a) whereby the economy, society, and environment are all recognized equally important (b) to knowing the economic system as the structure of a society as they are both reliant on environment and (c) assessing the economic system as a platform for arranging resources with the objective of preserving or improving socioeconomic well-being, environmental reliability, and economic progress [33].

Source: Copyright 2010. Reproduced with permission from Elsevier.

interactions range from each having equivalent significance to a hierarchy relationship with society and economy inherent in the environment.

These differences emerge throughout the execution and assessment of sustainable growth [40]. While the reported literature also provide a view of selective relationship between these dimensions, incorporating that one attribute should not deteriorate the others [41], others obviously call into query the requirement of inevitable economic growth [42]. Brundtland emerged to make a distinct development demand for impoverished countries from richer countries (WCED, 1987), with starvation elimination seen as major requirement before preserving harmony and preventing environmental calamity.

Brundtland also illustrated the interdependence of the nature, community, and economics, placing ecological resource management in a broader framework. Brundtland thought that while humans might transform the world, there might also be equilibrium among individuals and their surroundings while providing economic gains. It was retained in the conviction that economic expansion could coexist with the retention or even development of ecological resources, an idea that has achieved widespread support in the corporate sustainability community. Industrial growth, as well as ecological quality, were regarded as vital for poverty reduction. However, there was a possibility that economic progress would be accompanied with ecological damage, for which the circular economy claims to offer remedies. Overall, sustainable growth adopts an international and long-term view to ensure a successful, fair, and safe future (WCED, 1987) [43].

2.4.2 Equity between Generations

The development of economic pillars and the reduction of poverty, as well as the advancement of fairness among and between generations, are key aspects of environmental sustainability. Equitable accessibility to resources is essential for present and subsequent generations to have equal chances. For instance, our current loss of exhaustible resources has an impact on prospective generations' chances. Furthermore, instinctual resources (such as oil and bauxite) in emerging nations are being utilized as guarantee for loans from established countries, which are frequently used to charge for the aforementioned to create infrastructure in the original as part of global political endeavours to confirm the sustained flow of raw resource imports [44]. However, the circular economy has paid relatively minimal consideration to both intra and social equality [45].

2.4.3   An Integrated Viewpoint: Environmental, Economic, and Human Systems Connecting the Local and International Platforms

Sustainability studies and implementation must take a comprehensive approach [46], connecting globalization with localized environmental and socioeconomic aspects. Sustainable development has been described as a

> developing area of study relating with the interconnection between natural and socio-economic mechanisms, and with how that interconnection influence the challenge of sustainable development of meeting future needs while reducing poverty and conserving the planet's natural life-sustaining systems.

[47]

Sustainable development typically arises at the interfaces of subsystems, as the ecology offers the foundation for individual existence. The societal norms structured the individual's behavior that influences their surroundings [46]. Sustainable development is about understanding the fundamental nature of interactions between society and the environment, enabling essential advancement in the awareness of issues in sophisticated internal operations, and about (irrecoverable) responses of the landscape framework to various difficulties [47].

2.4.4   A Change in Social Principles Is Required to Maintain Equilibrium between Economy and Environment

A change in human values is necessary to prevent an ecological calamity because technical solutions alone are unlikely to be sufficient [48]. According to Brundtland's proposed remedy, environmental sustainability was being hampered by the conditions of innovation and energy socialization, both of which were deemed acceptable for socioeconomic advancement (WCED, 1987) [43]. The world's life-sustaining systems must be preserved, and resource-hungry lifestyles and the authorization mechanisms that support and facilitate them, particularly in wealthier countries, must be in line with this goal [47,49]. Social ideals and human behaviours should be changed through education, public engagement, and discussion.

2.4.5   Utilizing Resources Sustainably, Leaving the Linear Economy, and Instead Developing a "Resource Flowing Society"

The consumption of irreplaceable products and services is expanded upon by the changes in lifestyles and providing infrastructure mentioned in the previous section

[50]. Malthus first raised issues related to resource scarcity in 1798, and Osborn and Vogt helped bring them back to the forefront of public discussion in 1948 [51]. The Brundtland Report (WCED, 1987) mentioned sustainable use of environmental resources named more than 600 times, making it an essential element of sustainable development. The sequential designs of separation, manufacturing, utilization, and generation of waste are essentially irreconcilable with ecological sustainability [40,41]. The Brundtland commission implied there is "an increasing realisation in worldwide and intergovernmental organisations that it is unfeasible to detach financial advancement problems from ecological problems." Many aspects of advancement weaken the natural environment and social assets. Komiyama and Takeuchi (2006) addressed the significance of a "resource flowing society," which depicts straightforward comparison to the recycling and reuse, competent of producing and using goods, establishing manufacturing processes that flow assets, and circulating resource ways of life [46].

2.4.6 SUSTAINABLE ASSETS, OR THE SUM OF HUMAN-MADE AND ENVIRONMENTAL CAPITAL, DOES NOT DETERIORATE WITH TIME

Environmental capital is referred to as "a collection of productive resources supporting economic activities," including everything from life supporting systems to primary resource availability [52]. Within the sustainable sciences there is a contrast among moderate and powerful sustainability [53]. Inadequate sustainability believes that natural and human-made assets can indeed be replaced and constitute one stock of invested wealth, which would not decline across generations [54]. Strong sustainability, on the other hand, is an important natural investment that cannot be replaced by human-made investment and should not be depleted for subsequent generations [55]. Rockström et al. (2009) provided a perspective on the natural resources that we should see as essential for the upkeep of ecosystems that are supportive of society as a whole [56].

2.4.7 COOPERATIVE CHANGE

Worldwide cooperation and decision-making have changed as a result of the fact that sustainable development affects all nations (WCED, 1987). The most significant change in (inter)national government relates to ongoing initiatives to integrate economic strategies into environmental protection and development. Beyond governments, a broad spectrum of societal competitors collaborates towards sustainable development. For instance, collaborative activity study findings, which combines theory with practical expertise and changes in legislation, businesses, and the public at large, offers collectively "different ways of acquiring knowledge and understanding" in concurrent processes of research and deployment [47]. The critical nature of the problems being studied often forces analyses of phenomena to move towards problem-solving prior to the phenomena which is fully comprehended in research on circular economies and sustainable development. It is crucial to adopt a preventative approach in such situations [40,46]. To thoroughly analyse issues related to sustainability, it is required to identify interactions among challenges, and propose new solutions, so that the applied and fundamental studies from various disciplines be combined within academia [54]. Sustain. Sci. can provide a long-term perspective, analyse potential sustainable development scenarios, create transitional routes for implementation, and combine expertise from the academic world with expertise from other sectors within society through progressive methods [57].

2.4.8 EXECUTION IS CONTEXT DEPENDENT

Although the aforementioned points propose fundamental values and concepts which pertain to all sustainability growth initiatives, their execution can be customized to the specifics of various scenarios and sections of the globe. To be capable of reacting to difficulties in various ecological and cultural situations, a variety of approaches might be developed [46]. Risks can be reduced by pursuing a variety of pathways and approaches towards sustainable growth in the occasion that one does not produce the desired results. The establishment of sustainable development plans can integrate numerous aspects and ideas such as planetary boundaries [56] and Agenda 21 (UN, 1992) [58]. The goals for sustainability and targets are incorporated and inseparable, worldwide in essence, and uniformly relevant, incorporating separate national facts, abilities, and developmental stages and recognizing nationwide goals and strategies. The processes of translating global sustainable growth values and objectives into context-dependent strategy and actions come with an inherent risk. They are susceptible to interpretation from various perspectives, such as the more anthropocentric or ecocentric, value processes [59]. They can lead to various regions of the world taking diverging paths towards varying sustainability results, a few of which might prove less comprehensive.

2.5 CIRCULAR ECONOMY AS A PROMOTER OF SUSTAINABLE DEVELOPMENT IN BUSINESS MANAGEMENT

Across a wide range of business sectors, the circular economy is having an impact on the development of more sustainable businesses. A common issue in studies is sustainable business practices [60]. This effect is more pronounced and seen as a fundamental problem in certain regions, in others the true commercial implications of circular economy practices are not as apparent towards businesses. However, despite the fact that numerous practices have been in use for a long time, many activities or practices that lead to circularity may not have been recognized as such, leading to different names. A few areas of a business or organization where the circular economy can serve as a driver for more sustainability solutions include long-term planning, budget control, supply management, quality assurance, environmental conservation, operational processes, logistic support and chain management, customer service, and scientific and technological advancement. These activities can involve reciprocal, multifaceted changes in the aforementioned domains in addition to others, along with the possibility of one activity having a direct, one-way impact over the other. In an effort to summarize the important effects of the circular economy on each identified domain, Figure 2.7 presents an important impact map with ideas to the major implications of the circular economy towards a sustainable business management [5].

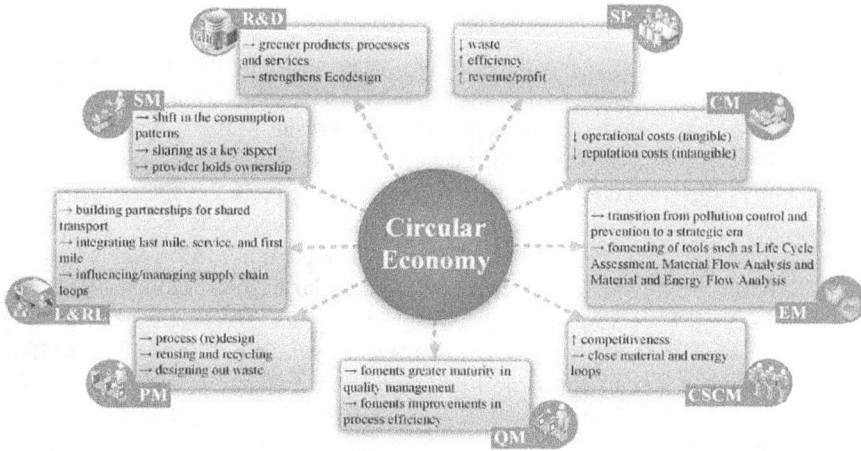

FIGURE 2.7 Benefits of the circular economy to different business areas [5].

Source: Copyright 2010. Reproduced with permission from Elsevier.

2.5.1 STRATEGIC PLANNING (SP)

Utilizing the circular economy's guiding principles to gain a competitive advantage allows businesses to minimize their environmental impact while also enhancing the financial aspects of their operations, which increases resource efficiency and promotes more sustainable economic growth [61]. Circular economy is viewed as a strategic challenge in the current business climate for value creation and profitability for businesses [62]. So, in terms of SP, executives need to fund more for sustainable environmental choices. Numerous experts have accepted the viewpoint that studying the circular economy helps the planning process and makes a business more resilient.

Regarding environmental considerations, organizations that follow circular economy concepts are better equipped to shut input loops and find ways to integrate more circular thinking into their long-term planning. So, the circular economy serves as an incentive for businesses to embrace more environmental friendly business strategies. For illustration, a business might describe how it manages its distribution network to improve resource productivity via reducing waste, reusing, and recycling while also pursuing ecological objectives [63].

Additionally, using circular economy ideas as the primary principles for strategy development gives businesses the ability to recognize and address various economic sources. Businesses that choose this route are capable of reaching various consumers as well as develop ways to lower operational expenses using asset reuse and reclamation [64]. In effort to establish circular business practices fully feasible, organizations can intentionally evolve their corporate models to achieve better circular economy inside their distribution networks. To accomplish this, they can encourage their vendors to embrace activities like goods regeneration and resource recycling [65].

2.5.2 Cost Management (CM)

Adopting circular economy practices and principles is important for CM because it enables businesses to use resources from products that are nearing the end of their usefulness to create new or different products. As a result, waste can be minimized [66] and less virgin material is required for input [62]. Additionally, resource scarcity results in higher and more uncertain prices, which has a detrimental effect on a company's ability to create and capture value [63]. Undoubtedly, every participant associated with the network of the company may need to make investments (of various kinds) in order to transition from linear to circular business models effectively and sustainably [67].

Alcatel's decision to implement electronic waste tracking and disposal is an effective illustration of how to apply the principles of the sustainable economy to CM. As Alcatel enters the Chinese market, it notes the obvious financial benefits that this practice offers, such as the cost savings that come from the recycle of components and, concurrently, the reduction of potential intangible expenses that are to be expected of poor environmental performers [65]. In Alcatel's case, recycling and reusing the waste it produces not only gave the company a definite economic advantage, but it also helped it avoid an intangible cost of reputational damage, which might be anticipated from businesses that are not considered favourably by the public.

2.5.3 Circular Supply Chain Management (CSCM)

The circular economy is frequently seen as offering opportunities for businesses to extend the economic lives of goods by collaborating with stakeholders along the supply chain. It also involves consumers to regain the worth of these goods throughout their life cycles. As a result, the possibility of value recovery presents excellent chances for the development of sustainable supply chain [68,69].

When it comes to supply chain management, the circular economy offers a way to incorporate its key ideas into the current management values. The arrangement and integration of organizational activities within and between company divisions in order to stop, slow down, or narrow the flow of energy and material. As a result, the system's resource input is minimized, and emissions and waste are also kept from disposing of the system, improving efficiency and competitiveness [70].

Companies must coincide their strategies to effectively accomplish particular goals in order to embrace more CSCM, which calls for proactive action from a variety of stakeholders as well as consideration of the financial, ecological, and societal bottom lines. In fact, achieving a balance between the company's ecological, financial, logistical, organizational, and marketing outcomes is essential in order to successfully incorporate circular economy into supply chain management [71]. Considering this, Tura et al. (2019) contend that more circular supply chains have the potential to be more independent as well as to prevent uncertain and high prices [72].

Businesses can achieve social goals in a variety of ways, such as by focusing on local job growth in their business models. Other businesses may choose to take socially conscious actions like pledging their loyalty to their clients and communities.

The circular economy provides businesses an organized method of thinking and enables them to incorporate it into a supply chain growth that is more environmentally friendly [73].

2.5.4 QUALITY MANAGEMENT (QM)

Reliability in goods and procedures is still essential when moving in a circular fashion. Businesses require a change in policies that permits environmental protection while supporting marketing approaches that ensure production and manufacturing quality. It also allows a firm to keep its viability, retaining customer oriented and segmentation approaches [67].

Tukker (2015) notes that the transition from a linear to a circular economy raises issues with the quality of the produced goods and the processes that weren't present or noticed before [74]. Companies need to deal with the possibility that their customers will view products made with retrieved input materials as being of lower quality. Therefore, in order to protect their reputation, organizations are compelled to adopt QM practices. Choosing the primary material moves in QM and the necessity of having a QM system that is more advanced and compatible with the circular economy are therefore crucial. As a result, a circular economy can promote process efficiency enhancement. The selection and approval of vitamins and minerals, soil, and water wastes, and other materials have to be done in accordance with a set of quality guidelines if the profitability of closed-loop systems is to be regarded as of high quality.

2.5.5 ENVIRONMENTAL MANAGEMENT (EM)

Growing environmental concerns have been raised in relation to sustainable and circular business models [75]. The studies have recommended that organizations implement environmental stewardship rather than advocating that they control the environment, which implicitly entails that they own the natural resources they have in their possession [76]. Additionally, it has been noted that the principles of the circular economy are closely related to EM and performance of a company [77]. On the other hand, it is asserted that more sustainable systems can help prevent harmful environmental effects [78]. On the other hand, one shouldn't disregard the implications of rebound effects, which, if not taken into account, could cancel out the effects that the intended strategy aimed to prevent and undermine the sustainability of more circular systems.

The life cycle assessment is one of the environmental evaluation tools that can help the circular economy [79]. Life cycle assessment is referred to as a tool by the International Organization for Standardization (2006a, 2006b) [80], which aims to assess the ecological aspects and possible effects of the life cycle of an entire process, goods, or services. According to Lofgren et al. (2011), it is the most significant tool for determining potential ecological impacts in contemporary EM [81].

By aiding in decision-making, choosing of environmental efficiency indicators, advertising tactics, selection, categorization, and administration decision support, the tool can direct enhancements on the environmental performance of products.

It can also be used to promote green practices and encourage sustainable decision-making [82]. A combination of inputs and results is used to perform the assessment, which helps EM recognize possibilities for production system upgrades. Therefore, the application of life cycle assessment enables a system's overall evaluation with the goal to quantify any prospective environmental effects.

2.5.6 PROCESS MANAGEMENT (PM)

In a circular economy, processes may be redesigned to improve financial outcomes, lessen environmental effects, or improve the life span of products [83]. Because of this, approaches to resource and recovery of products have received a lot of attention. As a result, many waste management procedures and approaches are built on the circularity concept. Numerous processes have undergone redesign in order to streamline aspects of recycling, reusing, and shipping, as well as transportation and reverse logistics [66]. Changing through one pair of processes to the other (manufacturing relying on virgin input data vs. manufacturing relying on non-input data) or even reworking manufacturing mechanisms to create them quite circular (e.g., transformation facilities) may have an impact on planning of processes [84].

The ideas of the recycling and reuse have influenced tactics and modifications in a variety of sectors, notably biochemical, biological, and a variety of manufacturing activities [85]. Modifications to increased circular economy might have an impact on elements like energy usage patterns and varieties of raw resources, waste control techniques (such as minimization, repurposing, composting, and final destruction), and even cooperation agreements to either supply or collect waste.

Finally, all of these interconnections between operations could result in industry symbiosis. This could occur between activities in the identical manufacturing facilities (internal recovery), as well as between enterprises that are connected or unrelated (external recovery). The results of one activity may serve as the intakes of the other, and this connection or link may be motivated by charity or monetary or commercial motivations. Whether or not they are intended to be environmental friendly, such relationships affect the ecological characteristics of activities, guiding businesses towards a more sustainable course.

2.5.7 LOGISTICS AND REVERSE LOGISTICS (L&RL)

Logistics management has spread round the globe, including all distribution network levels in different businesses and industries. This study field has grown to be an essential skill in contemporary distribution networks as well as a revenue-generating activity. A few consequences result from adopting circular economy principles to the domain of logistics. Sharing mobility resources to boost payload capacity (and hence decrease idleness) and eliminate wasteful transit is one of the concerns that stands out, as is encouraging and improving economic agglomeration so that different businesses can exchange facilities [86]. All of this helps to reduce cost of transport, as well as ecological implications and links to one fundamental principle of a circular economy (i.e. sharing).

L&RL governance helps close, slow down, and minimize distribution network loops in the framework of the recycling and reuse. In the interest of promoting recovery measures like recycling and reprocessing, the importance of forming collaborations to encourage backward logistics strategies is also emphasized. Additionally, backward logistics is a crucial part of how a green economy operates. It is based on the idea that post-use wastes need to be moved back downstream in order to be reprocessed again and so recover their utility that relies on the idea of take-back systems [87].

In this aspect, the ultimate client is the agency who regulates the end-of-life work flow, whether they would like to or not. Consumers are the ones who have the power to create or destroy such backward systems, so it is crucial to educate consumers and promote cultural transformation to become more adaptive for circular activities. Moreover, early mile (backward) logistics, final mile (delivery), and utility organization can all be combined to promote circularity [86]. If combined or coordinated, transports, grab, and interim logistic operations could improve transportation effectiveness and circularity. Again, these connections have good effects for company viability in terms of both the ecology and the economy.

2.5.8 SERVICE MANAGEMENT (SM)

Investigation on the circular economy has steadily moved from national policy to the control of the production chain and flow of materials, although it now has a more commercial focus. As a result of their key position among producers and consumers, professional organizations must incline with R imperatives [88]. In order to lessen a detrimental effect, significant modifications in the way items are processed are frequently made, but the comparable possibility in activities is sometimes overlooked. The implementation of product systems and services (PSS), a prototype that utilizes environmental assistance with the possibility to recreate and contend with the "quick" marketplace, has been progressively noticed [89]. These structures have indeed been noted as fantastic indicators of a sustainable society. Providing services instead of items is, in fact, among the most efficacious aspects to move forward into a circular economy. The PSS proposal also affects the creation of efficient take-back mechanisms and the creation of goods that are more robust and make it easier to reclaim value at the end of their useful life spans (for recycling, reprocessing, etc.) [90].

Despite the fact that large and medium-size organizations typically receive more attention; small and medium-sized businesses can also make a contribution to circularity models that involve both upstream and downstream collaborators. Obviously, there are obstacles to overcome, most of which are financial because comparatively small and medium-sized businesses may find it difficult to adapt to alterations in the short- to mid-term.

Additionally, a number of researchers endorse the idea that exchanging resources and providing skills are important measures in the direction of a sustainable future [90,91] support the assertion that PSSs, irrespective of whether they are product characteristics, act effectively, or natural consequence, increase circularity. These procedures might enable customers to get the goods and activities they desire but otherwise wouldn't be able to afford either financially or in terms of duration or location. It does, however, reduce the rate of utilization, which can occasionally devastate reserves.

2.5.9 RESEARCH AND DEVELOPMENT (R&D)

In order to succeed in a more globally interconnected world, technological advancement has been discussed as an effort to investigate strategic benefits and creativity. In this regard, development and investigation focused on eco-design and life cycle assessments which enables choosing different resources and ecological effectiveness throughout the entire product's life cycle [92]. Because of this, an item's development phase can encourage a more circular functioning. The use period of those items is prolonged by inventing long-lasting commodities and improving the life span of existing products (for example, providing services to prolong life of a product, such as repair and reprocessing), which slows the resource flow.

Furthermore, the role product design contributes in a circular business varies somewhat to that of businesses depending on a linear economy [93]. It has been demonstrated that considerate product design can give organizations a competitive advantage, so one could see the significance of committing to R&D activities as it affects economic performance [94]. It's crucial to remember that eco-design doesn't just apply to product growth or even to businesses that focus on selling products. Service-oriented businesses can incorporate the circular economy into their everyday activities by applying eco-design to their company models [88]. It claims that the effects of product design extend a variety of industries and contribute to increased sustainability in production chains by enhancing the life spans of goods, making them easier to handle, and introducing innovation [95].

2.6 CONCLUSIONS

The ultimate goal of this thorough analysis was to determine whether or not CE may help to mitigate the negative ecological effects of conventional economic structures. The concept of CE is only now beginning to be implemented on a global scale; it offers a solid structure for drastically enhancing the current business simulation. Besides, in order to promote preventive measures and regenerating environmental industrial growth it also helps to improved health based on restored integrity of the environment. Only a handful of nations have, however, made the first steps to CE, and more willingness is still needed. With a strong focus on advancements in technology in the form of greener technologies, CE is primarily founded on ecological economy and corporate ecology on both a conceptual and operational level. The latter is a crucial CE principle and needs to be given priority with appropriate regulations. Additionally, the focus on maximising resource utilisation is not entirely in line with the frequently stated desire to reduce the consumption of assets or the heavy dependence on resources that are not renewable. Because effectiveness is not the "prevailing card" in a growth-focused economic structures, where the rebounding effect and rivalry in the market are likely to reduce the potential advantages of improved efficiency, CE is not a suitable tool for such systems (i.e., it is unable to support further growth in the economy). Instead, CE performance and sustainability will emerge as important considerations that guide policies for adapting to novel manufacturing and consumption patterns, able to delay the fall and allow an easier adjustment to more and various sustainable ways of life and economic and social dynamics. This is true both in steady-state focused financial structures in addition to the potential future stage of descending of some global economies.

REFERENCES

[1] Sariatli, F. Linear economy versus circular economy: A comparative and analyzer study for optimization of economy for sustainability. *Visegr. J. Bioecon. Sustain. Dev.*, *6*(1), p. 31, 2017. https://doi.org/10.1515/vjbsd-2017-0005.

[2] Yuan, Z., Bi, J. and Moriguichi, Y. The circular economy: A new development strategy in China. *J. Ind. Ecol.*, *10*(1–2), p. 4, 2006. https://doi.org/10.1162/108819806775545321.

[3] Ogunmakinde, O.E. A review of circular economy development models in China, Germany and Japan. *Recycling*, *4*(3), 2019. https://doi.org/10.3390/recycling4030027.

[4] Diaz, A., Schöggl, J., Reyes, T., et al. Sustainable product development in a circular economy: Implications for products, actors, decision-making support and lifecycle information management. *Sustain. Prod. Consum.*, *26*, p. 1031, 2021. https://doi.org/10.1016/j.spc.2020.12.044.

[5] Barros, M.V., Salvador, R., do Prado, G.F., et al. Circular economy as a driver to sustainable businesses. *Clean. Environ. Syst.*, *2*, p. 100006, 2021. https://doi.org/10.1016/j.cesys.2020.100006.

[6] Ekins, P., Domenech, T., Drummond, P., et al. The circular economy: What, why, how and where. *Background Paper for an OECD/EC Workshop on 5 July 2019 within the Workshop Series "Managing Environmental and Energy Transitions for Regions and Cities,"* Paris.

[7] Allen, D., Davis, S. and Halloran, P. *Sustainable Materials Management: The Road Ahead*. United States Environmental Protection Agency, p. 1. www.epa.gov/smm/sustainable-materials-management-road-ahead, accessed on 25th May 2023.

[8] Blomsma, F. and Brennan, G. The emergence of circular economy: A new framing around prolonging resource productivity. *J. Ind. Ecol.*, *21*(3), p. 603, 2017. https://doi.org/10.1111/jiec.12603.

[9] Belmonte-Ureña, L.J., Plaza-Úbeda, A., Vazquez-Brust, D., et al. Circular economy, degrowth and green growth as pathways for research on sustainable development goals: A global analysis and future agenda. *Ecol. Econ.*, *185*(107050), p. 1, 2021. https://doi.org/10.1016/j.ecolecon.2021.107050.

[10] Antikainen, M., Valkokari, K. and McClelland, J. A framework for sustainable circular business model Innovation. *Technol. Innov. Manag. Rev.*, *6*(7), p. 5, 2016.

[11] Kirchherr, J., Reike, D. and Hekkert, M. Conceptualizing the circular economy: An analysis of 114 definitions. *Resour. Conserv. Recycl.*, *127*, p. 221, 2017. https://doi.org/10.1016/j.resconrec.2017.09.005

[12] Hoyos, D., Bermejo, R. and Arto, I. *Sustainable Development in the Brundtland Report and Its Distortion: Implications for Development Economics and International Cooperation*, p. 13, 2010. www.researchgate.net/publication/278036532_Sustainable_Development_in_the_Brundtland_Report_and_Its_Distortion_Implications_for_Development_Economics_and_International_Cooperation, accessed on 1st June 2023.

[13] Schoggl, J.P., Stumpf, L. and Baumgartner, R.J. The narrative of sustainability and circular economy—A longitudinal review of two decades of research. *Resour. Conserv. Recycl.*, *163*(105073), p. 1, 2020. https://doi.org/10.1016/j.resconrec.2020.105073.

[14] Soh, K.L. and Wong, W.P. Circular economy transition: Exploiting innovative eco-design capabilities and customer involvement. *J. Clean. Prod.*, *320*(128858), p. 1, 2021. https://doi.org/10.1016/j.jclepro.2021.128858.

[15] Calisto Friant, M., Vermeulen, W.J.V. and Salomone, R. A typology of circular economy discourses: Navigating the diverse visions of a contested paradigm. *Resour. Conserv. Recycl.*, *161*, p. 104917, 2020. https://doi.org/10.1016/j.resconrec.2020.104917.

[16] Fura, B., Stec, M. and Miś, T. Statistical evaluation of the level of development of circular economy in European union member countries. *Energies*, *13*(23), 2020. https://doi.org/10.3390/en13236401.

[17] MacArthur, E. Towards the circular economy vol. 1: An economic and business rationale for an accelerated transition." *Ellen Macarthur Foundation*, 2013, p. 1. https://ellenmacarthurfoundation.org/towards-the-circular-economy-vol-1-an-economic-and-business-rationale-for-an, accessed on 23rd May 2023.

[18] Ellen MacArthur. Growth within: A circular economy vision for a competitive Europe. *Ellen MacArthur Foundation*, 2015, p. 100. https://unfccc.int/sites/default/files/resource/Circular%20economy%203.pdf, accessed on 23rd May 2023.

[19] Andersson, D.E. and Andersson, Å.E. Sustainability and the built environment: The role of durability. *Sustainability (Switzerland)*, *11*(18), p. 1, 2019. https://doi.org/10.3390/su11184926.

[20] Ghisellini, P., Cialani, C. and Ulgiati, S. A review on circular economy: The expected transition to a balanced interplay of environmental and economic systems. *J. Clean. Prod.*, *114*, p. 11, 2016. https://doi.org/10.1016/j.jclepro.2015.09.007.

[21] Panchal, R., Singh, A. and Diwan, H. Does circular economy performance lead to sustainable development?—A systematic literature review. *J. Environ. Manage.*, *293*, p. 112811, 2021. https://doi.org/10.1016/j.jenvman.2021.112811.

[22] Jones, P. and Wynn, M.G. The circular economy, natural capital and resilience in tourism and hospitality. *Int. J. Contemp. Hosp. Manag.*, *31*(6), p. 2544, 2019. https://doi.org/10.1108/IJCHM-05-2018-0370.

[23] The world in transition, and Japan's efforts to establish a sound material—cycle society reduce reuse establish a sound material cycle society. *Ministry of the Environment*, Government of Japan, 2008, p. 1. www.env.go.jp/recycle/3r/en/approach/report_material-cycle/2008.pdf, accessed on 15th June 2023.

[24] Langen, V.S.K., Vassillo, C., Ghisellini, P., Restaino, D., Passaro, R. and Ulgiati, S. Promoting circular economy transition: A study about perceptions and awareness by different stakeholders groups. *J. Clean. Prod.*, *316*, p. 128166, 2021. https://doi.org/10.1016/j.jclepro.2021.128166.

[25] Potting, J., Hekkert, M., Worrell, E., et al. Circular economy: Measuring innovation in the product chain. *PBL Netherlands Environ. Assess. Agen.*, *2544*, p. 42, 2017.

[26] Morseletto, P. Targets for a circular economy. *Resour. Conserv. Recycl.*, *153*(December), 2020. https://doi.org/10.1016/j.resconrec.2019.104553.

[27] Alreahi, M., Bujdosó, Z., Dávid, L.D., et al. Green supply chain management in hotel industry: A systematic review. *Sustainability (Switzerland)*, *15*(7), 2023. https://doi.org/10.3390/su15075622.

[28] Car, T., Pilepic, L. and Šimunic, M. Mobile technologies and supply chain management—Lessons for the hospitality industry. *Tour. Hosp. Manag.*, *20*(2), pp. 207–219, 2014. https://doi.org/10.20867/thm.20.2.5.

[29] Carvalho, J.C.de, Vilas-Boas, J. and O'Neill, H. Logistics and supply chain management: An area with a strategic service perspective. *Am. J. Ind. Bus. Manag.*, *4*(1), p. 24, 2014. https://doi.org/10.4236/ajibm.2014.41005.

[30] Suárez-Eiroa, B., Fernández, E., Méndez-Martínez, G., et al. Operational principles of circular economy for sustainable development: Linking theory and practice. *J. Clean. Prod.*, *214*, p. 952, 2019. https://doi.org/10.1016/j.jclepro.2018.12.271.

[31] F. Blomsma and M. Tennant, "Circular economy: Preserving materials or products? Introducing the Resource States framework. *Resour. Conserv. Recycl.*, 156 (1292). pp. 1–4, 2020. doi: 10.1016/j.resconrec.2020.104698.

[32] Mwasilu, F. and Jung, J.W. Potential for power generation from ocean wave renewable energy source: A comprehensive review on state-of-the-art technology and future prospects. *IET Renew. Power Gene.* *13*(3), p. 363, 2019.

[33] Velenturf, A.P.M. and Purnell, P. Principles for a sustainable circular economy. *Sustain. Prod. Consum.* *27*(February), p. 1437, 2021. https://doi.org/10.1016/j.spc.2021.02.018.

[34] Bibri, S.E. A foundational framework for smart sustainable city development: Theoretical, disciplinary, and discursive dimensions and their synergies. *Sustain. Cities Soc.*, *38*, p. 758, 2018. https://doi.org/10.1016/j.scs.2017.12.032.

[35] Mentink, B. Circular business model innovation: A process framework and a tool for business model innovation in a circular economy. *Delft Univer. Technol.*, p. 168, 2014.

[36] Kazerooni Sadi, M., Abdullah, A., Navazandeh Sajoudi, M., Bin Mustaffa Kamal, M.F., Torshizi, F. and Taherkhani, R. Reduce, reuse, recycle and recovery in sustainable construction waste management. *Adv. Mat. Res.*, p. 446, 2012. https://doi.org/10.4028/scientific5/amr.446-449.937.

[37] Eddy, R. The circular economy: A new development strategy in China. *Build. Eng.*, *94*(11), p. 24, 2019.

[38] Ogunmakinde, O.E., Egbelakin, T. and Sher, W. Contributions of the circular economy to the UN sustainable development goals through sustainable construction. *Resour. Conserv. Recycl.*, *178*, p. 106023, 2022. https://doi.org/10.1016/j.resconrec.2021.106023

[39] Geng, Y. and Doberstain, B. Developing circular economy in China, challenges and opportunity. *J. Sustain. Dev.*, 2014, p. 37, 2008. https://doi.org/10.3843/SusDev.15.3.

[40] Sala, S., Ciuffo, B. and Nijkamp, P. A systemic framework for sustainability assessment. *Ecol. Econ.*, *119*, p. 314, 2015. https://doi.org/10.1016/j.ecolecon.2015.09.015.

[41] Millar, N., McLaughlin, E. and Börger, T. The circular economy: Swings and roundabouts? *Ecol. Econ.*, *158*, p. 11, 2019. https://doi.org/10.1016/j.ecolecon.2018.12.012.

[42] Schroder, P., Bengtsson, M., Cohen, M., Dewick, P., Hofstetter, J. and Sarkis, J. Degrowth within—aligning circular economy and strong sustainability narratives. *Resour. Conserv. Recycl.*, *146*, p. 190, 2019.

[43] WCED, S.W.S. World commission on environment and development. *Our Common. Future*, *17*(1), p. 1, 1987. https://idl-bnc-idrc.dspacedirect.org/bitstream/handle/10625/152/WCED_v17_doc149.pdf

[44] Bettencourt, L.M. and Kaur, J. Evolution and structure of sustainability science. *Proc. Natl. Acad. Sci.*, *108*(49), p. 19540, 2011. https://doi.org/10.1073/pnas.1102712108.

[45] Kirchherr, J., Reike, D. and Hekkert, M. Conceptualizing the circular economy: An analysis of 114 definitions. *Resour. Conserv. Recycl.*, *127*, p. 221, 2017. https://doi.org/10.1016/j.resconrec.2017.09.005.

[46] Komiyama, H. and Takeuchi, K. Sustainability science: Building a new discipline. *Sustain. Sci.*, *1*, p. 1, 2006. https://doi.org/10.1007/s11625-006-0007-4.

[47] Kates, R.W., Clark, W.C., Corell, R., Hall, J.M., Jaeger, C.C., Lowe, I., McCarthy, J.J., Schellnhuber, H.J., Bolin, B., Dickson, N.M. and Faucheux, S. Sustainability science. *Science*, *292*(5517), p. 641, 2011. https://doi.org/10.1007/s11625-010-0117-x.

[48] Meadows, D.H., Goldsmith, E.I. and Meadow, P. *The Limits to Growth*, vol. 381. London: Earth Island Limited, 1972. www.donellameadows.org/wp-content/userfiles/Limits-to-Growth-digital-scan-version.pdf, accessed on 21st May 2023.

[49] Wiedmann, T., Lenzen, M., Keyber, L.T. and Steinberger, J.K. Scientists' warning on affluence. *Nat. Commun.*, *11*(1), p. 3107, 2020. https://doi.org/10.1038/s41467-020-16941-y.

[50] Geissdoerfer, M., Savaget, P., Bocken, N.M. and Hultink, E.J. The circular economy—A new sustainability paradigm? *J. Clean. Prod.*, *143*, p. 757, 2017. https://doi.org/10.1016/j.jclepro.2016.12.048.

[51] Rome, A. Sustainability: The launch of spaceship earth. *Nature*, *527*, p. 443, 2015. https://doi.org/10.1038/527443a.

[52] Pearce, D. Economics, equity and sustainable development. *Futures*, *20*(6), p. 598, 1988. www.francoarchibugi.it/pdf/89-11a%20(E)%20The%20challenge%20of%20sustainable%20development.pdf, accessed on 25th May, 2023.

[53] Stilwell, F. and Jones, E. Weak versus strong sustainability: Exploring the limits of two opposing paradigms. *J. Aust. Political. Econ.*, *65*, p. 163, 2010. https://doi.org/10.4337/9781781007082.

[54] Kates, R.W. What kind of a science is sustainable science? *Proc. Natl. Acad. Sci.*, *108*(49), p. 19449, 2011. www.pnas.org/doi/pdf/10.1073/pnas.1116097108, accessed on 20th May 2023.

[55] Bond, A.J., Dockerty, T., Lovett, A., Riche, A.B., Haughton, A.J., Bohan, D.A., Sage, R.B., Shield, I.F., Finch, J.W., Turner, M.M. and Karp, A. Learning how to deal with values, frames and governance in sustainability appraisal. *Reg. Stud.*, *45*(8), p. 1157, 2011. https://doi.org/10.1080/00343404.2010.485181.

[56] Rockström, J., Steffen, W., Noone, K., Persson, Å., Chapin, F.S., Lambin, E.F., Lenton, T.M., Scheffer, M., Folke, C., Schellnhuber, H.J. and Nykvist, B. A safe operating space for humanity. *Nature*, *461*(7263), p. 472, 2009. https://doi.org/10.1038/461472a.

[57] Wise, C., Pawlyn, M. and Braungart, M. Eco-engineering: Living in a materials world. *Nature*, p. 172, 2013. https://doi.org/10.1038/494172a.

[58] Briant Carant, J. Unheard voices: A critical discourse analysis of the millennium development goals' evolution into the sustainable development goals. *Third World Q.*, *38*(1), p. 16, 2017. https://doi.org/10.1080/01436597.2016.1166944

[59] Barrett, J. and Scott, K. Link between climate change mitigation and resource efficiency: A UK case study. *Glob. Environ. Change*, *22*(1), p. 299, 2012. https://doi.org/10.1016/j.gloenvcha.2011.11.003.

[60] Dentchev, N., Rauter, R., Jóhannsdóttir, L., Snihur, Y., Rosano, M., Baumgartner, R., Nyberg, T., Tang, X., van Hoof, B. and Jonker, J. Embracing the variety of sustainable business models: A prolific field of research and a future research agenda. *J. Clean. Prod.*, *194*, p. 695, 2018. https://doi.org/ 10.1016/j.jclepro.2018.05.156.

[61] Haas, W., Krausmann, F., Wiedenhofer, D. and Heinz, M. How circular is the global economy?: An assessment of material flows, waste production, and recycling in the European Union and the world in 2005. *J. Ind. Ecol.*, *19*(5), p. 765, 2015. https://doi.org/10.1111/jiec.12244.

[62] Fonseca, L.M., Domingues, J.P., Pereira, M.T., Martins, F.F. and Zimon, D. Assessment of circular economy within Portuguese organizations. *Sustainability*, *10*(7), p. 2521, 2018. https://doi.org/10.3390/su10072521.

[63] Singh, M.P., Chakraborty, A. and Roy, M. Developing an extended theory of planned behavior model to explore circular economy readiness in manufacturing MSMEs, India. *Resour. Conserv. Recycl.*, *135*, p. 313, 2018. https://doi.org/10.1016/j.resconrec.2017.07.015.

[64] Park, J., Sarkis, J. and Wu, Z. Creating integrated business and environmental value within the context of China's circular economy and ecological modernization. *J. Clean. Prod.*, *18*(15), p. 1494, 2010. https://doi.org/10.1016/j.jclepro.2010.06.001.

[65] Geissdoerfer, M., Morioka, S.N., de Carvalho, M.M. and Evans, S. Business models and supply chains for the circular economy. *J. Clean. Prod.*, *190*, p. 712, 2018. https://doi.org/10.1016/j.jclepro.2018.04.159.

[66] Stahel, W.R. The circular economy. *Nature*, *531*(7595), p. 435, 2016. https://doi.org/10.1038/531435a.

[67] Lahti, T., Wincent, J. and Parida, V. A definition and theoretical review of the circular economy, value creation, and sustainable business models: Where are we now and where should research move in the future? *Sustainability*, *10*(8), p. 2799, 2018. https://doi.org/10.3390/su10082799.

[68] Hofmann, F. Circular business models: Business approach as driver or obstructer of sustainability transitions? *J. Clean. Prod.*, *224*, p. 361, 2019. https://doi.org/10.1016/j.jclepro.2019.03.115.

[69] Hofmann, F. and Jaeger-Erben, M. Organizational transition management of circular business model innovations. *Bus. Strategy Environ.*, *29*(6), p. 2770, 2020. https://doi.org/10.1002/bse.2542.

[70] Geisendorf, S. and Pietrulla, F. The circular economy and circular economic concepts—a literature analysis and redefinition. *Thunderbird Int. Bus. Rev.*, *60*(5), p. 771, 2018. https://doi.org/10.1002/tie.21924.

[71] Kazancoglu, Y., Kazancoglu, I. and Sagnak, M. A new holistic conceptual framework for green supply chain management performance assessment based on circular economy. *J. Clean. Prod.*, *195*, p. 1282, 2018. https://doi.org/10.1016/j.jclepro.2018.06.015.

[72] Tura, N., Hanski, J., Ahola, T., Stahle, M., Piiparinen, S. and Valkokari, P. Unlocking circular business: A framework of barriers and drivers. *J. Clean. Prod.*, *212*, p. 90, 2019. https://doi.org/10.1016/j.jclepro.2018.11.202.

[73] Leigh, M. and Li, X. Industrial ecology, industrial symbiosis and supply chain environmental sustainability: A case study of a large UK distributor. *J. Clean. Prod.*, *106*, p. 632, 2015. https://shura.shu.ac.uk/8796/7/Li%20Industrial%20ecology%2C%20industrial%20symbiosis%20and%20supply%20chain%20environmental%20sustainability.pdf, accessed on 17th May, 2023.

[74] Tukker, A. Product services for a resource-efficient and circular economy—a review. *J. Clean. Prod.*, *97*, p. 76, 2015. https://doi.org/10.1016/j.jclepro.2013.11.049.

[75] Salvador, R., Barros, M.V., da Luz, L.M., Piekarski, C.M. and de Francisco, A.C. Circular business models: Current aspects that influence implementation and unaddressed subjects. *J. Clean. Prod.*, *250*, p. 119555, 2020. https://doi.org/10.1016/j.jclepro.2019.119555.

[76] Cristoni, N. and Tonelli, M. Perceptions of firms participating in a circular economy. *Eur. J. Sustain. Dev.*, *7*(4), p. 105, 2018. https://doi.org/10.14207/ejsd.2018.v7n4p105.

[77] Jabbour, C.J.C., Seuring, S., de Sousa Jabbour, A.B.L., Jugend, D., Fiorini, P.D.C., Latan, H. and Izeppi, W.C. Stakeholders, innovative business models for the circular economy and sustainable performance of firms in an emerging economy facing institutional voids. *J. Environ. Manage.*, *264*, p. 110416, 2020. https://doi.org/10.1016/j.jenvman.2020.110416.

[78] Tura, N., Hanski, J., Ahola, T., Ståhle, M., Piiparinen, S. and Valkokari, P. Unlocking circular business: A framework of barriers and drivers. *J. Clean. Prod.*, *212*, p. 90, 2019. https://doi.org/10.1016/j.jclepro.2018.11.202.

[79] Scheepens, A.E., Vogtländer, J.G. and Brezet, J.C. Two life cycle assessment (LCA) based methods to analyse and design complex (regional) circular economy systems. Case: Making water tourism more sustainable. *J. Clean. Prod.*, *114*, p. 257, 2016. https://doi.org/10.1016/j.jclepro.2015.05.075.

[80] Kjaer, L.L., Pagoropoulos, A., Schmidt, J.H. and McAloone, T.C. Challenges when evaluating product/service-systems through life cycle assessment. *J. Clean. Prod.*, *120*, p. 95, 2016. https://doi.org/10.1016/j.jclepro.2016.01.048.

[81] Lofgren, B., Tillman, A.M. and Rinde, B. Manufacturing actor's LCA. *J. Clean. Prod.*, *19*(17–18), p. 2025, 2011. https://doi.org/10.1016/j.jclepro.2011.07.008.

[82] Morrissey, A.J. and Browne, J. Waste management models and their application to sustainable waste management. *Waste Manag.*, *24*(3), p. 297, 2004. https://doi.org/10.1016/j. wasman.2003.09.005.

[83] Lofthouse, V. and Prendeville, S. Human-centred design of products and services for the circular economy—a review. *Des. J.* *21*(4), p. 451, 2018. https://doi.org/10.1080/146 06925.2018.1468169.

[84] De los Rios, I.C. and Charnley, F.J. Skills and capabilities for a sustainable and circular economy: The changing role of design. *J. Clean. Prod.*, *160*, p. 109, 2017. https://doi. org/10.1016/j.jclepro.2016.10.130.

[85] Li, H., Bao, W., Xiu, C., Zhang, Y. and Xu, H. Energy conservation and circular economy in China's process industries. *Energy*, *35*(11), p. 4273, 2010. https://doi. org/10.1016/j.energy.2009.04.021.

[86] Van Buren, N., Demmers, M., Van der Heijden, R. and Witlox, F., 2016. Towards a circular economy: The role of Dutch logistics industries and governments. *Sustainability*, *8*(7), p. 647, 2016. https://doi.org/10.3390/su8070647.

[87] Stal, H.I. and Jansson, J., 2017. Sustainable consumption and value propositions: Exploring product—service system practices among Swedish fashion firms. *Sustain. Dev.*, *25*(6), p. 546, 2017. https://doi.org/10.1002/sd.1677.

[88] Heyes, G., Sharmina, M., Mendoza, J.M.F., Gallego-Schmid, A. and Azapagic, A. Developing and implementing circular economy business models in service-oriented technology companies. *J. Clean. Prod.*, *177*, p. 621, 2018. https://doi.org/10.1016/j. jclepro.2017.12.168.

[89] Bocken, N.M., Mugge, R., Bom, C.A. and Lemstra, H.J. Pay-per-use business models as a driver for sustainable consumption: Evidence from the case of HOMIE. *J. Clean. Prod.*, *198*, p. 498, 2018. https://doi.org/10.1016/j.jclepro.2018.07.043.

[90] Todeschini, B.V., Cortimiglia, M.N., Callegaro-de-Menezes, D. and Ghezzi, A. Innovative and sustainable business models in the fashion industry: Entrepreneurial drivers, opportunities, and challenges. *Bus. Horiz.*, *60*(6), p. 759, 2017. https://doi. org/10.1016/j.bushor.2017.07.003.

[91] Maffei, A., Grahn, S. and Nuur, C. Characterization of the impact of digitalization on the adoption of sustainable business models in manufacturing. *Procedia Cirp.*, *81*, p. 765, 2019. https://doi.org/10.1016/j.procir.2019.03.191.

[92] Ribeiro, I., Peças, P. and Henriques, E. A life cycle framework to support materials selection for Ecodesign: A case study on biodegradable polymers. *Mater. Des.*, *51*, p. 300, 2013. https://doi.org/10.1016/j.matdes.2013.04.043.

[93] Den Hollander, M.C., Bakker, C.A. and Hultink, E.J. Product design in a circular economy: Development of a typology of key concepts and terms. *J. Ind. Ecol.*, *21*(3), p. 517, 2017. https://doi.org/10.1111/jiec.12610.

[94] Bocken, N.M., De Pauw, I., Bakker, C. and Van Der Grinten, B. Product design and business model strategies for a circular economy. *J. Ind. Prod. Eng.*, *33*(5), p. 308, 2016. https://doi.org/10.1080/21681015.2016.1172124.

[95] Macarthur, E. and Heading, H. How the circular economy tackles climate change. *Ellen MacArthur Found*, *1*, p. 1, 2019. https://ellenmacarthurfoundation.org/completing-the-picture, accessed on 21st June 2023.

3 Lithium as New Gold for Advanced Energy Material, Availability in Nature, Recovery from Waste, and Sustainability

Majeti Narasimha Vara Prasad*
and Chanchal Kumar Mitra

3.1 INTRODUCTION

Population explosion and excessive utilization of natural resources are causing pollution and other serious environmental damages. Emissions from the consumption of natural resources, particularly fossil fuels, are also causing climate change. With increasing standard of life and quality of living, our energy demand is ever increasing. Several non-conventional sources, particularly renewable sources of energy, are now tapped to meet the increased energy demand as the quality of life improves. To address atmospheric and environmental pollution, new strategies and techniques to recover beneficial materials from waste have to be developed. Though exploitation of natural resources has improved the quality of the standard of living with enhanced material comfort, there is a need to generate energy and address related environmental problems such as carbon emissions. Conventional high-energy materials, like coal, petroleum, and natural gas, give us easily usable energy (for example electricity). This is a class of materials with a high amount of stored chemical energy that can be released for the production of usable energy (for example, by combustion). Centralized production of energy also encompasses a wide array of materials involved in energy transmission (like metals like copper and aluminum for electricity conductors or electric cables and steel used in transmission towers) and storage (like in rechargeable batteries). Further, some special categories of energy materials can play a role in reducing the power consumption or efficiency of existing devices. This field is rapidly growing, cutting across disciplines such as material science, engineering, environmental science, economics, biology, geology, etc. A few keywords in the broad area of energy materials (everything related to the production, distribution and storage) include:

* mnvsl@uohyd.ac.in

DOI: 10.1201/9781003269779-3

- Solar energy; solar systems integration; photovoltaic materials, devices, modules, and systems
- Renewable energy, geothermal, solar thermal, fossil fuels, natural gas, nuclear energy, and bioenergy
- Energy storage and grid modernization, batteries and fuel cells, electric grid, superconductors, and CO_2 capture, storage, and mineralization

It is in this direction that in 2015 UN proposed Sustainable Development Goals (SDGs) to be achieved by 2030 and the European Climate Law enshrining the 2050 climate neutrality; these are two global key objectives. The European Green Deal Roadmap highlights the climate ambition toward clean, affordable, and secure energy, an industrial strategy for a clean and circular economy, toward an almost zero-pollution ambition for a toxic-free environment [1]. The European Commission has already set out a clear vision of how to achieve climate neutrality by 2050. UN-SDG and EU-Green Deal have somewhat overlapping agendas. A Clean Planet for all—A European strategic long-term vision for a prosperous, modern, competitive, and climate-neutral economy is envisioned. UN-SDG and EU-Green Deal Roadmap (EU-GDR) highlight achieving a climate-neutral and circular economy for full mobilization of the industry. It takes 25 years—close to a generation—to transform an industrial energy-intensive sector and all the associated value or related supply chains. European Commission will adopt an EU industrial strategy to address the twin challenges of the Green Deal and the digital transformation. Europe must leverage the potential of digital transformation, which is a key enabler for reaching the Green Deal (EU-GDR) objectives. Together with the industrial strategy, a new circular economy action plan will help modernize the EU's economy and draw benefit from the opportunities of the circular economy domestically and globally. A key aim of the new policy framework will be to stimulate the development of lead markets for climate-neutral and circular products, both within the EU and beyond.

3.2 CIRCULAR ECONOMY

A circular economy is based on three principles, all driven by (1) design, (2) conservation, and (3) recycling and reuse. It is broadly a philosophy based on a conservative lifestyle. The salient features of a circular economy are as follows.

1. *Eliminate waste and reduce pollution.* Disposal of waste, particularly industrial waste, is a serious environmental concern. Chemical and biological waste disposal can cause severe pollution. Elimination and or reduction of industrial and/or domestic waste may be an energy-intensive process.
2. *Circulate products and reuse materials.* Metals like copper, silver, iron, aluminum, etc. are ideal candidates for reuse and recycling, and this too reduces the waste of natural resources. Economic considerations must favor reuse and recycling rather than fresh mining of raw materials. However, waste products from the biotechnology industry (say from a bioreactor or a

fermenter) are often tough to recycle. Modern synthetic items like plastics too are hard to recycle.

3. *Regenerate nature by conservation.* Both agriculture and agroforestry must practice the best ideas for conservation. For example, plantations for palm oil production can be reduced, new substitutions for palm oil found, or use of oil must be curtailed. Similarly, the use of wood as a structural material may be reduced and monoculture forestry may be avoided as far as practicable. The major focus should be on sustainability rather than immediate economic gain.

In brief, a circular economy promotes sustainability. Renewable energy use is one of the key pillars of this endeavor and energy storage using Li-ion batteries on a large scale is the current focus. Figure 3.1 shows three types of use of raw materials.

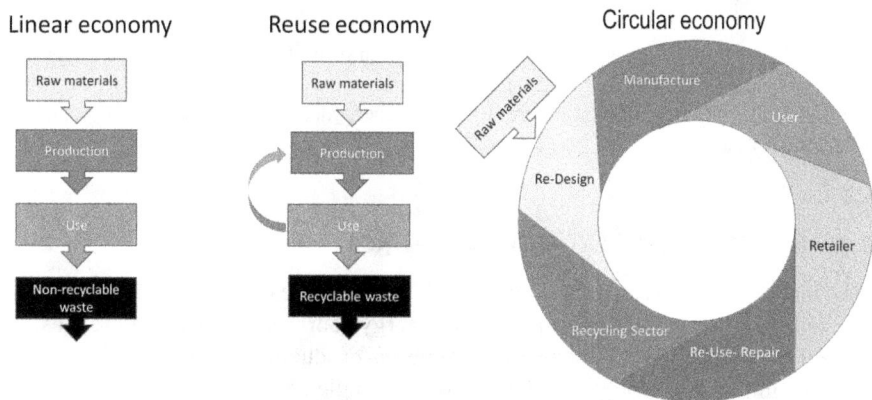

FIGURE 3.1   Three types of use of raw materials: linear economy, reuse economy, and circular economy.

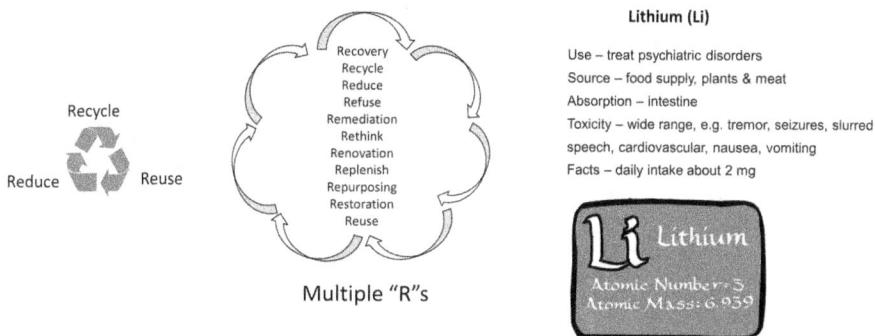

FIGURE 3.2   Conventional three Rs expanded to multiple Rs.

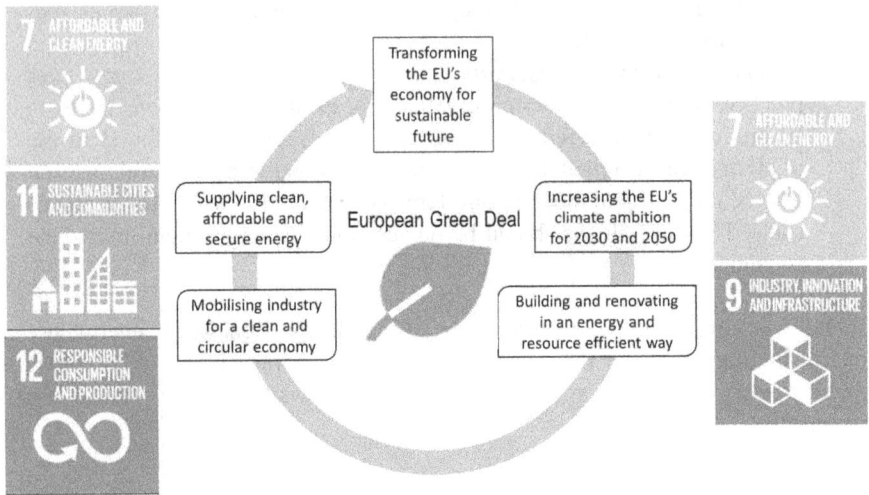

FIGURE 3.3 Interrelationships of objectives between several UN Sustainable Development Goals and the EU Green Deal roadmap. These goals are enumerated in detail in the references cited. In brief, the goals concern renewable energy, sustainable growth, and reducing waste and pollution.

3.3   ENERGY AS A MATERIAL RESOURCE

Like matter, energy is conserved, but usable energy (like electricity; also called the free energy) is mostly derived from fossil energy sources, like coal, natural gas, or natural oil (we can also include nuclear energy-producing materials like uranium and thorium in this group). Electricity, after a single use, is often converted into a form of energy (waste heat, for example) that is not usable. Therefore we need to supply energy (fossil fuel or otherwise) continuously to generate more and more electricity to meet our demands in modern lifestyle. Today most of the commercial electricity is produced from fossil fuel sources. However, electricity can also be produced from sunlight (solar energy) using the photovoltaic effect (solar panels), and these devices have become very competitive these days in terms of price and performance. Solar energy, in terms of overall cost and performance, generates fewer side products (pollution) during its life (cradle-to-grave) cycle. Solar energy is hence called a renewable resource, like wind energy or wave energy. Mineral extraction, for example, iron or lithium or cobalt, also uses energy for processing (iron extraction uses coke, a derivative product of coal, as a source of energy), and the per capita energy use is often used as a quality or mark of the standard of living.

3.4   COSTS ASSOCIATED WITH ENERGY PRODUCTION

Energy production (for example, electricity) is associated with some costs of production. There are several different ways these costs are computed, depending on the intended use. Some of these costs can be considered as mineral (material) costs, some energy is also used (energy costs) during the production process, and some are investment costs (rewards to the financiers). The calculations can be byzantine and hence we shall stick only to the basic ideas. Consider the example of a coal-fired electricity generation plant. The basic costs are as follows:

1. Plant setup costs
2. Plant running costs
3. Electricity distribution costs
4. Investment costs (profit, interest on capital, etc.).

As mentioned, computation of the individual costs can be daunting. For example, the plant setup can be considered a fixed cost with a built-in depreciation period of, say, 25 years. However, the plant may need upgradation during this period (say, a newer pollution control device needs to be installed) or may not perform well (at full capacity) during part of this period. These costs may be included here or in the running costs. The plant may contain lots of steel and the mining and processing of the steel involved considerable environmental damage with considerable costs associated with damage remediation. Similar consideration may be applied to cement and concrete used in the construction. These costs are often not explicitly included or specified. Similarly, the running cost includes the cost of input raw materials, cost towards depreciation and contingencies (for example, legal costs) and salaries plus profit. The running plant also produces pollution in the form of stack emissions (CO_2, SO_2, and NOx; in general, air pollution) and water pollution (from coal washeries). Once the plant reaches the end of its life cycle, it can be disposed of and part of the cost may be recovered. The overall cost calculations usually depend on local accounting practices. Some related common terms are shown in Figure 3.4. Usually, overall costs are represented in some fixed reference currency for ease of comparison (for example, in 1960 US dollars).

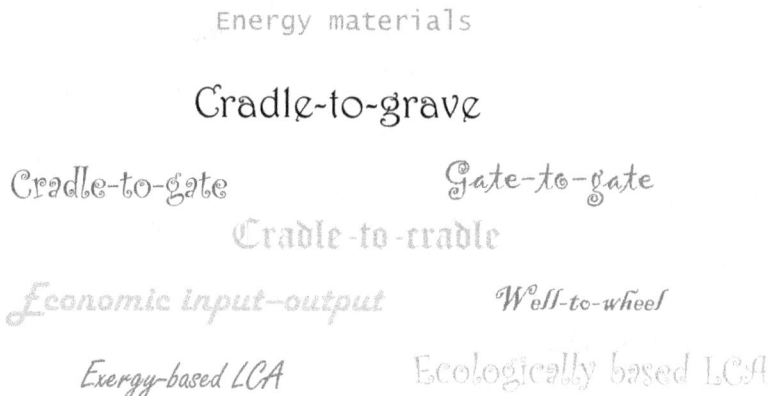

Energy materials

Cradle-to-grave

Cradle-to-gate   Gate-to-gate

Cradle-to-cradle

Economic input–output   Well-to-wheel

Exergy-based LCA   Ecologically based LCA

FIGURE 3.4 Life cycle analysis of energy materials. The terms are explained in the text.

Cradle-to-grave: Boundary covers the source or resource extraction (cradle) for the product creation throughout its use phase and to the end of the disposal phase (grave). Examination of all upstream and downstream processes and emissions from the surroundings. Assessment will assist in widening a holistic view of a product process but sometimes may be complicated to assess precisely.

Cradle-to-gate: Boundary covers the source or resource extraction (cradle) of the product factory gate (gate) before reaching the customers. Mainly makes the basis of environmental product declarations (EPD) normally termed business-to-business EDPs (Cao 2017) [2].

Gate-to-gate: The system boundary is set only to processing and linked with the production channel, which includes raw materials, transportation, disposal, recycling, etc. Mainly focuses on the relationship of the inventory to the already available information (Jiménez-González et al. 2000) [3].

Well-to-wheel: Involves analysis of all the stages in process development. Estimates emissions from processes through all the life cycle stages till the recovery process (Rahman et al. 2017) [4]. Provides methodology and policy-neutral technology to comprehend implications and issues with each technological pathway by taking into account the performance of emissions reduction and enhancement of energy efficiency (Wheel to Wheel 2016).

Cradle-to-cradle or closed-loop production: Cradle-to-grave assessment when the disposal phase is incorporated with the recycling process. Implies to minimize the environmental impacts toward sustainability (Smelror 2020) [5].

3.5 LITHIUM IN ENERGY STORAGE

Lithium salts have been used in medicine for a long time (in psychopharmacology and bipolar disorder) but it is not classified as a micronutrient and no specific daily doses are recommended. However, trace levels of lithium have considerable beneficial effects on the mental well-being of an individual and there are demands for food fortification with lithium (www.ncbi.nlm.nih.gov/pmc/articles/PMC6443601/) [6].

World production of the metal was around 82,000 metric tons in 2020 and the main uses are batteries, 71%; ceramics and glass, 14%; lubricating greases, 4%; continuous casting mold flux powders, 2%; polymer production, 2%; air treatment, 1%; and other uses, 6% (https://pubs.usgs.gov/periodicals/mcs2021/mcs2021-lithium.pdf) [7]. The demand for the metal lithium for use in batteries is expected to rise significantly in the coming years. Lithium batteries are potential candidates for energy storage devices in renewable electricity production systems (for example, energy produced from solar panels or wind energy, which do not deliver energy continuously and hence need a storage device as a buffer).

3.6 LITHIUM BATTERY

3.6.1 How Does a Typical Battery Work?

A common carbon-zinc cell (the ubiquitous AA and AAA cells, also called dry cells or zinc-carbon cells) is considered for comparison because they are so common as a portable source of power. Lithium-ion cells are modern inventions but the basic chemistry is broadly similar. There are several variants of lithium cells (lithium-ion, lithium polymer, lithium iron phosphate, lithium manganese, and others) but they are functionally very similar. Lithium cells provide higher energy density (more energy per unit weight) and also higher voltage (close to 4 V compared to 1.5 V for a regular carbon-zinc cell—the voltage does not depend on the size).

All electrochemical cells are redox systems, and oxidation takes place at one electrode and reduction takes place at the other. These two chambers (anode and cathode) are separated by a simple membrane that must allow the passage of ions. For a simple carbon-zinc cell the reactions are:

$$Zn \rightarrow Zn^{+2} + 2e$$

This reaction (oxidation of zinc) takes place at the negative electrode; the two electrons that are released travel via the external circuit.

$$2H^+ + 2MnO_2 + 2e \rightarrow H_2O + Mn_2O_3$$

This reaction (reduction of two protons) takes place at the positive electrode; this is a carbon-based indifferent electrode. The electrode is packed with manganese dioxide (mixed with graphite powder in the form of a paste; also called a depolarizer) that helps reduce hydrogen to water and prevent the formation of hydrogen gas. Without the manganese dioxide, the reaction would have been:

$$2H^+ + 2e \rightarrow H_2$$

The gas formed must be removed by some means.

The Li-ion battery is *functionally* very similar. Here both the electrodes contain lithium but the positive electrode contains CoO_2 (cobalt dioxide) that can form a compound with lithium ions. The reaction taking place at the positive electrode (called cathode; positive charge leaves this electrode during the discharge of the cell) is given by:

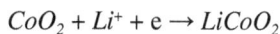

$$CoO_2 + Li^+ + e \rightarrow LiCoO_2$$

The electron is released at the negative electrode (called anode; positive charge enters this electrode during discharge) in the following reaction (this is a graphite electrode containing intercalated Li atoms):

$$C_6Li \rightarrow C_6 + Li^+ + e$$

The electron travels via the external circuit whereas the lithium ion moves in the internal circuit. Unlike the common carbon-zinc cells, the Li-ion cells are rechargeable. During charging, the above two reactions are reversed. In real life, a Li-ion cell can be charged and discharged hundreds of times whereas a typical carbon-zinc cell must be discarded after a single use and cannot be regenerated (a considerable amount of the active metal, zinc, is discarded without being used). We see below a

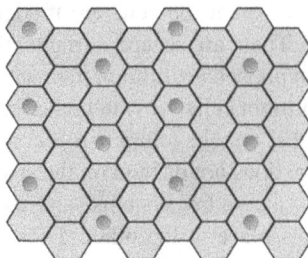

FIGURE 3.5   A pictorial diagram of graphite (one layer shown) intercalated with lithium ions (solid spheres inscribed within the hexagons are lithium ions). In reality, the number of lithium ions will be perhaps less than shown (one lithium ion per 10–20 carbon atoms). The ions can move in and out (of the matrix) rapidly and the electron will be shared by the graphite matrix. The graphite is not involved in electrochemistry but does affect the potential (the lithium reduction potential). This forms the negative electrode of the conventional Li-ion batteries (the reaction is usually written as $C_6Li \rightarrow C_6 + Li^+ + e$, but the material is usually non-stoichiometric). The material (usually in the form of a paste) is thinly painted on a metal foil that acts as the negative electrode.

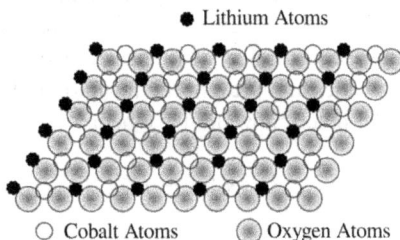

FIGURE 3.6   The positive electrode of a Li-ion cell consists of $LiCoO_2$ pasted on a metal foil that acts as the positive electrode. Although the structural formula is shown as $LiCoO_2$, the material is non-stoichiometric and often contains variable amounts of Ni, Mn, Al, or Fe. During discharge, the Li-ions move into the lattice (a process called intercalation) but it is the cobalt atom that undergoes the reduction from Co(IV) to Co(III) stage. The reduction is commonly represented as $Li^+ + CoO_2 + e \rightarrow LiCoO_2$.

diagram of Li-ion intercalated in between graphite sheets (Figure 3.5) that makes the negative electrode of typical Li-ion cells. Figure 3.6 shows the positive electrode and Figures 3.5 and 3.6 show the relative arrangement of these two electrodes.

3.6.2   WHY LITHIUM?

Lithium is widely used in modern batteries (particularly Li-ion batteries) because:

1. The batteries with high terminal voltages cannot use an aqueous medium (because water will undergo electrolysis); a non-aqueous solvent is essential and lithium salts are useful as ionic conductors in organic solvents.
2. We need an ionic carrier (like H^+) that can move easily across the separator and Li^+ is an ideal candidate because of its small ionic size.

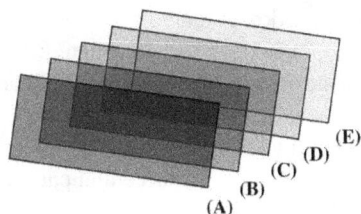

(E)
(D)
(C)
(B)
(A)

FIGURE 3.7 The organization of a typical lithium cell. The cells can be either cylindrical or rectangular (flat-pack) form. The cell is made up of several layers of foil. The top and bottom layers (A and E) are made of insulating plastic sheets. The next two layers are metal foils containing coatings of $LiCoO_2$ (positive electrode; B) and LiC_6 (negative electrode; D), respectively. These two foils have external connections for the charge/discharge function. The central layer (C) is a porous film (must allow passage of Li-ions) saturated (soaked with) a lithium salt in a suitable organic solvent; this acts as a separator between the anode and cathode chambers. The solvent and the solutes are not shown for clarity. The complete sandwich can be rolled to fit inside a cylindrical enclosure or can be folded flat for a flat battery (commonly used in cellular phones). The spacing between the layers is usually very small compared to the electrode surface area.

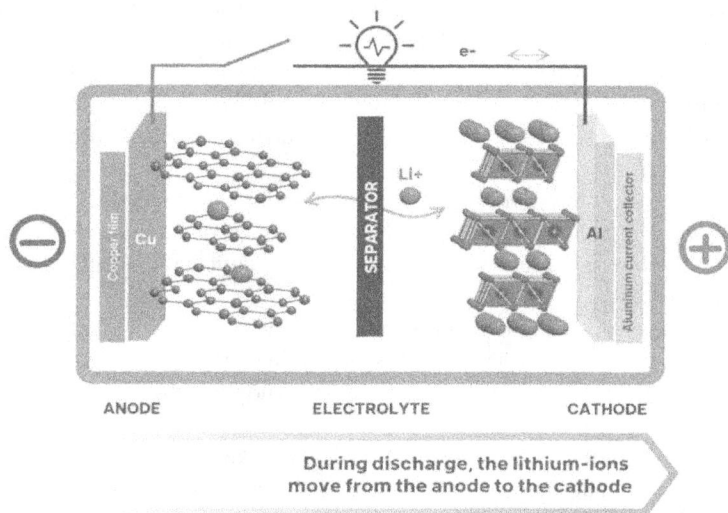

ANODE   ELECTROLYTE   CATHODE

During discharge, the lithium-ions
move from the anode to the cathode

FIGURE 3.8 Schematic of a lithium-ion battery. It is made of four components: (1) anode (negative electrode), (2) cathode (positive electrode) being separated by (3) a separator, and (4) electrolyte. The description of anode and cathode changes during the charging process. The anode is made of graphite that is deposited or pasted on a Cu foil current collector. The cathode is composed of a lithium metal oxide that is deposited on an Al current collector. When the battery is discharged, all the lithium ions are located on the cathode side. When the battery is charged, the lithium ions are forced to move from the cathode to the anode. The two electrodes are physically separated by the separator, which prevents electrical shorts. A liquid Li-ion–bearing electrolyte solution within the battery allows Li ions to move from one electrode to the other during charge or discharge.

Source: Reproduced with permission of the American Chemical Society from Goodenough and Park (2013) [8].

3. Lithium has the highest redox potential.
4. But lithium minerals are not widely distributed; most of today's lithium comes from waste brines from specific sources. Therefore alternative metals are actively being sought, for example, sodium.
5. Besides lithium, another metal, namely cobalt (Co), also used in Li-ion batteries, comes mostly from a single source and can pose supply chain problems. However, cobalt can be replaced with iron or manganese with some loss in performance. Replacing lithium in Li-ion batteries has proved to be much more difficult.

3.7   WHERE THE LITHIUM COMES FROM

According to the *CRC Handbook of Chemistry and Physics*, the abundance of lithium in the earth's crust is 20 ppm (ppm = mg/kg). To put in perspective, lithium is more abundant than lead (14 ppm), boron (10 ppm), or tin (2.3 ppm). In other words, lithium is less common compared to copper (60 ppm), zinc (70 ppm), nickel (84 ppm), or cobalt (25 ppm). The rarest elements in the earth's crust are ruthenium, Ru (1 ppb); rhodium, Rh (1 ppb); palladium, Pd (15 ppb); tellurium, Te (1 ppb); rhenium, Re (0.7 ppb); osmium, Os (1.5 ppb); iridium, Ir (1 ppb); platinum, Pt (5 ppb); and gold, Au (4 ppb) (these are siderophile elements; these are likely to be concentrated in the earth's core; 1 ppb = 1 μg/kg). On the other hand, the concentration of lithium in seawater is 100 ppb and several methods of isolation of the metal from the seawater have been proposed (https://web.stanford.edu/group/Urchin/mineral.html).

Lithium is commonly extracted from salt brines but few minerals are also known. These are all silicate minerals, Petalite, lepidolite, and spodumene are the most common lithium mineral. Silicate minerals need harsh chemical treatment for the extraction of the metal.

In 2020, Australia supplied approximately 50% of the world's demand for lithium (~40,000 tons) from mineral sources. Argentina and Chile produced more than 24,000 tons of the metal from brine sources (https://pubs.usgs.gov/periodicals/mcs2021/mcs2021-lithium.pdf).

Tables 3.1–3.3 contain relevant references and sources of this potentially very important metal. The demand for the metal is currently low but is expected to increase rapidly as electric vehicles take-off (the batteries in electric vehicles (EVs) contain considerable amounts of lithium).

Tables 3.1–3.3 clearly show that the metal is widely distributed but at a low concentration. Therefore the metal cannot be extracted from these sources (soil and water) economically at the current rates. The demand for the metal is expected to increase rapidly in the coming years and new sources must be made operational to meet the increased demand. Currently, there is an active interest in the extraction of lithium from the seawater, which contains about 100 ppb of lithium. It is also likely that the price of the metal may increase significantly due to the increased demand from the battery industry (used in electric vehicles).

TABLE 3.1
Lithium for Energy

Source	Author(s)	Year	Reference
Concentrations of lithium in Chinese coals	Sun et al.	2010	[9]
Minimum mining grade of associated Li deposits in coal seams	Sun et al.	2012	[10]
The Li-ion rechargeable battery: a perspective	Goodenough and Park	2013	[11]
The future of lithium availability for electric vehicle batteries	Speirs et al.	2014	[12]
Sustainable management of lithium-ion batteries after use in electric vehicles	Richa	2016	[13]
Energy-storage element lithium from seawater and spent lithium-ion batteries (LIBs)	Choubey et al.	2017	[14]
Small scale energy storage systems	Kokkotis	2017	[15]
Advance review on the exploitation of the prominent energy-storage element lithium	Choube et al.	2017	[14]
Lithium market research—global supply and future demand	Martin et al.	2017	[16]
How safe are the new green energy resources for marine wildlife? The case of lithium	Viana et al.	2020	[17]

TABLE 3.2
Lithium in Soil

Source	Author(s)	Reference	Year
Diagnostic criteria for plants and soils	Bradford	[18]	1973
Sources, amounts, and forms of alkali elements in soils	Scott and Smith	[19]	1987
The distribution of lithium in selected soils and surface waters of the southeastern USA	Anderson et al.	[20]	1988
Surface charge and solute interactions in soils	Bolan et al.	[21]	1999
Lithium, an emerging environmental contaminant, is mobile in the soil-plant system	Yalamanchali	[22]	2012
Leaching of lithium (Li), from plant growth media composed of coal fly ash (FA), and of FA amended with sphagnum peat moss and soil	Bilski	[23]	2013
Lithium distribution and isotopic fractionation during chemical weathering and soil formation in a loess profile	Tsai et al.	[24]	2014
Distribution of lithium in agricultural and grazing land soils at the European continental scale (GEMAS project)	Negrel et al.	[25]	2017
Lithium as an emerging environmental contaminant: mobility in the soil-plant system	Robinson et al.	[26]	2018

(*Continued*)

TABLE 3.2   (Continued)
Lithium in Soil

Source	Author(s)	Reference	Year
Impacts of molybdenum-, nickel-, and lithium-oxide nanomaterials on soil activity and microbial community	Avila-Arias et al.	[27]	2019
Growth and physiological response of spinach to various lithium concentrations in soil	Bakhat et al.	[28]	2019
Lithium in soils and plants of western Transbaikalia	Kashin	[29]	2019
Extraction of soils above concealed lithium deposits for rare metal exploration in Jiajika area: a pilot study	Xu	[30]	2019
Phytoremediation of soil contaminated with lithium-ion battery active materials—a proof-of-concept study	Henschel et al.	[31]	2020
Chemical removal of lithium in contaminated soils using promoted white eggshells with different catalysts	Abbas et al.	[32]	2021
Investigation of lithium application and effect of organic matter on soil health	Hayyat et al.	[33]	2021

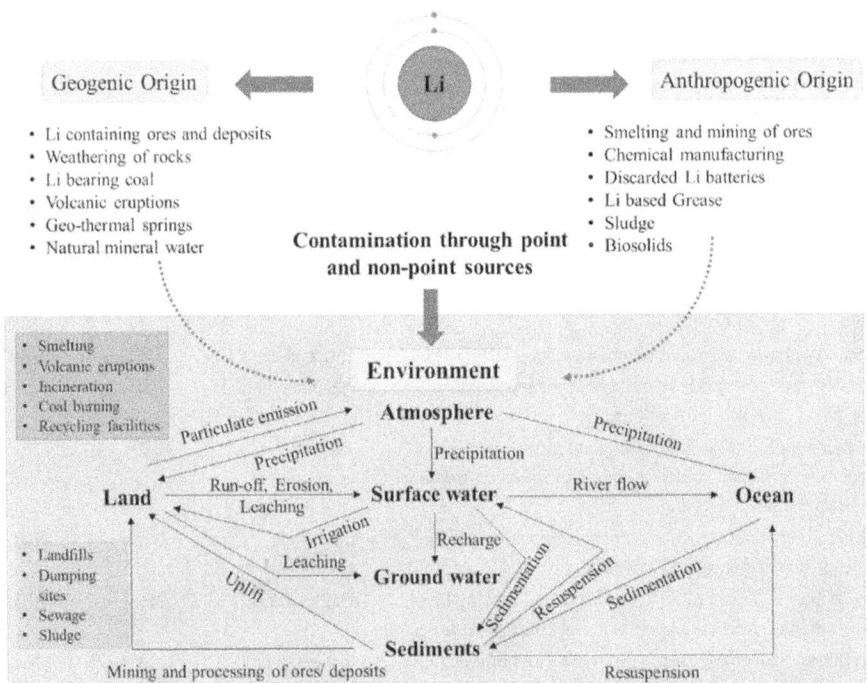

FIGURE 3.9   Sources and pathways of lithium contamination in the environment [47].

Source: Reproduced with permission from Elsevier 2021.

TABLE 3.3

Selected References on Lithium Concentrations in Soil, Water, and Sediment

A. Li Concentrations in Different Soils

Locations and Soil Depth	Soil Type	Total Li Concentration (mean) (ppm; mg kg^{-1})	Reference
Lincoln University dairy farm, New Zealand 0–20 cm	Templeton silt loam	31.8	Robinson et al. (2018)[26]
Jiajika rare metal mining area, China 10–25 cm	Not available	169.5	Xu et al. (2019) [30]
Weinan, China 0–14 cm	Not available	28.1	Tsai et al. (2014) [24]
Nearby desert areas 0–20 cm	Mud crust	24.3	
Transbaikal region, The Republic of Buryatia 0–20 cm	Gray forest soil	25.1	Kashin (2019) [29]
Cecil	Clayey, kaolinitic	11.49	Anderson et al. (1988) [20]
Iredell	Fine, montmorillonotic	25.38	
Madison	Clayey, kaolinitic	11.93	
Louisa	Fine montmorillonotic	33.29	
Japan	Coastal water	1.173	Choubey et al. (2017) [14]
Indian Ocean	Coastal water	0.160	
North Sea	Seawater	0.1	
Wakulla and Bonifay, USA, soil depth not available; Jordan Valley (JV) Northern JV; Middle JV; and South JV, Jordan 0–20 cm (topsoil) 20–40 cm (subsoil)	Sandy, siliceous, loamy, siliceous ustochreptic and ustollic camborthids and calciorthids, ustic torriorthents	5.82 3.74 Mean soluble Li concentration ranged from 1.04 to 2.68 mg L^{-1}; soluble Li concentration in subsoil layer was relatively higher than that of topsoil layer	Ammari et al. (2011) [34]

(Continued)

TABLE 3.3 *(Continued)*
Selected References on Lithium Concentrations in Soil, Water, and Sediment

B. Li Concentrations in Different Water Bodies

Location of Water Sample	Water Body	Li Concentration (mean) $1\ \mu mol\ L^{-1} = 7\ ppm$	Reference
Tibetan Plateau, China	Lake Donggi Cona	4.6–5.7 $\mu mol\ L^{-1}$	Weynell et al. (2017) [35]
	Dongqu River	1.2–1.8 $\mu mol\ L^{-1}$	
	Yellow River	8.3–8.6 $\mu mol\ L^{-1}$	
	Stream	2.8–5.2 $\mu mol\ L^{-1}$	
	Spring	2.9 $\mu mol\ L^{-1}$	
	Pond	5.7–9.3 $\mu mol\ L^{-1}$	
	Hot spring Wenquan	103.1–135.2 $\mu mol\ L^{-1}$	
Stillwater, wildlife management area in Nevada, USA	Wetland	>1000 $\mu g\ L^{-1}$	Hallock (1993) [36]
Abia and Imo States, Southeast Nigeria	Springs	2.49 $\mu g\ L^{-1}$	Ewuzie et al. (2020) [37]
	Streams	1.58 $\mu g\ L^{-1}$	
Public supply wells Domestic supply wells across USA	Groundwater	<1–396 $\mu g\ L^{-1}$	Lindsey et al. (2021) [38]
	Groundwater	<1–1700 $\mu g\ L^{-1}$	
Northeast Iceland	Groundwater	130–10,000 nmol L^{-1}	von Strandmann et al. (2016) [39]
Changjiang, China	River	116–237 nmol L^{-1}	Wang et al. (2015) [40]
	River	1260 nmol L^{-1}	
North Atlantic	Seawater	0.22 ppm	

C. Li Concentrations in Sediment

Location of Sediment Sample	Sediment	Li Concentration (range/ mean) (ppm; mg kg^{-1})	Reference
China	Top catchment sediment	5.37–400	Liu et al. (2020) [41]
	Deep catchment sediment	5.27–400	
Patos Lagoon, Brazil	Lagoon sediments	10.05–61.61	Niencheski et al. (2002) [42]
Aegean Sea, Greece	Coastal sediments	9.74–37.1	Aloupi and Angelidis (2001) [43]
Costa Rica	Subduction zone sediments	0.50–78.09	Chan and Kastner (2000) [44]
South Sandwich Island Arc		6.2–57.3	
East Sunda Island Arc		2.4–41.9	
Lesser Antilles Island Arc		35.2–74.3	
Mackenzie tributary	River sediments	57.8	Millot et al. (2010) [45]
Red Arctic tributary		56.8	
Liard tributary		46.1	
Slave tributary, Canada		41.2	
Loire River Basin, France	River sediments	41–73	Millot and Negrel (2021) [46]
Dongqu River	River sediments	14.7–44.9	Weynell et al. (2017) [35]
Lake Donggi Cona Tibetan Plateau, China	Lake sediments	52.2	

Source: Adapted with permission from Elsevier 2021 [47].

3.7.1 LITHIUM AND HEALTH

Lithium has been widely used in the treatment of mood disorders, and this simple drug is effective, not subject to patents, and very cheap. However, excessive amounts can cause toxic effects and some of these have been studied in detail. Although modern, expensive, and patented drugs for bipolar disorders are currently available, treatment with lithium is generally believed to be a better first option in most cases (McKnight et al. 2012) [48]. The authors report patients with clinical signs of lithium toxicity at concentrations of 1.5 mEq/L (approx. 10 ppm) or greater. The toxic effects include, according to the authors, increased risk of reduced urinary concentrating ability, hypothyroidism, hyperparathyroidism, and weight gain and no significant increased risk of congenital malformations, alopecia, or skin disorders, and little evidence for a clinically significant reduction in renal function in most patients.

It is interesting to note that a simple, small, monovalent cation (Li^+) can have such a great stabilizing effect on a disease that has baffled understanding for a long time. The exact nature of molecular action of the ion (Li^+) on the brain remains elusive even today. However, several important observations have been made in this direction (Williams et al. 2002) [49]. In addition, lithium has been reported as teratogenic: Klein and Melton report that lithium, one of the most effective drugs for the treatment of bipolar (manic-depressive) disorder, also has dramatic effects on morphogenesis in the early development of numerous organisms (Klein et al. 1996) [50].

Table 3.4 presents some information about the lithium content in water and its relation to health. Table 3.5 reports some selected references on the toxic effects of the lithium salts.

TABLE 3.4
Lithium in Water

Source	Author(s)	Year	Reference
Public water supplies of the 100 largest cities in the United States	Durfor et al.	1962	[51]
The lithium content of seawater	Riley and Tongudai	1964	[52]
High lithium concentrations in drinking water and plasma of exposed subjects	Zaldívar	1980	[53]
Incipient toxicity of lithium to freshwater organisms representing a salmonid habitat	Emery et al.	1981	[54]
The distribution of lithium in selected soils and surface waters of the southeastern USA	Anderson et al.	1988	
Toxicity of lithium to three freshwater organisms and the antagonistic effect of sodium	Kszos	2003	[55]
Review of lithium in the aquatic environment: distribution in the United States, toxicity and case example of groundwater contamination	Kszos et al.	2003	[55]

(Continued)

TABLE 3.4 (Continued)
Lithium in Water

Source	Author(s)	Year	Reference
Lithium and water reaction mechanisms, environmental impact and health effects	Lenntech	2007	[56]
The responses of rainbow trout gills to high lithium and potassium concentrations in water	Tkatcheva et al.	2007	[57]
Lithium levels in drinking water and risk of suicide	Ohgami et al.	2009	[58]
Lithium in the natural waters of the South East of Ireland	Kavanagh et al.	2017	[59]
Association of lithium in drinking water with the incidence of dementia	Kessing et al.	2017	[60]
Removal of lithium from water by aminomethyl phosphonic acid–containing resin	Çiçek et al.	2018	[61]
Electrochemical lithium recovery and organic pollutant removal from industrial wastewater of a battery recycling plant	Kim et al.	2018	[62]
The impact of anthropogenic inputs on lithium content in river and tap water	Choi et al.	2019	[63]
Lithium in drinking water sources in rural and urban communities in Southeastern Nigeria	Ewuzie et al.	2020	[37]
Lithium in groundwater used for drinking water supply in the United States	Lindsey et al.	2021	[38]

TABLE 3.5
Lithium Toxicity

Source	Author(s)	Year	Reference
Lithium toxicity in cattle	Johnson et al.	1980	[64]
Incipient toxicity of lithium to freshwater organisms representing a salmonid habitat	Emery et al.	1981	[54]
Teratogenic effects of lithium in mice	Smithberg and Dixit	1982	[65]
Lithium toxicity in yeast is due to the inhibition of RNA processing enzymes	Dichtl et al.	1997	[66]
Lithium chloride: a flow-through embryo-larval toxicity test with the fathead minnow, Pimephales promelas Rafinesque	Long et al.	1998	[67]
Lithium intoxication	Timmer and Sands	1999	[68]
Review of lithium in the aquatic environment: distribution in the United States, toxicity and case example of groundwater contamination	Kszos et al.	2003	[55]
Lithium-induced renal toxicity in rats: protection by a novel antioxidant caffeic acid phenethyl ester	Oktem et al.	2005	[69]

(Continued)

TABLE 3.5 *(Continued)*
Lithium Toxicity

Source	Author(s)	Year	Reference
The syndrome of irreversible lithium-effectuated neurotoxicity	Munshi and Thampy	2005	[70]
Lithium toxicity and expression of stress-related genes or proteins in A549 cells	Allagui et al.	2007	[71]
Sinoatrial block in lithium toxicity	Goldberger	2007	[72]
Toxicity of lithium to humans and the environment—a literature review	Aral and Vecchio-Sadus	2008	[73]
Lithium nephrotoxicity revisited	Grünfeld and Rossier	2009	[74]
Lithium toxicity: the importance of clinical signs	Dunne et al.	2010	[75]
Lithium: environmental pollution and health effects	Aral, and Vecchio-Sadus	2011	[73]
Lithium toxicity profile: a systematic review and meta-analysis	McKnight et al.	2012	[48]
The safety of lithium	Goodwin	2015	[76]
Lithium and nephrotoxicity	Davis et al.	2018	[77]
Lithium in the environment and potential targets to reduce lithium toxicity in plants	Tanveer et al.	2019	[78]
The toxicity of lithium to human cardiomyocytes	Shen	2020	[79]

3.7.2 LITHIUM PRODUCTION

A major part of lithium is isolated from the waste brines (only brines from specific salt lakes in Argentina, Bolivia, and Chile) and the major (70%–80% of the world's reserves) part of lithium comes from this source (Murodjon, Yu, Li, Duo & Deng 2020) [80]. For each ton of $LiCO_3$ (containing slightly more than 10% of the metal), about half a million liters of brine (about 500 tons) need to be evaporated (mostly using solar energy), but the process is slow and cannot be scaled up easily (Flexer, Baspineiro & Galli 2018) [81]. The process also produces large amounts of NaCl, KCl, and $MgCl_2$ among other products that can also be sold profitably.

Lithium can also be isolated from minerals that may contain 1%–3% of lithium by weight (pure spodumene contains 8% of Li_2O or about 3.3% of Li). The mineral (rocks) are treated with hot sulfuric acid (to break down the rock) and crushed and treated with sodium hydroxide. The process is more expensive (compared to extraction from brines) but is scalable but the resources (minerals) are limited. Seawater contains only trace amounts of lithium (180 ppb; the Na:Li proportion is approximately 60,000:1 by weight). However, newer technologies are being tried out to isolate Li from seawater because the total reserve of lithium in seawater is estimated as 200 billion tons (Jacoby) [82].

Lithium is widely used in batteries and the retired batteries can be recycled efficiently. Both lithium and cobalt can be recovered with approximately >90% efficiency. Using automated processes, the damage to environment can be kept at a minimum. Table 3.6 shows several references that point to various sources of lithium and the possible modes of extraction. The list is not exhaustive and only suggestive.

TABLE 3.6
Lithium Availability

Source	Author(s)	Year	Reference
Lithium availability—market economy countries	Bleiwas and Coffman	1986	[83]
Recycling of batteries: a review of current processes and technologies	Bernardes	2004	[84]
Removal and recovery of lithium using various microorganisms	Tsuruta	2005	[85]
Using the cumulative availability curve to assess the threat of mineral depletion: the case of lithium	Yaksic	2009	[86]
Global lithium availability: a constraint for electric vehicles?	Gruber et al.	2011	[87]
Recovery of lithium from waste materials	Jandova et al.	2012	[88]
The future of lithium availability for electric vehicle batteries	Speirs et al.	2014	[89]
Hydrometallurgical processing of spent lithium-ion batteries (LIBs) in the presence of a reducing agent	Meshram et al.	2015	[90]
Coal as a promising source of lithium	Qin et al.	2015	[91]
Sustainable management of lithium-ion batteries after use in electric vehicles	Richa	2016	[92]
Material flow analysis of lithium in China	Hao et al.	2017	[93]
Trade-linked material flow analysis of lithium	Sun et al.	2017	[94]
Lithium market research—global supply, future demand, and price development	Martin et al.	2017	[16]
Induced plant accumulation of lithium—agro-farming	Kavanagh et al.	2018	[59]
Spodumene: the lithium market, resources and processes	Dessemond et al.	2019	[95]

3.8 LITHIUM RECYCLING

Currently, there is little incentive to recycle Li-ion batteries, but this may change soon because of the extensive use of Li batteries in electric vehicles. Gruber et al. suggest that the current lithium reserve is sufficient to meet the global demand for electric vehicles (Gruber, Medina, Keoleian, Kesler, Everson, and Wallington, 2011) [87]. Kushnir and Sandén have estimated that the future lithium demand can easily be met with reasonable recycling efforts (Kushnir and Sandén 2012) [96]. Different types of Li-ion batteries must be segregated and recycled and not lumped together in one bin (Harper et al. 2019) [97]. The authors also suggest robotic tools to disassemble batteries for more effective separation. Table 3.7 lists recent efforts towards lithium recovery and recycling.

As a major part of lithium and cobalt are produced from the waste products of refining other metals (magnesium and copper/nickel, respectively), environmental concerns associated with extensive mining are relatively less. However, this also means the production cannot be stepped up to meet any sudden increased demand. Therefore sourcing of lithium from seawater is under active consideration and development. Substitutes for cobalt in Li-ion batteries are also being investigated.

TABLE 3.7
Lithium Recovery

Source	Author(s)	Year	Reference
Removal and recovery of lithium using various microorganisms	Tsuruta	2005	[85]
Recovery of lithium from waste materials	Jandova et al.	2012	[88]
Recovery of lithium from Urmia Lake by a nanostructure MnO_2 ion sieve	Zandevakili et al.	2014	[98]
Recovery of lithium from wastewater using the development of Li ion-imprinted polymers	Luo et al.	2015	[99]
Recovery of lithium ions from sodium-contaminated lithium bromide solution by using electrodialysis process	Parsa et al.	2015	[100]
Electrochemical lithium recovery and organic pollutant removal from industrial wastewater of a battery recycling plant	Kim et al.	2018	
Recovery methods and regulation status of waste lithium-ion batteries in China: a mini-review	Siqi	2019	[101]
Investigation of solution chemistry to enable efficient lithium recovery from low-concentration lithium-containing wastewater	Zhao	2020	[102]

A suggested flowchart for the recovery of lithium and cobalt from spent Li-ion batteries (from electric vehicles) is seen in Figure 3.7. The batteries must be fully discharged and dismantled using robotic technology. Maximum efficiency is obtained if the batteries are properly segregated with respect to their internal chemistry. The recovery can be very high (>90%).

3.9   CONCLUSIONS AND OTHER CONSIDERATIONS

Another important element that is need in the production of Li-ion batteries is cobalt (Co). Cobalt often appears with copper and nickel and is usually isolated from the ores of copper and nickel during the isolation of copper and nickel. Most cobalt produced today is a byproduct in the refining of copper and nickel. However, major sources of cobalt are highly localized and can be a problem for a robust supply chain. Today, more than 70% of the world's supply of the metal comes from the mines of the Democratic Republic of Congo (DRC) mainly operated by multinational corporations. Most of the world's cobalt production is a byproduct of copper and nickel production. A significant part of the cobalt mining in the DRC includes child labour, human rights abuses, unsafe working conditions, and worker exploitation   (www.theverge.com/2022/2/15/22933022/cobalt-mining-ev-electriv-vehicle-working-conditions-congo). Cobalt has many uses but the major use is in the lithium-ion battery. Another component used in lithium-ion batteries is graphite. Graphite is an allotrope of carbon and is widely distributed. More than 80% of the world's production of graphite in 2021 came from China. Graphite has wide uses in

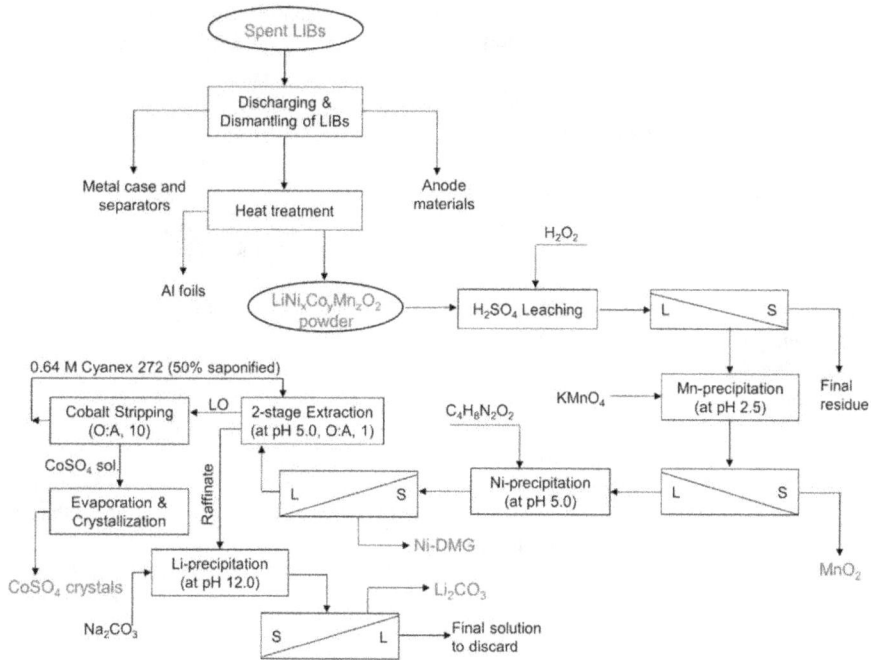

FIGURE 3.10 Resource recovery of critically rare metals by hydrometallurgical recycling of spent lithium-ion batteries (with permission from Sattar et al. 2019) [103]. For details, please see the original reference. Lithium is recovered as Li_2CO_3, cobalt is recovered as $CoSO_4$, nickel is recovered as the organic salt nickel-dimethylglyoxime, and manganese is recovered as the dioxide.

Source: Reproduced with permission from Elsevier 2019.

metallurgy (high-temperature crucibles), lubrication (high-temperature lubricants), battery and other electrochemical applications, and also in writing tools (pencils). Graphite can be produced synthetically, but artificial (synthetic) graphite is expensive. However, artificial (synthetic) graphite can be extremely pure and is suitable for semiconductor use.

REFERENCES

[1] Shevchenko, H., Petrushenko, M., Burkynskyi, B., Khumarova, N., 2021. SDGs and the ability to manage change within the European green deal: The case of Ukraine. *Probl. Perspect. Manag.* 19(1), 53.

[2] Cao, C., 2017. Sustainability and life assessment of high strength natural fibre composites in construction. In: *Advanced high strength natural fibre composites in construction.* Woodhead Publishing, pp. 529–544.

[3] Jiménez-González, C., Kim, S., Overcash, M.R., 2000. Methodology for developing gate-to-gate life cycle inventory information. *Int. J. Life Cycle Assess.* 5, 153–159.

[4] Rahman, S.M., Handler, R.M., Mayer, A.L., 2016. Life cycle assessment of steel in the ship recycling industry in Bangladesh. *J. Clean. Prod.* 135, 963–971.

[5] Smelror, M., 2020. Geology for society in 2058: Some down-to-earth perspectives. *Geol. Soc. London Spec. Publ.* 499(1), 17–47.

[6] www.ncbi.nlm.nih.gov/pmc/articles/PMC6443601/.

[7] https://pubs.usgs.gov/periodicals/mcs2021/mcs2021-lithium.pdf.

[8] Bibienne, T., Magnan, J.-F., Rupp, A., Laroche, N., 2020. From mine to mind and mobiles: Society's increasing dependence on lithium. *Elements* 16(4), 265–270.

[9] Sun, Y., Li, Y., Zhao, C., Lin, M., Wang, J., Qin, S., 2010. Concentrations of lithium in Chinese coals. *Energy Explor. Exploit.* 28, 97–104.

[10] Sun, Y., Yang, J., Zhao, C., 2012. Minimum mining grade of associated Li deposits in coal seams. *Energy Explor. Exploit.* 30, 167–170.

[11] Goodenough, J.B., Park, K.-S., 2013. The Li-ion rechargeable battery: A perspective. *J. Am. Chem. Soc.* 135, 1167–1176.

[12] Speirs, J., Contestabile, M., Houari, Y., Gross, R., 2014. The future of lithium availability for electric vehicle batteries. *Renew. Sustain. Energy Rev.* 35, 183–193.

[13] Richa, K., 2016. Sustainable management of lithium-ion batteries after use in electric vehicles. *Resour. Conserv. Recycl.* 168, 105249.

[14] Choubey, P.K., Chung, K.-S., Kim, M.-S., Lee, J.-C., Srivastava, R.R., 2017. Advance review on the exploitation of the prominent energy-storage element lithium. Part II: From sea water and spent lithium ion batteries (LIBs). *Miner. Eng.* 110, 104–121.

[15] Kokkotis, P.I., Psomopoulos, C.S., Ioannidis, G.C., Kaminaris, S.D., 2017. Small scale energy storage systems. A short review in their potential environmental impact. *Fresenius Environ. Bull.* 26, 5658–5665.

[16] Martin, G., Rentsch, L., Höck, M., Bertau, M., 2017. Lithium market research—global supply, future demand and price development. *Energy Stor. Mater.* 6, 171–179.

[17] Viana, T., Ferreira, N., Henriques, B., Leite, C., De Marchi, L., Amaral, J., Freitas, R., Pereira, E., 2020. How safe are the new green energy resources for marine wildlife? The case of lithium. *Environ. Pollut.* 267, 115458.

[18] Bradford, G.R., 1973. Lithium. In: Chapman, H.D. (Ed.), *Diagnostic criteria for plants and soils.* Second Printing. Quality Printing Co., pp. 218–224.

[19] Scott, A., Smith, S., 1987. Sources, amounts, and forms of alkali elements in soils. In: *Advances in soil science.* Springer, pp. 101–147.

[20] Anderson, M.A., Bertsch, P.M., Miller, W.P., 1988. The distribution of lithium in selected soils and surface waters of the southeastern USA. *Appl. Geochem.* 3, 205–212.

[21] Bolan, N.S., Naidu, R., Syers, J.K., Tillman, R.W., 1999. Surface charge and solute interactions in soils. *Adv. Agron.* 67, 87–140.

[22] Yalamanchali, R., 2012. *Lithium, an emerging environmental contaminant, is mobile in the soil-plant system.* Lincoln University.

[23] Bilski, J., Kraft, C., Jacob, D., Soumaila, F., Farnsworth, A., 2013. Leaching of selected trace elements from plant growth media composed of coal fly ash (FA), and of FA amended with sphagnum peat moss and soil. Part 1: Leaching of trace elements from group 1: Cesium (Cs) and lithium (Li), and from group 2: Beryllium (Be), strontium (Sr), and barium (Ba). *Res. J. Chem. Environ. Sci.* 1, 7.

[24] Tsai, P.-H., You, C.-F., Huang, K.-F., Chung, C.-H., Sun, Y.-B., 2014. Lithium distribution and isotopic fractionation during chemical weathering and soil formation in a loess profile. *J. Asian Earth Sci.* 87, 1–10.

[25] Negrel, P., Millot, R., Brenot, A., Bertin, C., 2010. Lithium isotopes as tracers of groundwater circulation in a peat land. *Chem. Geol.* 276, 119–127.

[26] Robinson, B.H., Yalamanchali, R., Reiser, R., Dickinson, N.M., 2018. Lithium as an emerging environmental contaminant: Mobility in the soil-plant system. *Chemosphere* 197, 1–6.

[27] Avila-Arias, H., Nies, L.F., Gray, M.B., Turco, R.F., 2019. Impacts of molybdenum-, nickel-, and lithium-oxide nanomaterials on soil activity and microbial community structure. *Sci. Total Environ.* 652, 202–211.

[28] Bakhat, H.F., Rasul, K., Farooq, A.B.U., Zia, Z., Fahad, S., Abbas, S., Shah, G.M., Rabbani, F., Hammad, H.M., 2019. Growth and physiological response of spinach to various lithium concentrations in soil. *Environ. Sci. Pollut. Control. Ser.* 1–9.

[29] Kashin, V., 2019. Lithium in soils and plants of western Transbaikalia. *Eurasian Soil Sci.* 52, 359–369.

[30] Xu, Z., Liang, B., Geng, Y., Liu, T., Wang, Q., 2019. Extraction of soils above concealed lithium deposits for rare metal exploration in Jiajika area: A pilot study. *Appl. Geochem.* 107, 142–151.

[31] Henschel, J., Peschel, C., Klein, S., Horsthemke, F., Winter, M., Nowak, S., 2020. Clarification of decomposition pathways in a State-of-the-art lithium ion battery electrolyte through 13C-labeling of electrolyte components. *Angew. Chem. Int. Ed.* 59(15), 6128–6137.

[32] Abbas, M.N., Al-Tameemi, I.M., Hasan, M.B., Al-Madhhachi, A.-S.T., 2021. Chemical removal of cobalt and lithium in contaminated soils using promoted white eggshells with different catalysts. *S. Afr. J. Chem. Eng.* 35, 23–32.

[33] Hayyat, M.U., Nawaz, R., Siddiq, Z., Shakoor, M.B., Mushtaq, M., Ahmad, S.R., Ali, S., Hussain, A., Irshad, M.A., Alsahli, A.A., 2021. Investigation of lithium application and effect of organic matter on soil health. *Sustainability* 13, 1705.

[34] Ammari, T.G., Al-Zu'bi, Y., Abu-Baker, S., Dababneh, B., Tahboub, A., 2011. The occurrence of lithium in the environment of the Jordan Valley and its transfer into the food chain. *Environ. Geochem. Health* 33, 427–437.

[35] Gruber, P.W., Medina, P.A., Keoleian, G.A., Kesler, S.E., Everson, M.P., Wallington, T.J., 2011. Global lithium availability: A constraint for electric vehicles? *J. Ind. Ecol.* 15, 760–775.

[36] Hallock, L.L., 1993. Detailed study of irrigation drainage in and near wildlife management areas, west-central Nevada, 1987–90; Part B, Effect on biota in Stillwater and Fernley Wildlife Management Areas and other nearby wetlands. In: *US Geological survey.* Books and Open-File Reports Section [distributor]. U.S. Geological Survey 333 West Nye Lane, Room 203 Carson City, NV 89706-0866.

[37] Ewuzie, U., Nnorom, I.C., Eze, S.O., 2020. Lithium in drinking water sources in rural and urban communities in Southeastern Nigeria. *Chemosphere* 245, 125593.

[38] Lindsey, B.D., Belitz, K., Cravotta III, C.A., Toccalino, P.L., Dubrovsky, N.M., 2021. Lithium in groundwater used for drinking-water supply in the United States. *Sci. Total Environ.* 767, 144691.

[39] von Strandmann, P.A.P., Burton, K.W., Opfergelt, S., Eiríksdóttir, E.S., Murphy, M.J., Einarsson, A., Gislason, S.R., 2016. The effect of hydrothermal spring weathering processes and primary productivity on lithium isotopes: Lake Myvatn, Iceland. *Chem. Geol.* 445, 4–13.

[40] Wang, Q.-L., Chetelat, B., Zhao, Z.-Q., Ding, H., Li, S.-L., Wang, B.-L., Li, J., Liu, X.-L., 2015. Behavior of lithium isotopes in the Changjiang River system: Sources effects and response to weathering and erosion. *Geochem. Cosmochim. Acta* 151, 117–132.

[41] Liu, H., Wang, X., Zhang, B., Wang, W., Han, Z., Chi, Q., Zhou, J., Nie, L., Xu, S., Yao, W., 2020. Concentration and distribution of lithium in catchment sediments of China: Conclusions from the China Geochemical Baselines project. *J. Geochem. Explor.* 215, 106540.

[42] Niencheski, L.F.H., Baraj, B., Franca, R.G., Mirlean, N., 2002. Lithium as a normalizer for the assessment of anthropogenic metal contamination of sediments of the southern area of Patos Lagoon. *Aquat. Ecosys. Health Manag.* 5, 473–483.

[43] Aloupi, M., Angelidis, M., 2001. Normalization to lithium for the assessment of metal contamination in coastal sediment cores from the Aegean Sea, Greece. *Mar. Environ. Res.* 52(1), 1–12.

[44] Chan, L.-H., Kastner, M., 2000. Lithium isotopic compositions of pore fluids and sediments in the Costa Rica subduction zone: Implications for fluid processes and sediment contribution to the arc volcanoes. *Earth Planet Sci. Lett.* 183, 275–290.

[45] Millot, R., Vigier, N., Gaillardet, J., 2010. Behaviour of lithium and its isotopes during weathering in the Mackenzie Basin, Canada. *Geochem. Cosmochim. Acta* 74, 3897–3912.

[46] Millot, R., Négrel, P., 2021. Lithium isotopes in the Loire River Basin (France): Hydrogeochemical characterizations at two complementary scales. *Appl. Geochem.* 125, 104831.

[47] Bolan, N., Hoang, S.A., Tanveer, M., Wang, L., Bolan, S., Sooriyakumar, P., Robinson, B., Wijesekara, H., Wijesooriya, M., Keerthanan, S., Vithanage, M., 2021. From mine to mind and mobiles—Lithium contamination and its risk management. *Environ. Pollut.* 290, 118067.

[48] McKnight, R.F., Adida, M., Budge, K., Stockton, S., Goodwin, G.M., Geddes, J.R. 2012. Lithium toxicity profile: A systematic review and meta-analysis. *Lancet* 379(9817), 721–728.

[49] Williams, R.S.B., Cheng, L., Mudge, A.W., Harwood, A.J. 2002. A common mechanism of action for three mood-stabilizing drugs. *Nature* 417(6886), 292–295.

[50] Klein, P.S., Melton, D.A. 1996. A molecular mechanism for the effect of lithium on development. *Proc. Natl. Acad. Sci.*, 93(16), 8455–8459.

[51] Durfor, C.N., Becker, E., 1964. *Public water supplies of the 100 largest cities in the United States, 1962*. US Government Printing Office.

[52] Riley, J., Tongudai, M., 1964. The lithium content of seawater. In: *Deep sea research and oceanographic abstracts*. Elsevier, pp. 563–568.

[53] Zaldívar, R., 1980. High lithium concentrations in drinking water and plasma of exposed subjects. *Arch. Toxicol.* 46, 319–320.

[54] Emery, R., Klopfer, D., Skalski, J., 1981. Incipient toxicity of lithium to freshwater organisms representing a salmonid habitat. In: Battelle Pacific Northwest Labs. *Environ. Sci. Water Res. Technol.* 4, 175–182.

[55] Kszos, L.A., Beauchamp, J.J., Stewart, A.J., 2003. Toxicity of lithium to three freshwater organisms and the antagonistic effect of sodium. *Ecotoxicology* 12, 427–437.

[56] Lenntech, B., 2007. *Lithium and water reaction mechanisms, environmental impact and health effects*. Retrieved from www.lenntech.com/periodic/water/lithiu m/lithium.

[57] Tkatcheva, V., Franklin, N.M., McClelland, G.B., Smith, R.W., Holopainen, I.J., Wood, C.M., 2007a. Physiological and biochemical effects of lithium in rainbow trout. *Arch. Environ. Contam. Toxicol.* 53, 632–638.

[58] Ohgami, H., Terao, T., Shiotsuki, I., Ishii, N., Iwata, N., 2009. Lithium levels in drinking water and risk of suicide. *Br. J. Psychiatr.* 194, 464–465.

[59] Kavanagh, L., Keohane, J., Cleary, J., Garcia Cabellos, G., Lloyd, A., 2017. Lithium in the natural waters of the South East of Ireland. *Int. J. Environ. Res. Publ. Health.* 14, 561.

[60] Kessing, L.V., Gerds, T.A., Knudsen, N.N., Jørgensen, L.F., Kristiansen, S.M., Voutchkova, D., Ernstsen, V., Schullehner, J., Hansen, B., Andersen, P.K., 2017. Association of lithium in drinking water with the incidence of dementia. *JAMA Psychiatr.* 74, 1005–1010.

[61] Çiçek, A., Yılmaz, O., Arar, O., 2018. Removal of lithium from water by aminomethylphosphonic acid containing resin. *J. Serb. Chem. Soc.* 83, 1059–1069.

[62] Kim, S., Kim, J., Kim, S., Lee, J., Yoon, J., 2018. Electrochemical lithium recovery and organic pollutant removal from industrial wastewater of a battery recycling plant. *Environ. Sci. Water Res. Technol.* 4, 175–182.

[63] Choi, H.-B., Ryu, J.-S., Shin, W.-J., Vigier, N., 2019. The impact of anthropogenic inputs on lithium content in river and tap water. *Nat. Commun.* 10, 1–7.

[64] Johnson, J., Crookshank, H., Smalley, H., 1980. Lithium toxicity in cattle. *Vet. Hum. Toxicol.* 22, 248–251.

[65] Smithberg, M., Dixit, P.K., 1982. Teratogenic effects of lithium in mice. *Teratology* 26, 239–246.

[66] Dichtl, B., Stevens, A., Tollervey, D., 1997. Lithium toxicity in yeast is due to the inhibition of RNA processing enzymes. *EMBO J.* 16, 7184–7195.

[67] Long, K., Brown Jr., R., Woodburn, K., 1998. Lithium chloride: A flow-through embryo-larval toxicity test with the fathead minnow, Pimephales promelas Rafinesque. *Bull. Environ. Contam. Toxicol.* 60, 312–317.

[68] Timmer, R.T., Sands, J.M., 1999. Lithium intoxication. *J. Am. Soc. Nephrol.* 10, 666–674.

[69] Oktem, F., Ozguner, F., Sulak, O., Olgar, S., Akturk, O., Yilmaz, H.R., Altuntas, I., 2005. Lithium-induced renal toxicity in rats: Protection by a novel antioxidant caffeic acid phenethyl ester. *Mol. Cell. Biochem.* 277, 109–115.

[70] Munshi, K.R., Thampy, A., 2005. The syndrome of irreversible lithium-effectuated neurotoxicity. *Clin. Neuropharmacol.* 28, 38–49.

[71] Allagui, M., Vincent, C., Gaubin, Y., Croute, F., 2007. Lithium toxicity and expression of stress-related genes or proteins in A549 cells. *Biochim. Biophys. Acta Mol. Cell Res.* 1773, 1107–1115.

[72] Goldberger, Z.D., 2007. Sinoatrial block in lithium toxicity. *Am. J. Psychiatr.* 164, 831–832.

[73] Aral, H., Vecchio-Sadus, A., 2008. Toxicity of lithium to humans and the environment—a literature review. *Ecotoxicol. Environ. Saf.* 70, 349–356.

[74] Grünfeld, J.-P., Rossier, B.C., 2009. Lithium nephrotoxicity revisited. *Nat. Rev. Nephrol.* 5, 270.

[75] Dunne, F.J., 2010. Lithium toxicity: The importance of clinical signs. *Br. J. Hosp. Med.* 71, 206–210.

[76] Goodwin, G.M., 2015. The safety of lithium. *JAMA Psychiat.* 72, 1167–1169.

[77] Davis, J., Desmond, M., Berk, M., 2018. Lithium and nephrotoxicity: A literature review of approaches to clinical management and risk stratification. *BMC Nephrol.* 19, 1–7.

[78] Tanveer, M., Hasanuzzaman, M., Wang, L., 2019. Lithium in environment and potential targets to reduce lithium toxicity in plants. *J. Plant Growth Regul.* 38, 1574–1586.

[79] Shen, J., Li, X., Shi, X., Wang, W., Zhou, H., Wu, J., Wang, X., Li, J., 2020. The toxicity of lithium to human cardiomyocytes. *Environ. Sci. Eur.* 32, 1–12.

[80] Murodjon, S., Yu, X., Li, M., Duo, J., Deng, T., 2020. *Lithium recovery from brines including seawater, salt lake brine, underground water and geothermal water.* IntechOpen.

[81] Flexer, V., Baspineiro, C.F., Galli, C.I. 2018. Lithium recovery from brines: A vital raw material for green energies with a potential environmental impact in its mining and processing. *Sci. Total Environ.* 639, 1188–1204.

[82] Jacoby, M., 2021. Can seawater give us the lithium to meet our battery needs? *Chemical & Engineering News*, 19–21.

[83] Bleiwas, D., Coffman, J., 1986. Lithium availability-market economy countries. *Bureau Mines Infor. Circ.* 9102, 23.

[84] Bernardes, A.M., Espinosa, D.C.R., Tenório, J.S., 2004. Recycling of batteries: A review of current processes and technologies. *J. Power Sources* 130, 291–298.

[85] Tsuruta, T., 2005. Removal and recovery of lithium using various microorganisms. *J. Biosci. Bioeng.* 100, 562–566.

[86] Yaksic, A., Tilton, J.E., 2009. Using the cumulative availability curve to assess the threat of mineral depletion: The case of lithium. *Resour. Pol.* 34, 185–194.

[87] Gruber, P.W., Medina, P.A., Keoleian, G.A., Kesler, S.E., Everson, M.P., Wallington, T.J., 2011. Global lithium availability: A constraint for electric vehicles? *J. Ind. Ecol.* 15, 760–775.

[88] Jandova, J., Dvorak, P., Kondas, J., Havlak, L., 2012. Recovery of lithium from waste materials. *Ceramics* 56, 50–54.

[89] Speirs, J., Contestabile, M., Houari, Y., Gross, R., 2014. The future of lithium availability for electric vehicle batteries. *Renew. Sustain. Energy Rev.* 35, 183–193.

[90] Meshram, P., Pandey, B., Mankhand, T., 2015. Hydrometallurgical processing of spent lithium ion batteries (LIBs) in the presence of a reducing agent with emphasis on kinetics of leaching. *Chem. Eng. J.* 281, 418–427.

[91] Qin, S., Zhao, C., Li, Y., Zhang, Y., 2015. Review of coal as a promising source of lithium. *Int. J. Oil Gas Coal Technol.* 9, 215–229.

[92] Richa, K., 2016. Sustainable management of lithium-ion batteries after use in electric vehicles. *Resour. Conserv. Recycl.* 168, 105249.

[93] Hao, H., Liu, Z., Zhao, F., Geng, Y., Sarkis, J., 2017. Material flow analysis of lithium in China. *Resour. Pol.* 51, 100–106.

[94] Sun, X., Hao, H., Zhao, F., Liu, Z., 2017. Tracing global lithium flow: A trade-linked material flow analysis. *Resour. Conserv. Recycl.* 124, 50–61.

[95] Dessemond, C., Lajoie-Leroux, F., Soucy, G., Laroche, N., Magnan, J.-F., 2019. Spodumene: The lithium market, resources and processes. *Minerals* 9, 334.

[96] Kushnir, D., Sandén, B.A., 2012. The time dimension and lithium resource constraints for electric vehicles. *Resour. Policy* 37, 93–103.

[97] Harper, G., Sommerville, R., Kendrick, E., Driscoll, L., Slater, P., Stolkin, R., Walton, A., Christensen, P., Heidrich, O., Lambert, S., Abbott, A., Ryder, K., Gaines, L., Anderson, P., 2019. Recycling lithium-ion batteries from electric vehicles. *Nature* 575, 75–86.

[98] Zandevakili, S., Ranjbar, M., Ehteshamzadeh, M., 2014. Recovery of lithium from Urmia Lake by a nanostructure MnO2 ion sieve. *Hydrometallurgy* 149, 148–152.

[99] Luo, X., Guo, B., Luo, J., Deng, F., Zhang, S., Luo, S., Crittenden, J., 2015. Recovery of lithium from wastewater using development of Li ion-imprinted polymers. *ACS Sustain. Chem. Eng.* 3, 460–464.

[100] Parsa, N., Moheb, A., Mehrabani-Zeinabad, A., Masigol, M.A., 2015. Recovery of lithium ions from sodium-contaminated lithium bromide solution by using electrodialysis process. *Chem. Eng. Res. Des.* 98, 81–88.

[101] Siqi, Z., Guangming, L., Wenzhi, H., Juwen, H., Haochen, Z., 2019. Recovery methods and regulation status of waste lithium-ion batteries in China: A mini review. *Waste Manag. Res.* 37, 1142–1152.

[102] Zhao, C., He, M., Cao, H., Zheng, X., Gao, W., Sun, Y., Zhao, H., Liu, D., Zhang, Y., Sun, Z., 2020. Investigation of solution chemistry to enable efficient lithium recovery from low-concentration lithium-containing wastewater. *Front. Chem. Sci. Eng.* 14, 639–650.

[103] Sattar, R., Ilyas, S., Bhatti, H.N., Ghaffar, A., 2019. Resource recovery of critically-rare metals by hydrometallurgical recycling of spent lithium ion batteries. *Sep. Purif. Technol.* 209, 725–733.

4 Photocatalytic Materials for Production of Hydrogen from Water Using Solar Energy

Nandana Chakinala, Praveen K. Surolia*,
Raksh Vir Jasra, and Anand G. Chakinala

4.1 INTRODUCTION

Global energy consumption is steadily increasing, with fossil fuels currently accounting for 80% of the energy demand and contributing to environmental pollution through greenhouse gas emissions [1,2]. To achieve sustainable development, the United Nations implemented the Sustainable Development Goals (SDGs) in 2015. Meeting these goals requires the development of clean energy systems that utilize renewable resources [3].

Hydrogen, a clean and abundant energy source, shows promise for various applications, particularly as a vehicle fuel. However, current industrial demands of hydrogen requirement are met through steam methane reformers and coal gasification processes which result in CO_2 emissions. As we seek sustainable solutions for hydrogen generation through water splitting, harnessing solar energy emerges by way of a highly advantageous alternative. The generation of green hydrogen from solar-driven water splitting stands out as a capable method, meeting the ever-growing global energy demand while minimizing environmental impact. Efforts have been applied to design efficient photocatalytic substances for water-based hydrogen production, following the groundbreaking work of Honda and Fujishima [4]. This chapter delivers a comprehensive outline of the fundamental and thermodynamic aspects concerning to hydrogen production through water splitting. Numerous photocatalytic methods and catalyst design strategies are discussed, along with the existing limitations of the process. Moreover, the chapter explores potential avenues for enhancing conversion efficiency to ensure the technical viability of harnessing solar energy aimed at green hydrogen production. Emphasizing its significance, an integration of solar-driven water splitting into hydrogen generation is seen as a pivotal step towards a sustainable future for both society and industry, promising a transformative shift towards a greener and more environmentally conscious energy landscape.

* praveenkumar.surolia@jaipur.manipal.edu

DOI: 10.1201/9781003269779-4

77

4.2   FUNDAMENTAL AND THERMODYNAMIC ASPECTS

In the realm of water splitting for hydrogen production, the theoretical minimum Gibbs free energy stands at 237 kJ/mol (Eq. 4.2). However, practical implementation faces the challenge of surpassing the kinetic energy barrier inherent in both the oxidation and reduction processes occurring at catalyst surface. Notably, the water oxidation reaction demands a larger overpotential compared to proton reduction. Nonetheless, through strategic catalyst modifications, it becomes possible to diminish the activation energy barrier, effectively overcoming this limitation [5,6].

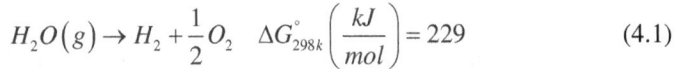

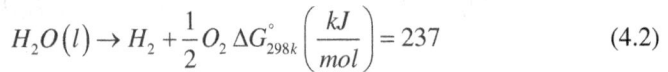

$$H_2O(g) \rightarrow H_2 + \frac{1}{2}O_2 \quad \Delta G^\circ_{298k}\left(\frac{kJ}{mol}\right) = 229 \tag{4.1}$$

$$H_2O(l) \rightarrow H_2 + \frac{1}{2}O_2 \quad \Delta G^\circ_{298k}\left(\frac{kJ}{mol}\right) = 237 \tag{4.2}$$

The photocatalytic water splitting process unfolds through a sequential series of three fundamental steps:

- *Photoexcitation:* When a photocatalyst is exposed to solar irradiation, it takes up photons having energies (hυ) equal to or greater than its bandgap (E_g). This absorption energizes the photocatalyst, generating electron-hole duos in both its conduction band and valence band, respectively (Eq. 4.3).
- *Charge carrier separation and transfer:* The newly produced electron-hole duos are efficiently separated within the photocatalyst and subsequently directed towards the reaction sites of the catalyst.
- *Simultaneous redox reactions:* Once the charge carriers reach the catalyst's surface, simultaneous oxidation and reduction reactions occur (Eqs. 4.4 and 4.5). The adsorbed water molecules undergo transformation, leading to the production of oxygen and hydrogen, respectively.

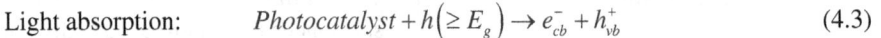

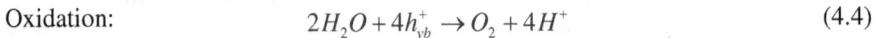

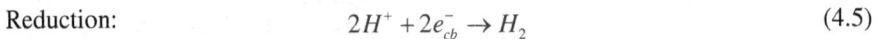

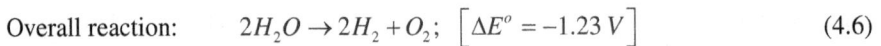

Light absorption: $Photocatalyst + h\left(\geq E_g\right) \rightarrow e^-_{cb} + h^+_{vb}$ $\tag{4.3}$

Oxidation: $2H_2O + 4h^+_{vb} \rightarrow O_2 + 4H^+$ $\tag{4.4}$

Reduction: $2H^+ + 2e^-_{cb} \rightarrow H_2$ $\tag{4.5}$

Overall reaction: $2H_2O \rightarrow 2H_2 + O_2; \; \left[\Delta E^\circ = -1.23\,V\right]$ $\tag{4.6}$

The thermodynamics governing the photocatalytic water splitting process are explored in terms of various factors, including light-induced excitation, bandgap characteristics, and electron states, such as quasi-equilibrium states and quasi-Fermi levels. Under dark conditions, an electron within the photocatalyst tends to attain internal equilibrium at the ground state energy level, primarily due to their extremely short relaxation time, which is shorter than the time essential to traverse the bandgap. The chemical potentials or quasi-Fermi levels of both electrons and holes are defined as follows (Eqs. 4.7 and 4.8) [6].

$$F_n = E_c + k_B T ln \frac{n}{N_c} \tag{4.7}$$

$$F_p = E_v - k_B T ln \frac{p}{N_v} \tag{4.8}$$

where E_c and E_v are the energy levels of conduction band minimum (CBM) and valency band maximum (VBM), T is the reaction temperature, k_B is the Boltzmann constant, n and p are the concentrations of e^- and h^+, and N_c and N_v are the effective densities of states in CB and VB.

As explained in various steps of photocatalytic process, when a photocatalyst is irradiated with a photon energy ≥ bandgap, the charge carriers e^-/h^+ will be generated by excitation, causing a disturbance in the quasi-equilibrium states and quasi-Fermi levels of the photocatalyst and hence also causing a change in the chemical potentials of electron and hole. In such case, the reaction becomes spontaneous because of negative DG, resulting from a higher value of electron chemical potential as compared to that of the hole. The calculation of ΔG is given in Eq. 4.9. From this perspective, electron transfer becomes spontaneous and can either contribute in the desired reaction of water splitting or in undesired e^-/h^+ recombination reaction (Eq. 4.10), leading to energy losses. The photon energy of irradiation is not completely utilized by the photocatalyst due to unavoidable entropy loss ($T\Delta S_{mix}$) from the CB electron and VB hole charge transferring. An energy difference between the absorbed photon energy (E_g) and the entropy loss represents the maximum energy utilized for driving the reaction as given in Eqs. 4.11 and 4.12.

$$\Delta G = -\left|F_n - F_p\right| = -E_g - k_B T ln \frac{np}{N_c N_v} \tag{4.9}$$

$$r_{recombination} = k_{recombination} \left[e_{CB}^-\right]\left[h_{VB}^+\right] \tag{4.10}$$

$$T\Delta S_{mix} = k_B T \left(ln \frac{N_c}{n} + ln \frac{N_v}{p} \right) = k_B T ln \frac{N_c N_v}{np} \tag{4.11}$$

$$E_g - T\Delta S_{mix} = E_g - k_B T ln \frac{N_c N_v}{np} = -\Delta G \tag{4.12}$$

Overall, from a thermodynamics perspective, it can be summarized that, a photocatalyst upon light irradiation absorbs photon energy $> E_g$ to produce e^-/h^+ pairs, making the charge transfer spontaneous and attaining a lower Gibbs free energy. The generated charge carriers split water to produce hydrogen and oxygen simultaneously recombining resulting in an entropy loss thereby reducing a low utilization of absorbed photons. Hence, this intrinsic thermodynamic limit of photocatalytic water splitting is responsible for the lower solar to hydrogen (STH) efficiency of lower than 1%–2%.

4.3 PERFORMANCE MEASUREMENT OF AN EFFICIENT PHOTOCATALYST

The efficient utilization of solar energy is very crucial for the photocatalytic water splitting process as photons having lower energies than catalyst's bandgap go off unused. Majority of the existing photocatalysts developed till date utilizes only UV and high-energy upper portion of visible spectrum that typically account for about only 4% of total solar spectrum. Therefore, a photocatalyst for this application must fulfil the following attributes for enhanced and efficient hydrogen production

- *Bandgap*: The bandgap of the photocatalyst should be lower than 3 eV but higher than 1.23 eV, preferably between 2 and 2.4 eV for efficient light absorption. In addition, it should possess more negative CB bottom level and a more positive VB top level.
- *Charge separation*: It is important that the photocatalyst is effective in enhanced charge carrier (e^-/h^+) separation and boost the surface reactions as the recombination of e^-/h^+ has a negative impact on the process efficiency.
- *Quantum yield*: The performance of photocatalyst can also be compared in the way of quantum yield, which is described as the number of electrons reacted per absorbed/incident photon as specified in Eq. 4.13.

$$Quantum\,Yield = \frac{Number\,of\,electrons\,reacted}{Number\,of\,incident\,photons} * 100$$
$$= \frac{2 * Number\,of\,evolved\,H_2\,molecules}{Number\,of\,incident\,photons} * 100$$

(4.13)

- *Solar to hydrogen efficiency*: STH efficiency is another standard of assessing the photocatalyst performance, that is defined as energy generated in the form of H_2 gas to the energy of incident light and calculated as given in Eq. 4.14.

$$STH(\%) = \frac{Output\,energy\,as\,H_2\,gas}{energy\,of\,incident\,light} = \frac{r_{H_2} * \Delta G}{P_{sun} * S} * 100$$

(4.14)

where r_{H2} = rate of H_2 production (mmol/s), P_{sun} = sunlight energy flux (mW/cm^2), S = reactor area (cm^2), and ΔG = Gibbs free energy (J/mol).

4.4 PHOTOCATALYTIC WATER-SPLITTING METHODS

To achieve hydrogen production with minimal environmental impact, numerous studies have been carried out to develop efficient and effective technologies. These include electrolysis, thermolysis, thermochemical processes, photocatalysis, photoelectrochemical process, biochemical process, and photo electrolysis among others. However, these technologies face certain limitations such as high capital costs, dependence on rare materials, greenhouse gas emissions (e.g., "gray hydrogen" production), and limited efficiency and durability at the commercial scale [7].

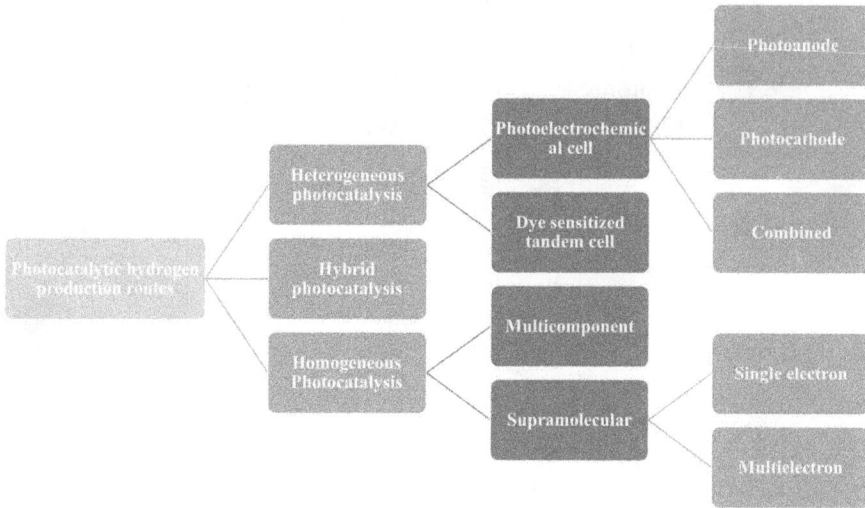

FIGURE 4.1 Classification of different hydrogen production routes by photocatalytic water splitting process.

Light-assisted water splitting can be accomplished in various ways, including the association with electricity, such as the photoelectrochemical process or sensitized tandem electrolysis cell, and direct utilization of photons without involving electricity, such as photolysis. Additionally, it can also involve homogeneous photocatalysis systems with supra-molecular devices. Based on these combinations, photocatalytic water splitting processes can be categorized into heterogeneous photocatalytic processes, homogeneous photocatalytic processes, and hybrid processes (see Figure 4.1).

4.4.1 Heterogeneous Photocatalytic Water Splitting

Heterogeneous photocatalytic systems are classified into two types such as photo-electrochemical process and dye-sensitized tandem electrolysis cells. In photoelectrochemical process, a photoelectrochemical cell (PEC) is utilized, consisting of two electrodes—an anode and a cathode—and at least one of them being a semiconductor photocatalyst electrode. The choice of the electrolyte, whether acidic or basic, relies on the characteristics of the electrode material. When semiconductor electrode is exposed to photons with energies larger than its bandgap energy, it generates electron/hole duos, which can then either oxidize or reduce water.

Several semiconductor materials, such TiO_2, ZnO, Fe_2O_3, $BiVO_4$, and WO_3 have been studied for use in PEC. Among these, TiO_2 has proven to be promising so far [8,9]. The efficiency of the PEC is influenced by whether the semiconductor is applied to one or both electrodes. Applying the semiconductor to both electrodes typically results in higher efficiency (Figure 4.3(a) and (b)) [10]. The efficiency of PEC with single $SrTiO_3$ electrode was 1.2% whereas 6% when applied at both electrodes [11].

An alternative process, heterogeneous photocatalysis, has been explored using a sulphite/sulphide electrolyte [11,12]. In this case, the semiconductor catalyst captures incident photons, exciting electrons from the VB to the CB (Figure 4.2(a)). These

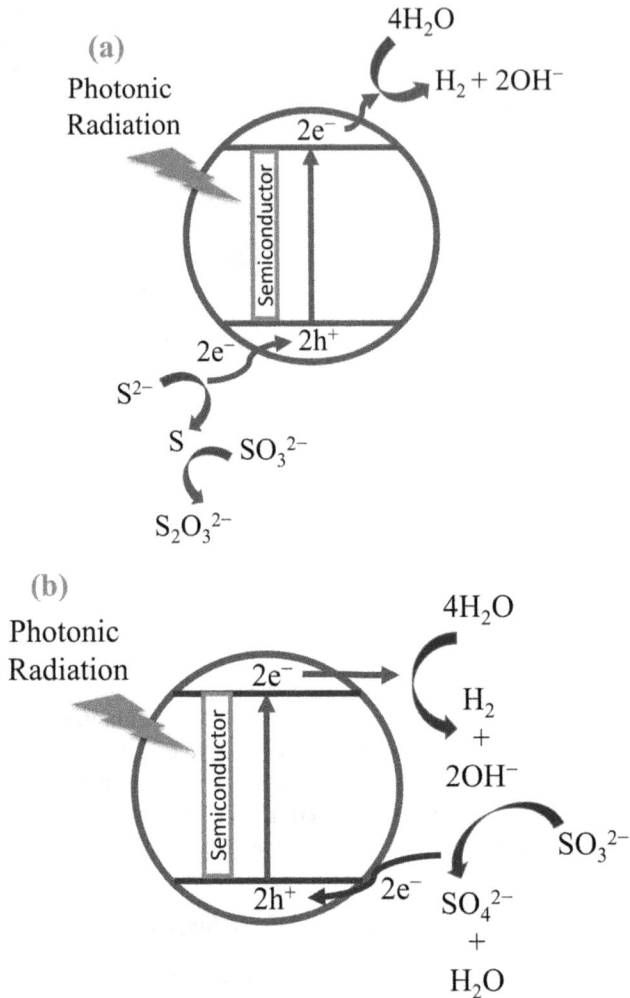

FIGURE 4.2 Representation of (a) heterogeneous photocatalysis process and (b) alternative photocatalysis process with sulphite/sulphide.

excited electrons then participate in the water-splitting process. The catalyst further would be regenerated through accepting electrons from sulphide ions, converting them into sulphur. The generated sulphur subsequently reacts with sulphite ions to produce thiosulfate. Additionally, when the hydroxyl ion concentration becomes significant, an alternative route of the photochemical process is triggered, leading to the conversion of sulphite ions into sulphate ions. This alternate process helps mitigate photo corrosion of the catalyst (Figure 4.2(b)). PECs generally do not require additional power supplies; however, they can be linked with photovoltaic (PV) arrays to provide part of the energy needed to run the process, in combination with the photon-driven energy input at the semiconductor electrode (Figure 4.3(c)) [10].

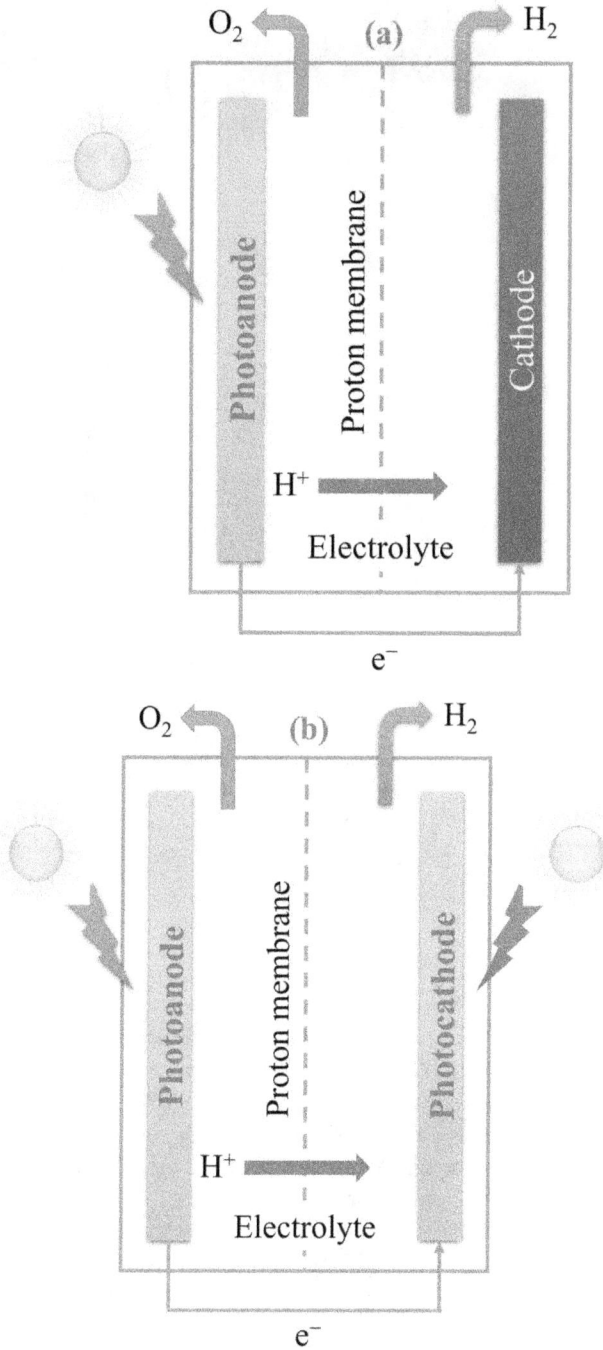

FIGURE 4.3 Different photoelectrochemical cells for water electrolysis consisting of (a) photoanode and metallic cathode, (b) combined photoanode/photocathode cell, and (c) PV array-assisted cell.

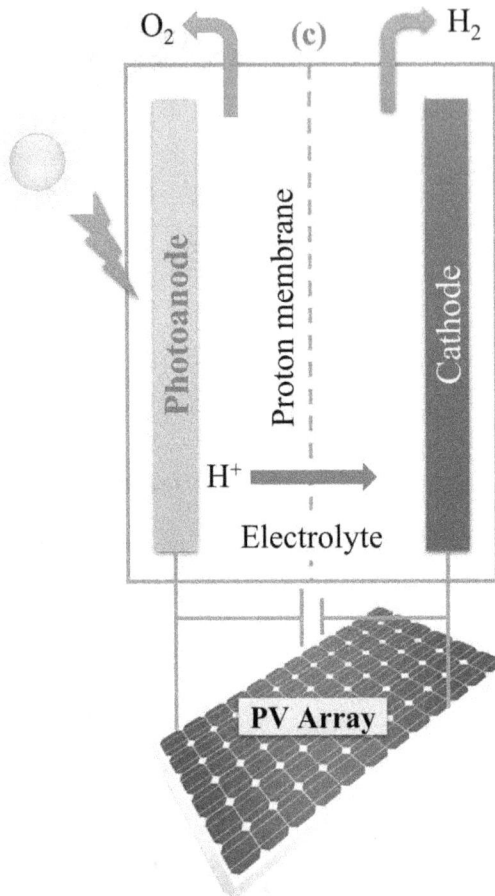

FIGURE 4.3 (Continued)

Sensitized tandem electrolysis cells includes both photocatalytic electrode and a photosensitizer such as dye (Figure 4.4). These tandem cells are divided into two sections—one with an aqueous electrolyte for the electrolysis process and the other with an iodine/iodide redox electrolyte [13]. The first photoanode, usually made of a transparent semiconductor like WO_3, absorbs light and transmits it to the second electrode, which is made of another semiconductor (e.g., TiO_2) sensitized with a suitable dye, such as a Ru-complex dye molecule. This setup allows the tandem cell to cover a broad light spectrum from 477 nm to 744 nm.

Hydrogen production using water splitting can also be done on semiconductor particles by applying a similar concept of PEC (Figure 4.5). Semiconductor particles produce e^-/h^+ pairs under irradiation which further cause reduction and oxidation reactions. To attain complete water splitting, the semiconductor must have a bottom of the conduction band more negative than the reduction potential of H^+ to H_2 (0 V vs. NHE at pH 0) and a top of the valence band more positive than the oxidation

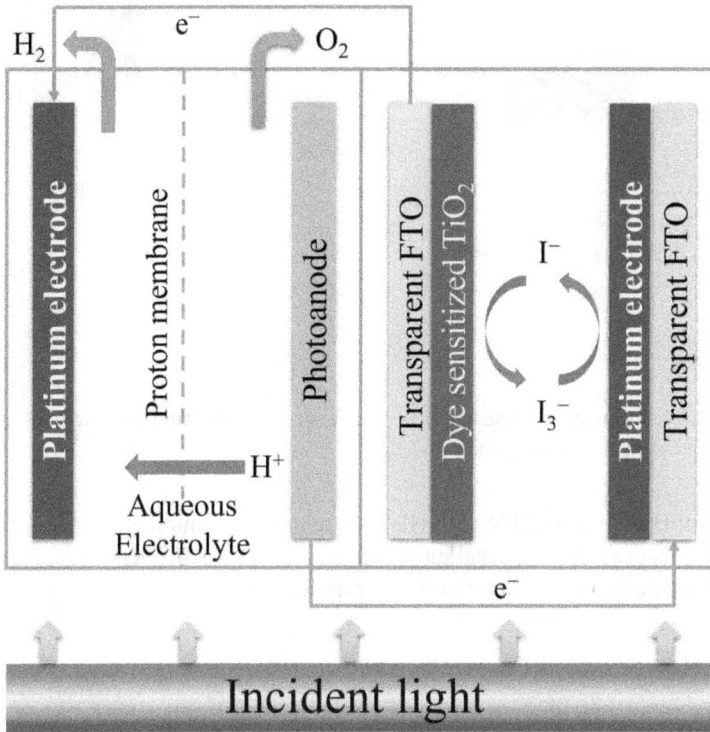

FIGURE 4.4   Representation of dye-sensitized tandem photo electrolysis cell.

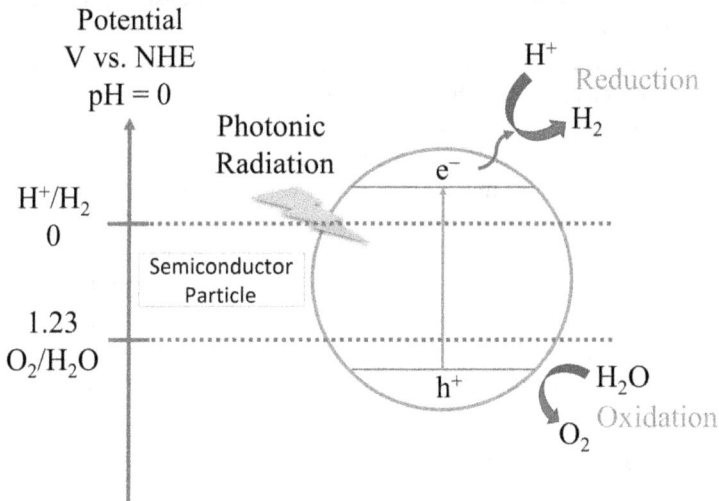

FIGURE 4.5   Schematic representation of basic processes in photocatalytic water splitting on semiconductor particles.

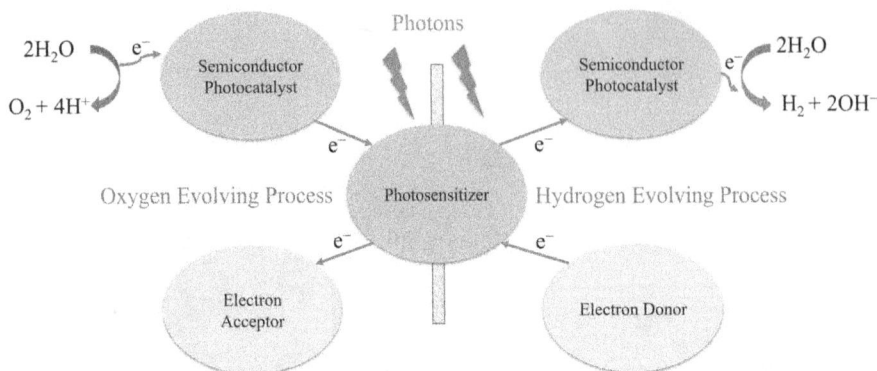

FIGURE 4.6 Schematic representation of homogeneous photocatalysis system for hydrogen-evolving and oxygen-evolving processes.

potential of H_2O to O_2 (1.23 V vs. NHE), requiring a minimum energy of 1.23 eV [7]. However, this process poses challenges, such as activation barriers in charge transfer process between semiconductor and water molecules.

4.4.2 HOMOGENEOUS PHOTOCATALYSIS

Hydrogen production through homogeneous photocatalysis involves a liquid medium with a dissolved catalyst. This process relies on multiple molecular complexes, each playing a distinct role such as photon capture, charge separation and transfer, electron acceptance or release, and catalytic water splitting [12]. Within homogeneous photocatalysis, hydrogen evolving processes (HEP) and oxygen evolving processes (OEP) can be distinguished (Figure 4.6). In HEP, a photosensitizer absorbs photons of appropriate energy, becoming excited and interacting with a semiconductor catalyst to transfer electrons. The catalyst, in turn, utilizes these electrons to decompose water. The photosensitizer regains its electrons from dissolved electron donor in the solution, completing the reaction in four cycles. OEP, on the other hand, involves the transfer of electrons in the opposite direction, requiring an electron acceptor. This process generates hydrogen and hydroxyl from water, thereby completing the water splitting cycle. In both HEP and OEP processes, sensitizer plays an important role and ruthenium tris-2,2′-bipyridyne, that is $[Ru(bpy)_3]^{2+}$, is mostly used. This molecule excites to generate singlet state by absorbing the photon of 450 nm (2.75eV) which further coverts to triplet excited state by losing 0.65 eV energy through an intersystem crossing. The life of excited state is notably long at 890 nanoseconds, allowing for efficient induction of water oxidation and reduction reactions.

4.4.3 HYBRID PHOTOCATALYSIS

Hybrid systems consist of interconnected reactors to enhance the generation of gaseous products [14]. The process comprises two main steps: the OEP and the HEP. In the first reactor, the OEP takes place. This reactor contains a photoanode, and when incident photons strike the photoanode, the water oxidation process is initiated. An electron acceptor existing in system captures the electrons produced during the reaction and undergoes reduction. This reduced electron acceptor then proceeds to the second reactor to facilitate the HEP. The second reactor contains an electron acceptor, and in this step, an electron donor transfers its electrons to the electron acceptor and generate hydrogen gas. Baniasadi et al., established a dual-cell hybrid photocatalytic system, based on a similar concept [15]. Their setup involved two interconnected cells separated by an anion exchange membrane. Further, they applied sulfur-based catholyte, while the photocatalyst employed was ZnS nanoparticles. The electron donor used in their experiments was sodium sulfide. Both photoreactors and dual cells operated within an alkaline aqueous solution. This system achieved enhanced gas production and overall efficiency in generating hydrogen gas through photocatalysis.

4.5 STRATEGIES FOR THE MODIFICATION OF PHOTOCATALYSTS

Most of the photocatalysts utilize photons of high-energy UV light and in very few cases upper part of visible light. Thus, it is realized that their electronic structure is required to be tuned to increase the absorption of wider part of solar spectrum. Further, recombination of the charge particles electron and hole is a great challenge and needs to be minimized for better output of the process.

4.5.1 DOPING OF METALS AND NON-METALS

Impregnation of semiconductor with metals (Au, Pt, Fe, Co, Zn, Ni, Ag, etc.) and non-metals (N, C, S, Sn, etc.) is done to tune the bandgap, minimize the charge recombination and alter its optical properties [16–21]. The doping of metal creates a Schottky barrier at metal-semiconductor interface and trap to photogenerated electron (Figure 4.7). The introduced Schottky barrier allows an easy transfer of electron

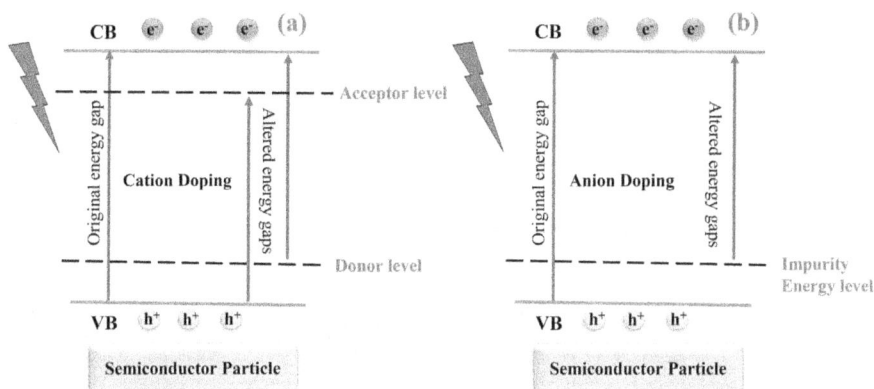

FIGURE 4.7 Modification of energy gaps in semiconductor with (a) cation doping and (b) anion doping.

from CB of semiconductor to metal, which ultimately acts as electron sink, thus minimizing the charge recombination and improving the photocatalytic efficiency. For example, Pt doped [20] and Cu doped TiO_2 [21] have been reported for higher hydrogen production as compared to undoped TiO_2. Metal doping also enhance the photocatalytic performance via surface plasma resonance (SPR), where metal particle absorbs the photon when irradiated with their plasmon resonance frequency and a strong local electric field is created near to metal surface. This will generate hot electrons and holes. The hot electron will further be transferred to semiconductor to perform photocatalytic process. Liu et al. applied Ag doped TiO_2 for photocatalytic hydrogen evolution under both UV and visible light and observed high efficiency compared to undoped TiO_2 [22]. Non-metal doping introduces new energy level just above the VB due to the impurities. This extra energy level makes the photocatalyst active in visible range also, making it overall active in both UV and visible light. Further, non-metal dopants create oxygen vacancies and inhibit the charge recombination, ultimately enhance the performance. Higher absorption of visible light was observed with the doping of N, S, Cl, and Br in TiO_2 [19,23].

4.5.2 SEMICONDUCTOR COMPOSITE FORMATION

Enhanced photocatalytic performance can also be achieved through the combination or coupling of two different semiconductors by creating type I, II, and III heterojunctions (Figure 4.8). These heterojunctions construct Z-scheme and S-type photocatalytic systems [5,24,25]. In a study, type-I heterojunction was constructed by employing g-Pt/C_3N_4/Fe_2O_3 to TiO_2 [26]. The charge carrier electrons and holes could transport towards Fe_2O_3, resulting a better charge separation. This led to 1150 times greater hydrogen generation as compared to pristine TiO_2.

A type-II composite g-C_3N_4/TiO_2 was applied where high performance for hydrogen evolution was observed [27]. In this case photogenerated electrons were

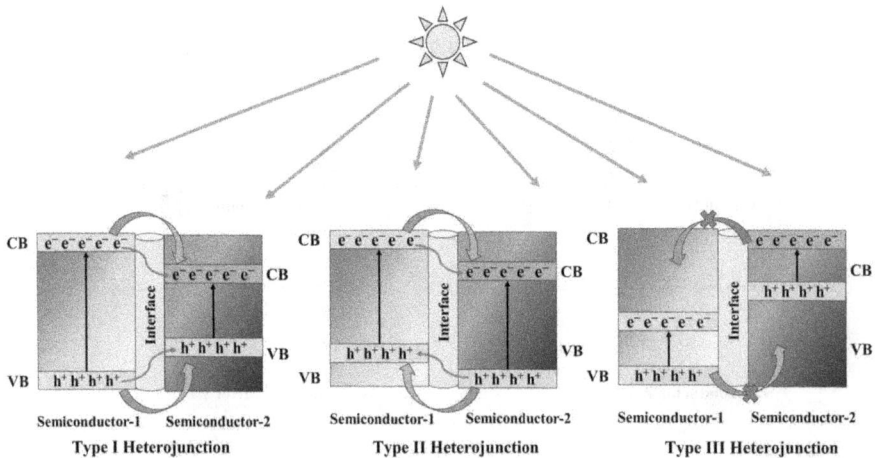

FIGURE 4.8 Different types of heterojunctions in composite semiconductors.

transported to TiO_2 and holes were transported to g-C_3N_4, minimizing the electron-hole recombination. An efficient charge separation was observed in Z-scheme hetero-junction of TiO_2/NiO [28]. The better charge separation was because of the mid-gap state by Ti^{3+} and oxygen vacancies. An easy Z-scheme charge transfer would be possible between the VB of NiO and Ti^{3+}/oxygen vacancy state.

4.5.3 INTRODUCTION OF POROSITY

Designing of photocatalyst with porosity (particularly mesoporosity) has an impact on its performance as this can change its photon harvesting capacity [29–31]. There are multiple reasons due to which porosity enhance the performance. Macro porous network improves the radiation transfer rate and stimulate multiple reflections to improve the photon harvesting ability. Cascade effect is another factor for the higher photocatalytic performance while illuminating the active sites present in the pores [32]. Another study has shown that high performance for hydrogen evolution can be achieved by structural alignment in mesoporous C_3N_4 [33]. Further, mesoporous transition metal oxides present a higher number of active sites due to higher surface area as compared to bulk symmetry and perform better for hydrogen evolution.

4.5.4 CONTROLLED SIZE AND SHAPE SYNTHESIS

With the alteration in the dimension of the semiconductor particle size, quantum confinement effect (Q-effect) occurs due to the 3-D confinement of electrons [34]. In the quantum-sized semiconductor particles, photogenerated charges electrons and holes are confined in a potential well which restricts their delocalization. This is the property of bulk materials. With the reduction in particle size, bandgap gets wider. The surface vacancies and modified band positions both are generated due to Q-effect that act to trap the charges to enhance the charge separation [35]. Many semiconductors have been investigated for Q-effect such as TiO_2, ZnO, Fe_2O_3, ZnS, CdS, etc. Further, by modifying the facets of nanocrystals, hydrogen evolution capacity can be enhanced. In a study, TiO_2 nanoparticles with facet {101} were observed more efficient than facet {001}[36].

4.5.5 DEFECT ENGINEERING

Some defects like step edges and oxygen vacancies are reported to support in enhanced photocatalytic performance for hydrogen production [9,37,38]. These defects help in better charge separation. Furthermore, oxygen vacancies improve optical properties, electrical conductivity, reductive property, and dissociative adsorption of metal oxides. These oxygen vacancies can be created by applying non-metal doping, thermal treatment of metal oxide under reducing condition, or by metals through bombardment of high energetic photons [38]. In a study, hydrogen evolution using ultrathin In_2O_3 photo-electrode was superior compared to bulk material [37]. This was due to the poor oxygen vacancy and narrow bandgap property of bulk material.

4.5.6 Bionic Engineering

In nature, there are hydrogenases enzymes which metabolize hydrogen in cyano-bacteria and other blue-green algae. Some of them are [NiFe]-H_2ase, [FeFe]-H_2ase and [Fe]-H_2ase [39]. These enzymes catalyse the reduction of H^+ to hydrogen gas at a high rate. In view of this, efforts were made to couple these enzymes with semi-conductors for water splitting. A turnover frequency (TOF) of 380–900 cm^{-1} was observed over CdS hybrid with [FeFe]-H_2ase [8]. Similarly, with of [NiFeSe]-H_2ases coated on Ru dye–sensitized TiO_2 generated hydrogen with TOF of 50 s^{-1} at 25°C and pH 7, in visible region [40].

4.5.7 Sensitization of Semiconductor

The primary purpose of the sensitization is to make photoelectrode active in lower energy part as well for photon harvesting [17,41]. Semiconductors such as TiO_2 are active mainly in UV light region and the application of sensitizer, for example dye may allow the photon absorption in visible light. Out of many possible sensitizers, organic dyes and ruthenium complex dyes are quite popular for the purpose. Upon light irradiation, dye molecule gets excited by π–π^* transition from HOMO level of dye to LUMO level. The photoelectron generated in the process is transferred to CB of semiconductor which is further utilized in the reduction process of hydro-gen evolution. Rhodamine B [42] and coumarin dye [43] were applied as sensitizers for Pt loaded TiO_2 for an efficient hydrogen evolution. Similarly, sensitized Pt/TiO_2 catalyst with Pt(dcbpy)Cl_2 have shown greater efficiency for hydrogen production in visible light [44].

4.6 CHALLENGES OF PHOTOCATALYTIC WATER SPLITTING

The photocatalytic water splitting process has to cross over many challenges in order to reach the real-world large-scale applications. Even after several years of research progress in photocatalytic water splitting, still some major challenges are related with process, that needs to be addressed to make the process economically feasible from a commercial viewpoint.

4.6.1 Energy Losses, Low Quantum Efficiency

The faster recombination rate as compared to the charge generation rate and the inevitable entropy losses associated with the process during recombination leads to the low quantum efficiency of the process. Several approaches to inhibit the recombi-nation rate, include co-catalyst addition that enhances the H_2/O_2 production, particle size reduction of photocatalysts, development of single-layer materials that shorten the migration path of charge carriers from bulk to surface [45], and development of Z-scheme water splitting system consisting of hydrogen generative photocatalyst and oxygen generative photocatalyst [46]. Despite several research attempts towards the enhanced charge separation, the overall quantum efficiency of the process is still less than 5%.

4.6.2 PRODUCT SEPARATION

Product separation poses a major challenge in solar-driven water splitting, as simultaneous H_2 and O_2 generation within the same reactor could lead to hazardous conditions. Various solutions have been proposed, including the use of flammability suppressants like N_2 or CO_2, employing a Z-scheme configuration with separate compartments, or developing heterojunction catalysts from two semiconductors. However, these solutions impact the process's economic viability. To improve the economics, the cost of H_2/O_2 separation must be minimized with energy-efficient systems. Thermodynamically, the minimum Gibbs energy required for separation at room temperature is 2.4 kJ, only 1.1% of the overall energy change in water splitting reactions.

4.6.3 MASS TRANSFER LIMITATIONS

Several mass transfer resistances are present in the heterogenous photocatalytic water splitting process such as transport of water molecules from bulk to surface, internal diffusion of water molecules to catalyst pores, diffusion of products from pores to the surface, and transport of products from surface to the bulk that are aligned with the solid-liquid heterogenous catalytic reactions. The diffusional transport of molecules and the surface reaction were reported to be much slower than the transmission of photon and photoexcitation. These transport limitations can be overcome with turbulent mixing that decreases the thickness of solid-liquid boundary layer over the catalyst. Therefore, the batch scale/lab scale operations are carried under continuous stirring to homogenously disperse the catalyst, to shorten the interdiffusional path with the use of nanoparticles. The usage of slurry reactors for large-scale operations faces the challenge of the catalyst recovery from large volumes of water. Hence, for continuous operations, immobilized photocatalysts can be used where photocatalyst powders fixes thinly on to the substrates using silica binders. The catalyst layer should be thin, to alleviate the mass transfer resistances associated with the transport of reactants and products. The hydrophilicity and the porosity of these layers are quite significant to control for protecting the activity of these layers [2]. These layers can be used for commercial scale operations and the reactor configuration are to be cautiously engineered to have a high surface to volume ratio to overcome the pronounced mass transfer limitations.

4.6.4 UTILIZING INCIDENT PHOTON ENERGY EFFICIENTLY

The solar spectrum comprises 5% UV (<400 nm), 43% visible (400–700 nm) and 52% near-infrared light regions (700–2500 nm). The photon energy irradiated over the photocatalytic reactor transmits through multiple mediums of reactor, water, and photocatalyst surface with notable reflection, scatters, and finally absorbs on the photocatalyst surface. The solar spectrum energy utilization remains a challenge, as wider bandgap energies lead to the underutilization of irradiated photon energy. Hence, novel photocatalytic materials with lower bandgaps and controlled electronic and optical properties are essential for better visible light response, accounting for

43% of the solar spectrum. The maximum allowed bandgap of photocatalysts should be considered (2.36 eV, corresponding to a minimum wavelength of 526 nm) to achieve high STH efficiency for large-scale applications [6].

4.6.5 MINIMIZING THE USE OF SACRIFICIAL REAGENTS

To mitigate the undesired $e^+/h-$ recombination and to boost the photocatalyst performance, sacrificial reagents that serve as hole scavengers/electron donors are required to be incorporated in the process. A sacrificial reagent scavenges the photogenerated hole and conserves the electron that is required for the reduction reaction of water to yield H_2. Attempts were made to enhance the production of H_2 by the addition of several sacrificial reagents such as methanol, ethanol, and triethanol amine. However, these reagents may undergo photo reforming reactions, producing secondary pollutants that harm the environment. Additionally, the cost of treating these secondary pollutants and using high-valued, irrecoverable reagents poses an economic challenge [2,46].

4.6.6 ADDRESSING FRESH WATER AVAILABILITY

While photocatalytic water splitting predominantly uses fresh water, in regions where fresh water is scarce, the utilization of seawater appears more reasonable. Some studies have reported enhanced H_2 generation in seawater environments due to impurities acting as sacrificial agents. However, the deactivation of catalyst active sites caused by inorganic salt deposition remains a significant challenge. Energy-efficient methods for producing fresh water from seawater are required to overcome this hurdle [6].

4.6.7 REACTOR CONFIGURATIONS

Conventional photocatalytic water splitting involves particulate photocatalytic systems where photocatalysts disperse in water to produce hydrogen under light irradiation. In these processes, the redox reaction of water occurs on each individual catalyst particle surface, acting as a micro-photoelectrode. To enhance water splitting without substrate requirements or film deposition, researchers have explored catalysts with more active sites and larger surface areas. However, scaling up these processes for large-scale operations with significant STH conversion efficiency remains challenging.

Lab-scale flask reactors have been commonly used for studies with particulate photocatalytic systems. Unfortunately, scaling them up for practical applications is difficult due to limitations in STH conversion efficiency. Additionally, in such systems, gravity tends to cause catalysts to settle at the bottom of the reactor, leading to inefficient light absorption unless the reactor is perfectly stirred. Tube-shaped reactors with parabolic concentrators were investigated, using sacrificial reagents, to improve light absorption. However, maintaining turbulent flows to prevent catalyst sedimentation and subsequent catalyst recovery through energy-intensive methods like filtration or centrifugation remain significant drawbacks [47]. Thin photocatalyst layers on substrates using drop cast methods with silica binders have also been explored. In these systems, controlling the hydrophilicity and porosity of the layers is crucial for preserving the intrinsic activity of the catalysts. However, these systems generally exhibit lower activity compared to suspension-type systems. An alternative approach involves panel-type reactors, which offer promise for large-scale operations. Some studies utilizing highly active $SrTiO_3:Al$ photocatalyst sheets indicate that these reactors can sustain high gas evolution rates with a potential for scale-up to square-meter scale and beyond without mass transfer limitations [48]. Notably, these panel-type reactors can also be integrated with Z-scheme type photocatalytic sheets, which effectively harvest sunlight.

4.7   INTEGRATING CIRCULAR ECONOMY PRINCIPLES IN PHOTOCATALYTIC WATER SPLITTING

Photocatalytic water splitting has emerged as a promising and cost-effective technology for harnessing solar energy to produce hydrogen, setting it apart from other solar water splitting methods such as photoelectrochemical and photovoltaic electrochemical processes. However, while this approach proves to be environmentally attractive, it faces challenges related to charge separation efficiency, charge recombination issues, and a relatively low STH efficiency, hindering its scalability for large-scale adoption within a circular economy framework.

Techno-economic analyses conducted on photocatalytic water splitting processes have uncovered significant shortcomings, presenting a major barrier to the widespread implementation of this sustainable technology. Current estimates suggest hydrogen production costs to be around 10.36 \$/kg, a stark contrast to the 2–4 \$/kg achieved through conventional steam methane reforming units and coal gasification plants. To enhance the circularity of the hydrogen production process, it is imperative to bridge this cost gap and align it with circular economy principles, such as resource efficiency and waste reduction.

Maeda and Domen [49] projected an infeasibly extensive plantation area of 250,000 km^2, comprising 10,000 units operating at 10% efficiency to meet approximately one-third of the global energy demand by 2050. This underscores the urgent need for not only enhancing efficiency but also achieving cost-effectiveness and scalability. Hisatomi and Domen [2] estimated the maximum feasible capital cost for photocatalytic water splitting to be 102 \$/m^2, considering certain assumptions such as a fixed production cost of 3.5 \$/kg of H_2, 7.6 hours of sunlight irradiation with a constant power density of 240 W/m^2, a plant life of 10 years, and 10% STH efficiency. However, the current low STH efficiency (<3%) significantly inflates actual production costs, surpassing the anticipated 3.5 \$/kg mark by up to fivefold. To overcome these barriers and establish a sustainable STH production process, it is imperative to explore innovative materials and systems to meet the targeted cost goals.

In pursuit of economic viability and circularity, innovative materials and systems should be explored to improve charge separation efficiency and STH efficiency, thus minimizing production costs while maximizing the utilization of solar energy. Circular design strategies, such as cradle-to-cradle approaches, should be applied to ensure the efficient recycling of materials and minimize waste generation during the manufacturing and operation of photocatalytic water splitting units. Furthermore, the development of large-scale demonstration units is critical for a comprehensive and accurate estimation of the system's total cost, encompassing hard and soft balance of system (BOS) costs, which are subject to fluctuations with the progress of research and development. The aspect of technical safety concerning hydrogen and oxygen separation remains a key challenge, as the existing enclosed reaction systems prove impractical for large-scale operations. Moreover, within the circular economy paradigm, collaboration among researchers, policymakers, and industry stakeholders becomes essential to drive innovations and share best practices for sustainable and affordable hydrogen production. Leveraging circular supply chains and integrated energy-water systems will create synergies that foster cost-effective scalability and circularity, ensuring economic, social, and environmental benefits. As part of this circular transition, a crucial aspect to consider is the technical safety surrounding hydrogen and oxygen separation. Integrating circular economy principles with robust safety measures will pave the way for large-scale and secure photocatalytic water splitting operations.

4.8 FUTURE OUTLOOK

Photocatalytic water splitting, a promising technology that utilizes semiconductor materials and solar radiation to produce hydrogen, faces challenges in achieving widespread implementation due to high hydrogen cost and low STH efficiency (<10%). To enhance the process, efforts must focus on optimizing solar light utilization by expanding spectral receptive regions, suppressing e^-/h^+ pair recombination, and equalizing the overpotential with water. This presents a unique opportunity for extensive research and development in this field.

- *Enhancing photon absorption*: Innovative design strategies using low-cost, stable, and panchromatic responsive materials, like cation/anion doping and plasmonic photocatalysts, can significantly boost light-harvesting capacity.
- *Improving charge separation*: Novel approaches are needed to develop highly efficient photocatalytic processes, such as synthesizing well-crystallized materials with low bulk defects and passivating surface modifications.
- *Understanding photoexcited carriers*: Thoroughly understanding photoexcited carrier's dynamic behaviour under non-equilibrium conditions using advanced technologies like operando analysis can accelerate the discovery of superior photocatalytic materials.

While significant progress has been made in comprehending photocatalytic systems and advancements in analytical technologies, continued research and acceleration are essential for sustainable energy production. Resolving the challenges associated with large-scale photocatalytic water splitting demands relentless, extensive research to develop efficient, cost-effective, robust, and scalable carbon-free technologies. Only through such advancements can we secure a greener and more sustainable energy future. The journey towards achieving this ambitious vision relies on ongoing dedication and ingenuity from researchers and stakeholders alike.

REFERENCES

[1] IEA World Energy Investment, 2023. [Online]. Available: www.iea.org/reports/world-energy-investment-2023, accessed on 15th August 2023.

[2] Hisatomi, T., and Domen, K., Reaction systems for solar hydrogen production via water splitting with particulate semiconductor photocatalysts, *Nat. Catal.*, 2, 387, 2019.

[3] Department of Economic and Social Affairs of United Nation, THE 17 GOALS | Sustainable Development, *Sustainable Development*, 2022. Available: https://sdgs.un.org/goals, accessed on 20th April 2023.

[4] Fujishima, A., and Honda, K., Electrochemical photolysis of water at a semiconductor electrode, *Nature*, 238, 37, 1972.

[5] Ishaq, T., et al., A perspective on possible amendments in semiconductors for enhanced photocatalytic hydrogen generation by water splitting, *Int. J. Hydrog. Energy*, 46, 39036, 2021.

[6] Ng, K. H., et al., Photocatalytic water splitting for solving energy crisis: Myth, fact or busted?, *Chem. Eng. J.*, 417, 128847, 2021.

[7] Yue, M., et al., Hydrogen energy systems: A critical review of technologies, applications, trends and challenges, *Renew. Sustain. Energy Rev.*, 146, 111180, 2021.

[8] Li, X., et al., Engineering heterogeneous semiconductors for solar water splitting, *J. Mater. Chem. A*, 3, 2485, 2015.

[9] Wang, G., Ling, Y., and Li, Y., Oxygen-deficient metal oxide nanostructures for photoelectrochemical water oxidation and other applications, *Nanoscale*, 4, 6682, 2012.

[10] Dincer, I., and Zamfirescu, C., Sustainable hydrogen production options and the role of IAHE, *Int. J. Hydrog. Energy*, 37, 16266, 2012.

[11] Zamfirescu, C., Naterer, G.F., and Dincer, I., Solar light—Based hydrogen production systems, in *Encyclopedia of Energy Engineering and Technology*, 2nd edn, CRC Press, 2014, 1722–1744.

[12] Dincer, I., and Zamfirescu, C., Hydrogen production by photonic energy, in *Sustainable Hydrogen Production*, 2016, 309–391. Elsevier. https://doi.org/10.1016/B978-0-12-801563-6.00005-4

[13] Arachchige, S. M., et al., High turnover in a photocatalytic system for water reduction to produce hydrogen using a Ru,rh,ru photoinitiated electron collector, *ChemSusChem*, 4, 514, 2011.

[14] Minggu, L. J., Wan Daud, W. R., and Kassim, M. B., An overview of photocells and photoreactors for photoelectrochemical water splitting, *Int. J. Hydrog. Energy*, 35, 5233, 2010.

[15] Baniasadi, E., Dincer, I., and Naterer, G. F., Measured effects of light intensity and catalyst concentration on photocatalytic hydrogen and oxygen production with zinc sulfide suspensions, *Int. J. Hydrog. Energy*, 38, 9158, 2013.

[16] Azam, M. U., et al., Engineering approach to enhance photocatalytic water splitting for dynamic H_2 production using La_2O_3/TiO_2 nanocatalyst in a monolith photoreactor, *Appl. Surf. Sci.*, 484, 1089, 2019.

[17] Chen, X., and Mao, S. S., Titanium dioxide nanomaterials: Synthesis, properties, modifications and applications, *Chem. Rev.*, 107, 2891, 2007.

[18] Udayabhanu, et al., One-pot synthesis of Cu—TiO_2/CuO nanocomposite: Application to photocatalysis for enhanced H2 production, dye degradation & detoxification of Cr (VI), *Int. J. Hydrog. Energy*, 45, 7813, 2020.

[19] Luo, H., et al., Photocatalytic activity enhancing for titanium dioxide by co-doping with bromine and chlorine, *Chem. Mater.*, 16, 846, 2004.

[20] Guo, N., et al., Crumpled and flexible cotton-fiber-like TiO_2 with Pt anchored and its notable photocatalytic activity facilitate by Schottky junction interface, *Mater. Lett.*, 221, 183, 2018.

[21] Christoforidis, K. C., and Fornasiero, P., Photocatalytic hydrogen production: A rift into the future energy supply, *ChemCatChem*, 9, 1523, 2017.

[22] Liu, E., et al., Plasmonic Ag deposited TiO_2 nano-sheet film for enhanced photocatalytic hydrogen production by water splitting, *Nanotechnology*, 25, 165401, 2014.

[23] Piskunov, S., et al., C-, N-, S-, and Fe-Doped TiO_2 and $SrTiO_3$ Nanotubes for visible-light-driven photocatalytic water splitting: Prediction from first principles, *J. Phys. Chem. C*, 119, 18686, 2015.

[24] Fajrina, N., and Tahir, M., Monolithic Ag-Mt dispersed Z-scheme pCN-TiO_2 heterojunction for dynamic photocatalytic H_2 evolution using liquid and gas phase photoreactors, *Int. J. Hydrog. Energy*, 45, 4355, 2020.

[25] Sharma, K., et al., Photocatalytic process for oily wastewater treatment: A review, *Int. J. Environ. Sci. Technol.*, 20(4), 4615–4634, 2023.

[26] Mohamed, R. M., Kadi, M. W., and Ismail, A. A., A Facile synthesis of mesoporous α-Fe_2O_3/TiO_2 nanocomposites for hydrogen evolution under visible light, *Ceram. Int.*, 46, 15604, 2020.

[27] Tiwari, A., and Pal, U., Effect of donor-donor-π-acceptor architecture of triphenylamine-based organic sensitizers over TiO_2 photocatalysts for visible-light-driven hydrogen production, *Int. J. Hydrog. Energy*, 40, 9069, 2015.

[28] Liu, J., et al., Facial construction of defected NiO/TiO_2 with Z-scheme charge transfer for enhanced photocatalytic performance, *Catal. Today*, 335, 269, 2019.

[29] Jin, Y., et al., Preparation of mesoporous Ni_2P nanobelts with high performance for electrocatalytic hydrogen evolution and supercapacitor, *Int. J. Hydrog. Energy*, 43, 3697, 2018.

[30] Onsuratoom, S., Chavadej, S., and Sreethawong, T., Hydrogen production from water splitting under UV light irradiation over Ag-loaded mesoporous-assembled TiO_2–ZrO_2 mixed oxide nanocrystal photocatalysts, *Int. J. Hydrog. Energy*, 36, 5246, 2011.

[31] Surolia, P. K., and Jasra, R. V., Photocatalytic degradation of p-nitrotoluene (PNT) using TiO_2-modified silver-exchanged NaY zeolite: Kinetic study and identification of mineralization pathway, *Desalin. Water Treat.*, 57, 22081, 2016.

[32] Vaudreuil, S., et al., Synthesis of macrostructured silica by sedimentation-aggregation, *Adv. Mater.*, 13, 1310, 2001.

[33] Han, Q., et al., Atomically thin mesoporous nanomesh of graphitic C_3N_4 for high-efficiency photocatalytic hydrogen evolution, *ACS Nano*, 10, 2745, 2016.

[34] Yang, P., et al., Generalized syntheses of large-pore mesoporous metal oxides with semicrystalline frameworks, *Nature*, 396, 152, 1998.

[35] Hoffmann, M. R., et al., Environmental applications of semiconductor photocatalysis, *Chem. Rev.*, 95, 69, 1995.

[36] Gordon, T. R., et al., Nonaqueous synthesis of TiO_2 nanocrystals using TiF_4 to engineer morphology, oxygen vacancy concentration, and photocatalytic activity, *J. Am. Chem. Soc.*, 134, 6751, 2012.

[37] Lei, F., et al., Oxygen vacancies confined in ultrathin indium oxide porous sheets for promoted visible-light water splitting, *J. Am. Chem. Soc.*, 136, 6826, 2014.

[38] Pan, X., et al., Defective TiO_2 with oxygen vacancies: Synthesis, properties and photocatalytic applications, *Nanoscale*, 5, 3601, 2013.

[39] Lubitz, W., et al., Hydrogenases, *Chem. Rev.*, 114, 4081, 2014.

[40] Reisner, E., et al., Visible light-driven H_2 production by hydrogenases attached to dye-sensitized TiO_2 nanoparticles, *J. Am. Chem. Soc.*, 131, 18457, 2009.

[41] Joseph, M., and Haridas, S., Recent progresses in porphyrin assisted hydrogen evolution, *Int. J. Hydrog. Energy*, 45, 11954, 2020.

[42] Duonghong, D., Borgarello, E., and Grätzel, M., Dynamics of light-induced water cleavage in colloidal systems, *J. Am. Chem. Soc.*, 103, 4685, 1981.

[43] Abe, R., et al., Photocatalytic activity of R_3MO_7 and $R_2Ti_2O_7$ (R = Y, Gd, La; M = Nb, Ta) for water splitting into H_2 and O_2, *J. Phys. Chem. B*, 110, 2219, 2006.

[44] Li, J., et al., Visible light induced dye-sensitized photocatalytic hydrogen production over platinized TiO_2 derived from decomposition of platinum complex precursor, *Int. J. Hydrog. Energy*, 38, 10746, 2013.

[45] Villa, K., et al., Photocatalytic water splitting: Advantages and challenges, *Sustain. Energy Fuels*, 5, 4560, 2021.

[46] Lin, L., et al., Visible-light-driven photocatalytic water splitting: Recent progress and challenges, *Trends Chem.*, 2, 813, 2020.

[47] Jing, D., et al., Photocatalytic hydrogen production under direct solar light in a CPC based solar reactor: Reactor design and preliminary results, *Energy Convers. Manag.*, 50, 2919, 2009.

[48] Goto, Y., et al., A particulate photocatalyst water-splitting panel for large-scale solar hydrogen generation, *Joule*, 2, 509, 2018.

[49] Maeda, K., and Domen, K., Photocatalytic water splitting: Recent progress and future challenges, *J. Phys. Chem. Lett.*, 1, 2655, 2010.

5 Valorization of Lignocellulosic Biomass into High-Value Chemicals
Forging a Route towards the Circular Economy

Mangat Singh, Sahil Kumar, Hadi Ali,
Surinder Singh, and Sushil Kumar Kansal*

5.1 INTRODUCTION

Most of the energy sources utilized in the present, such as coal, natural gas, and oil, are derived from fossil fuels. Nevertheless, due to the millions of years required for the formation of these resources, they are considered finite and not replenishable [1–2]. One of the most urgent challenges confronting contemporary society involves substituting limited fossil fuels with renewable resources to satisfy the worldwide need for oil and chemicals [3]. The sole carbon-based renewable resource abundantly available in nature, with the potential to effectively substitute for these exhaustible sources, is biomass. Biomass stands as the most prevalent form of renewable energy at present. It ranks as the world's fourth-largest energy source [4], trailing only coal and oil. Biomass encompasses organic matter sourced from living organisms or those that have recently lived. Although biomass is often linked with plants, its scope encompasses both animal- and plant-derived organic matter. This encompasses a diverse array of materials such as agricultural leftovers, forestry resources, wood processing byproducts, food waste, municipal refuse, animal waste, and more (as depicted in Figure 5.1) [5]. Biomass plays a substantial role in curbing net carbon emissions. The carbon dioxide released during the burning and utilization of biomass is absorbed by plants for growth and other metabolic activities. Consequently, there is no net increase in atmospheric carbon dioxide levels [6].

Vast reserves of biomass are stored by plants, spanning from terrestrial to aquatic environments. The total biomass on land globally is estimated to reach 1.8 trillion

* goyalsahil12@gmail.com

DOI: 10.1201/9781003269779-5

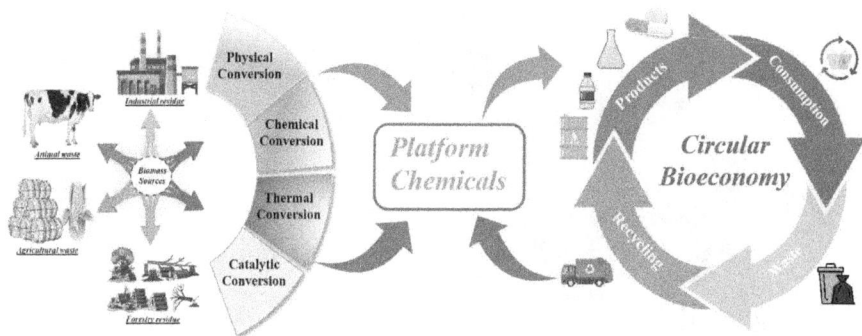

FIGURE 5.1 Various origins of biomass and its utilization within the framework of the circular economy.

tons, while aquatic biomass accounts for around 4 billion tons [7]. Biomass is chemically composed of cellulose, lignin, and hemicellulose, and thus it is termed as lignocellulosic biomass (LCB) or simply lignocellulose (LC). A varying proportions of these components are found in plants remains, whereas cereals and animal waste are considered to be a rich sources of starch and proteins respectively [8]. Utilizing biomass holds paramount significance in addressing the challenges of contemporary society.

LCB assumes a crucial role within the framework of the circular economy due to its capacity for application in diverse, ecologically sustainable processes. The circular economy concept (Figure 5.1), aiming to minimize waste, enhance resource efficiency, and optimize material reuse, recycling, and rejuvenation, can effectively incorporate lignocellulosic biomass—encompassing materials such as wood, agricultural residues, and specific plant types—through various avenues: (1) harnessing bioenergy and biofuels, (2) creating bioproducts, (3) establishing biorefineries, (4) developing bio-based materials, (5) valorizing waste, and (6) advancing carbon neutrality and sustainability [9]. The US Department of Energy has established a target suggesting that by 2030, 20% of the fuel used in the transportation sector [10] should originate from biomass sources. The European Union has likewise formulated an objective to achieve a minimum of 10% of the energy used in the transport sector [11] from renewable resources.

The US Department of Energy has additionally identified the foremost value-enhancing chemicals, which encompass 5-hydroxymethyl furfural, sorbitol, levulinic acid, 1,4-diacids (such as succinic, fumaric, and malic acids), 2,5-furan dicarboxylic acid, 3-hydroxy propionic acid, aspartic acid, glucaric acid, glutamic acid, itaconic acid, 3-hydroxybutyrolactone, glycerol, and xylitol/arabinitol [12]. These compounds are derivable from LCB and hold substantial applications in the pharmaceutical, chemical, and agrochemical sectors [13]. Nonetheless, all the crucial value-added chemicals enumerated earlier are currently sourced from fossil fuels [14], underscoring the necessity for sustainable methods of their production.

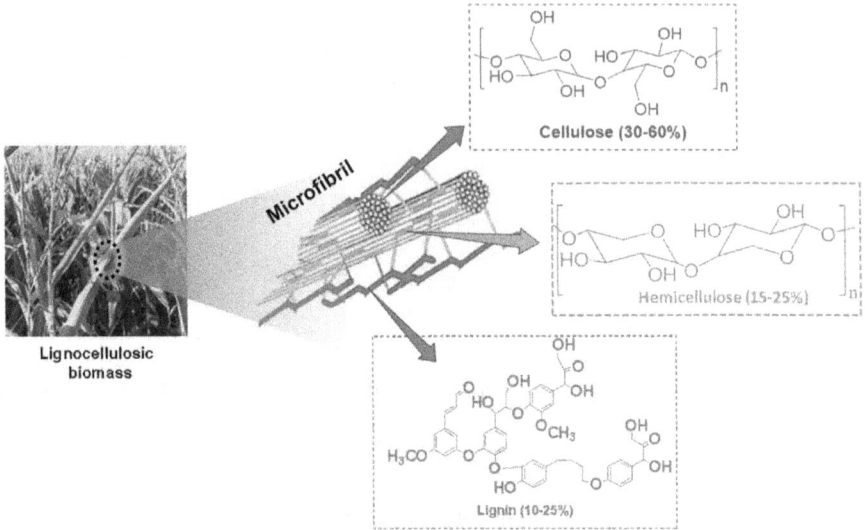

FIGURE 5.2   Structure of plant cell wall.

5.1.1   STRUCTURE OF PLANT CELL WALL

The plant's cell wall exhibits a layered arrangement, consisting of cellulose microfibrils, hemicellulose, lignin, pectin, and soluble proteins (Figure 5.2). These components are structured into three primary layers: the primary cell wall, middle lamella, and secondary cell wall. Among these layers, the outermost one is the middle lamella, composed mainly of pectin, which plays a role in cell adhesion. Moreover, the middle lamella functions as a binding agent between two adjacent primary cell walls. Meanwhile, the primary cell wall primarily consists of cellulose microfibrils, enveloped by pectin and hemicellulose fibers forming the matrix. It serves essential roles, such as providing structural strength, and flexibility, and safeguarding the plant.

The third layer, known as the secondary cell wall, is positioned between the primary cell wall and the plasma membrane. It encompasses lignin, cellulose, hemicellulose, and certain structural proteins. The presence of lignin in the secondary cell walls facilitates water conduction within plant tissue cells. Notably, while some plants exhibit all three layers—middle lamella, primary cell wall, and secondary cell wall—others may lack the secondary wall, resulting in a composition of only the middle lamella and primary cell wall [15,16].

5.2   COMPOSITION OF LIGNOCELLULOSIC BIOMASS

5.2.1   CELLULOSE

Cellulose, an immensely abundant organic polymer on Earth's crust, is annually produced at a staggering 7.5×10^{10} tons [17]. It is primarily composed of carbon (44.44%), hydrogen (6.17%), and oxygen (49.39%). Its distribution spans earthly

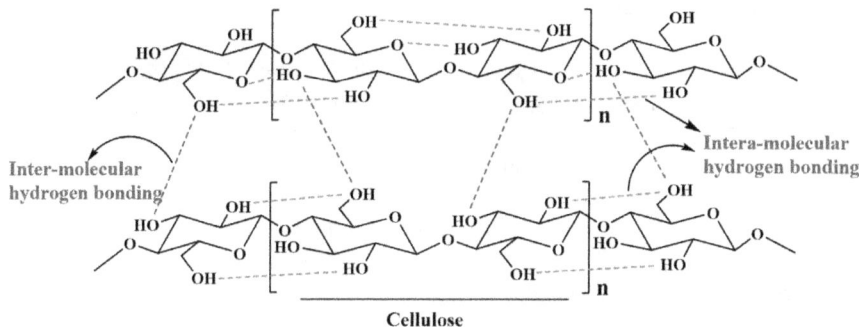

FIGURE 5.3 Molecular structure and illustration of hydrogen bonds in cellulose.

plants, bacteria, and marine algae. As a renewable natural biopolymer, cellulose offers an eco-friendly alternative to petrochemical-based polymers. Its diverse applications have been recognized since ancient times, particularly in clothing, paper, and construction material creation. Its vital role extends to maintaining plant cell wall structures. Cellulose is a linear polymer composed of glucose units, also known as D-glucopyranose units, interconnected through β (1–4) linkages as shown in Figure 5.2. With a degree of polymerization (DP) ranging from 2000 to 6000, cellulose bears the chemical formula $(C_6H_{10}O_5)_n$ and a molecular weight (MW) of 342.3 g/mol, where n signifies the total glucose units in the molecule. Typically, a lengthy cellulose polymer chain encompasses around 7000 to 15,000 glucose units [18–22]. Native cellulose is insoluble in water but dissolves in strong acidic or alkaline conditions. It contains various functional groups like hydroxyl, methoxy, and ether groups within its structure. Methoxy and hydroxyl groups are significant, existing in all repeating cellulose units.

Cellulose polymorph contains two distinct types of hydroxyl groups: primary hydroxyl ($-CH_2OH$) at the C-6 position, and secondary –OH at the C-3 and C-4 positions. These hydroxyl groups contribute to both intramolecular and intermolecular hydrogen bonds [23] within the cellulose polymorph. Intermolecular bonds link D-glucopyranose units within the same polymer chains, while intermolecular bonds form between different polymer chains. The extent of these hydrogen bonds influences the degree of crystallinity within cellulose fibrils. The molecular structure and illustration of hydrogen bonds in cellulose are depicted in Figure 5.3. The three-dimensional crystal structure resembles a web when intermolecular and intramolecular hydrogen bonds fill all available sites in the polymer. When glucose units cross-link, a less ordered para-crystalline cellulose structure is formed. Despite varying with plant species, cellulose biosynthesis initially generates glucose units, followed by their combination to form cellulose chains. These chains aggregate to produce fibrils through intermolecular hydrogen bonding between hydroxyl groups of adjacent monomeric units. These fibrils are further arranged into microfibrils, typically ranging in diameter from 5–50 nm. Within the cellulose polymer, regions exist in either highly ordered (crystalline) or disordered (amorphous) arrangements [17], classifying it as a semi-crystalline polymer.

5.2.2 HEMICELLULOSE

Hemicellulose, the second most abundant biopolymer after cellulose, is present in lignocellulosic biomass. It primarily consists of pentose sugars like xylose and arabinose, with smaller amounts of C-6 sugars such as glucose, galactose, and mannose, as well as sugar acids like glucuronic acid and galacturonic acid. As a heterogeneous biopolymer, it possesses an amorphous structure and a lower degree of polymerization (ranging from 50 to 200) [24]. Its composition makes up around 20–30 wt.% of total biomass in dry wood. Due to its lower polymerization degree, it is prone to degradation in dilute acidic or hot aqueous conditions. Hemicellulose has a shorter chain

TABLE 5.1
Major Constituents' Part of Lignocellulosic Biomasses [64]

Lignocellulosic Biomass	Cellulose (wt%)	Hemicellulose (wt%)	Lignin (wt%)	Extractives (wt%)	Ash (wt%)
Albizzia wood	58.3	8.1	33.2	1.9	1.1
Almond shell	50.7	28.9	20.4	2.5	3.3
Aspen wood	60.7	19.1	14.8	–	0.4
Barley straw	32.5	25.7	2	18.4	9.8
Birch	54.7	24.8	12.2	–	0.8
Canola straw	42.4	16.4	14.2	–	2.1
Coconut shell	20	48.8	30	–	0.5
Coffee pulp	35	46.3	18.8	–	8.2
Corn cob	45	35	15	–	–
Corn stover	40	22	18	–	–
Esparto grass	33–38	27–32	17–19	–	–
Flax straw	28.7	26.8	22.5	19.5	3
Hazelnut shell	26.8	30.4	42.9	3.3	1.4
Hemp hurds	44.5	32.8	21	3.6	3
Miscanthus straw	47.5	20.9	21.4	6.4	3.4
Oak	53.9	29.0	9.43	–	0.2
Oat straw	37.6	23.3	12.9	–	2.2
Olive husk	24	23.6	48.4	9.4	3.3
Pinecone	32.7	37.6	24.9	4.8	0.9
Pinewood	38.8	23.6	20.4	15.7	1.5
Poplar	43.8	14.8	29.1	–	–
Rice straw	32.1	24	18	–	–
Salix straw	43.8	14.6	22.5	2.7	1.2
Spruce straw	49.4	4.7	27.7	3.5	0.3
Sugarcane bagasse	42.4	35.3	20.8	–	1.6
Sunflower shell	48.4	34.6	17	2.7	4
Switch grass	30–50	10–40	5–20	–	4.5
Timothy grass	34.2	30.1	18.1	16.5	1.1
Walnut shell	25.6	22.1	52.3	2.8	2.8
Wheat straw	39.1	24.1	16.3	19.2	1.3

length compared to cellulose, typically comprising 500 to 3000 sugar units [25]. The predominant sugar in hemicellulose is xylose ($C_5H_{10}O_5$, MW 150.1 g/mol). The xylan polysaccharides in hemicellulose form a polymeric chain of D-xylose monomers which are linked through β(1,4) linkages [26]. The composition of xylan polysaccharides varies across biomass sources and depends on plant species (Table 5.1). For example, birch wood (*Betula*) biomass xylan contains xylose (89.3 wt.%), glucose (1.4 wt.%), arabinose (1.0 wt.%), and anhydrobiotic acid (8.3 wt.%) [27]. On the other hand, corn fiber xylan comprises xylose (48–54 wt.%), arabinose (33–35 wt.%), galactose (5–11 wt.%), and glucuronic acid (3–6 wt.%) [28]. Similarly, wheat arabinoxylan polysaccharides include xylose (65.8 wt.%), arabinose (33.5 wt.%), glucose (0.3 wt.%), mannose (0.1 wt.%), and galactose (0.1 wt.%) [29]. Another crucial component of hemicellulose is xyloglucans, composed of D-glucopyranose monomer units linked through β-(1,4)-glycosidic bonds. Xyloglucans contain xylose, glucose, and galactose in a ratio of 3:4:1 [30]. Additionally, hemicellulose includes mannose compounds such as glucomannan, galactomannan, glucuronic acid, and galacturonic acid. Mannose units are linked by β-1,4 bonds, while galactomannan comprises mannose and galactose connected through α-1,6 bonds. Glucomannan features glucose and mannose chains with a 1:3 ratio [30]. In non-woody biomass like agricultural crop residues, arabino-4-O-methylglucuronoxylans (arabino-glucurono-xylans) are prevalent, while in softwoods, they are minor components [31]. Hemicellulose content in softwoods, hardwoods, and herbaceous plants typically ranges from 18–23 wt.%, 10–15 wt.%, and 20–25 wt.%, respectively [32]. Carbohydrate polysaccharides in softwood hemicellulose consist of xyloglucan, galacto-gluco-mannan, and arabinose-glucuronic-xylan, while hardwood hemicellulose includes glucuronoxylan, glucomannan, and xyloglucan. In herbaceous plants, hemicellulose comprises xyloglucan and glucurono-arabinoxylan [31,33,34]. Figure 5.4 depicts the molecular structure of xylan polysaccharides and Figure 5.5 illustrates the monomer units found in xylan polysaccharides.

5.2.3 Lignin

Lignin ranks as the second most abundant renewable biopolymer on Earth, following cellulose. It stands as a hetero-polymer composed of phenyl propene sub-units, characterized by its amorphous nature. These phenyl propene sub-units within the lignin structure

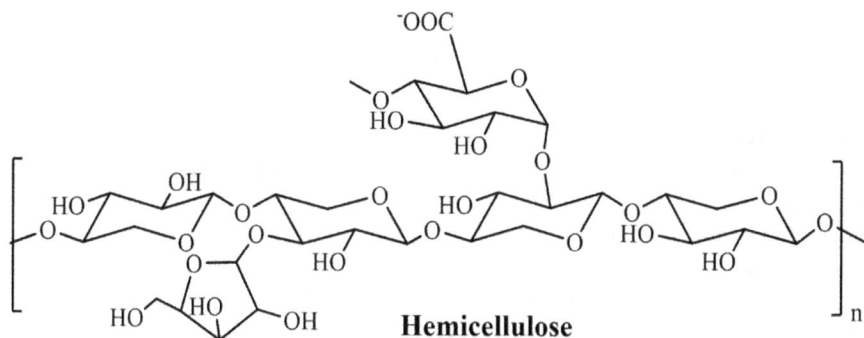

FIGURE 5.4 Molecular structure of xylan polysaccharides.

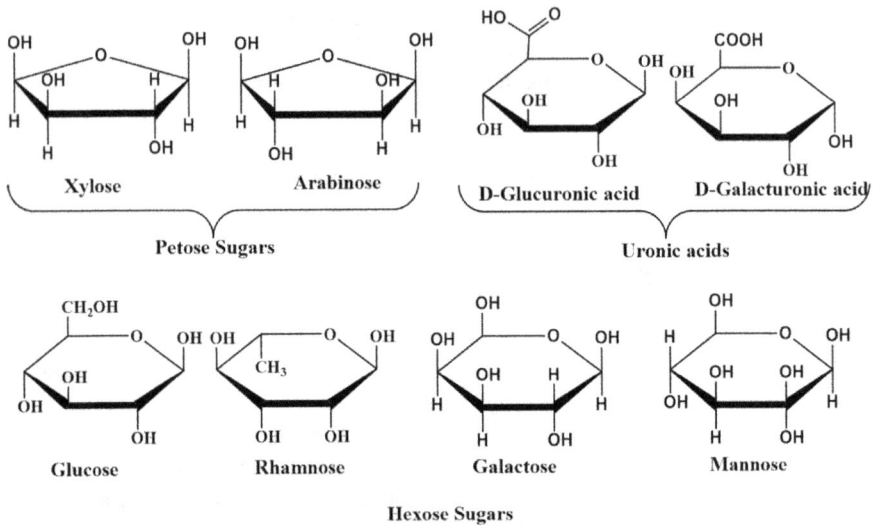

FIGURE 5.5 Monomer units found in xylan polysaccharides.

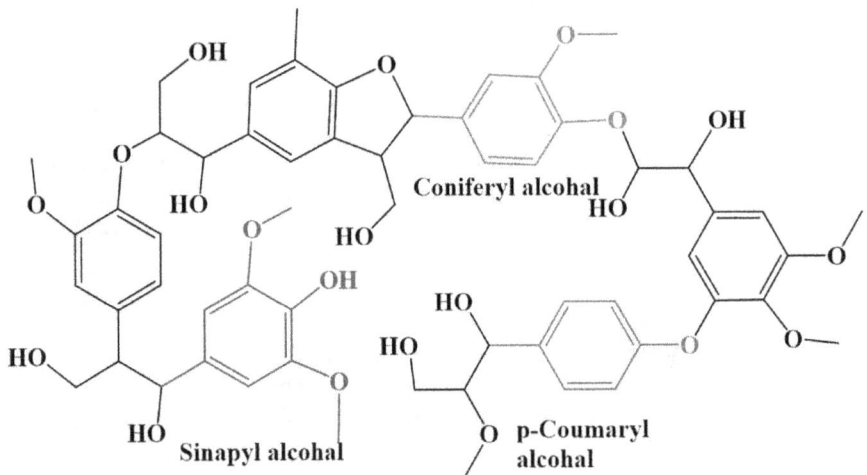

FIGURE 5.6 Representative chemical structure of lignin and its major subunits.

stem from two fundamental building blocks: tyrosine and phenylalanine [35,36]. In their structure, these phenylpropane monomeric units are intricately interconnected through ester bonds in a non-linear and random manner. Lignin contributes durability and firmness to terrestrial plants by forming cross-links with cellulose and hemicellulose polymers [37,38]. Notably, the major phenyl propane subunits present in lignin encompass p-coumaric alcohol (4-hydroxycinnamic alcohol), coniferyl alcohol (3-methoxy-4-hydroxycinnamyl alcohol), and sinapyl alcohol (3,5-dimethoxy-4-hydroxycinnamyl alcohol) [39]. These significant units are also known as p-hydroxyphenyl (H), guaiacyl (G), and syringyl (S) units, respectively. They are termed as such due to their origins as polymeric forms of hydroxyphenyl propane, guaiacyl propane, and syringyl propane [24,31]. The structure of lignin and its principal monomeric units is illustrated in Figure 5.6.

The composition and quantity of lignin in lignocellulosic biomasses varies depending on the type of biomass and its origin (Table 5.1). In general, the lignin content follows the order of softwood > hardwood > grasses, with softwood containing the highest amount [39]. Reports indicate that hardwood lignin consists of a significant number of methoxyl ($-OCH_3$) groups due to approximately equal proportions of syringyl (S) and guaiacyl (G) units. On the other hand, softwood lignin is primarily composed of around 90 wt.% guaiacyl (G) units. Lignin macromolecules feature various functional groups on their side chains, including methoxyl, hydroxyl, carboxyl, and carbonyl. The molecular weight of lignin is closely linked to branching, cross-linkages, and its polymeric nature, generally ranging from 1 to 20 kg/mol. Its calorific value falls within the range of 23.26–25.59 MJ/kg, approximately 30% higher than other existing polymers like cellulose and hemicellulose [40]. It stands as a hydrophobic and thermally stable constituent of plants, insoluble in water under ambient conditions but readily soluble in alkaline solutions (e.g., NaOH, KOH) [41], ionic liquids (e.g., 1-butyl-3-methylimidazolium [BMIM], 1-ethyl-3-methylimidazolium [EMIM]) [42], and appropriate organic solvents.

The true structure of untreated lignin, referred to as proto-lignin, found in plants remains uncertain. The isolated lignin structure consistently differs from proto-lignin. Therefore, various spectroscopy techniques such as ultraviolet (UV) spectrophotometry [43], Fourier-transform infrared (FT-IR) spectroscopy [44], Raman spectroscopy [19], powder x-ray diffraction (PXRD) [45], elemental dispersive spectroscopy (EDS) [46], x-ray photoelectron spectroscopy (XPS) [47], nuclear magnetic resonance (NMR) spectroscopy [48], and mass spectrometry (MS) have been employed to unravel lignin's exact structure [49]. The type of functional groups present in lignin also depends on the extraction method. For instance, lignin extracted via the Kraft process, known as Kraft lignin, has lower sulfur impurities compared to lignin from the lignosulfonate process. Obtaining the purest and unaltered form of lignin is achievable through the organosolv process, which employs organic solvents like ethanol, butanol, methanol, ethylene glycol, phenol, etc. Occasionally, Lewis acid catalysts like $AlCl_3$, $FeCl_3$, $FeCl_2$, and $CuCl_2$ are used to enhance delignification rates or lower the reaction temperature [50,51].

5.2.4 EXTRACTIVES

The non-structural components, both inorganic and organic, within lignocellulosic biomass are referred to as extractives. These substances are soluble in both water and various organic solvents such as ethanol, acetone, benzene, hexane, dichloromethane, and toluene. Extractives encompass a broad spectrum of biopolymers, including fats, waxes, lipids, proteins, terpenes, terpenoids, steroids, gums, resins, starches, essential oils, pectin, mucilage, glycosides, saponins, fatty acids, sterols, flavonoids, phenolic compounds such as lignans, tannins, and stilbenes, and more [52]. Mohan et al. (2006) conducted significant research into the role of extractives in wood biomass and discovered their negative impact on bio-oil production and levoglucosan formation. Removing extractives from biomass leads to a reduction in the hydrogen and oxygen content of solid char [53]. Table 5.1 illustrates the varying levels of extractives in different lignocellulosic biomass.

Terpenes and their oxygenated derivatives, known as terpenoids, are of notable commercial significance due to their wide applications in food flavoring, perfumery, and medicine. A terpene is composed of isoprene (2-methylbuta-1,3-diene) units and is synthesized within plant cell cytoplasm through the mevalonic acid pathway. The condensation of isoprene units in a head-to-tail manner results in a multitude of higher terpenes, including monoterpenes (C10 skeleton), diterpenes (C20 skeleton), sesquiterpenes (C15 skeleton), and triterpenes (C30 skeleton), among others. Terpenoids, on the other hand, are a diverse group formed through biochemical modifications like rearrangement and oxidation of terpenes. Over 40,000 terpene and terpenoid molecules have been identified to date. Terpenes serve to safeguard biomass against microbial and fungal attacks [54,55].

Pectin is another constituent considered an extractive, present in the microfibrils of cellulose polysaccharides within plant cell walls. It is classified into pectic acid and protopectin. Water-soluble pectic acid forms a gel of calcium pectate when exposed to calcium. This calcium pectate consists of a linear chain of around 100 galacturonic acids linked by α-1,4 bonds [24]. In contrast, protopectin, commonly found in the primary plant cell wall, is water insoluble. Protopectin can be converted to pectin with the assistance of the proto-pectinase enzyme or dilute acids. A network with cross-links is formed between two pectin molecules through the creation of a calcium bridge, and this network structure of pectin significantly influences water movement within cells [24].

5.2.5 ASH AND MINERALS

The powdery residue or non-combustible inorganic material left after the complete burning of lignocellulosic biomass is termed ash. This powdery residue comprises a huge number of elements and minerals such as silicates, CO_4^{2-}, PO_4^{3-}, Na, Ca, K, Mg, Cr, and Ni. Among all components of biomass ash, the major portion of it is silica, which in plants occurs in the form of silicates e.g., quartz, plagioclase, chlorite, feldspars, opal, and kaolinite. These silicates provide strength to the plants by supporting the plant's tissues such as bark [56,57]. Different lignocellulosic biomass contains different amounts of ash as described in Table 5.1.

In comparison to agricultural residues, woody biomass contains a relatively small amount of ash. However, it does contain some unburnt carbon, a result of incomplete combustion and inefficient fuel utilization. Research by Demirbas et al. (2005) revealed that the ash produced from boilers can contain more than 50% unburnt carbon, rendering it unsuitable for forest recycling. The presence of unburnt carbon contributes to increased ash volume and can hinder chemical hardening [58,59]. Understanding the physical and chemical properties of biomass ashes is crucial for identifying deposits that form in gasification boilers. Ash resulting from the thermo-chemical degradation of biomass is categorized as bottom and fly ash, with distinct physicochemical properties determined by operating conditions, biomass type, and conversion systems employed. Poor management or improper disposal of volatile heavy metals generated during ash formation can lead to significant environmental repercussions. The composition of heavy metals in ashes is typically contingent on factors such as biomass source, processing conditions during biomass conversion, and the system used. Fly ash tends to exhibit a higher concentration of heavy metals, while bottom ash displays a lower concentration [59,60–62].

Vamvuka et al. (2009) conducted a study on the thermal behavior and environmental impact of ashes obtained from olive kernel combustion in fixed and fluidized bed combustors [50]. The investigation revealed that ash produced from the fixed bed combustion of olive kernel exhibited elevated concentrations of transition elements like chromium (Cr), copper (Cu), zinc (Zn), nickel (Ni), and manganese (Mn), alongside minimal concentrations of toxic metals such as selenium (Se) and lead (Pb) [63,64].

5.3   VALORIZATION OF LIGNOCELLULOSIC BIOMASS

Valorization is the process of enhancing the value of materials derived from various biogenic sources, including food crops (such as rice and corn husks), aquatic plants (such as algae), lignocellulosic plants (such as rice straw), municipal waste, and animal waste. With declining fossil fuel reserves and fluctuating fuel prices, only LCB has the potential to generate opportunity to be a valuable source of chemicals, including reducing sugars, furfural, ethanol, and various other high-value products. Thus, valorization of biomass into biofuels and high-value chemicals is a necessity. However, complexity of lignocellulose biomass's recalcitrance structure for converting it to platform chemicals is a challenging task [65]. Consequently, diverse approaches have been explored to effectively valorize it into bio-based chemicals, solvents, fuels, and products. In particular chemical applications, it is essential and necessary to isolate individual components of biomass such as cellulose, hemicellulose, and lignin. In numerous cases, biorefineries prioritize the utilization of cellulose or hemicellulose from lignocellulosic biomass to produce chemicals and materials, often resulting in lignin being generated as a byproduct [66].

5.3.1   PRE-TREATMENT OF BIOMASS

Lignocellulosic biomass has a very complex intricate structure due to the existence of cellulose, hemicellulose, and lignin polysaccharides. Biomass pre-treatment not only overcomes its crystallinity but also allows these polysaccharides, locked in the intricacy of the cell walls, to be easily accessible and amenable so that they can be transformed efficiently into higher-value chemicals and products [64,67]. Pre-treatment of biomass can be carried out with a variety of methods such as physical, physicochemical, chemical, biological, and electrical methods for altering its size, structure, and chemical composition. Cellulose is the key polysaccharide of biomass; it holds 50 wt.% and can be used to produce various value-added chemicals including sorbitol, hydroxymethylfurfural (HMF), and 2,5-furan dicarboxylic acid (FDCA). However, the hemicellulose extracted from LCB can be transformed into various fine chemicals such as furfural, furfuryl alcohol, xylitol and γ-valerolactone. Thus, for the effective valorization of biomass to platform value chemicals and products, pre-treatment of biomass is necessary for process development [68,69].

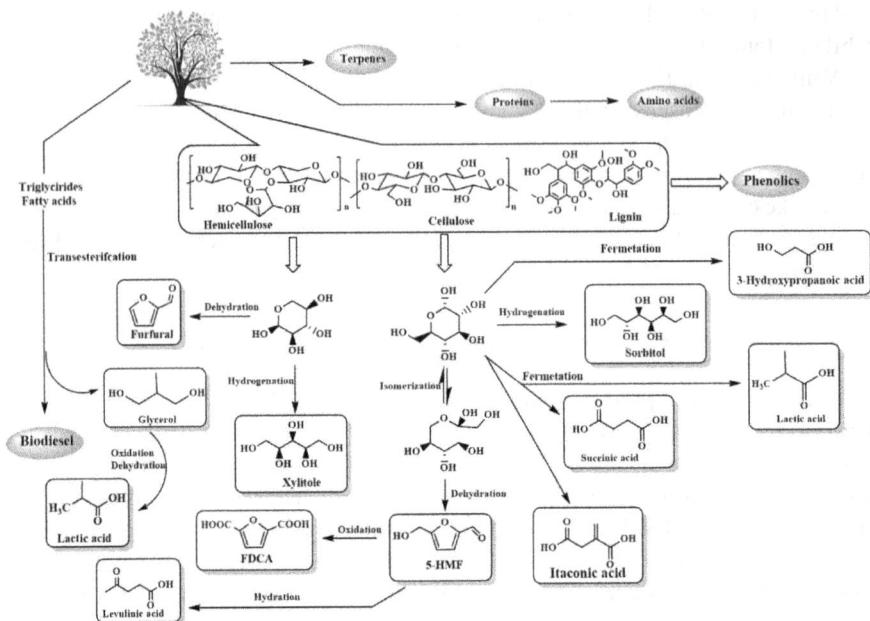

FIGURE 5.7 Schematic representation of the top 11 value-added chemicals.

5.4 PLATFORM CHEMICALS FROM BIOMASS

A chemical substance is classified as a platform chemical when it can serve as a building block to create various other high-value products. The US Department of Energy (DOE) identified a total of 11 value-added chemicals (Figure 5.7) which encompass, Furfural 5-hydroxymethylfurfural, sorbitol, levulinic acid, 1,4-diacids (such as succinic), 2,5-furan dicarboxylic acid, 3-hydroxy propionic acid, lactic acid, itaconic acid, glycerol, and xylitol.

All such molecules can be derived from LCB via chemical and biological routes. Back in 2004, the DOE coined the term "platform chemicals" for these substances, recognizing their potential to be transformed into a range of other valuable chemicals [70]. In 2012, McKinsey & Co. (BIO, 2016) conducted an estimation that the sales of bio-based products amounted to a total of $252 billion. Among these, approximately 9% of the global chemical sales were accounted for by bio-based chemicals. This figure is projected to rise, reaching a range of $375 to $441 billion by the year 2020 [71].

5.4.1 FURFURAL

Furfural, is a heterocyclic aldehyde that serves as a versatile chemical, finding applications in the synthesis of petrochemical products and solvents such as 5-membered heterocycles [72], furfuryl amine [73], succinic acid [74], levulinic acid [75], methyl furan [76], tetrahydrofuran, furfuryl alcohol [77,79], and furoic acid [80]. Its traditional uses range from adhesives and flavor enhancers in food and beverages to transportation fuel, gasoline additives, and decolorizing agents. Moreover, furfural derivatives contribute to insecticides, lubricants, resins, bioplastics, medicines, and agricultural products [81].

Typically, furfural is produced from xylan polysaccharides or xylose using an acid catalyst. Commercially, it is derived either by the dehydration of C-5 sugars such as xylose and arabinose or through the acid hydrolysis of agricultural residues, energy crops, and hardwoods. The latter method involves a two-step process: the release of pentose sugars, mainly xylose, from lignocellulosic biomass's xylan polysaccharide, followed by the cyclodehydration of pentose sugars to yield furfural as the primary product [82]. The production of furfural via direct hydrolysis of agro-residue is a promising process as it entails the reduced cost of the process and the potential to process the remaining solid residues into other valuable chemicals like glucose, phenols, and ethanol [83,84]. Global furfural production is estimated at around 300,000 tons per year, with approximately 70% of it being utilized for furfuryl alcohol manufacturing. China is a major global producer, primarily sourcing furfural from corn cobs, followed by the Dominican Republic using sugarcane bagasse [85,86].

Historically, the Quaker Oats company achieved large-scale furfural production (<50%) in 1922 from agricultural byproducts, mainly corn and sugarcane bagasse [87]. Furfural extraction from sources such as paddy straw [88], sunflower husks [89], and eucalyptus wood [90] using varying concentrations of mineral acids has also been documented. Ultrasound-assisted acid hydrolysis of renewable materials like sugarcane straw, rice husk, grass, wood waste, and yerba mate waste has also been attempted for furfural production [91]. The combination of ionic liquids (e.g., [BMIM][Cl]) with various minerals or solid acids (e.g., HCl, H_2SO_4, $FeCl_3$, $CrCl_3$, $AlCl_3$, HCl CuCl, H_3PO_4, Amberlyst-15, LiCl) leads to notable furfural yields (range 40%–70%). However, the inefficient separation of ionic liquids was a major drawback associated with the process. In addition, biphasic systems comprising a water-organic solvent mixture (e.g., toluene, MTHF, GVL) have also been investigated to enhance the furfural yield [92–95]. However, the higher cost and complexity of these systems remains a challenge.

5.4.2 Hydroxymethylfurfural

Hydroxymethylfurfural (5-HMF), a multifunctional compound, incorporates an aromatic aldehyde, aliphatic alcohol, and a furan ring within its structure. 5-HMF finds extensive utility in generating high-value organic chemicals, fuel additives, biofuel precursors, flavoring agents, polymers, and pharmaceutical intermediates. Notably, essential intermediates and chemicals like 2,5-dimethyl tetrahydrofuran (DMHTF), 2,5-diformylfuran (DFF), dimethylfuran (DMF), 2,5-furan dicarboxylic acid (FDCA), and levulinic acid (LA) can all be derived from this versatile molecule [96–99].

Beyond its role in valuable chemicals, 5-HMF also holds significance in traditional medicinal systems. For instance, it is a key component of processed steamed Rehmanniae Radix, a Chinese medicine renowned for treating conditions like anaemia and diabetes [100].

Historically, its synthesis originated from D-glucose or cellulose. Its synthesis from glucose typically occurs in three steps. First, hydrolysis of glucan to glucose is performed using a Brønsted acid (e.g., HCl, H_3BO_3, NaH_2PO_4, H_2MoO_4, propyl sulfonic acid, propionic acid, acrylic acid, benzoic). This is followed by glucose isomerization to fructose with a Lewis acid (e.g., $ZnCl_2$, $AlCl_3$, $FeCl_3$, $SnCl_4$, $CrCl_3$, LiCl) [101–107]. Last, fructose is dehydrated to produce HMF in the presence of a Brønsted acid. Implication of Brønsted acids and metals chlorides under aqueous media limits the process efficacy not only due to high processing cost but also pause difficulty in product separation. However, a wide array of renewable materials, such as sugarcane bagasse, paddy straw, grass, and wood waste, have also been employed [108]. But low HMF yield due to the interference of side reactions and the requirement of drastic reaction conditions such as high temperature, longer reaction time, and highly acidic reaction media, etc. creates hurdles for their industrial scale implications A one-pot transformation has been reported for the direct conversion of lignocellulosic biomass to HMF [109]. This approach is not only more environmentally friendly with reduced CO_2 emissions but also offers energy savings, making it a cost-effective and sustainable technology. However, the direct conversion of lignocellulosic biomass to HMF is more intricate due to its complex and heterogeneous nature when compared to model carbohydrates.

5.4.3 GLYCEROL

Glycerol, also known as propane-1,2,3-triol, is a colorless, odorless liquid with a sweet taste typically obtained from bio-diesel industries as a byproduct. It is around 10% of bio-diesel industries, with approximately 90% of total glycerol stemming from other sources. Pure glycerol, or glycerine, finds extensive applications in the food, cosmetics, and pharmaceutical industries [110]. It is a key ingredient in products like toothpaste, mouthwashes, shaving cream, lotions, and soaps, contributing to smoothness and lubrication, and acting as a humectant to prevent drying [111]. The pharmaceutical field incorporates glycerol in cough syrups, suppositories, elixirs, expectorants, and cardiac medications [112]. It is estimated that glycerol production reached approximately 2.8×10^6 tons in 2020 and is projected to grow to 4.0×10^6 tons by 2027, with a compound annual growth rate of about 5.2% between 2020 and 2027 [113].

Several rational approaches, both chemical and biological, have been explored for the efficient utilization of this abundant waste from bio-diesel industries. Biological conversion of glycerol offers advantages over chemical methods, as it is target specific, yields high-purity products, and is environmentally friendly. Microorganisms like *Clostridium acetobutylicum* and *Escherichia coli* have also been explored to generate products such as 1,3-propanediol and 1,2-propanediol, respectively [114]. Anaerobic fermentation with bacteria like *Klebsiella*, *Enterobacter*, and *Clostridium* yields organic acids like oxalic acid, and citric acid. Polyols like mannitol, erythritol, and arabitol can be produced in significant yield through yeast and filamentous fungi fermentation. Other derivatives of glycerol, such as glycidol and epichlorohydrin, are used in the production of epoxy resins, polyurethanes, and other compounds. Glycerol carbonate, synthesized from glycerol, finds applications in the preparation of glycidol, polymers, coatings, adhesives, and lubricants [115].

Chemical transformation includes hydrogenation with catalysts like Ru/C or Pt/C, glycerol can lead to compounds like ethylene glycol, propylene glycol, and acetol [116]. Dehydration of glycerol yields hydroxy acetone, 3-hydroxypropyl, and acrolein [117]. Oxidation of glycerol produces various products, including glyceraldehyde, 1,3-dihydroxyacetone, and organic acids like lactic acid, tartronic acid, and acrylic acid [118,119]. Lactic acid production using glycerol as raw material is now gaining more attention as it is a precursor in several chemical transformations. Yang et al. (2016) converted glycerol into lactic acid by incorporating a heterogeneous catalyst, namely 30%CuO/ZrO_2, and achieved nearly 100% glycerol conversion with a product selectivity of ~94.6% [120]. $LaMnO_3$ perovskite–supported Au catalysts have been devoted by C.D. Evans et al. (2022) to the synthesis of tartronic acid from glycerol [121]. Selective glycerol oxidation to acrylic acid (yield; 87%) with Au/CeO_2 catalysts was reported by Minsu Kim and Hyunjoo Lee in 2017 [122]. Hydrogenolysis (HG) of glycerol leads to commercially important alcohols such as 1,3-propanediol, 1,2-propanediol, and allyl alcohol [123,124]. Since the demand for 1,3-propanediol-derived polymer materials is increasing constantly, significant reports have also been published in the literature to implicate glycerol in 1,3-PDO production. In this direction, Tabah B. et al. (2016) were the first to study instant baker's yeast (*Saccharomyces cerevisiae*) for the production of 1,3-propanediol using glycerol as substrate [124]. A highly energetic, selective, and stable Cu/MgO catalyst (10 wt%) was devoted to the hydrogenolysis of glycerol to 1,2-propanediol (yield 95.5%) by Pandhare et al. (2014) [125]. Allyl alcohol (AAL) is a pivotal intermediate for several organic transformations and it can be derived from renewable and surplus glycerol. Zhao et al. (2020) developed a bimetallic CoFe-ZIF-derived CoFe alloy for the selective synthesis of AAL directly from glycerol [126]. Monoclinic zirconia (m-ZrO_2) supported metals including Ru, Rh, Pt, and Pd-based nanoparticles were scrutinized in glycerol hydrogenolysis to propylene glycol and ethylene glycol by Wang et al. (2013) [127]. Nevertheless, both processes come with their own challenges and techno-economic status. Biological routes are products selective, easily operational, but economically not feasible due to high enzyme cost, and are prone to environmental conditions, etc. whereby chemical route is cost-effective but emission of hazardous chemicals, low reusability of the catalyst, difficulty in catalyst separation from the reaction mixture, and formation of undesired products, etc. creates barriers for their multiscale implication. Therefore, further assessment of both pathways is still needed for efficient glycerol utilization.

5.4.4 2,5-FURANDICARBOXYLIC ACID

2,5-Furandicarboxylic acid (FDCA) has garnered extensive attention due to its role as a crucial building block for petrochemical-derived adipic acid and terephthalic acid. It can be efficiently produced with high yields (90%–99%) through the oxidation of HMF using various homo/heterogeneous catalytic systems [128]. Another approach involves synthesizing FDCA directly from fructose through two-step dehydration followed by oxidation, eliminating the need to isolate the HMF intermediate [129]. FDCA finds versatile applications in the production of polyesters, polyamides, and plasticizers. Particularly, the creation of polyethylene furanoate (PEF) polymers using FDCA and ethylene glycol has yielded materials with similar properties to conventional petroleum-based polymers like polyethylene terephthalate [128].

Companies like Avantium and BASF have invested in the production of FDCA and PEF polymers, with pilot plants and commercial production facilities in operation [130,131]. Dupont and Archer Daniels Midland (ADM) have also developed a high-yielding process for FDCA methyl ester production directly from fructose [132]. Beyond its role in bio-based polymer synthesis, FDCA serves as a versatile synthetic intermediate in various domains such as organic synthesis, pharmaceuticals, and the creation of metal-organic framework materials [133,134]. Market projections anticipate FDCA's value to grow significantly, reaching around €770 million by 2025 with a compound annual growth rate of approximately 11% [135].

The applications of FDCA extend to the production of biochemicals like succinic acid, fungicides, isodecylfuran-2,5-dicarboxylate, and poly(ethylenedodecanedioate-2,5-furandicarboxylate). Additionally, FDCA finds use as corrosion inhibitors, pharmaceutical intermediates, cross-linking agents, and medicinal products [136].

5.4.5 LACTIC ACID

Lactic acid, also known as 2-hydroxypropanoic acid, is a significant commodity chemical widely utilized in the cosmetic, food, and pharmaceutical industries. It has diverse applications for producing oxygenated chemicals, plant growth regulators, and specialized chemical intermediates [137,138]. Lactic acid plays a pivotal role in creating alkyl lactates [139], propylene glycol [140], propylene oxide [141], acrylic acid [142], and poly-lactic acid (PLA) [143]. The biopolymers derived from lactic acid, such as PLA, are of considerable interest due to their biodegradability, biocompatibility, and eco-friendly nature.

Lactic acid is primarily produced through carbohydrate fermentation processes involving glucose, sucrose, cellulose, and lactose as substrates. However, these traditional methods often yield lactic acid with impurities and low productivity. Companies like Corbion have developed improved fermentation processes to produce L-lactic acid, with the primary byproduct being calcium lactate. Innovative purification techniques like nanofiltration, electrodialysis, and ion exchange resins have been explored to address the challenges posed by calcium sulfate ($CaSO_4$) byproducts [144–146]. Additionally, efforts have been made to engineer yeast to convert pure xylose into lactate. These advancements are driving global lactic acid production, which currently stands at approximately 270,000 tons/year, with projections of substantial growth in the coming years. It is also expected that the consumption of PLA in Europe alone may reach 650,000 t by 2025, which is almost 3 times its current value [147].

Lactic acid's derivatives, such as lactate esters, hold potential as green solvents in organic synthesis [148]. The use of lactic acid in the production of biodegradable polylactic acid (PLA) is particularly noteworthy. PLA has emerged as a valuable material in the textile and packaging industries due to its admirable barrier properties and thermal stability. Several companies are currently scaling up PLA production to meet the growing demand [149].

Lactic acid exists in two isomeric forms, L-lactic acid and D-lactic acid, each possessing distinct industrial applications. L-lactic acid is crucial for the synthesis of poly L-lactic acid (PLLA), a semi-crystalline, biodegradable, and thermally stable polymer. PLLA finds applications in medical products such as orthopedic fixation devices, cardiovascular applications, dental products, intestinal applications, and sutures [150,151]. On the other hand, D-lactic acid is employed in producing poly D-lactic acid (PDLA). The properties and biodegradability of polylactic acid (PLA) can be tailored by combining different ratios of L- and D-lactic acid. The resulting stereo complexes exhibit higher melting points and enhanced biodegradability compared to their individual constituents.

5.4.6 SUCCINIC ACID

Succinic acid, with the chemical formula $(CH_2)_2(CO_2H)_2$, is a dicarboxylic acid that naturally occurs in both plant and animal tissues. It is found in various foods such as broccoli, rhubarb, sugar beets, fresh meat extracts, cheeses, and sauerkraut [152,153]. In the realm of food products, succinic acid serves as a sequestrant, buffer, and neutralizing agent. The global production of succinic acid was estimated to be around 16,000 to 18,000 metric tons in 1990 [154]. Traditional methods involve producing succinic acid through the hydrogenation of petrochemical-derived maleic acid or the oxidation of butanediol. However, modern industrial processes, as employed by companies like Roquette/DSM and Bio-Amber, use fermentation of glucose or biorefinery sugars with *E. coli* to produce succinic acid on a large scale [155–158].

Renewable sources such as cassava bagasse, sugarcane bagasse, agave, cheese whey, and sake lees (source/route to produce succinic acid) have been used to synthesize bio-based succinic acid [159]. Succinic acid is a valuable precursor for various products, including succinate esters, which are essential building blocks for 1,4-butanediol, adipic acid, maleic anhydride, γ-butyrolactone, N-methyl pyrrolidinone, and tetrahydrofuran (THF) synthesis [160,161]. The water solubility of succinic acid makes it suitable for preparing lacquers, resins, perfume esters, coating chemicals, medicines, surfactants, and biodegradable plastics. Additionally, succinic acid is utilized as a flavoring agent and additive in numerous food products and beverages [162]. Succinic anhydride, derived from the dehydrogenative cyclization of succinic acid, serves as a starting material for producing fumaric acid and maleic acid. Companies like NatureWorks and Bio-Amber have harnessed succinic acid and 1,4-butanediol to create bio-renewable polyesters. Succinic acid is crucial for the preparation of polyethylene succinate (PES) [163], a fundamental material in the plastics industry. Biodegradable polyesters like PPS (polypropylene succinate) [164] and PBS (polybutylene succinate) [165] can also be synthesized from succinic acid, with poly(propylene succinate) and poly(butylene succinate) having achieved commercialization success [166].

As of 2022, the market value of succinic acid was approximately $181.0 million, and this figure is projected to reach $359.8 million by 2032, exhibiting a compound annual growth rate (CAGR) of 7.3% [167].

5.4.7 LEVULINIC ACID

Levulinic acid (LA) is a highly versatile chemical with a wide range of applications, e.g., resin, plasticizer, textile, animal feed, coating, and antifreeze [168]. Generally,

LA is synthesized by two different reaction pathways. In the first pathway, the catalytic hydrolysis of furfuryl alcohol (a hydrogenated product of furfural) is being carried out followed by acid treatment in the presence of suitable homo/heterogeneous catalyst for a series of chemical transformations such as ring-opening, hydrolysis, and rearrangement [169–172]. The second pathway comprises the catalytic hydrolysis of lignocellulosic biomass in the presence of mineral acids such as sulfuric acid, hydrochloric acid, and trifluoroacetic acid [173–177]. Initially, LA was produced from simple sugars, e.g., glucose, fructose, etc. [178]; however, in view of high cost of monosaccharides, newer methods were developed that rely on renewable sources such as polymeric sugars (e.g., starch, cellulose, etc.) [179,180], starch material (e.g., corn starch, food waste) [181], forestry residues (e.g., pine sawdust, wood chips) [182], industrial waste (e.g. paper and pulp) [183], and agro-residues (e.g., bagasse, paddy straw) [184–186].

In literature reports, a variety of homogeneous catalysts (e.g., HCl, H_3PO_4, HCHO, H_2SO_4, etc.) and heterogeneous catalysts (e.g., H-ZSM-5, $Al_2(SO_4)_3$, Cu–Nb_2P_5, etc.) were employed for LA production from renewable sources. Sorghum was the first raw material used for LA production using H_2SO_4 as a catalyst [187]. Chang et al. (2007) reported the use of wheat straw as a substrate for LA production in ~19.86 mol% yield [188]. Yang et al. (2013) reported the synthesis of LA from glucose in the presence of a mixture of acids such as $CrCl_3$ and H_3PO_4 [189]. Besides the use of Brønsted acids and Lewis acids, the incorporation of transition metal salts like Cu^{2+} and Fe^{3+} under homogeneous conditions was found to accelerate the defibrillation rate of cellulose to glucose causing a high LA yield (~74 mol% with D-fructose) [190,191].

Solid acid catalysts [192] have been also tested for LA synthesis due to many advantages, e.g., eco-friendly nature, high product selectivity, recyclability, cost-effective down streaming, etc. Omari et al. (2012) discussed the effect of $SnCl_4.5H_2O$ concentration on LA yield when chitosan (23.9%) and glucosamine hydrochloride (27.4%) were heated at 200°C [193]. The solid super acid, a type of acid superior to the liquid acids, was also tested for the production of LA from rice straw [194]. Commercial resin such as Amberlyst 70 bearing reactive $-SO_3H$ groups was found efficient for LA production [195]. Despite the high efficacy of $-SO_3H$-bearing solid catalysts, it was expected that leaching of the $-SO_3H$ group during the process is a major obstacle that limits the use of such catalysts for large commercial scale production. In addition, processes devoted to LA production using mineral acids result in high product yield, with high efficiency in less reaction time. The maximum levulinic acid yield by acid catalysis of lignocellulose was 76% mole and 62.1% reported by Girisuta et al. 2007 [196] and Hurst et al. 2019 [197], respectively. However, some shortcomings [198] associated with the use of mineral acids are the generation of acidic waste, high acid recovery cost, difficulties in product separation, requirement of specialty equipment, etc. Therefore, further assessments are still required for the efficient production of this versatile chemical from renewable materials.

5.4.8   3-HYDROXYPROPIONIC ACID

3-Hydroxypropionic acid (3-HP) is a highly sought-after platform chemical with significant global market demand. This nonchiral carboxylic acid possesses both hydroxyl and carboxyl groups in its structure, making it a valuable precursor for the synthesis of more complex and valuable chemicals. Although numerous methods for producing 3-HP have been reported, commercial processes for its production from renewable sources are not yet established. One key challenge in the biological production of 3-HP is the toxicity of its aldehyde form, 3-hydroxypropionaldehyde (3-HPA), from which it is derived [199]. Currently, 3-HPA is obtained through the fermentation of glycerol, which is then oxidized to yield 3-HP. Some bacterial strains, such as *Klebsiella pneumoniae, Lactobacillus reuteri,* and *E. coli,* are capable of converting glycerol to 3-HP through oxidative or reductive pathways [200]. Glucose and other sugars like xylose and sucrose are also valuable sources for 3-HP production. Additionally, lignocellulosic biomass, agricultural byproducts, forest biomass, energy crops, and municipal waste can serve as renewable sources for 3-HP production [201].

The range of high-value compounds derived from both 3-hydroxypropionaldehyde and 3-hydroxypropionic acid is extensive. These compounds include 1,3-propanediol (1,3-PDO), malonic acid, acrolein, acrylic acid (AA), acrylonitrile, acrylamide, propiolactone, acrylic acid esters, 3-hydroxypropionic esters, homopolymers, heteropolymers, and bioplastics [202]. These compounds find applications in the production of adhesives, polymers, plastic packaging, fibers, cleaning agents, and resins. Acrylic monomers, important for polymer synthesis, are manufactured by companies like BASF, AkzoNobel, Dow Chemical Company, and Lubrizol through the polymerization of acrylic acid derived from 3-hydroxypropionic acid [203]. Acrylamides, obtained from 3-hydroxy-propionaldehyde, can be converted to non-toxic polyacrylamide for various uses, including in mining, oil recovery, water treatment, papermaking, and electrophoresis gels. Acrolein and acrylonitrile, derived from 3-HP and 3-HPA, are essential for the synthesis of polymers [204].

5.4.9 SORBITOL

Sorbitol is a naturally occurring sugar alcohol that finds wide applications in various industries such as food, beverages, drugs, cosmetics, textiles, pharmaceuticals, healthcare, and more [205]. The global consumption of sorbitol was around 2.66 million metric tons in 2020, and its market is projected to continue growing, with estimates indicating a volume of 2.91 million metric tons by 2028 [206], whereas another report illustrated that the global sorbitol market will be reached at approximately $1.77 billion by the end of 2023 and is expected to exhibit a CAGR of 6.7% from 2023 to 2030 [207]. Sorbitol is commercially produced through the raney nickel metal-catalyzed hydrogenation of D-glucose. Renewable sources like corncob, paddy straw, cassava, and wheat are also used as raw materials for sorbitol production. In this process, enzymatic hydrolysis of these renewable sources yields glucose, which is then hydrogenated to form sorbitol, primarily using Ru-based catalysts. Roquette Freres is a major producer of sorbitol, holding a substantial market share, alongside companies like Cargill and SPI Polyols. Recent developments include the preparation of sorbitol from cellulose and cellobiose through enzymatic hydrolysis followed by hydrogenation [208,209]. Sorbitol production from cellulose can also be achieved using mineral acids or supported transition metal catalysts like

Pt or Ru [210–213]. Sorbitol is employed in the production of ascorbic acid through fermentation processes, constituting about 15% of total sorbitol usage. Sorbitol can undergo dehydration with a Cu catalyst to yield isosorbide, which is precursor for making non-ionic surfactant [214,215]. Additionally, hydrogenolysis of sorbitol with various catalysts can generate a range of low-degree alcohols, including methanol, ethanol, glycerol, propylene glycol, and ethylene glycol, which can further be converted into high-value chemicals or products [216]. Sorbitol is also utilized in the preparation of biodegradable polymers, which have applications in bio-composites and biomedicines. A notable example is poly (isosorbide carbonate), recognized as a promising alternative to petrochemical-based BPA polycarbonate due to its superior properties. In summary, sorbitol's diverse applications, including its use as a sugar substitute, its role in the synthesis of various chemicals, and its contribution to the development of biodegradable polymers, make it a valuable and versatile compound in multiple industries [217].

5.4.10 XYLITOL

Xylitol is a sugar alcohol with a 5-carbon structure and is used as an artificial sweetener due to its sweetness being about 20% higher than sucrose. It has gained popularity as a sugar substitute and is particularly appealing to individuals with diabetes since its metabolism is independent of insulin. This makes it a suitable option for those managing their blood sugar levels [218]. Commercially, xylitol is produced through chemical methods, involving the catalytic reduction, H_2/Ni, of pure xylose obtained from hardwood hemicellulosic hydrolysate [219]. Metal catalysts like Ru, Pd, and Pt are utilized under high temperature and pressure conditions. However, biochemical reductive methods using enzymes or microorganisms have also been developed as an alternative production approach. Among these methods, the biochemical production of xylitol from D-xylose is considered more efficient and cost-effective [220].

Recent research has explored innovative production methods, such as a one-pot process using acid in combination with a Ru/C catalyst to produce xylitol from hemicellulose [221]. Additionally, about 60% xylitol yield can be obtained from corncob-derived hemicellulose when treated with a Ru-catalyst supported on carbon nanotubes (Ru/CNT) [222]. Xylitol's properties, including its cooling effect, sweet taste, and flavor, have led to its incorporation in food products. The xylitol benefits have also extended to the pharmaceutical industry, where it is valued as a good sugar substitute with added health advantages. These benefits include being anti-cariogenic (helping prevent tooth decay), promoting tooth rehardening, gastrointestinal tolerance, low energy content, and potential prevention against ear and upper respiratory infections [223,224]. The consumption of xylitol has been on the rise, reaching around 160,000 metric tons in 2013. It's anticipated that xylitol consumption will continue to grow, potentially reaching 242,000 metric tons by the end of 2020. The combination of its sweetness, health benefits, and versatility has positioned xylitol as a valuable sugar alternative with applications in various industries [225]. Techno-economic analysis of processes devoted to platform chemical production are given in Table 5.2.

TABLE 5.2

Techno-Economic Analysis of Processes Devoted to Platform Chemical Production*

Chemical	Reagents	Feedstocks	Process Economy	References
Levulinic acid	HCOOH	Wheat straw	MSP: 9.59–12.8 \$/kg Feed rate: 100 kg/h Return rate: 6.1% Payback period: 20 years	226
		Banana bunches	Process capacity: 15,175.60 kg/h Production rate: 3883.13 kg/h Production cost: 0.178 \$/kg Annual cost: \$29,163,638.95	227
5-HMF	H_2SO_4	Spruce waste	Expected investment: 257 M€ Production capacity: 5 kt/year MSP: 1.93 €/kg Return rate: 15.90% Loss probability: 17%	228
FDCA	AuPd/Mg $(OH)_2$ K_2CO_3/Cs_2CO_3	Furfural	MSP: 2000 ± 500 \$/ton Market price: 1400 \$/ton Minimal return on investment: 10% Product yield: 83 mol% Energy consumption: 24 MJ/kg	229
Furfural	2-sec-butylphenol	Xylose or birchwood	MSP: 1.62 €/kg Product yield: 54–59 mol% Production cost:14 M€ Payback period: 5 years IRR: ~20.7%	230
	ChCl, MIBK	Switchgrass	MSP: 625 \$/t Expected production: 17.9% Feed rate: 40% Cost reduction: 37%	231
Glycerol	–	Biodiesel waste	Capital cost: 71 M\$ Operating costs: 303 M\$/year Gross profit: 60.5 M\$/year Net profit value: 235 M\$ Payback period: 1.7 years Market price: 900 \$/ton	232
Succinic acid	Bio-tech	Sugar-rich residual streams Biogas	Capital investment: €5,211,000 Production cost: 2,339,000 €/year Expected production: 1000 tons/year Expected revenue: 2,811,000 €/year Return on investment: 11.68% Return on payback period: 8.56 years IRR: 11.11%	233
	DMSO	Glycerol	Expected profit: \$190 million IRR: 33.3% Estimated payout period: 4.48 years Reduction of GHG: 26%	234

(Continued)

TABLE 5.2   (*Continued*)

Techno-Economic Analysis of Processes Devoted to Platform Chemical Production*

Chemical	Reagents	Feedstocks	Process Economy	References
Sorbitol	Mineral acid Ammonia fiber explosion (AFEX)	Sargassum (Brown algae)	Expected production: 116 kt/yr Investment: 120.5 M€ Production costs: 0.25 €/kg	235
		Switchgrass, corn stover, miscanthus	Expected production: 157.6–202 kt/yr Total investment cost: ~112–120 M€ Total production cost: 0.22–0.28 €/kg	
Xylitol	–	Sugarcane bagasse	MSP: 3000 $/t IRR: 12.3% Hurdle rate: 9.7% MSP: 1687 $/t IRR: 21.3%–59.8%	236
3-HPA	*Escherichia coli*	Glycerol	Production: 19.7 g/L Production rate: 9.1 g/L/h Product yield: 77 mol%	237
Lactic acid	Fermentation using bacteria, fungi, and yeast	Corn stover and miscanthus	Production capacity: 100,000 MT/y MSP: $993–1392 IRR: 10%	238
Itaconic acid	Biotech route using *Aspergillus terreus*	Glycerol	MSP: ND Production rate: 171 kg/h Operational cost: 42 $/kg	239

* Listed as per market potentials. MSP, minimum selling price; IRR, internal rate of return; GHG, greenhouse gases; ND, not disclosed.

5.4.11   ITACONIC ACID

Itaconic acid (IA), also known as methylene succinic acid, is an unsaturated organic acid with a unique structure consisting of two carboxyl groups and one methylene group. This structural arrangement, characterized by carbonyl functional groups and a conjugated double bond, has attracted significant attention from researchers, academia, and industries due to its potential for diverse chemical transformations. Although it was first synthesized in 1837 through the thermal decarboxylation of citric acid [240], economic production methods were challenging until biotech routes based on carbohydrates emerged in the mid-20th century. These pathways significantly improved the yield and feasibility of IA production [241].

Currently, IA is industrially synthesized through fermentation using the *Aspergillus terreus* strain, with a production capacity of around 80 g L^{-1}. The global production of IA is estimated to be around 80,000 tons per year, with a selling price of approximately $2 per kg [242]. The Asia Pacific region contributes the majority of the global IA supply (over 54%), followed by Europe and North America. IA is used in various applications, including the production of styrene-butadiene rubber (SBR) latex, methyl methacrylate, and poly-itaconic acid.

The IA market has experienced growth, with its value reaching close to $95.4 million in 2021. This growth is expected to continue, potentially reaching around $108.4 million by the end of 2026 [243,244]. Structural similarities between IA and acrylic acid or methacrylic acid, along with their esters such as dimethyl itaconate and dibutyl itaconate, have led to intensive research on their use as co-monomers in the synthesis of poly(meth)acrylates. These polymers find applications in a range of fields from corrosion inhibition to dental materials, elastomers, and drug delivery systems [245].

IA itself is considered a precursor for various polymeric products, including paints, adhesives, plastics, resins, textiles, and super-absorbent polymers [246,247]. Despite its progress and growing demand for sustainable chemicals, the utilization of IA in everyday products still lags behind petroleum-based alternatives. For IA-based products to become economically competitive, the manufacturing cost should not exceed $0.5/kg. Achieving this goal requires the development of efficient fermentation routes using low-cost substrates. Several approaches have been explored for large-scale IA production, with a focus on renewable sources [248,249]. For instance, *Scenedesmus* sp. [250] has been employed to produce IA from lignocellulosic biomass through direct transesterification. Glycerol has also been used as a carbon source to produce IA through fermentation using *Aspergillus terreus* MJL05. Researchers have investigated the use of various hexose sugars and pentose sugars for IA production through fermentation, as well as polysaccharides like starch. Agrowaste materials such as husk, straw, bran, and sugarcane bagasse have also shown promise as feedstocks for IA production [251,252]. The development of sustainable and cost-effective IA production processes from renewable sources is crucial for its widespread adoption and integration into a range of products and industries.

5.5   CONCLUSION AND PERSPECTIVE

Efficiently utilizing lignocellulosic materials is a vital aspect of promoting the circular economy within the contemporary chemical industry. This entails the conversion of lignocellulosic biomass into valuable chemicals, a subject that has been thoroughly examined in accordance with the guidelines of the US Department of Energy (USDOE) and its current applicability in market production. Various industrial processes (Table 5.2) have been developed for the manufacture of these chemicals using sustainable raw materials, as illustrated by the commercial production of FDCA from fructose in a one-pot process with excellent yields. However, challenges related to the use of biphasic solvents remain, primarily due to cost factors at the pilot scale. Commercial furfural production also involves the use of sulfuric acid as a catalyst whereby this process is inefficient, resulting in lower furfural yields (less than 50%), and the requirement of specialty equipment, due to the corrosive nature of sulfuric acid. Technological advancements in renewable material processing unlock the potential of previously underutilized waste materials as valuable feedstock for chemical production. Substantial strides have been taken not only in processing agricultural waste but also in converting it into a wide range of everyday products and industrial chemicals. Nonetheless, concerns have arisen regarding the suitability of lignocellulosic biomass to produce bulk chemicals, given its uneven distribution, geographical variability, and complex composition. However, as an alternative, it proves to be a suitable feedstock for specialty and high-value chemicals. Conducting techno-economic analyses of current processing methods becomes crucial for industrial implementation when emphasizing sustainability. As a result, to effectively incorporate lignocellulosic biomass into the existing supply chain, it is essential to strategically position processing facilities near waste-producing sources and high-value chemical manufacturing sites. Additionally, the development of agricultural waste processing techniques should be customized to suit the physicochemical characteristics of locally generated waste materials and align with the specific requirements of the local chemical industry. In summary, to streamline the connection between molecules derived from agricultural waste and manufacturing hubs producing high-value chemicals, the future could benefit from the utilization of data mining and automated chemical route search techniques.

REFERENCES

[1] Sharma, A., Sharma, P., Sharma, A., Tyagi, R., & Dixit, A. Hazardous effects of petrochemical industries: A review. *Recent Adv. Petrochem. Sci.*, 3, 555607, 2017.

[2] Wenpo, R., Chaoha, Y., & Honghong, S. Characterization of average molecular structure of heavy oil fractions by 1H nuclear magnetic resonance and X-ray diffraction. *China Pet. Process. Petrochemical. Technol.*, 13, 1–7, 2011.

[3] Megia, P. J., Vizcaino, A. J., Calles, J. A., & Carrero, A. Hydrogen production technologies: From fossil fuels toward renewable sources. A mini review. *Energy Fuels*, 35, 16403–16415, 2021.

[4] Shahzad, U., Elheddad, M., Swart, J., Ghosh, S., & Dogan, B. The role of biomass energy consumption and economic complexity on environmental sustainability in G7 economies. *Bus. Strategy Environ.*, 32, 781–801, 2023.

[5] Kumar, A., Kumar, N., Baredar, P., & Shukla, A. A review on biomass energy resources, potential, conversion and policy in India. *Renew. Sust. Energ. Rev.*, 45, 530–539, 2015.

[6] Chen, L., Msigwa, G., Yang, M., Osman, A. I., Fawzy, S., Rooney, D. W., & Yap, P. S., Strategies to achieve a carbon neutral society: A review. *Environ. Chem. Lett.*, 20, 2277–2310, 2022.

[7] Tursi, A., A review on biomass: Importance, chemistry, classification, and conversion. *Biofuel Res. J.*, 6, 962–979, 2019.

[8] Zoghlami, A., & Paës, G. Lignocellulosic biomass: Understanding recalcitrance and predicting hydrolysis. *Front. Chem.*, 7, 874, 2019.

[9] Velvizhi, G., Balakumar, K., Shetti, N. P., Ahmad, E., Kishore Pant, K., & Aminabhavi, T. M., Integrated biorefinery processes for conversion of lignocellulosic biomass to value added materials: Paving a path towards circular economy. *Bioresour. Technol.*, 343, 2023.

[10] Gielen, D., Boshell, F., Saygin, D., Bazilian, M. D., Wagner, N., & Gorini, R., The role of renewable energy in the global energy transformation. *Energy Strategy Rev.*, 24, 38–50, 2019.

[11] European Commission: Renewable Energy Targets. https://energy.ec.europa.eu/topics/renewable-energy/renewable-energy-directive-targets-and-rules/renewable-energy-targets_en#:~:text=The%202020%20targets

[12] Kohli, K., Prajapati, R., & Sharma, B. K., Bio-based chemicals from renewable biomass for integrated biorefineries. *Energies*, 12, 233, 2019.

[13] Andrade, M. C., Gorgulho Silva, C. de O., de Souza Moreira, L. R., & Ferreira Filho, E. X., Crop residues: Applications of lignocellulosic biomass in the context of a biorefinery. *Front. Energy*, 16, 224–245, 2022.

[14] Chaturvedi, T., Hulkko, L. S. S., Fredsgaard, M., & Thomsen, M. H., Extraction, isolation, and purification of value-added chemicals from lignocellulosic biomass. *Processes*, 10, 2022.

[15] Alberts, B., Johnson, A., & Lewis, J., *Molecular Biology of the Cell*. 4th edition. New York: Garland Science; 2003. The Plant Cell Wall. 91, 401.

[16] Zhang, X., Li, L., & Xu, F., Chemical characteristics of wood cell wall with an emphasis on ultrastructure: A mini-review. *Forests*, 13, 2022.

[17] Habibi, Y., Lucia, L. A., & Rojas, O. J., Cellulose nanocrystals: Chemistry, self-assembly, and applications. *Chem. Rev.*, 110, 3479–3500, 2010.

[18] Blanco, A., Nanocellulose for industrial use: Cellulose nanofibers (CNF), cellulose nanocrystals (CNC), and bacterial cellulose (BC). In Blanco, A., Monte, M. C., Campano, C., Balea, A., Merayo, N., & Negro, C. (Eds.), *Handbook of Nanomaterials for Industrial Applications*. Madrid, Spain: Elsevier, 2018, Chapter 5, 74–126.

[19] Nanda, S., Maley, J., Kozinski, J. A., & Dalai, A. K., Physico-chemical evolution in lignocellulosic feedstocks during hydrothermal pretreatment and delignification. *J. Biobased. Mater. Bioenergy.* 9, 295–308, 2015.

[20] Hallac, B. B., & Ragauskas, A. J., Analyzing cellulose degree of polymerization and its relevancy to cellulosic ethanol. *Biofuel. Bioprod. Biorefin.* 5, 215–225, 2011.

[21] Pubchem. Cellulose. *National Library of Medicine*. https://pubchem.ncbi.nlm.nih.gov/compound/CELLULOSE, accessed on 4 May 2020.

[22] Moon, R. J., Martini, A., Nairn, J., Simonsen, J., & Youngblood, J., Cellulose nanomaterials review: Structure, properties and nanocomposites. *Chem. Soc. Rev.*, 40, 3941–3994, 2011.

[23] Kadla, J., & Gilbert, R. D., Cellulose structure: A review. *Cellul. Chem. Technol.*, 34, 197–216, 2000.

[24] Chen, H., Chemical composition and structure of natural lignocellulose. In: Chen, H. Z. (Ed.), *Biotechnology of Lignocellulose: Theory and Practice*. Dordrecht: Springer, 2014, 25–71.

[25] Gibson, L. J., & Soc, J. R., The hierarchical structure and mechanics of plant materials. *Interface*, 9, 2749–2766, 2012.

[26] Pubchem. D-Xylose. *National Library of Medicine*. https://pubchem.ncbi.nlm.nih.gov/compo und/135191, accessed on 4 May 2020.

[27] Kormelink, F. J. M., & Voragen, A. G. J., Degradation of different [(glucurono)arabino]xylans by a combination of purified xylan degrading enzymes. *Appl. Microbiol. Biotechnol.*, 38, 688–695, 1993.

[28] Doner, L. W., & Hicks, K. B., Isolation of hemicellulose from corn fiber by alkaline hydrogen peroxide extraction. *Cereal Chem.* 74, 176–181, 1997.

[29] Saha, B. C., Hemicellulose bioconversion. *J. Ind. Microbiol. Biotechnol.*, 30, 279–291, 2003.

[30] Yang, S. H., *Plant Fiber Chemistry*. Beijing: Light Industry Press, 2008.

[31] Gírio, F. M., Fonseca, C., Carvalheiro, F., Duarte, L. C., Marques, S., & Bogel-Łukasik, R., Hemicelluloses for fuel ethanol: A review. *Bioresour. Technol.*, 101, 4775–4800, 2010.

[32] Wang, S., Dai, G., Yang, H., & Luo, Z., Lignocellulosic biomass pyrolysis mechanism: A state-of-the-art review. *Prog. Energy Combust. Sci.* 62, 33–86, 2017.

[33] Zhou, X., Li, W., Mabon, R., & Broadbelt, L. J., A critical review on hemicellulose pyrolysis. *Energy Technol.*, 5, 52–79, 2017.

[34] Ragauskas, A. J., Nagy, M., Kim, D. H., Eckert, C. A., Hallett, J. P., & Liotta, C. L., From wood to fuels: Integrating biofuels and pulp production. *Ind. Biotechnol.* 2, 55–65, 2006.

[35] Buranov, A. U., & Mazza, G., Lignin in straw of herbaceous crops. *Ind. Crops Prod.* 28, 237–259, 2008.

[36] Reinoso, F. A. M., Rencoret, J., Gutiérrez, A., Milagres, A. M. F., del Río, J. C., & Ferraz, A., Fate of *p*-hydroxycinnamates and structural characteristics of residual hemicelluloses and lignin during alkaline-sulfite chemithermomechanical pretreatment of sugarcane bagasse. *Biotechnol. Biofuels*, 11, 153, 2018.

[37] Fougere, D., Nanda, S., Clarke, K., Kozinski, J. A., & Li, K., Effect of acidic pretreatment on the chemistry and distribution of lignin in aspen wood and wheat straw substrates. *Biomass Bioenerg.* 91, 56–68, 2016.

[38] Zakzeski, J., Bruijnincx, P. C. A., Jongerius, A. L., & Weckhuysen, B. M., The catalytic valorization of lignin for the production of renewable chemicals. *Chem. Rev.* 110, 3552–3599, 2010.

[39] Pandey, M. P., & Kim, C. S., Lignin depolymerization and conversion: A review of thermochemical methods. *Chem. Eng. Technol.* 34, 29–41, 2011.

[40] Novaes, E., Kirst, M., Chiang, V., Winter-Sederoff, H., & Sederoff, R., Lignin and biomass: A negative correlation for wood formation and lignin content in trees. *Plant Physiol.* 154, 555–561, 2010.

[41] Kim, J. S., Lee, Y. Y., & Kim, T. H., A review on alkaline pretreatment technology for bioconversion of lignocellulosic biomass. *Bioresour. Technol.* 199, 42–48, 2016.

[42] Szalaty, T. J., Klapiszewski, Ł., & Jesionowski, T., Recent developments in modification of lignin using ionic liquids for the fabrication of advanced materials—A review. *J. Mol. Liq.*, 301, 2020.

[43] Chen, Z., Naderi Nasrabadi, M., Staser, J. A., & Harrington, P. B., Application of generalized standard addition method and ultraviolet spectroscopy to quantify electrolytic depolymerization of lignin. *J. Anal. Test.*, 4, 35–44, 2020.

[44] Tamaki, Y., & Mazza, G., Rapid determination of lignin content of straw using Fourier transform mid-infrared spectroscopy. *J. Agric. Food Chem.*, 59, 504–512, 2011.

[45] Singh, M., Pandey, N., Negi, P., Jyoti, Larroche, C., & Mishra, B. B., Solvothermal conversion of spent aromatic waste to ethyl glucosides. *Chemosphere*, 292, 133428, 2022.

[46] Singh, M., Pandey, N., & Mishra, B. B., A divergent approach for the synthesis of (hydroxymethyl)furfural (HMF) from spent aromatic biomass-derived (chloromethyl) furfural (CMF) as a renewable feedstock. *RSC Adv.*, 10, 45081–45089, 2020.

[47] Ju, X., Engelhard, M., & Zhang, X., An advanced understanding of the specific effects of xylan and surface lignin contents on enzymatic hydrolysis of lignocellulosic biomass. *Bioresour. Technol.* 132, 137–145, 2013.

[48] Lu, Y., Lu, Y. C., Hu, H. Q., Xie, F. J., Wei, X. Y., & Fan, X., Structural characterization of lignin and its degradation products with spectroscopic methods. *J. Spectr.*, 2017, 1–15, 2017.

[49] Banoub, J., Delmas Jr., G. H., Joly, N., Mackenzie, G., Cachet, N., Benjelloun-Mlayah, B., & Delmas, M., A critique on the structural analysis of lignins and application of novel tandem mass spectrometric strategies to determine lignin sequencing. *J. Mass Spec.*, 50, 5–48, 2015.

[50] Chundawat, S. P. S., Balan, V., Sousa, L. D. C., & Dale, B. E., Thermochemical pretreatment of lignocellulosic biomass. In: Waldron, K. (Ed.), *Bioalcohol Production*, Michigan: Woodhead Publishing, 2010, 24–72.

[51] Constant, S., Basset, C., Dumas, C., Di Renzo, F., Robitzer, M., Barakat, A., & Quignard, F., Reactive organosolv lignin extraction from wheat straw: Influence of Lewis acid catalysts on structural and chemical properties of lignins. *Ind. Crops Prod.*, 65, 180–189, 2015.

[52] Okolie, J. A., Rana, R., Nanda, S., Dalai, A. K., & Kozinski, J. A., Supercritical water gasification of biomass: A state-of-the-art review of process parameters, reaction mechanisms and catalysis. *Sustain. Energy Fuel.*, 3, 578–598, 2019.

[53] Mohan, D., Pittman, C. U., & Steele, P. H., Pyrolysis of wood/biomass for bio-oil: A critical review. *Energy Fuels*, 20, 848–889, 2006.

[54] Aharoni, A., Jongsma, M. A., & Bouwmeester, H. J., Volatile science? Metabolic engineering of terpenoids in plants. *Trends Plant Sci.* 10, 594–602, 2005.

[55] Wang, X., Ort, D. R., & Yuan, J. S., Photosynthetic terpene hydrocarbon production for fuels and chemicals. *Plant Biotechnol. J.* 13, 137–146, 2015.

[56] Vassilev, S. V., Baxter, D., Andersen, L. K., Vassileva, C. G., & Morgan, T. J., An overview of the organic and inorganic phase composition of biomass. *Fuel*, 94, 1–33, 2012.

[57] Nanda, S., Mohanty, P., Pant, K. K., Naik, S., Kozinski, J. A., & Dalai, A. K., Characterization of North American lignocellulosic biomass and biochars in terms of their candidacy for alternate renewable fuels. *Bioenerg. Res.*, 6, 663–677, 2013.

[58] Demirbas, A., Potential applications of renewable energy sources, biomass combustion problems in boiler power systems and combustion related environmental issues. *Prog. Energy Combust. Sci.*, 31, 171–192, 2005.

[59] James, A. K., Thring, R. W., Helle, S., & Ghuman, H. S., Ash management review-applications of biomass bottom ash. *Energies*, 5, 3856–3873, 2012.

[60] Mohanty, P., Nanda, S., Pant, K. K., Naik, S., Kozinski, J. A., & Dalai, A. K., Evaluation of the physiochemical development of biochars obtained from pyrolysis of wheat straw, timothy grass and pinewood: Effects of heating rate. *J. Anal. Appl. Pyrolysis*, 104, 485–493, 2013.

[61] Nanda, S., & Abraham, J., Remediation of heavy metal contaminated soil. *Afr. J. Biotechnol.* 12, 3099–3109, 2013.

[62] Khan, A. A., de Jong, W., Jansens, P. J., & Spliethoff, H., Biomass combustion in fluidized bed boilers: Potential problems and remedies. *Fuel Process. Technol.* 90, 21–50, 2009.

[63] Vamvuka, D., Comparative fixed/fluidized bed experiments for the thermal behaviour and environmental impact of olive kernel ash. *Renew. Energy* 34, 158–164, 2009.

[64] Okolie, J. A., Nanda, S., Dalai, A. K., & Kozinski, J. A., Chemistry and specialty industrial applications of lignocellulosic biomass. *Waste Biomass Valori.*, 12, 145–2169, 2021.

[65] Nanda, S., Azargohar, R., Dalai, A. K., & Kozinski, J. A., An assessment on the sustainability of lignocellulosic biomass for biorefining. *Renew. Sustain. Energy Rev.*, 50, 925–941, 2015.

[66] Parakh, P. D., Nanda, S., & Kozinski, J. A., Eco-friendly transformation of waste biomass to biofuels. *Curr. Biochem. Eng.* 6, 120–134, 2020.

[67] Singh, M., Pandey, N., Dwivedi, P., Kumar, V., & Mishra, B. B., Production of xylose, levulinic acid, and lignin from spent aromatic biomass with a recyclable Brønsted acid synthesized from *d*-limonene as renewable feedstock from citrus waste. *Bioresour. Technol.*, 293, 122105, 2019.

[68] Kassaye, S., Pant, K. K., Jain, S., Synergistic effect of ionic liquid and dilute sulphuric acid in the hydrolysis of microcrystalline cellulose. *Fuel Process. Technol.*, 148, 289–294, 2016.

[69] Mankar, A. R., Pandey, A., Modak, A., & Pant, K. K., Pretreatment of lignocellulosic biomass: A review on recent advances. *Bioresour. Technol.*, 334, 2021.

[70] Werpy, T., & Petersen, G., *Top Value Added Chemicals from Biomass: Volume I -Results of Screening for Potential Candidates from Sugars and Synthesis Gas.* Washington, D.C.: NREL, 2004.

[71] Takkellapati, S., Li, T., & Gonzalez, M. A., An overview of biorefinery-derived platform chemicals from a cellulose and hemicellulose biorefinery. *Clean Technol. Environ. Policy*, 20, 1615–1630, 2018.

[72] Kabbour, M., & Luque, R., Furfural as a platform chemical: From production to applications. In *Biomass, Biofuels, Biochemicals.* Cordoba: Elsevier B.V., Ed., 2020, 283–297, 1, 218–224, 2012.

[73] Chatterjee, M., Ishizaka, T., & Kawanami, H., Reductive amination of furfural to furfurylamine using aqueous ammonia solution and molecular hydrogen: An environmentally friendly approach. *Green Chem.*, 18, 487–496, 2016.

[74] Thubsuang, U., Chotirut, S., Nuithitikul, K., Payaka, A., Manmuanpom, N., Chaisuwan, T., & Wongkasemjit, S., Oxidative upgrade of furfural to succinic acid using SO_3H-carbocatalysts with nitrogen functionalities based on polybenzoxazine. *J. Colloid Interface Sci.*, 565, 96–109, 2020.

[75] Shao, Y., Hu, X., Zhang, Z., Sun, K., Gao, G., Wei, T., Zhang, S., Hu, S., Xiang, J., & Wang, Y., Direct conversion of furfural to levulinic acid/ester in dimethoxymethane: Understanding the mechanism for polymerization. *Green Energy Environ.*, 4(4), 400–413, 2019.

[76] Barranca, A., Agirrezabal-Tellería, I., Rellán-Piñeiro, M., Ortuño, M. A., & Gandarias, I., Selective furfural hydrogenolysis towards 2-methylfuran by controlled poisoning of Cu-Co catalysts with chlorine Contents Page. *Sustain. Energy Fuels*, 5, 1379–1393, 2021.

[77] Villaverde, M. M., Bertero, N. M., Garetto, T. F., & Marchi, A. J., Selective liquid-phase hydrogenation of furfural to furfuryl alcohol over Cu-based catalysts. *Catal. Today*, 213, 87–92, 2013.

[78] Bonita, Y., Jain, V., Geng, F., O'Connell, T. P., Wilson, W. N., Rai, N., & Hicks, J. C., Direct synthesis of furfuryl alcohol from furfural: Catalytic performance of monometallic and bimetallic Mo and Ru phosphides. *Catal. Sci. Technol.*, 9, 3656–3668, 2019.

[79] Gong, W., Chen, C., Zhang, Y., Zhou, H., Wang, H., Zhang, H., Zhang, Y., Wang, G., & Zhao, H., Efficient synthesis of furfuryl alcohol from H_2-hydrogenation/transfer hydrogenation of furfural using sulfonate group modified Cu catalyst. *ACS Sustain. Chem. Eng.*, 5, 2172–2180, 2017.

[80] al Ghatta, A., Perry, J. M., Maeng, H., Lemus, J., & Hallett, J. P., Sustainable and efficient production of furoic acid from furfural through amine assisted oxidation with hydrogen peroxide and its implementation for the synthesis of alkyl furoate. *RSC Sustain*, 1, 303–309, 2023.

[81] Mathew, A. K., Abraham, A., Mallapureddy, K. K., & Sukumaran, R. K., Lignocellulosic biorefinery wastes, or resources? In Mathew, A. K., Abraham, A., Mallapureddy, K. K., & Sukumaran, R. K. (Eds.), *Waste Biorefinery: Potential and Perspectives*, Thiruvananthapuram: Elsevier, 2018, Chapter 9, 267–297.

[82] Mariscal, R., Maireles-Torres, P., Ojeda, M., Sadaba, I., & Lopez Granados, M., Furfural: A renewable and versatile platform molecule for the synthesis of chemicals and fuels. *Energy Environ. Sci.*, 9, 1144–1189, 2016.

[83] Zhang, T., Li, W., Xiao, H., Jin, Y., & Wu, S., Recent progress in direct production of furfural from lignocellulosic residues and hemicellulose. *Bioresour. Technol.*, 354, 2022.

[84] Yong, K. J., Wu, T. Y., Lee, C. B. T. L., Lee, Z. J., Liu, Q., Jahim, J. M., Zhou, Q., & Zhang, L., Furfural production from biomass residues: Current technologies, challenges and future prospects. *Biomass Bioenerg.* 161, 2022.

[85] Yan, K., Wu, G., Lafleur, T., & Jarvis, C., Production, properties and catalytic hydrogenation of furfural to fuel additives and value added chemicals. *Renew. Sust. Energy Rev.*, 38, 663–676, 2014.

[86] Gravitis, J., Vedernikov, N., Zandersons, J., & Kokorevics, A., Furfural and levoglucosan production from deciduous wood and agricultural wastes. Chemicals and materials from renewable resources. *J. Am. Chem. Soc.*, 784, 110–122, 2001.

[87] Dashtban, M., Production of furfural: Overview and challenges. *J-FOR*, 2, 2012.

[88] Sherif, N., Gadalla, M., & Kamel, D., Acid—hydrolysed furfural production from rice straw bio-waste: Process synthesis, simulation, and optimisation. *S. Afr. J. Chem. Eng.*, 38, 34–40, 2021.

[89] Ambalkar, V. U., & Talib, M., Synthesis and identification of furfural from sunflower husk. *Int. Adv. Res. J. Sci. Eng. Technol.*, 4, 2017.

[90] Peleteiro, S., Santos, V., Garrote, G., & Parajó, J. C., Furfural production from Eucalyptus wood using an Acidic Ionic Liquid. *Carbohydr. Polym.*, 146, 20–25, 2016.

[91] Bizzi, C. A., Santos, D., Sieben, T. C., Motta, G. V., Mello, P. A., & Flores, E. M. M., Furfural production from lignocellulosic biomass by ultrasound-assisted acid hydrolysis. *Ultrason. Sonochem.*, 51, 332–339, 2019.

[92] Lyu, X., & Botte, G. G., Investigation of factors that inhibit furfural production using metal chloride catalysts. *Chem. Eng. J.*, 403, 2021.

[93] Zhao, Y., Xu, H., Lu, K., Qu, Y., Zhu, L., & Wang, S., Dehydration of xylose to furfural in butanone catalyzed by Brønsted-Lewis acidic ionic liquids. *Energy Sci. Eng.*, 7, 2237–2246, 2019.

[94] Hu, L., Zhao, G., Hao, W., Tang, X., Sun, Y., Lin, L., & Liu, S., Catalytic conversion of biomass-derived carbohydrates into fuels and chemicals via furanic aldehydes. *RSC Adv.*, 2, 11184–11206, 2012.

[95] Ghosh, S., Falyouna, O., Malloum, A., Othmani, A., Bornman, C., Bedair, H., Onyeaka, H., Al-Sharify, Z. T., Jacob, A. O., Miri, T., Osagie, C., & Ahmadi, S., A general review on the use of advance oxidation and adsorption processes for the removal of furfural from industrial effluents. *Microporous Mesoporous Mater.*, 331, 2022.

[96] Mascal, M., & Nikitin, E. B., High-yield conversion of plant biomass into the key value-added feedstocks 5-(hydroxymethyl) furfural, levulinic acid, and levulinic esters via 5-(chloromethyl)furfural. *Green Chem.*, 12, 370–373, 2010.

[97] Ventura, M., Dibenedetto, A., & Aresta, M., Heterogeneous catalysts for the selective aerobic oxidation of 5-hydroxymethylfurfural to added value products in water. *Inorganica Chim. Acta*, 470, 11–21, 2018.

[98] Zhang, Z., & Deng, K., Recent advances in the catalytic synthesis of 2,5-furandicarboxylic acid and its derivatives. *ACS Catal.*, 5, 6529–6544, 2015.

[99] Bangalore Ashok, R. P., Oinas, P., & Forssell, S., Techno-economic evaluation of a biorefinery to produce γ-valerolactone (GVL), 2-methyltetrahydrofuran (2-MTHF) and 5-hydroxymethylfurfural (5-HMF) from spruce. *Renew. Energy*, 190, 396–407, 2022.

[100] Zhang, W., Cui, N., Su, F., Wang, Y., Yang, B., Sun, Y., Guan, W., Kuang, H., & Wang, Q., Comprehensive metabolomics and network pharmacology to explore the mechanism of 5-hydroxymethyl furfural in the treatment of blood deficiency syndrome. *Front. Pharmacol.*, 12, 2022.

[101] Kumar, A., Chauhan, A. S., Bains, R., & Das, P., Rice straw (Oryza sativa L.) biomass conversion to furfural, 5-hydroxymethylfurfural, lignin and bio-char: A comprehensive solution. *J. Ind. Eng. Chem.*, 104, 286–294, 2021.

[102] Ito, R., Miyafuji, H., Miyazaki, Y., & Kawai, T., Production of 5-hydroxymethylfurfural from wood by ionic liquid treatment. *J. Wood Sci.*, 62, 349–355, 2016.

[103] Dallas Swift, T., Nguyen, H., Anderko, A., Nikolakis, V., & Vlachos, D. G., Tandem Lewis/Brønsted homogeneous acid catalysis: Conversion of glucose to 5-hydoxymethylfurfural in an aqueous chromium(III) chloride and hydrochloric acid solution. *Green Chem.*, 17, 4725–4735, 2015.

[104] Enslow, K. R., & Bell, A. T., SnCl₄-catalyzed isomerization/dehydration of xylose and glucose to furanics in water. *Catal. Sci. Technol.*, 5, 2839–2847, 2015.

[105] Li, F., Shi, G., Wang, G., Guo, T., & Lei, X., Catalytic conversion of raw *Dioscorea composita* biomass to 5-hydroxymethylfurfural using a combination of metal chlorides in N,N-dimethylacetamide solvent containing lithium chloride. *Res. Chem. Intermed.*, 42, 6757–6767, 2016.

[106] Shirai, H., Ikeda, S., & Qian, E. W., One-pot production of 5-hydroxymethylfurfural from cellulose using solid acid catalysts. *Fuel Process. Technol.*, 159, 280–286, 2017.

[107] Li, M., Jiang, H., Zhang, L., Yu, X., Liu, H., Yagoub, A. E. G. A., & Zhou, C., Synthesis of 5-HMF from an ultrasound-ionic liquid pretreated sugarcane bagasse by using a microwave-solid acid/ionic liquid system. *Ind. Crops Prod.*, 149, 2020.

[108] Kumar, A., Chauhan, A. S., Bains, R., & Das, P., Catalytic transformations for agrowaste conversion to 5-hydroxymethylfurfural and furfural: Chemistry and scale-up development. *Green Chem.*, 25, 849–870, 2022.

[109] Haldar, D., Dey, P., Thomas, J., Singhania, R. R., & Patel, A. K., One pot bioprocessing in lignocellulosic biorefinery: A review. *Bioresour. Technol.*, 365, 2022.

[110] Izyan, N., Azelee, W., Nor, A., Ramli, M., Hasmaliana, N., Manas, A., Salamun, N., Man, R. C., & Enshasy, H. E., Glycerol in food, cosmetics and pharmaceutical industries: Basics and new applications. *Int. J. Sci. Technol. Res.*, 8, 2019.

[111] Glycerol: Application: Pharmaceutical and Personal Care Applications. www.wikidoc.org/index.php/Glycerol#:~:text=Glycerol%20is%20used%20in%20medical,hair%20care%20products%2C%20and%20soaps, accessed on 9 August 2012.

[112] Chilakamarry, C. R., Sakinah, A. M. M., Zularisam, A. W., & Pandey, A., Glycerol waste to value added products and its potential applications. *Syst. Microbiol. Biomanufacturing*, 1, 378–396, 2021.

[113] Globe Newswire: Global Glycerol Industry. www.globenewswire.com/news-release/2020/10/02/2102707/0/en/Global-Glycerol-Industry.html, accessed on 1 October 2020.

[114] Maris, E. P., & Davis, R. J., Hydrogenolysis of glycerol over carbon-supported Ru and Pt catalysts. *J. Catal.*, 249, 328–337, 2007.

[115] Pethan Rajan, N., Srinivasa Rao, G., Putrakumar, B., & Chary, K. V., Vapour phase dehydration of glycerol to acrolein over Vanadium Phosphorous Oxide (VPO) catalyst. *RSC Adv.*, 4, 53419–53428, 2014.

[116] Sanchez, A., Velasquez, M., Batiot-Dupeyrat, C., Espinal, J. F., & Santamaría, A., Mechanism of glycerol dehydration and dehydrogenation: An experimental and computational correlation. *DYNA*, 86, 126–135, 2019.

[117] Venkataseetharaman, A., Mishra, G., Ghosh, M. K., & Das, G. K., Role of glycerol oxidation pathways in the reductive acid leaching kinetics of manganese nodules using glycerol. *ACS Omega*, 6, 14903–14910, 2021.

[118] Wang, Y., Zhou, J., & Guo, X., Catalytic hydrogenolysis of glycerol to propanediols: A review. *RSC Adv.*, 5, 74611–74628, 2015.

[119] Sun, D., Yamada, Y., Sato, S., & Ueda, W., Glycerol hydrogenolysis into useful C3 chemicals. *Appl. Catal. B: Environ.*, 193, 75–92, 2016.

[120] Yang, G. Y., Ke, Y. H., Ren, H. F., Liu, C. L., Yang, R. Z., & Dong, W. S., The conversion of glycerol to lactic acid catalyzed by ZrO₂-supported CuO catalysts. *Chem. Eng. J.*, 283, 759–767, 2016.

[121] Evans, C. D., Bartley, J. K., Taylor, S. H., Hutchings, G. J., & Kondrat, S. A., Perovskite supported catalysts for the selective oxidation of glycerol to tartronic acid. *Catal. Letters*, 153, 2026–2035, 2023.

[122] Kim, M., & Lee, H., Highly selective production of acrylic acid from glycerol via two steps using au/CeO2 catalysts. *ACS Sustain. Chem. Eng.*, 5, 11371–11376, 2017.

[123] Jia, R., Zhao, B., Xue, W., Liu, L., & Dong, J., Production of high-purity allyl alcohol by the salting-out method from formic acid-mediated deoxydehydration of glycerol. *J. Chem. Eng. Data*, 63, 3874–3880, 2018.

[124] Tabah, B., Varvak, A., Pulidindi, I. N., Foran, E., Banin, E., & Gedanken, A., Production of 1,3-propanediol from glycerol via fermentation by Saccharomyces cerevisiae. *Green Chem.*, 18, 4657–4666, 2016.

[125] Pandhare, N. N., Pudi, S. M., Biswas, P., & Sinha, S., Selective hydrogenolysis of glycerol to 1,2-propanediol over highly active and stable Cu/MgO catalyst in the vapor phase. *Org. Process Res. Dev.*, 20, 1059–1067, 2016.

[126] Zhao, H., Jiang, Y., Liu, H., Long, Y., Wang, Z., & Hou, Z. Direct synthesis of allyl alcohol from glycerol over CoFe alloy. *Appl. Catal. B: Environ.*, 277, 2020.

[127] Wang, S., Yin, K., Zhang, Y., & Liu, H., Glycerol hydrogenolysis to propylene glycol and ethylene glycol on zirconia supported noble metal catalysts. *ACS Catal.*, 3, 2112–2121, 2013.

[128] Wadaugsorn, K., Lin, K. Y., Kaewchada, A., & Jaree, A., Production of 2,5-furandicarboxylic acid *via* oxidation of 5-hydroxymethylfurfural over Pt/C in a continuous packed bed reactor. *RSC Adv.*, 12, 18084–18092, 2022.

[129] Prasad, S., Khalid, A. J., Narishetty, V., Kumar, V., Dutta, S., & Ahmad, E., Recent advances in the production of 2,5-furandicarboxylic acid from biorenewable resources. *Mater. Sci. Energy Technol.*, 6, 502–521, 2023.

[130] de Jong, E., Visser, H. A., Dias, A. S., Harvey, C., & Gruter, G. J. M., The road to bring FDCA and PEF to the market. *Polymers*, 14, 2022.

[131] Avantium, Press Release, 2016. www.avantium.com/press-releases/synvina-joint-venture-basf-avantium-established/.

[132] DuPont Press Release, 2016. www.dupont.com/products-and-services/industrial-bio-technology/press-releases/dupont-admannounce-platform-technology-for-long-sought-after-molecule.html, accessed on 22 June 2017.

[133] Yusuf, V. F., Malek, N. I., & Kailasa, S. K., Review on metal-organic framework classification, synthetic approaches, and influencing factors: Applications in energy, drug delivery, and wastewater treatment. *ACS Omega*, 7, 44507–44531, 2022.

[134] Rose, M., Weber, D., Lotsch, B. V., Kremer, R. K., Goddard, R., & Palkovits, R., Biogenic metal-organic frameworks: 2,5-Furandicarboxylic acid as versatile building block. *Micropor. Mesopor. Mater.*, 181, 217–221, 2013.

[135] Albonetti, S., Hu, C., & Saravanamurugan, S., Preface to special issue on green conversion of HMF. *ChemSusChem*, 15, 2022.

[136] Rajesh, R. O., Godan, T. K., Sindhu, R., Pandey, A., & Binod, P., Bioengineering advancements, innovations and challenges on green synthesis of 2, 5-furan dicarboxylic acid. *Bioengineered*, 11, 19–38, 2020.

[137] Alsaheb, R. A. A., Aladdin, A., Othman, Z., Malek, R. A., Leng, O. M., Aziz, R., & Enshasy, H. A. El. Lactic acid applications in pharmaceutical and cosmeceutical industries. *J. Chem. Pharm. Res.*, 7, 729–735, 2015.

[138] Kim, J., Kim, Y. M., Lebaka, V. R., & Wee, Y. J., Lactic acid for green chemical industry: Recent advances in and future prospects for production technology, recovery, and applications. *Fermentation*, 8, 2022.

[139] de Clippel, F., Dusselier, M., van Rompaey, R., Vanelderen, P., Dijkmans, J., Makshina, E., Giebeler, L., Oswald, S., Baron, G. v., Denayer, J. F. M., Pescarmona, P. P., Jacobs, P. A., & Sels, B. F., Fast and selective sugar conversion to alkyl lactate and lactic acid with bifunctional carbon-silica catalysts. *J. Am. Chem. Soc.*, 134, 10089–10101, 2012.

[140] Jang, H., Kim, S. H., Lee, D., Shim, S. E., Baeck, S. H., Kim, B. S., & Chang, T. S., Hydrogenation of lactic acid to propylene glycol over a carbon-supported ruthenium catalyst. *J. Mol. Catal. A: Chem.*, 380, 57–60, 2013.

[141] Sobuś, N., & Czekaj, I., Lactic acid conversion into acrylic acid and other products over natural and synthetic zeolite catalysts: Theoretical and experimental studies. *Catal. Today*, 387, 172–185, 2022.

[142] Aida, T. M., Ikarashi, A., Saito, Y., Watanabe, M., Smith, R. L., & Arai, K., Dehydration of lactic acid to acrylic acid in high temperature water at high pressures. *J. Supercrit. Fluids*, 50(3), 257–264, 2009.

[143] Naser, A. Z., Deiab, I., & Darras, B. M., Poly(lactic acid) (PLA) and polyhydroxyal-kanoates (PHAs), green alternatives to petroleum-based plastics: A review. *RSC Adv.*, 11, 17151–17196, 2021.

[144] Yang, P. B., Tian, Y., Wang, Q., & Cong, W., Effect of different types of calcium carbonate on the lactic acid fermentation performance of Lactobacillus lactis. *Biochem. Eng. J.*, 98, 38–46, 2015.

[145] Ghaffar, T., Irshad, M., Anwar, Z., Aqil, T., Zulifqar, Z., Tariq, A., Kamran, M., Ehsan, N., & Mehmood, S., Recent trends in lactic acid biotechnology: A brief review on production to purification. *J. Radiat. Res. Appl. Sci.*, 7, 222–229, 2014.

[146] Datta, R., & Henry, M., Lactic acid: Recent advances in products, processes and technologies-a review. *J. Chem. Technol. Biotechnol.*, 81, 1119–1129, 2006.

[147] Alexandri, M., Schneider, R., Mehlmann, K., & Venus, J., Recent advances in d-lactic acid production from renewable resources: Case studies on agro-industrial waste streams. *Food Technol. Biotechnol.*, 57(3), 293–304, 2019.

[148] Pereira, C. S. M., Silva, V. M. T. M., & Rodrigues, A. E., Ethyl lactate as a solvent: Properties, applications and production processes-a review. *Green. Chem.*, 13, 2658–2671, 2011.

[149] Erickson, B., Nelson, J. E., & Winters, P., Perspective on opportunities in industrial biotechnology in renewable chemicals. *Biotechnol. J.*, 7, 176–185, 2012.

[150] Simamora, P., & Chern, W., Poly-L-lactic acid: An overview. *J. Drugs Dermatol.*, 5(5), 436–440, 2006.

[151] Senthamaraikannan, C., Akash, K., Amanullah, S., Barath, M., Manojkumar, R., & Jagadeeshwaran, J., Overview of Polylactic acid and its derivatives in medicinal applications. *IOP Conference Series: Mater. Sci. Eng.*, 988, 2020.

[152] Saxena, R. K., Saran, S., Isar, J., & Kaushik, R., Production and applications of succinic acid. In *Current Developments in Biotechnology and Bioengineering: Production, Isolation and Purification of Industrial Products*, New Delhi: Elsevier, 2016, 601–630.

[153] Vaghela, S. S., Jethva, A. D., & Gohil, M. S., Cyclic voltammetric and galvanostatic electrolysis studies on the reduction of maleic acid in buffered and unbuffered solutions. *Bull. Electrochem.*, 18, 237–240, 2002.

[154] Nghiem, N. P., Kleff, S., & Schwegmann, S., Succinic acid: Technology development and commercialization. *Ferment.*, 3, 2017.

[155] Orozco-Saumell, A., Mariscal, R., Iglesias, J., Maireles-Torres, P., & López Granados, M., Aqueous phase hydrogenation of maleic acid to succinic acid mediated by formic acid: The robustness of the Pd/C catalytic system. *Sustain. Energy Fuels*, 6, 5160–5176, 2022.

[156] Becker, J., Lange, A., Fabarius, J., & Wittmann, C., Top value platform chemicals: Bio-based production of organic acids. *Curr. Opin. Biotechnol.*, 36, 168–175, 2015.

[157] Roquette/DSM. Press Release, 2008. www.dsm.com/corporate/about /business-enti-ties/dsm-biobased-productsandservices/reverdia.html, accessed on 22 June 2017.

[158] Bioamber, Press Release. www.bio amber.com/bioamber/en/news/2015/bioamber-now-shipping-bio-succinic-acid to-customers, accessed on 13 October 2015.

[159] Song, H., & Lee, S. Y., Production of succinic acid by bacterial fermentation. *Enzym. Microb. Technol.*, 39, 352–361, 2006.

[160] Delhomme, C., Weuster-Botz, D., & Kuhn, F. E., Succinic acid from renewable resources as a C4 building-block chemical—a review of the catalytic possibilities in aqueous media. *Green Chem.*, 11, 13–26, 2009.

[161] Luque, R., Clark, J. H., Yoshida, K., & Gai, P. L., Efficient aqueous hydrogenation of biomass platform molecules using supported metal nanoparticles on Starbons. *Chem. Commun.*, 35, 5305–5307, 2009.

[162] Global Bonds in Chemistry: Scuccinic Acid. https://thechemco.com/chemical/succinic-acid/

[163] Morales-Huerta, J. C., de Ilarduya, A. M., & Muñoz-Guerra, S. A green strategy for the synthesis of poly(ethylene succinate) and its copolyesters via enzymatic ring opening polymerization. *Eur. Polym. J.*, 95, 514–519, 2017.

[164] Bikiaris, D. N., Papageorgiou, G. Z., Papadimitriou, S. A., Karavas, E., & Avgoustakis, K., Novel biodegradable polyester poly(propylene succinate): Synthesis and application in the preparation of solid dispersions and nanoparticles of a water-soluble drug. *AAPS Pharm. Sci. Tech.*, 10, 138–146, 2009.

[165] Aliotta, L., Seggiani, M., Lazzeri, A., Gigante, V., & Cinelli, P., A brief review of poly (butylene succinate) (PBS) and its main copolymers: Synthesis, blends, composites, biodegradability, and applications. *Polym*, 14, 2022.

[166] Isikgor, F. H., & Becer, C. R., Lignocellulosic biomass: A sustainable platform for the production of bio-based chemicals and polymers. *Poly. Chem.*, 6, 4497–4559, 2015.

[167] GLOBE Newswire: Succinic Acid Market. www.globenewswire.com/en/newsre-lease/2023/05/09/2664413/0/en/Succinic-Acid-Market-Size-and-Value-to-Reach-USD-359-8-Million-in-2032-Growing-at-CAGR-of-7-3-Market-us-Study.html, accessed on 9 May 2023.

[168] Cha, J. Y., & Hanna, M. A., Levulinic acid production based on extrusion and pressur-ized batch reaction. *Ind. Crops Prod.*, 16, 109–118, 2002.

[169] Shao, Y., Hu, X., Zhang, Z., Sun, K., Gao, G., Wei, T., Zhang, S., Hu, S., Xiang, J., & Wang, Y., Direct conversion of furfural to levulinic acid/ester in dimethoxymethane: Understanding the mechanism for polymerization. *Green Energy Environ.*, 4, 400–413, 2019.

[170] Velaga, B., & Peela, N. R., Levulinic acid production from furfural: Process develop-ment and techno-economics. *Green Chem.*, 24, 3326–3343, 2022.

[171] Fang, C., Liu, Y., Wu, W., Li, H., Wang, Z., Zhao, W., Yang, T., & Yang, S., One pot cascade conversion of bio-based furfural to levulinic acid with cu-doped niobium phos-phate catalysts. Waste Biomass Valorization, 10(5), 1141–1150, 2019.

[172] Gong, W., Chen, C., Fan, R., Zhang, H., Wang, G., & Zhao, H., Transfer-hydrogenation of furfural and levulinic acid over supported copper catalyst. *Fuel*, 231, 165–171, 2018.

[173] Kamm, B., Gruber, P. R., & Kamm, M., Biorefineries-industrial processes and prod-ucts: Ullmann's Encycl. Ind. Chem., 1–2, 1–959, 2008.

[174] Galletti, A. M. R., Antonetti, C., De Luise, V., Licursi, D., & Di Nasso, N. N. O., Levulinic acid production from waste biomass. *BioResour*, 7, 1824–1834, 2012.

[175] Aliko, K., Doudin, K., Osatiashtiani, A., Wang, J., Topham, P. D., & Theodosiou, E., Microwave-assisted synthesis of levulinic acid from low-cost, sustainable feedstocks using organic acids as green catalysts. *J. Chem. Technol. Biotechnol.*, 95, 2110–2119, 2020.

[176] Dong, D., Sun, J., Huang, F., Gao, Q., Wang, Y., & Li, R., Using trifluoroacetic acid to pretreat lignocellulosic biomass. *Biomass Bioenergy*, 33, 1719–1723, 2009.

[177] Muranaka, Y., Suzuki, T., Sawanishi, H., Hasegawa, I., & Mae, K., Effective production of levulinic acid from biomass through pretreatment using phosphoric acid, hydrochloric acid, or ionic liquid. *Ind. Eng. Chem. Res.*, 53, 11611–11621, 2014.

[178] Pyo, S. H., Glaser, S. J., Rehnberg, N., & Hatti-Kaul, R., Clean production of levulinic acid from fructose and glucose in salt water by heterogeneous catalytic dehydration. *ACS Omega*, 5, 14275–14282, 2020.

[179] Weingarten, R., Conner, W. C., & Huber, G. W., Production of levulinic acid from cellulose by hydrothermal decomposition combined with aqueous phase dehydration with a solid acid catalyst. *Energy Environ. Sci.*, 5, 7559–7574, 2012.

[180] Mukherjee, A., & Dumont, M. J., Levulinic acid production from starch using microwave and oil bath heating: A kinetic modeling approach. *Ind. Eng. Chem. Res.*, 55, 8941–8949, 2016.

[181] Dutta, S., Yu, I. K. M., Fan, J., Clark, J. H., & Tsang, D. C. W., Critical factors for levulinic acid production from starch-rich food waste: Solvent effects, reaction pressure, and phase separation. *Green Chem.*, 24, 163–175, 2022.

[182] Lappalainen, K., & Dong, Y., Simultaneous production of furfural and levulinic acid from pine sawdust via acid-catalysed mechanical depolymerization and microwave irradiation. *Biomass Bioenergy*, 123, 159–165, 2019.

[183] Chen, S. S., Wang, L., Yu, I. K. M., Tsang, D. C. W., Hunt, A. J., Jérôme, F., Zhang, S., Ok, Y. S., & Poon, C. S., Valorization of lignocellulosic fibres of paper waste into levulinic acid using solid and aqueous Brønsted acid. *Bioresour. Technol.*, 247, 387–394, 2018.

[184] Hurst, G., Brangeli, I., Peeters, M., & Tedesco, S., Solid residue and by-product yields from acid-catalysed conversion of poplar wood to levulinic acid. *Chem. Pap.*, 74, 1647–1661, 2020.

[185] Victor, A., Pulidindi, I. N., & Gedanken, A., Levulinic acid production from Cicer arietinum, cotton, Pinus radiata and sugarcane bagasse. *RSC Adv.*, 4, 44706–44711, 2014.

[186] Yan, L., Yang, N., Pang, H., & Liao, B., Production of levulinic acid from bagasse and paddy straw by liquefaction in the presence of hydrochloride acid. *Clean—Soil, Air, Water*, 36, 158–163, 2008.

[187] Fang, Q., & Hanna, M. A., Experimental studies for levulinic acid production from whole kernel grain sorghum. *Bioresour. Technol.*, 81, 187–192, 2002.

[188] Chang, C., Cen, P., & Ma, X., Levulinic acid production from wheat straw. *Bioresour. Technol.*, 98, 1448–1453, 2007.

[189] Yang, F., Fu, J., Mo, J., & Lu, X., Synergy of Lewis and Brønsted acids on catalytic hydrothermal decomposition of hexose to levulinic acid. *Energy & Fuels*, 27, 6973–6978, 2013.

[190] Tao, F., Song, H., & Chou, L., Hydrolysis of cellulose by using catalytic amounts of $FeCl_2$ in Ionic liquids. *ChemSusChem*, 3, 1298–1303, 2010.

[191] Fachri, B. A., Abdilla, R. M., Bovenkamp, H. H. V. De, Rasrendra, C. B., & Heeres, H. J., Experimental and kinetic modeling studies on the sulfuric acid catalyzed conversion of D-fructose to 5-hydroxymethylfurfural and levulinic acid in water. *ACS Sustain. Chem. Eng.*, 3, 3024–3034, 2015.

[192] Hegner, J., Pereira, K. C., DeBoef, B., & Lucht, B. L., Conversion of cellulose to glucose and levulinic acid *via* solid-supported acid catalysis. *Tetrahedron Lett.*, 51, 2356–2358, 2010.

[193] Omari, K. W., Besaw, J. E., & Kerton, F. M., Hydrolysis of chitosan to yield levulinic acid and 5-hydroxymethylfurfural in water under microwave irradiation. *Green Chem.*, 14, 1480–1487, 2012.

[194] Chen, H., Yu, B., & Jin, S., Production of levulinic acid from steam exploded rice straw via solid superacid, $S_2O_8^{2-}/ZrO_2-SiO_2-Sm_2O_3$. *Bioresour. Technol.*, 102, 3568–3570, 2011.

[195] Weingarten, R., Conner, W. C., & Huber, G. W., Production of levulinic acid from cellulose by hydrothermal decomposition combined with aqueous phase dehydration with a solid acid catalyst. *Energy Environ. Sci.*, 5, 7559, 2012.

[196] Girisuta, B., Janssen, L. P. B. M., & Heeres, H. J., Kinetic study on the acid-catalyzed hydrolysis of cellulose to levulinic acid. *Ind. Eng. Chem. Res.*, 46, 1696–1708, 2007.

[197] Hurst, G., Brangeli, I., Peeters, M., & Tedesco, S., Solid residue and by-product yields from acid-catalysed conversion of poplar wood to levulinic acid. *Chem. Pap.*, 74, 1647–1661, 2020.

[198] Zheng, X., Gu, X., Ren, Y., Zhi, Z., & Lu, X. Production of 5-hydroxymethyl furfural and levulinic acid from lignocellulose in aqueous solution and different solvents. *Biofuels, Bioprod. Biorefin.*, 10, 917–931, 2016.

[199] Zheng, P., Wereath, K., Sun, J. B., van den Heuvel, J., & Zeng, A., Overexpression of genes of the *dha* regulon and its effects on cell growth, glycerol fermentation to 1,3-propanediol and plasmid stability in *Klebsiella pneumoniae*. *Process Biochem.*, 41, 2160–2169, 2006.

[200] Kumar, V., Ashok, S., & Park, S., Recent advances in biological production of 3-hydroxypropionic acid. *Biotechnol Adv.*, 31, 945–961, 2013.

[201] Son, J., Baritugo, K. A., Lim, S. H., Lim, H. J., Jeong, S., Lee, J. Y., Choi, J. L., Joo, J. C., Na, J. G., & Park, S. J., Microbial cell factories for the production of three-carbon backbone organic acids from agro-industrial wastes. *Bioresour. Technol.*, 349, 2022.

[202] Corma, A., Iborra, S., & Velty, A., Chemical routes for the transformation of biomass into chemicals. *Chem. Rev.*, 107, 2411–2502, 2007.

[203] Huang, S. Y., Lipp, D. W., & Farinato, R. S., Acrylamide polymers. In: *Encyclopedia of Polyer Science and Technology.* Hoboken: Wiley, 2001.

[204] Bhagwat, S. S., Li, Y., Cortés-Peña, Y. R., Brace, E. C., Martin, T. A., Zhao, H., & Guest, J. S., Sustainable production of acrylic acid via 3-hydroxypropionic acid from lignocellulosic biomass. *ACS Sustain. Chem. Eng.*, 9, 16659–16669, 2021.

[205] Banu, M., Venuvanalingam, P., Shanmugam, R., Viswanathan, B., & Sivasanker, S., Sorbitol hydrogenolysis over Ni, Pt and Ru supported on NaY. *Top Catal.*, 55, 897–907, 2012.

[206] EMR: Global Sorbitol Market. www.expertmarketresearch.com/reports/sorbitol-market.

[207] Grand View Research: Market Report Analysis. www.grandviewresearch.com/industry-analysis/sorbitol-market.

[208] Chen, J., Wang, S., Huang, J., Chen, L., Ma, L., & Huang, X., Conversion of cellulose and cellobiose into sorbitol catalyzed by ruthenium supported on a polyoxometalate/metal—organic framework hybrid. *ChemSusChem*, 6, 1545–1555, 2013.

[209] Ribeiro, L. S., Órfão, J. J. M., & Pereira, M. F. R., Enhanced direct production of sorbitol by cellulose ball-milling. *Green Chem.*, 17, 2973–2980, 2015.

[210] Zhang, J., Li, J., Wu, S., & Liu, Y., Advances in the catalytic production and utilization of sorbitol. *Ind. Eng. Chem. Res.*, 52, 11799–11815, 2013.

[211] Zhu, W., Yang, H., Chen, J., Chen, C., Guo, L., Gan, H., Zhao, X., & Hou, Z., Efficient hydrogenolysis of cellulose into sorbitol catalyzed by a bifunctional catalyst. *Green Chem.*, 16, 1534–1542, 2014.

[212] Zhang, X., Durndell, L. J., Isaacs, M. A., Parlett, C. M. A., Lee, A. F., & Wilson, K., Platinum-catalyzed aqueous-phase hydrogenation of D-glucose to D-sorbitol. *ACS Catal.*, 6, 7409–7417, 2016.

[213] Wang, G., Tan, X., Lv, H., Zhao, M., Wu, M., Zhou, J., Zhang, X., & Zhang, L., Highly selective conversion of cellobiose and cellulose to hexitols by Ru-based homogeneous catalyst under acidic conditions. *Ind. Eng. Chem. Res.*, 55, 5263–5270, 2016.

[214] Kobayashi, H., & Fukuoka, A., Synthesis and utilisation of sugar compounds derived from lignocellulosic biomass. *Green Chem.*, 15, 1740–1763, 2013.

[215] Rose, M., & Palkovits, R., Isosorbide as a renewable platform chemical for versatile applications—quo vadis? *ChemSusChem*, 5, 167–176, 2012.

[216] Isikgor, F. H., & Becer, C. R., Lignocellulosic biomass: A sustainable platform for the production of bio-based chemicals and polymers. *Poly. Chem.*, 6, 4497–4559, 2015.

[217] Roquette. www.roquette.com/industries/performance-materials/polycarbonates/, accessed on 22nd June 2017.

[218] Lugani, Y., Oberoi, S., & Sooch, B. S., Xylitol: A sugar substitute for patients of diabetes mellitus. *World J. Pharm. Pharm. Sci.*, 6, 741–749, 2017.

[219] Rafiqul, I. S. M., & Sakinah, A. M. M., Processes for the production of xylitol—a review. *Food Rev. Int.*, 29, 127–156, 2013.

[220] Chen, X., Jiang, Z., Chen, S., & Qin, W., Microbial and bioconversion production of d-xylitol and its detection and application. *Int. J. Biol. Sci.*, 6, 834–844, 2010.

[221] Dietrich, K., Hernandez-Mejia, C., Verschuren, P., Rothenberg, G., & Shiju, N. R., One-pot selective conversion of hemicellulose to xylitol. *Org. Process Res. Dev.*, 21, 165–170, 2017.

[222] Ribeiro, L. S., Delgado, J. J., de Melo Órfão, J. J., & Pereira, M. F. R., A one-pot method for the enhanced production of xylitol directly from hemicellulose (corncob xylan). *RSC Adv.*, 6, 95320–95327, 2016.

[223] Umai, D., Kayalvizhi, R., Kumar, V., & Jacob, S., Xylitol: Bioproduction and applications: A review. *Front. Sustain.*, 3, 2022.

[224] Janakiram, C., Deepan Kumar, C. V., & Joseph, J., Xylitol in preventing dental caries: A systematic review and meta-analyses. *J. Nat. Sci. Biol. Med.*, 8, 16–21, 2017.

[225] Liu, C.-J., Zhu, N.-N., Ma, J.-G., & Cheng, P., Toward green production of chewing gum and diet: Complete hydrogenation of xylose to xylitol over ruthenium composite catalysts under mild conditions. *Research*, 5178573, 2019.

[226] Ukawa-Sato, R., Hirano, N., & Fushimi, C., Design and techno—economic analysis of levulinic acid production process from biomass by using co-product formic acid as a catalyst with minimal waste generation. *Chem. Eng. Res. Des.*, 192, 389–401, 2023.

[227] Meramo Hurtado, S. I., Puello, P., & Cabarcas, A., Technical evaluation of a levulinic acid plant based on biomass transformation under techno-economic and exergy analyses. *ACS Omega*, 6, 5627–5641, 2021.

[228] Bangalore Ashok, R. P., Oinas, P., & Forssell, S., Techno-economic evaluation of a biorefinery to produce γ-valerolactone (GVL), 2-methyltetrahydrofuran (2-MTHF) and 5-hydroxymethylfurfural (5-HMF) from spruce. *Renew. Energ.*, 190, 396–407, 2022.

[229] Dubbink, G. H. C., Geverink, T. R. J., Haar, B., Koets, H. W., Kumar, A., van den Berg, H., van der Ham, A. G. J., & Lange, J. P., Furfural to FDCA: Systematic process design and techno-economic evaluation. *Biofuels Bioprod. Biorefin.*, 15, 1021–1030, 2021.

[230] Gómez Millán, G., Bangalore Ashok, R. P., & Oinas, P., Furfural production from xylose and birch hydrolysate liquor in a biphasic system and techno-economic analysis. *Biomass Conv. Bioref.*, 11, 2095–2106, 2021.

[231] Zang, G., Shah, A., & Wan, C., Techno-economic analysis of an integrated biorefinery strategy based on one-pot biomass fractionation and furfural production. *J. Clean. Prod.*, 260, 2020.

[232] Pandit, K., Jeffrey, C., Keogh, J., Tiwari, M. S., Artioli, N., & Manyar, H. G., Techno-economic assessment and sensitivity analysis of glycerol valorization to biofuel additives via esterification. *Ind. Eng. Chem. Res.*, 62, 9201–9210, 2023.

[233] Kim, H., Sang, B. I., Tsapekos, P., Angelidaki, I., & Alvarado-Morales, M., Techno-economic analysis of succinic acid production from sugar-rich wastewater. *Energies*, 16, 2023.

[234] Thanahiranya, P., Charoensuppanimit, P., Sadhukhan, J., Soottitantawat, A., Arpornwichanop, A., Thongchul, N., & Assabumrungrat, S., Succinic acid production from glycerol by *Actinobacillus succinogenes*: Techno-economic, environmental, and exergy analyses. *J. Clean. Prod.*, 404, 2023.

[235] Galán, G., Martín, M., & Grossmann, I. E., Integrated renewable production of sorbitol and xylitol from switchgrass. *Ind. Eng. Chem. Res.*, 60, 5558–5573, 2021.

[236] Morakile, T., Mandegari, M., Farzad, S., & Görgens, J. F., Comparative techno-economic assessment of sugarcane biorefineries producing glutamic acid, levulinic acid and xylitol from sugarcane. *Ind. Crops Prod.*, 184, 2022.

[237] Zaushitsyna, O., Dishisha, T., Hatti-Kaul, R., & Mattiasson, B., Crosslinked, cryostructured Lactobacillus reuteri monoliths for production of 3-hydroxypropionaldehyde, 3-hydroxypropionic acid and 1,3-propanediol from glycerol. *J. Biotechnol.*, 241, 22–32, 2017.

[238] Manandhar, A., & Shah, A., Techno-economic analysis of the production of lactic acid from lignocellulosic biomass. *Fermentation*, 9, 2023.

[239] Nieto, L., Rivera, C., & Gelves, G., Economic assessment of itaconic acid production from aspergillus terreus using superpro designer. *J. Phys. Conf. Ser.*, 1655, 2020.

[240] Baup, S., Ueber eine neue pyrogen-citronensäure, und über benennung der pyrogensäuren überhaupt. *Ann. Chim. Phys.*, 39–41, 1837.

[241] Teleky, B. E., & Vodnar, D. C., Biomass-derived production of itaconic acid as a building block in specialty polymers. *Polymers*, 11, 2019.

[242] Robert, T., & Friebel, S., Itaconic acid-a versatile building block for renewable polyesters with enhanced functionality. *Green Chem.*, 18, 2922–2934, 2016.

[243] Market Data Forecast, 2021. Itaconic Acid Market. www.marketdataforecast.com/market-reports/itaconic-acid-market, accessed on 26 May 2021.

[244] Gopaliya, D., Kumar, V., & Khare, S. K. Recent advances in itaconic acid production from microbial cell factories. *Biocatal. Agric. Biotechnol.*, 36, 2021.

[245] Liu, Y., Yuan, J., Ma, H., Zhu, C., Zhang, D., Ding, Y., Gao, C., & Wu, Y., A type of itaconic acid modified polyacrylate with good mechanical performance and biocompatibility. *React. Funct. Polym.*, 143, 2019.

[246] Kumar, S., Krishnan, S., Samal, S. K., Mohanty, S., & Nayak, S. K., Itaconic acid used as a versatile building block for the synthesis of renewable resource-based resins and polyesters for future prospective: A review. *Polym. Int.*, 66, 1349–1363, 2017.

[247] Sano, M., Tanaka, T., Ohara, H., & Aso, Y., Itaconic acid derivatives: Structure, function, biosynthesis, and perspectives. *Appl. Microbiol. Biotechnol.*, 104, 9041–9051, 2020.

[248] Klement, T., & Büchs, J., Itaconic acid—a biotechnological process in change. *Bioresour. Technol.*, 135, 422–431, 2013.

[249] Yang, J., Xu, H., Jiang, J., Zhang, N., Xie, J., Wei, M., & Zhao, J., Production of itaconic acid through microbiological fermentation of inexpensive materials. *J. Bioresour. Bioprod.*, 4, 135–142, 2019.

[250] Sivaramakrishnan, R., Suresh, S., & Incharoensakdi, A., Improvement of methyl ester and itaconic acid production utilizing biorefinery approach on *Scenedesmus sp. Renew. Energ.*, 215, 2023.

[251] Saha, B. C., Kennedy, G. J., Qureshi, N., & Bowman, M. J., Production of itaconic acid from pentose sugars by *Aspergillus terreus*. *Biotechnol. Prog.*, 33, 1059–1067, 2017.

[252] Saha, B. C., Kennedy, G. J., Bowman, M. J., Qureshi, N., & Dunn, R. O., Factors affecting production of itaconic acid from mixed sugars by *Aspergillus terreus*. *Appl. Biochem. Biotechnol.*, 187, 449–460, 2019.

6 Sustainable Electrode Materials/Composites for Batteries and Fuel Cells for Achieving Supercapacitance and Environmentally Benign Mobility

Naveen Kumar, Surinder Singh, Neena Mehta, Alex Ibhadon, and S.K. Mehta*

ABBREVIATIONS

ALD	Atomic layer deposition
CBD	Chemical bath deposition
GO	Graphene oxide
GOSC	Graphene oxide supercapacitor
H$_2$SO$_4$	Sulphuric acid
KOH	Potassium hydroxide
MMOs	Mixed metal oxides
MOFs	Metal organic frameworks
NaOH	Sodium hydroxide
NiO	Nickel oxide
PANI	Polyaniline
PVA	Polyvinyl alcohol
PVDF	Polyvinylidene fluoride
rGO	Reduced graphene oxide
TMOs	Transition metal oxides
TM(OH)s	Transition metal hydroxides
VGN	Vertical graphene nanosheets

* surinder.sk1961@gmail.com

DOI: 10.1201/9781003269779-6

6.1 INTRODUCTION

Energy consumption has evidently grown over the past decades, mainly sourced from fossil fuel combustion that produce byproducts in the form of harmful pollutants. Byproducts including CO_2, NO_x, and SO_x cause deteriorating effects on the environment [1]. Although there are many other forms of world energy development as shown in Figure 6.1, the dependency on the use of fossil fuels is highest among all available options. Other concerning factor is that these fossil fuel sources of energy reserves are depleting at an alarming rate (Figure 6.1). Efforts have been undertaken in recent years to mitigate the difficulties associated with the usage of traditional energy sources [2, 3].

Since the sources other than fossil fuels, such as hydropower, wind energy, geothermal energy, and solar energy generators have been in use, the lack of efficient, cost-effective setups and requirement of huge land area for their establishments have been some of the major limiting factors. Another area of research that has received substantial consideration for an alternative green energy source is the electrochemical energy production. Batteries, fuel cells, and electrochemical supercapacitors are the devices for the conversion and storage of electrochemical energy [4]. The energy generation in case of batteries as well as in fuel cells is monitored by redox processes that occur at the anode and cathode of the system. Energy storage and conversion differs between batteries and fuel cells; for batteries, both electrodes (anode and cathode) operate as charge transfer medium and participate as active masses in redox processes. In case of fuel cells, the electrodes act only as the medium of charge transfer and active masses are provided from the external environment/source. Hence, the storage and conversion locations are locally separated. Apart from these two electrochemical sources of energy, a source having advanced electrochemical characteristics, encompassing high energy density, rapid charge-discharge rates, specific

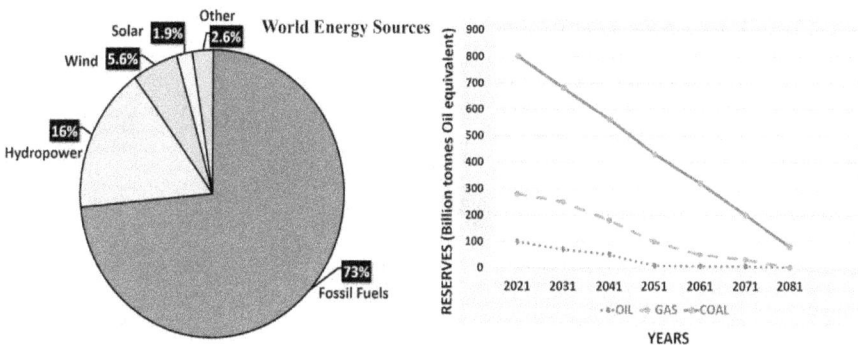

FIGURE 6.1 World energy generation from different sources and depleting rate of the sources.

capacitance, and exceptional cyclic stability is being researched in recent decades and that is where the inception of supercapacitors take place. After all these positive characteristics, the bottleneck for successful application of these supercapacitors has been their low energy density [5]. With the introduction of innovative electrode materials, a significant amount of research has been directed towards improving the electrochemical performance of supercapacitors.

6.1.1 ADVANCES IN ELECTRODE MATERIALS OF BATTERIES AND FUEL CELLS

Fuel cell and rechargeable batteries or secondary batteries are regarded as the oldest electrochemical energy systems which are used to store chemical energy [6, 7]. The most widespread application of batteries is their use as storage devices for electric power systems hence, batteries are built with different capacities ranging from less than 100 W to several megawatts [8, 9]. Different from batteries and fuel cells, supercapacitors generally show high power density (Figure 6.2) and may harvest energy in a relatively short amount of time before providing a burst of energy when needed [16]. Owing to the excellent electrochemical properties of supercapacitors (SCs) such as superior cycling life, rapid charging-discharging rates, and high-power density, they are also referred to as ultra or electrochemical capacitors (ECs) [10, 11]. The basic energy storage mechanism, i.e., charging-discharging at the electrode-electrolyte interference, in ECs is similar to that in traditional capacitors, however the charging and discharging process is much faster [12, 13].

The three different electrochemical capacitor groups, based on the different charge storage principles, are (a) EDLCs [14] (b) pseudocapacitors [15] and asymmetric supercapacitors (ASSC) or hybrid supercapacitors [16] (Figure 6.3).

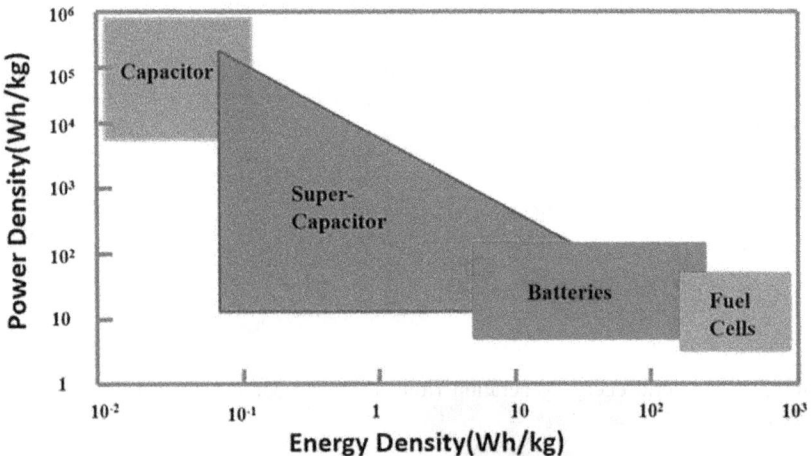

FIGURE 6.2   Energy density and power density range of the different sources.

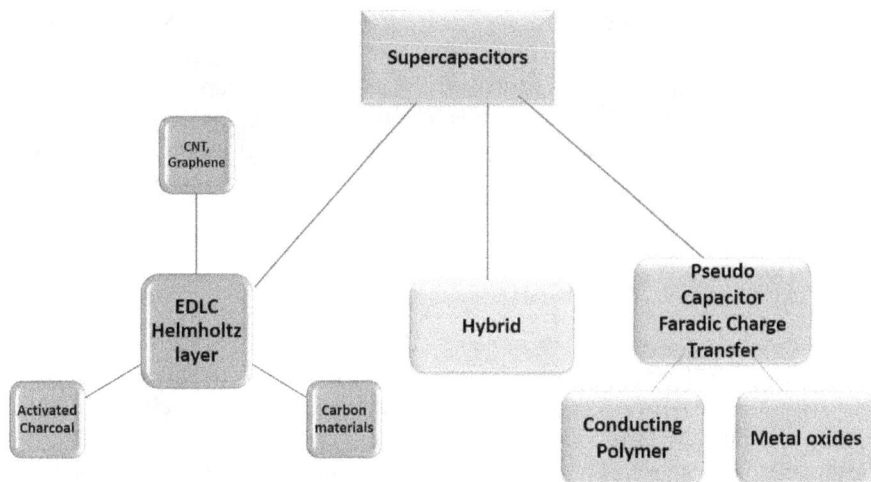

FIGURE 6.3 Different types of supercapacitor electrode materials.

6.1.2 ELECTRICAL DOUBLE LAYER CAPACITORS

The most popular form of electrochemical capacitor is the electrical double layer capacitor (EDLC), which stores energy via an electrostatic mechanism with a dielectric layer on the electrode electrolyte interphase. EDLCs employ liquid electrolytes that are mostly aprotic in nature, such as dimethyl carbonate (DME), propylene carbonate (PC), ethylene carbonate (EC), and diethyl carbonate (DEC), among others [17]. These supercapacitors store energy in Helmholtz double layers that develop at the phase interface between the electrodes' surface and the electrolyte. The creation of double-layer capacitance [18] is caused by the potential dependence of the surface energy at the interface of the capacitor electrodes. There is no electron exchange or redox reaction in EDLCs, and the energy is stored in a non-faradaic manner. With the enormous surface area of the electrodes and the Helmholtz layer thickness [19] these supercapacitors may achieve extraordinarily high capacitance. EDLC supercapacitors have a high cyclability and a long life span. Because of its high specific area, activated carbon (AC) is commonly utilized as an electrode material for EDLCs. At the moment, there is a lot of emphasis on developing novel electrode materials to improve the electrolytic performance of EDLCs [17].

6.1.3 PSEUDOCAPACITORS

Pseudocapacitors differ from EDLCs as they involve fast quasi-reversible faradaic processes at the electrodes during charging and discharging, primarily due to the occurrence of redox reactions, unlike EDLCs, which do not rely on redox reactions [1]. The electrochemical characteristics of pseudocapacitive materials lie in between purely capacitive behavior and bulk faradaic processes. The capacitance of pseudocapacitive-electrode material exceeds the capacitance of electrodes material in the EDLC supercapacitors as in case of the pseudocapacitors, they have large specific surface area and their capacitance is resultant of both the EDL capacitance and faradaic reaction pseudocapacitance [20–22]. Materials utilized for pseudocapacitive electrodes include polymers which are electrically conducting such as polythiophene, polyaniline, polypyrrole, and others, as well as metal oxides with varying degrees of oxidation such as RuOx, MnOx, etc. [23].

6.2   ELECTRODE MATERIAL

To achieve the supercapacitance in battery and fuel cells, electrode materials play a vital role. The electrode material/composite employed must provide high specific surface area, extended cyclic stability, good thermal stability, excellent electrical conductivity, corrosion resistance and appropriate chemical stability [24–29]. For sustainable and environmentally benign mobility, the electrode material must be cost-effective as well as efficient and ecologically friendly. Beside the specific surface area, the capacity to transmit the faradic charge is influenced by morphology, pore size, pore shape, and their availability for the electrolyte [30, 31].

Thus, the main requirements for the super-capacitive electrode material for battery and fuel cells are (1) improving the electrochemical active sites and (2) modification of the pore size and shape [25, 32]. The electrode materials/composites are categorized into two categories: (1) based on chemical composition and (2) based on the dimensional network of the material. Based on chemical composition, electrode materials can further be classified in four categories: carbonaceous material, transition metal oxides (TMOs), metal carbon hybrid materials, and conducting polymers (CPs) [14, 15, 33–35]. Based on dimensional network, electrode material are divided in the four categories: 0D, 1D, 2D, and 3D materials [36] (Figure 6.4).

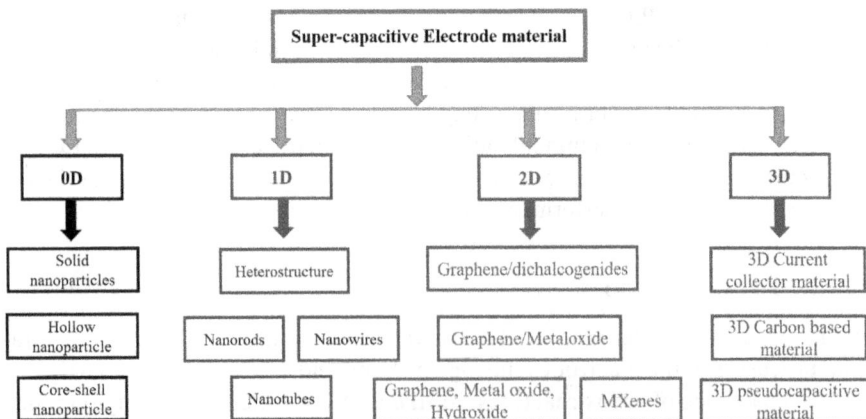

FIGURE 6.4   Commonly used supercapacitive electrode material.

6.3 CARBON-BASED MATERIAL

Carbon electrode materials are one of the greatest choices for sustainable and environmentally friendly benign mobility due to their low cost and environmental friendliness. The first-ever carbon-based EDLC was developed by Becker in 1975 [36, 37]. Carbon-based electrode materials can perform evenly in every chemical state (neutral, acidic, or basic) due to their strong thermal and chemical stability. Carbon-based materials are ideal for supercapacitive electrode materials due to the rectangular form of the cyclic voltammetry (CV) curve and symmetrical galvanostatic charge-discharge (GCD) profile [38]. The charge storage capacity of carbon-based materials primarily depends on the specific surface area and morphology (pore size and shape). This leads to the formation of thin layer, called the Helmholtz layer, in between electrolyte material, which serves as mechanism for storing charges in the carbon-based electrode materials [39–41]. The introduction of redox-active moieties, chemical doping, and combination of the aromatic isomers can increase the capacitance of the carbon-based electrode materials.

Other advantages of carbon-derived electrode materials are their abundance, facile synthesis, and ease of modification. Out of the several example of carbon generated electrode materials, activated carbon fiber, activated carbon derived from biomass, graphene, carbon nanotubes, carbon cloth and hybrid/composite material are some the types which are under focus in the recent times [33, 42–44]. These examples are discussed in detail in the following sections.

6.3.1 BIOMASS-DERIVED MATERIALS

The use of biomass for the electrode material is eco-friendly and therefore can play a crucial role in achieving sustainable and environmental benign mobility. Based on this key factor it has gained much interest in the field of electrochemical energy conversion and storage [45–47]. The charge storage mechanism involved is EDLC, in case of activated carbon materials derived from biomass. To employ the materials having EDLC mechanism as supercapacitive electrode material they must have large specific surface area, low resistance and high capacitance, and long lifetime stability [26, 40]. These characteristics in biomass-derived electrode materials can be attained by various carbonization, chemical activation and oxidation techniques [48] (Figure 6.5).

6.3.1.1 Precursor

The biomass-based carbon material can be synthesized by any biological precursors which have high carbon content. Generally the precursors derived from plant, animal, micro-organisms, and fruits are used extensively for the synthesis. The biopolymers like cellulose, lignin, and hemicellulose are derived from the lignocellulose plant biomass [49, 50]. The most prevalent plant biopolymer cellulose is composed of β-glucose molecules and the second most available plant biopolymer lignin is aromatic and mainly composed of alcohols, p-courmyl, conifryl, and sinapyl. Hemicellulose is composed of different short chain sugar molecules. Out of these three plant biopolymers, lignin has the highest thermal stability followed by the cellulose and hemicelluloses [51–53].

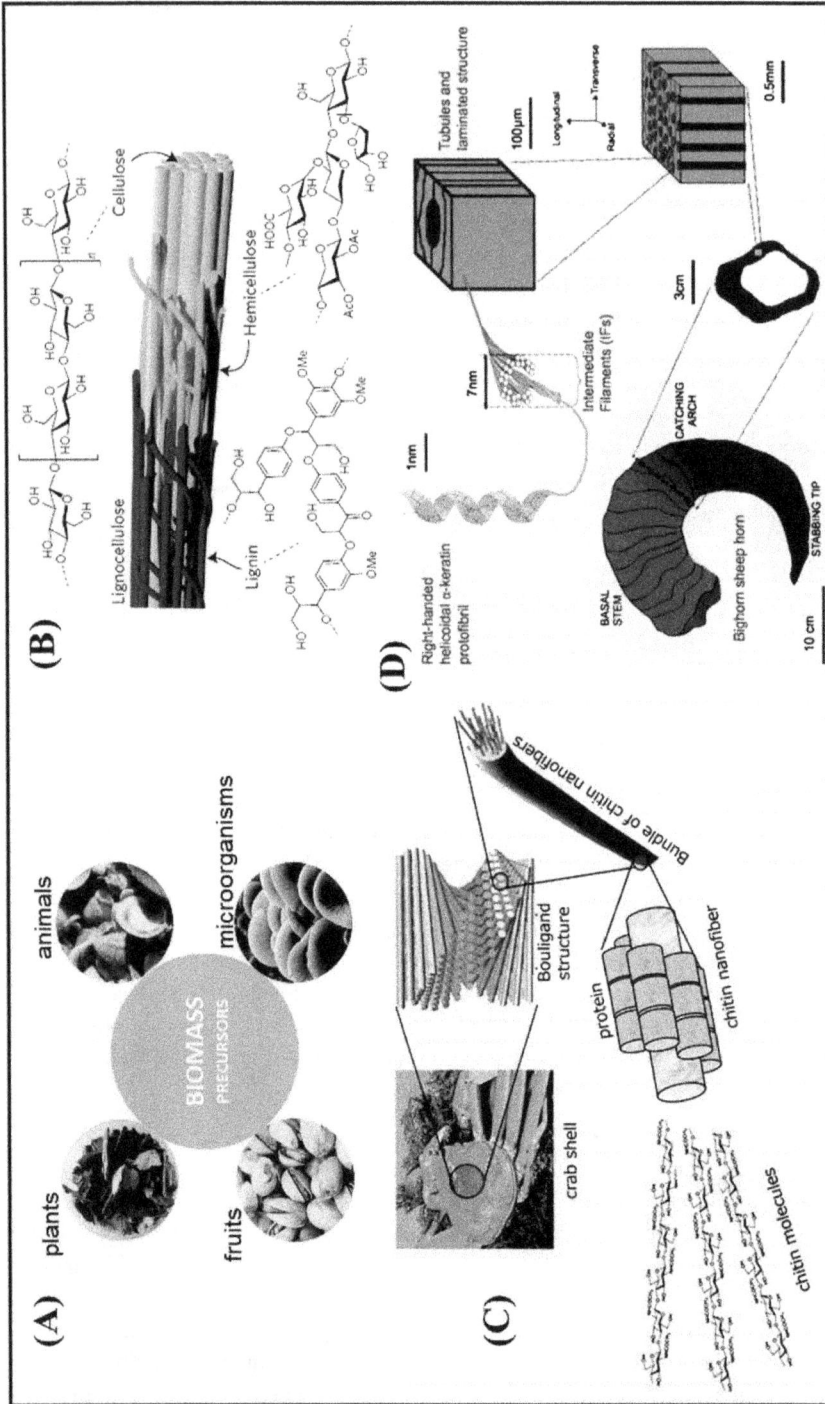

FIGURE 6.5 (A) Sources of plant material precursor. Examples of (B) plant biomass (lignin, cellulose and hemicellulose), (C) crab shell biomass (chitin nanofiber), and (D) sheep horn biomass structure (α-keratin).

Source: Copyright 2016. Reproduced with permission from Royal Society of Chemistry.

Fruit- and animal-based biomass contains different types of biomolecules/polymers depending upon the sources from which they have been obtained. Fruit- and animal-based biomass produces low yield of carbon material due the presence of high amount protein and lipids.

Despite the low yield, fruit-based biomass is still being used extensively for the production of carbon materials due the presence heteroatoms (nitrogen, sulphur, and phosphorus) in their biomass [48]. Keratin being the exception, produces high-quality carbon material with good yield due to the existence of resilient covalent bonding in its molecular structure [51].

6.3.1.2 Synthesis Strategies

Sustainable, eco-friendly, and cost-effective carbon-based electrode material can be manufactured by employing the various carbonization processes such as hydrothermal, direct conversion, pyrolysis, etc. [54]. All these strategies have significant relevance with morphology, surface area, mesoporous pore size, and shape of obtained carbon material. Raw biomass generally contains 30%–60% of carbon, 20%–30% of oxygen, around 5% of hydrogen, and about 1% of nitrogen, sulphur, phosphorous, etc. Carbonization will help to obtain the carbon material with a high content of oxygen or heteroatom doping and it can be carried out in two ways [51] (Figure 6.6): hydrothermal carbonization (HTC) and pyrolysis.

In HTC, carbon material is obtained by self-generated hydrothermal vessel which contains biomass and aqueous media. The temperature varies from 120–250°C in case of HTC. Pyrolysis is a high-temperature carbonization and usually it varies from 500°C to 1000°C. Pyrolysis is performed in an inert environment and the carbon material obtained by pyrolysis generally has high specific surface area [48, 55]. The carbon material obtained by the above carbonization methods generally have low surface area, low porosity, and low electrical conductivity, and therefore have room for improvement. This scope of betterment can be achieved by activating the carbon materials. Activation of carbonaceous material can be carried out by two methods: (1) chemical activation and (2) physical/thermal activation (Figure 6.6).

For the chemical activation, temperature varies from 450°C to 900°C; in this process carbon material is heated in the presence of various reagents such as KOH, $ZnCl_2$, H_3PO_4, and H_3BO_3; rarely, K_2CO_3, NaOH, H_2SO_4, and their mixtures are also used. The inert atmosphere is the key requirement in case of physical/thermal activation, and the temperature for this varies from 800°C to 1200°C. The high temperature treatment eliminates nanocarbon material which leads to creation of porosity in carbon materials [54].

Chemical activation technique is superior to physical/thermal activation method in such a way that chemical activation method usually takes place in a single step and consumes less energy as compared to the physical/thermal activation method. The chemical activation process takes place with an uncertain and complex mechanism. In case of KOH as activating agent (Figure 6.7), the proposed mechanism is as follows; dehydration of KOH takes place at 450°C (Eq. 1), then carbon material reacts with H_2O and release of H_2 takes place (Eq. 2). In the temperature that lies between 450°C and 650°C, the evolution of CO and CO_2 takes place (Eq. 3). K_2O produced in

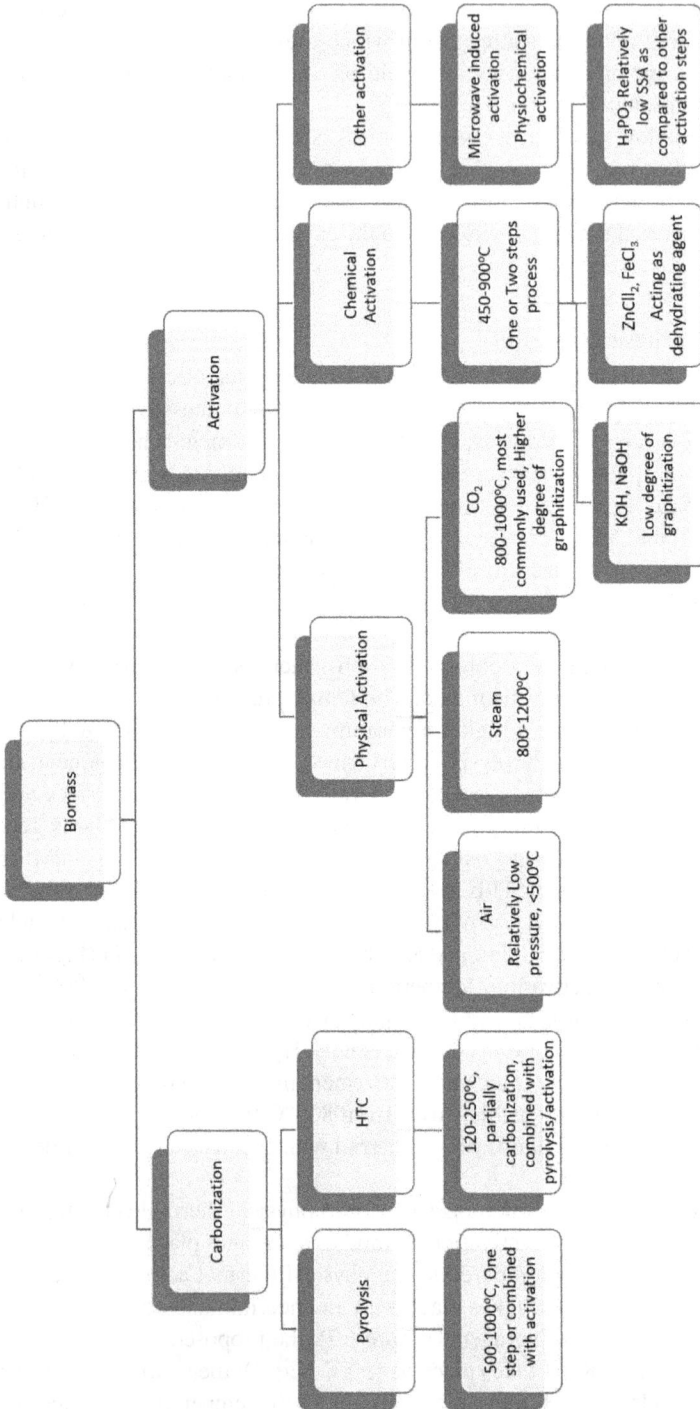

FIGURE 6.6 Schematic diagram of common ways for converting and activating biomass (raw plant material) to carbon materials.

$2KOH$	$\longrightarrow$ $K_2O + H_2O$	(1)
$C + H_2O$	$\longrightarrow$ $CO + H_2$	(2)
$CO + H_2O$	$\longrightarrow$ $CO_2 + H_2$	(3)
$CO_2 + K_2O$	$\longrightarrow$ K_2CO_3	(4)
K_2CO_3	$\longrightarrow$ $K_2O + CO_2$	(5)
$CO2_2 + C$	$\longrightarrow$ $2CO$	(6)
$K_2CO_3 + 2C$	$\longrightarrow$ $2K + 3CO$	(7)
$C + K_2O$	$\longrightarrow$ $2K + CO$	(8)

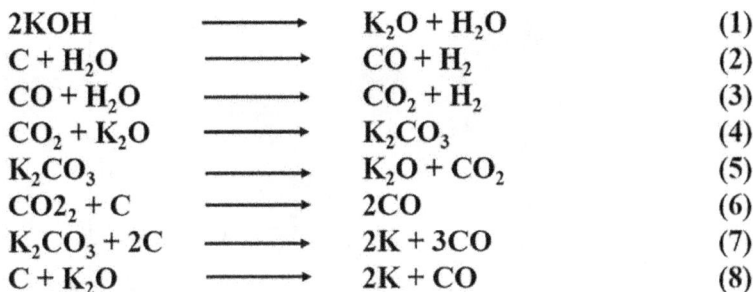

FIGURE 6.7 Mechanism for the chemical activation, KOH as activating agent.

Eq. 1 reacts with the CO_2 (Eq. 4); the decomposition or reduction of produced K_2CO_3 by carbon (Eqs. 5–8) takes place above 700°C [38, 56].

Various synthesized and activated carbon-based supercapacitive electrode materials with their electrochemical properties (current density, specific capacitance) are summarized in Table 6.1.

6.3.2 GRAPHENE-BASED ELECTRODE MATERIAL

Ever since its discovery by Geim and Novoselov [64], graphene has dominated various fields of science and technology. This dominance is associated with the unprecedented and astonishing properties of the graphene. Graphene is a 2D layered structure of one atom thick sp^2-linked carbon atoms. The single layer of graphene has a honeycomb-like, crystal structure. Excellent electrical conductivity, excellent carrier mobility, and high chemical and thermal stability make graphene the best candidate for the supercapacitive electrode material [64–66]. Unlike the carbon-based electrode material, graphene does not have much dependency over its solid state morphology. Graphene offers high surface area of 2400 m^2/g with excellent specific conductance of 5880 S/m, which in turn produces high specific conductance of 120 F/g, excellent energy density of 26 Wh/Kg, and about 10 W/g power density. Graphene oxide, reduced graphene oxide, heteroatom (nitrogen, boron, sulphur, and phosphorus etc.) doped graphene, functionalized graphene are some of the commonly known forms of the grapheme [28, 65]. Along with these forms, some of the newly discovered forms of graphene such as graphene quantum dots, graphene fibers, graphene nano-ribbons, pillared graphene, zig-zag graphene and 3D printed graphene have also been extensively investigated for the supercapacitive electrode materials [28, 65, 67, 68].

6.3.2.1 Synthesis Strategies

There are several methods to synthesis the above mentioned forms of graphene. The most often employed methods are microwave plasma chemical vapor deposition, chemical reduction, electrophoretic deposition, micromechanical exfoliation, arch discharge, unzipping of carbon nanotubes, laser pyrolysis technique, epitaxial growth, and intercalation technique in graphite [65, 66]. All these methods have their own unique advantages over the other, but the chemical reduction method is the most successful, profitable, economical, sustainable, and eco-friendly for the industrial scale production

TABLE 6.1

Carbon-Based Supercapacitive Electrode Material Synthesized from Various Sources

Source	SSA (m^2/g)	Activating Agent	Scan Rate/Current Density	C_{sp} (F/g)	Electrolyte	Electrode System (2E/3E)	Ref.
Bamboo	2352	KOH	1 A/g	268	6 M KOH	3E	[57]
Banana leaves	1097	$ZnCl_2$	0.5 A/g	74	1 M Na_2SO_4	3E	[58]
Fraxinus Chinensis fallen leaves	1078	KOH	1 A/g	268	6 M KOH	3E	[59]
Human hair	1306	KOH	1 A/g	340	6 M KOH	3E	[60]
Rice Husk	2996	NaOH	100 mA/g	147	6 M KOH	2E	[61]
Coconut shell	1874	$ZnCl_2$	5 mV/s 1 A/g	268	6 M KOH	3E	[62]
Membrane (egg)	1575	KOH	200 mA/g	203	1 M KOH	3E	[63]
Batata leaves	3114	KOH	1 A/g	532	1 M H_2SO4	3E	[40]
Potato	2342	KOH	50 mA/g	335	6 M KOH	2E	[41]
Silkworm	2523	KOH	1 A/g	304	6 M KOH	3E	[19]
Cotton stalk	1481	H_3PO_4	500 mA/g	114	1 M Et_4NBF_4	2E	[42]

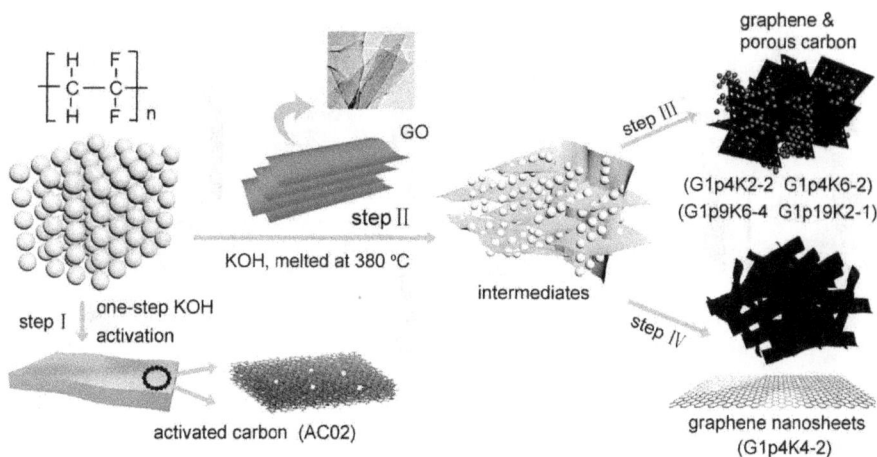

FIGURE 6.8 A schematic diagram for the synthesis of graphene nanosheets and activation of carbon. The material is represented as $Ga_1pa_2Ka_{3-4}$, where G = GO; p = PVDF; K = KOH; a_1 and a_2 are the ratios of PVDF to graphene oxide; a_3 is the ratio of base (KOH):(PVDF + GO); and a_4 is the reaction duration.

Source: Copyright 2013. Reproduced with permission from Royal Society of Chemistry.

the graphene and its various forms towards the electrochemical applications [36, 68]. Figure 6.8 illustrates the facile synthesis of porous nanocarbon and graphene composite [69]. Briefly, the mixture of polyvinylidene fluoride (PVDF) and graphene oxide powder was hand-grinded and was placed in a nickel crucible along with the KOH. Graphene oxide was reduced by elevating the temperature up to 380°C. The different reaction times, temperatures, and reagent ratios produced the different morphologies. The resultant electrode material shows the specific capacitance of 185 F/g.

In another study, Caixia Zhou et al. [70] reported the synthesis of asymmetric flexible micro-supercapacitor material based on graphene fiber. The electrode material was produced by the hydrothermal technique. In a glass capillary homogenous combination of graphene oxide and 1, 4-naphtoqunone was placed at 180°C for 10 h to obtain the graphene fiber.

The obtained electrode material shows the specific capacitance of 223 F/g with current density of 210 mA/g and sustains the 88% of original capacitance even after 1000 cycles. When reduced graphene oxide and graphene are combined, they can provide some good electrochemical characteristics.

Interestingly, Chikako Ogata et al. [71] reported the design of a device with tunable battery and supercapacitor behaviors based only on graphene oxide. The rGO/GO/rGO device was fabricated by the single step photo-irradiation of GO under specific conditions, which results in the development of rGO/GO/rGO; the two sides of rGO act as electrode and exhibit the electrical conductivity and GO which separates the rGO act as solid state electrode. The device made up of rGO/GO/rGO is represented as a graphene oxide supercapacitor (GOSC-t), where t is the irradiation time. Figure 6.9 shows the fabrication of the GOSC device which

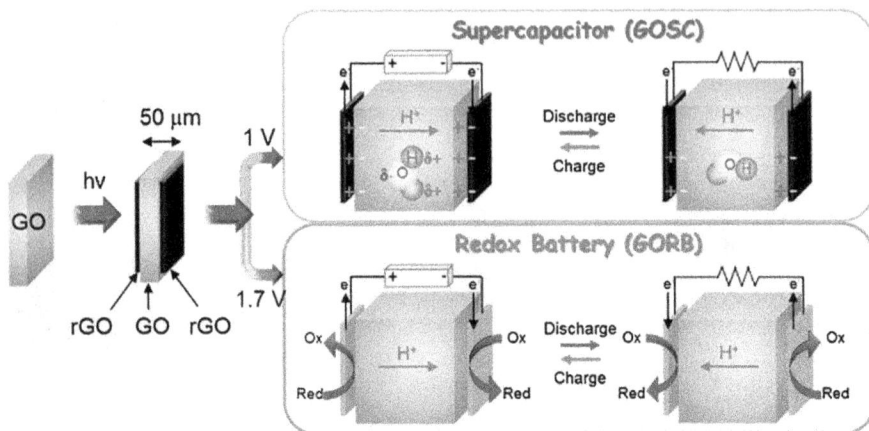

FIGURE 6.9 Schematic illustration of the rGO/GO/rGO battery and supercapacitor.

Source: Copyright 2013. Reproduced with permission from Royal Society of Chemistry.

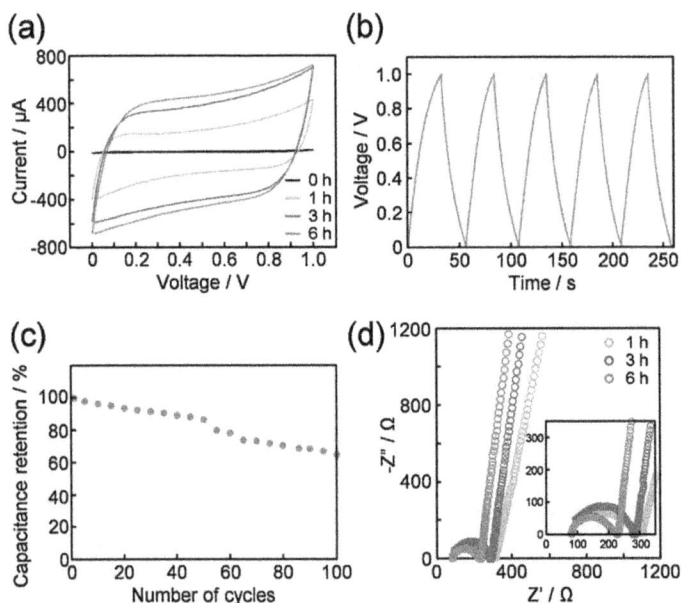

FIGURE 6.10 (a) Cyclic voltammetric profiles of GOSC produced at various temperatures at a fix scan rate of 0.3 V s^{-1}. (b) CC of GOSC-6 h at 0.1 mA. (c) Stability study of GOSC-6 h at 0.1 mA. (d) Impedance spectra of GOSC-1,3 and 6.

Source: Copyright 1996. Reproduced with permission from Royal Society of Chemistry.

works as supercapacitor (above 1.5 V) as well as a battery (below 1.2 V) depending upon the voltage window. The fabricated GOSC-6 shows the energy density of 1.1×10^{-4} Wh/cm^3 and power density of 0.12 W/cm^3. Figure shows the rectangular cyclic voltammetry curves, good stability up to 80% after 100 cycles (Figure 6.10).

Chemical vapor deposition assisted with enhanced microwave plasma is one the important technique to grow the graphene materials with increased wettability. Subrata Ghosh et al.[18, 72] has synthesized vertical graphene nano-sheets (VGN) using the same technique. Ultra-pure CH_4 was selected as the source of carbon and pure argon was used as carrier gas.

The reaction was carried out at 800°C temperature, 1.2×10^{-3} mbar pressure, and 300 W plasma power. Synthesized VGN was activated using KOH solution. The VGN electrode material exhibits superior capacitance of 3.32 mF/cm^2 in H_2SO_4.

6.4   TRANSITION METAL–DERIVED ELECTRODE MATERIALS

Redox active substances are the best-known electrode materials for energy conversion and storage devices [18, 72–74]. Transition metal–based electrode materials are the most often employed for charge storage devices. Transition metal–based materials can provide very specific conductance and capacitance along with very long cyclic stability. Noble metals such as Pt, Ir, and Ru have outstanding electrochemical characteristics among all transition metals [75, 76]. The scarcity and high price of these metals does not uphold the cost-effective property for sustainable development which ultimately limits their large-scale use. For sustainable and environmentally benign mobility, either we have to replace these precious metals with non-precious metals or reduce the overall content of these metals in the electrode material. To tackle this issue, oxides, hydroxide, ferrites, carbides, and nitrides of various affordable transition metals like nickel (Ni), iron (Fe), cobalt (Co), titanium (Ti), molybdenum (Mo), manganese (Mn), and vanadium (V) have been investigated extensively [24, 25, 43, 77]. Furthermore, mixed metals oxides have also gained interest as supercapacitor electrode material in recent times. Several examples of mixed metal oxides (such as $NiFe_2O_4/Fe_2O_3$, $ZnO/ZnFe_2O_4$, CuO/Cu_2O, $NiO@ZnO$, $Fe_2O_3@TiO_2$, Fe_2O_3/Co_3O_4, $Fe_2O_3/NiCo_2O_4$, $Co_3O_4/ZnFe_2O_4$, $Co_3O_4/NiCo_3O_4$, and $Cr_2O_3@TiO_2$) [30, 31, 78–85] have been derived from different precursors. These electrode materials provide higher conductivity and lower resistance as compared to carbon-based materials [86].

6.4.1   SYNTHETIC STRATEGIES

The low-cost supercapacitive electrode materials based on transition metals can be synthesized from various precursors such as metal nitrate/hydroxide/chloride/acetate [87], metal organic frameworks (MOFs), co-ordination complexes/polymers etc. Generally, synthesis can be carried out by different methods like sonication, chemical reduction, hydrothermal, solvothermal, annealing, calcination, and several other methods. Each method and precursor have their unique advantages, e.g. when we chose MOFs as precursor [32, 88, 89], we usually end up getting the material having very high specific surface area (Figure 6.11) as compared to other precursors. transition metal–based electrode materials generally shows the battery-type behavior for the charge storage mechanism. Some examples of transition metals, including oxide/hydroxide, mixed metal oxide, and metal carbide/nitride–based materials, are discussed briefly in the following section.

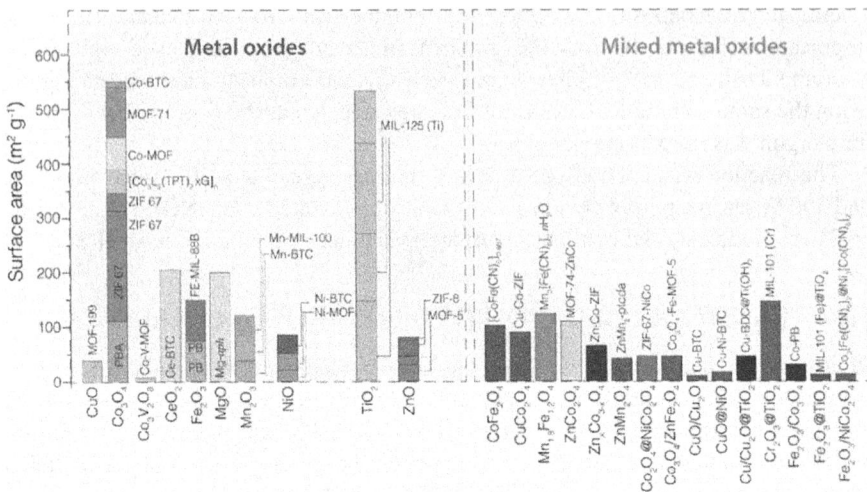

FIGURE 6.11   The surface of diverse metal and mixed metal oxides generated from distinct MOF precursors is shown schematically.

6.4.2 TRANSITION METAL OXIDE/HYDROXIDE–BASED ELECTRODE MATERIALS

The extraordinary properties of TMOs and TM(OH)s, such as high thermal and chemical stability, superior metallic conductivity, shorter diffusion path, high theoretical capacitance, and large potential window, make them an excellent choice for the supercapacitor electrode materials [39, 90]. Because of the low cost and large abundance, till date a vast number of reports have been reported on the TMOs and TM(OH)s–based electrode materials. Among various supercapacitive electrode materials based on MnO_2 [91–93], Abhilash Pullanchiyodan et al. [94] fabricated an interesting stretchable supercapacitor based on MnO_x synthesized by the electrodeposition technique. The electrode material showed the functional specific conductance of 108 mF/g, capacitance of 580 mF/cm^2, and energy density of 0.051 mWh/cm^2.

Electrode materials which possess the pseudocapacitive charge storage mechanism have also been largely investigated. Among several promising pseudocapacitive materials, Ni oxide and Ni hydroxide–based materials outperform many other available electrode materials [95–97].

Reziwanguli A. et al. [44] reported the synthesis of Ni oxide nanofibers by electro-spun strategy. Briefly, a homogenous solution of Ni nitrate and polyvinyl alcohol (PVA) was placed in the NANON-01 electrospinning setup. The optimization of Ni nitrate and PVA concentrations were carried out separately to achieve the best results. Finally, 0.1 mol/L and 0.2 mol/L concentrations of Ni nitrate and PVA were used in the synthesis. The as synthesized product was calcined at 650°C to obtain the NiO nanofibers. In a report by Xiaona Wang et al. [98] an excellent supercapacitive material was designed based on Ni hydroxide. The fabrication was carried out in multiple steps (Figure 6.12), in which hydrothermal annealing and chemical vapor deposition processes were involved. The electrode material showed the areal capacitance of 291.9 mF/cm^2 at an operational voltage of 1.6 V with the energy density of 37.7 mW/cm^3.

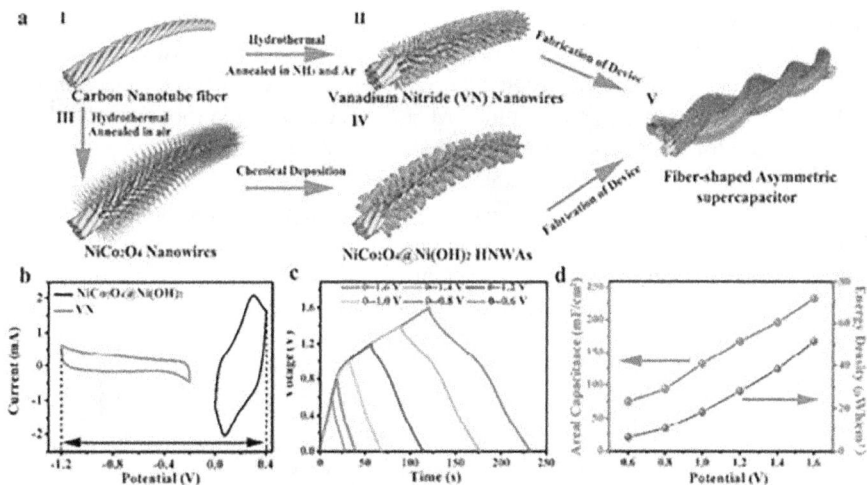

FIGURE 6.12 (a) Asymmetric supercapacitor device, stepwise fabrication. (b) Cyclic voltammetric curves at a scan rate of 5 mV/s of VN/CNTF and NiCo$_2$O$_4$@Ni(OH)$_2$/CNTF in a 3E system and 1 M KOH electrolyte. (c) GCD profiles of the fiber-shaped asymmetric supercapacitor device at 0.4 to 1.6 V and 3 mA/cm^2 current density. (d) Calculations of specific capacitance and energy density.

Source: Copyright 2019. Reproduced with permission from American Chemical Society.

The material possess the extraordinary cyclic stability of 90% even after 3000 cycles which ultimately enhanced the practical use of the fabricated electrode material. The most commonly used metal oxide materials for the supercapacitive electrode material are mixed metal oxides (MMOs) [99]. As the different oxidation states of mixed metal oxides provides better redox activity. Therefore, various efficient MMOs and their corresponding composite electrode materials have been found as promising supercapacitive materials. The various MMOs can be synthesized by using different chemical techniques including hydrothermal synthesis, sol-gel processing, solvothermal synthesis, electrochemical/electropolymerization synthesis, atomic-layer deposition (ALD), microwave-assisted synthesis, and chemical bath deposition (CBD) [20, 21, 100, 101]. Presently, the principle interest of researchers among MMOs is metal ferrites (Mfe$_2$O$_4$, where M = Cu, Co, Zn, Ni, Mn, etc.) because of their synergetic activity, low price, widespread availability and low hazards [102–105]. All these metal ferrites are normal spinals in which Fe takes the octahedral position and metal (Cu, Co, Zn and Mn) occupies the tetrahedral sites, except Ni ferrite which is an inverse spinal [106, 107]. The composites of TMOs, TM(OH)s, and MMOs with different carbon supports like GO, rGO, carbon nanotubes (CNTs), carbon cloth, conductive polymers, etc. possess the features of both carbon-based electrode materials which are highly stable and of metals which are highly conductive [108, 109].

Such composite materials provides the outstanding specific capacitance and a long cyclic stability. These hybrid materials have much wider range of operation as compare to the EDLC (generally carbon-based electrode materials) and pseudocapacitors (metal-based electrode materials) [21, 22]. Sherif A. El-Khodary and co-workers [110] reported the fabrication of supercapacitive electrode material based of Fe and polyaniline (PANI). The composite was obtained by one-pot synthesis followed by freeze-drying procedure and the composite was designated as Fe-PANI@S-MWCNTs. The composite electrode material shows excellent specific capacitance of 2105.64 F/g at 1 A/g in acidic medium (1 M H_2SO_4). Figure 6.13a shows the synthesis of the ASC with FeCo2O4@PPy as cathode and AC as anode. In another report, Xinyi He et al. [111] synthesized the flexible solid state asymmetric supercapacitive material based on the Fe, Co, and polypyrrole supported on the carbon cloth. The synthesis was carried out by chemical liquid process and obtained material was designated as $FeCo_2O_4$@polypyrrole. The material have the core shell nanowire like morphology and act as anode for the asymmetric supercapacitor having 2269 Fg^{-1}

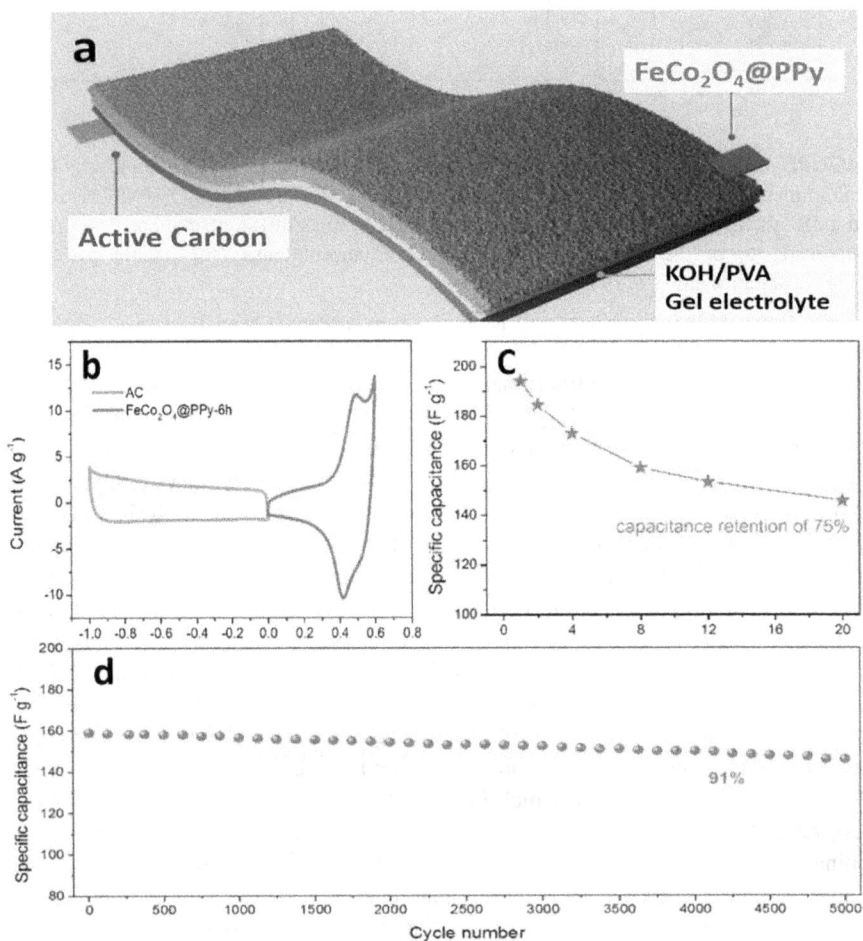

FIGURE 6.13　(a) Assembling of asymmetric supercapacitor based on $FeCo_2O_4$@PPy and AC; (b) cyclic voltammetry curves of $FeCo_2O_4$@PPy and AC; (c) specific capacitance retention up to 20 hl and (d) stability by cyclic efficiency.

Source: Copyright 2018. Reproduced with permission from American Chemical Society.

TABLE 6.2

Transition Metal–Based Supercapacitive Electrode Material with Their Electrochemical Performance

Composite	Energy Density (W/m²)	Power Density (W/m²)	C_{sp}(F/g)	Electrolyte	Ref.
MnO_2/CNTs	177 W/m²	250 W/m²	1980 at 1 A/g	1 M Na_2SO_4	[112]
SnMO-O@P-HPC-700	114 Wh/Kg	515 W/Kg	1272.8 at 1 A/g	6 M KOH	[113]
Graphene/Mn_3O_4	30.1 Wh/Kg	475 W/Kg	208.3	1 M Na_2SO_4	[114]
Ni–Co–O@CF	16.3 Wh/Kg	600 W/Kg	863.3 at 3 Ag^{-1}	3 M KOH	[115]
Zn_2SnO_4 /SnO_2 /CNT	98 Wh/Kg	100 W/Kg	702 at 1 A/g	6 M KOH	[33]
$NiCo_2O_4$/GCNF	48.6 Wh/Kg	749 W/Kg	1416 at 1 A/g	6 M KOH	[116]
Graphene wrapped $CuCo_2O_4$	42.5 Wh/Kg	15 KW/Kg	1813 at 2 A/g	3 M KOH	[117]
Ni-Co-Fe LDH/Fe_2O_3/rGO	101 Wh/Kg	92 KW/Kg	3130 at 1 A/g	3 M KOH	[118]
WS_2@$NiCo_2O_4$/CC	45.6 Wh/Kg	992 W/Kg	2449 mF/cm² at 1 mA/cm²	3 M KOH	[34]
CNT-Fe_2O_3	37 Wh/Kg	10 KW/Kg	373 at 10 mV/s	1 M Na_2SO_4	[35]
$CuMoO_4$/$ZnMoO_4$	34.8Wh/Kg	472 W/Kg	186 at 1.5 A/g	1 M KOH	[16]

specific capacitance at 1 Ag^{-1}. Figure 6.13a depicts an asymmetric supercapacitor fabricated by FeCo2O4@PPy anode and activated carbon as cathode. The assembly showed the excellent energy density of 68.8 Wh Kg^{-1} (0.8 KW Kg^{-1}) having long-term stability (Figure 6.13d, 91% after 5000 cycles) (Figure 6.13). Various electrode materials based on TMOs, TM(OH)s, and composite materials have been listed in Table 6.2.

6.5 CONCLUSION

With the unprecedented dependency of humans on the conventional sources of energies like fossil fuels, the smooth functionality towards the growth of mankind cannot be achieved without the replacement of these exhaustible resources of energy. Efforts have been made and are being made to explore other options of renewable, sustainable and clean sources of energy as the demands for energy is unlimited. In this book, Chapter 1 discusses such clean and green energy sources in terms of supercapacitive electrochemical electrode materials. With various supercapacitive electrode materials such as carbonaceous material, transition metal oxides (TMOs), metal carbon hybrid materials, and composite materials are discussed in detail with their electrochemical performances. The electrode materials based on carbon are mostly worked with due to their abundance, facile synthesis, ease of modification and can be sourced from environment. Activated carbon fibers, activated carbon derived from biomass, graphene, carbon nanotubes, carbon cloth, and hybrid/composite material are in the main focus in the recent times. Low surface area, low porosity, and low electrical conductivity in the obtained materials therefore have room for improvement. This scope of betterment can be achieved by activating the carbon materials and have been discussed. Another branch of supercapacitive materials of transition metal oxide/hydroxide–based materials has also been covered which replaces the scarce and valuable metals like Pt, Ir, and Ru with affordable metals, and also these electrode materials exhibit superior conductivity and lower resistance when compared to carbon-based materials. Material composites of TMOs, TM(OH)s, and MMOs with different carbon supports like GO, rGO, carbon nanotubes (CNTs), carbon cloth, conductive polymers, etc. have been covered that possess the features of both carbon-based electrode material which are highly stable and that of metals which have high conductivity. These composite materials have a high specific capacitance as well as a long cycle stability. As a result, a wide range of such complex composites must be investigated in order to successfully capture green and clean energy in the long run for mankind's sustainable growth and development.

REFERENCES

[1] Chen, T.W., Anushya, G., Chen, R., et al., Recent advances in nanoscale based electro-catalysts for metal-air battery, fuel cell and water-splitting applications: An overview. *Materials*, 15, 458, 2022.

[2] Zhou, M., Chi, M., Luo, J., et al., An overview of electrode materials in microbial fuel cells. *J. Power Sources*, 196, 4427, 2011.

[3] Gielen, D., Boshell, F., Saygin, D., et al., The role of renewable energy in the global energy transformation. *Energy Strategy Rev.*, 24, 38, 2019.

[4] Winter, M., and Brodd, R., What are batteries, fuel cells, and supercapacitors? *Chem. Rev.*, 104, 4245, 2004.

[5] Ando, K., Matsuda, T., and Imamura, D., Degradation diagnosis of lithium-ion batteries with a $LiNiO.5CoO.2MnO.3O_2$ and $LiMn_2O_4$ blended cathode using dV/dQ curve analysis. *J. Power Sources*, 390, 278, 2018.

[6] Baker, J.N., and Collinson, A., Electrical energy storage at the turn of the millennium. *Power Eng. J.*, 13, 107, 1999.

[7] Cairns, E.J., *Energy Conversion & Storage Program. 1995 Annual Report* (No. LBL-38350). Lawrence Livermore National Lab. (LLNL), Livermore, CA (United States), 1996.

[8] Aneke, M., and Wang, M., Energy storage technologies and real life applications—A state of the art review. *Appl. Energy*, 179, 350, 2016.

[9] Divya, K.C., and Østergaard, J., Battery energy storage technology for power systems–An overview. *Electr. Power. Syst. Res.*, 79, 511, 2009.

[10] Hornikx, J.M.A., and Kelly-Holmes, H., Advertising as multilingual communication Basingstoke, *Palgrave Macmillan*, 45, 1, 2006.

[11] Winter, M., Barnett, B., and Xu, K., Before Li ion batteries. *Chem. Rev.*, 118, 11433, 2018.

[12] Kakaei, K., Esrafili, M.D., and Ehsani, A., Graphene-based electrochemical supercapacitors. *Interface Sci. Technol.*, 27, 339, 2019.

[13] Zhang, L.L., and Zhao, X.S., Carbon-based materials as supercapacitor electrodes. *Chem. Soc. Rev.*, 38, 2520, 2009.

[14] Lee, J.W., Ahn, T., Kim, J.H., et al., Nanosheets based mesoporous NiO microspherical structures via facile and template-free method for high performance supercapacitors. *Electrochim. Acta*, 56, 4849, 2011.

[15] Xing, Z., Chu, Q., Ren, X., et al., Ni_3S_2 coated ZnO array for high-performance super-capacitors. *J. Power Sources*, 245, 463, 2014.

[16] Appiagyei, A.B., Bonsu, J.O., and Han, J.I., Robust structural stability and performance-enhanced asymmetric supercapacitors based on $CuMoO_4/ZnMoO_4$ nanoflowers prepared via a simple and low-energy precipitation route. *J. Mater. Sci.: Mater. Electron.*, 32, 6668, 2021.

[17] Libich, J., Máca, J., Vondrák, J., et al., Supercapacitors: Properties and applications. *J. Energy Storage*, 17, 224, 2018.

[18] Ghosh, S., Sahoo, G., Polaki, S.R., Krishna, N.G., et al., Enhanced supercapacitance of activated vertical graphene nanosheets in hybrid electrolyte. *J. Appl. Phys.*, 122, 21, 2017.

[19] Gong, C., Wang, X., Ma, D., et al., Microporous carbon from a biological waste-stiff silkworm for capacitive energy storage. *Electrochim. Acta*, 220, 331, 2016.

[20] Moyseowicz, A., and Gryglewicz, G., Hydrothermal-assisted synthesis of a porous polyaniline/reduced graphene oxide composite as a high-performance electrode material for supercapacitors. *Compos. B Eng.*, 159, 4, 2019.

[21] Buldu-Akturk, M., Toufani, M., Tufani, A., et al., ZnO and reduced graphene oxide electrodes for all-in-one supercapacitor devices. *Nanoscale*, 14, 3269, 2022.

[22] Fu, M., Chen, W., Zhu, X., et al., Crab shell derived multi-hierarchical carbon materials as a typical recycling of waste for high performance supercapacitors. *Carbon*, 141, 748, 2019.

[23] Volfkovich, Y.M., Electrochemical supercapacitors (a review). *Russ. J. Electrochem.*, 57, 311, 2021.

[24] Zhong, Y., Xia, X., Shi, F., et al., Transition metal carbides and nitrides in energy storage and conversion. *Adv. Sci.*, 3, 1500286, 2016.

[25] Rakhi, R.B., Ahmed, B., Anjum, D., et al., Direct chemical synthesis of MnO_2 nanowhiskers on transition-metal carbide surfaces for supercapacitor applications. *ACS Appl. Mater. Interfaces*, 8, 18806, 2016.

[26] Lai, L., Yang, H., Wang, L., et al., Preparation of supercapacitor electrodes through selection of graphene surface functionalities. *ACS Nano.*, 6, 5941, 2012.

[27] Zhao, N., Wu, S., He, C., et al., Hierarchical porous carbon with graphitic structure synthesized by a water soluble template method. *Mater. Lett.*, 87, 77, 2012.

[28] Kumar, R., Joanni, E., Sahoo, S., et al., An overview of recent progress in nanostructured carbon-based supercapacitor electrodes: From zero to bi-dimensional materials. *Car.*, 193, 298, 2022.

[29] Zhao, Y., Liu, F., Zhu, K., et al., Three-dimensional printing of the copper sulfate hybrid composites for supercapacitor electrodes with ultra-high areal and volumetric capacitances. *Adv. Compos. Hybrid. Mater.*, 5, 1537, 2022.

[30] Huang, G., Zhang, L., Zhang, F., et al., Metal—organic framework derived $Fe_2O_3@$ $NiCo_2O_4$ porous nanocages as anode materials for Li-ion batteries. *Nanoscale*, 6, 5509, 2014.

[31] Hu, L., Huang, Y., Zhang, F., et al., CuO/Cu_2O composite hollow polyhedrons fabricated from metal—organic framework templates for lithium-ion battery anodes with a long cycling life. *Nanoscale*, 5, 4186, 2013.

[32] Du, M., He, D., Lou, Y., et al., Porous nanostructured $ZnCo_2O_4$ derived from MOF-74: High-performance anode materials for lithium ion batteries. *J. Energy Chem.*, 26, 673, 2017.

[33] Samuel, E., Kim, T.G., Park, C.W., et al., Supersonically sprayed $Zn_2SnO_4/SnO_2/CNT$ nanocomposites for high-performance supercapacitor electrodes. *ACS Sustain Chem. Eng.*, 7, 14031, 2019.

[34] Li, L., Gao, J., Cecen, V., et al., Hierarchical WS2@ $NiCo_2O_4$ core—shell heterostructure arrays supported on carbon cloth as high-performance electrodes for symmetric flexible supercapacitors. *ACS Omega.*, 5, 4657, 2020.

[35] Kumar, A., Sarkar, D., Mukherjee, S., et al., Realizing an asymmetric supercapacitor employing carbon nanotubes anchored to Mn_3O_4 cathode and Fe_3O_4 anode. *ACS Appl. Mater. Interfaces*, 10, 42484, 2018.

[36] Forouzandeh, P., Kumaravel, V., and Pillai, S.C., Electrode materials for supercapacitors: A review of recent advances. *Catalysts*, 10, 969, 2020.

[37] Becker, H.I., General Electric Co, Low voltage electrolytic capacitor. *U.S. Patent*, 2, 800, 616, 1957.

[38] Marichi, R.B., Sahu, V., Sharma, R.K., et al., Efficient, sustainable, and clean energy storage in supercapacitors using biomass-derived carbon materials. In *Handbook of Ecomaterials*, 1, 2017. Springer, Cham.

[39] Zhao, C., and Zheng, W., A review for aqueous electrochemical supercapacitors. *Front. Energy Res.*, 3, 23, 2015.

[40] Wei, X., Li, Y., and Gao, S., Biomass-derived interconnected carbon nanoring electrochemical capacitors with high performance in both strongly acidic and alkaline electrolytes. *J. Mater. Chem. A*, 5, 181, 2017.

[41] Zhao, S., Wang, C.Y., Chen, M.M., et al., Potato starch-based activated carbon spheres as electrode material for electrochemical capacitor. *J. Phys. Chem. Solids*, 70, 2009.

[42] Chen, M., Kang, X., Wumaier, T., et al., Preparation of activated carbon from cotton stalk and its application in supercapacitor. *J Solid State Electrochem.*, 17, 2013.

[43] Wan, K., Li, Y., Wang, Y., et al., Recent advance in the fabrication of 2D and 3D metal carbides-based nanomaterials for energy and environmental applications. *Nanomaterials*, 11, 1, 2021.

[44] Aihemaitituoheti, R., Alhebshi, N.A., and Abdullah, T., Effects of precursors and carbon nanotubes on electrochemical properties of electrospun nickel oxide nanofibers-based supercapacitors. *Molecules*, 26, 2021.

[45] Simon, P., and Gogotsi, Y., Materials for electrochemical capacitors. *Nat. Mater.*, 7, 845, 2008.

[46] Endo, M., Takeda, T., Kim, Y.J., et al., High power electric double layer capacitor (EDLC's); from operating principle to pore size control in advanced activated carbons. *Carbon Lett.*, 1, 117, 2001.

[47] Helmholtz, H., Ueber einige Gesetze. *Ann. Phys. Chem.*, 165, 211, 1853.

[48] Ruiz-Montoya, J.G., Quispe-Garrido, L.V., Gómez, J.C., Baena-Moncada, A.M., and Gonçalves, J.M., Recent progress in and prospects for supercapacitor materials based on metal oxide or hydroxide/biomass-derived carbon composites. *Sustain. Energy Fuels*, 5, 5332, 2021.

[49] Tombolato, L., Novitskaya, E.E., Chen, P.Y., et al., Microstructure, elastic properties and deformation mechanisms of horn keratin. *Acta Biomater.*, 6, 319, 2010.

[50] Ifuku, S., and Saimoto, H., Chitin nanofibers: preparations, modifications, and applications. *Nanoscale*, 4, 3308, 2012.

[51] Deng, J., Xiong, T., Wang, H., et al., Effects of cellulose, hemicellulose, and lignin on the structure and morphology of porous carbons. *ACS Sustain Chem. Eng.*, 4, 3750, 2016.

[52] Liu, M., Yang, J., Liu, Z., et al., Cleavage of covalent bonds in the pyrolysis of lignin, cellulose, and hemicellulose. *Energy Fuels*, 29, 5773, 2015.

[53] Rubin, E.M., Genomics of cellulosic biofuels. *Nature*, 454, 841, 2008.

[54] Lu, H., and Zhao, X.S., Biomass-derived carbon electrode materials for supercapacitors. *Sustain. Energy Fuels*, 1, 1265, 2017.

[55] Takaya, C.A., Fletcher, L.A., Singh, S., et al., Phosphate and ammonium sorption capacity of biochar and hydrochar from different wastes. *Chemosphere*, 145, 518, 2016.

[56] Wang, J., and Kaskel, S., KOH activation of carbon-based materials for energy storage. *J. Mater. Chem.*, 22, 23710, 2012.

[57] Li, J., and Wu, Q., Water bamboo-derived porous carbons as electrode materials for supercapacitors. *New J. Chem.*, 39, 3859, 2015.

[58] Subramanian, V., Luo, C., Stephan, A.M., et al., Supercapacitors from activated carbon derived from banana fibers. *J. Phys. Chem. C*, 111, 7527, 2007.

[59] Li, Y.T., Pi, Y.T., Lu, L.M., et al., Hierarchical porous active carbon from fallen leaves by synergy of K2CO3 and their supercapacitor performance. *J Power Source*, 299, 519, 2015.

[60] Qian, W., Sun, F., Xu, Y., et al., Human hair-derived carbon flakes for electrochemical supercapacitors. *Energy Environ. Sci.*, 7, 379, 2014.

[61] Teo, E.Y.L., Muniandy, L., Ng, E.P., et al., High surface area activated carbon from rice husk as a high performance supercapacitor electrode. *Electrochim. Acta*, 192, 110, 2016.

[62] Sun, L., Tian, C., Li, M., et al., From coconut shell to porous graphene-like nanosheets for high-power supercapacitors. *J. Mater. Chem. A*, 1, 6462, 2013.

[63] Li, Z., Zhang, L., Amirkhiz, S.B., et al., Carbonized chicken eggshell membranes with 3D architectures as high-performance electrode materials for supercapacitors. *Adv. Energy Mater.*, 2, 431, 2012.

[64] Geim, A.K., and Novoselov, K.S., The rise of graphene. *Nat. Mater.*, 6, 11, 2007.

[65] Bokhari, S.W., Siddique, A.H., et al. Advances in graphene-based supercapacitor electrodes. *Energy Rep.*, 6, 2768, 2020.

[66] Huang, X., Zeng, Z., Fan, Z., et al., Graphene-based electrodes. *Adv. Mater.*, 24, 5979, 2012.

[67] Arvas, M.B., Gencten, M., and Sahin, Y., One-step synthesized N-doped graphene-based electrode materials for supercapacitor applications. *Ionics*, 27, 2241, 2021.

[68] Ci, S., Cai, P., Wen, Z., et al., Graphene-based electrode materials for microbial fuel cells. *Sci. China Mater.*, 58, 496, 2015.

[69] Zhang, H., Wang, K., Zhang, X., et al., Self-generating graphene and porous nanocarbon composites for capacitive energy storage. *J. Mater. Chem. A*, 3, 11277, 2015.

[70] Zhou, C., Gao, T., Liu, Q., et al., Preparation of quinone modified graphene-based fiber electrodes and its application in flexible asymmetrical supercapacitor. *Electrochim. Acta*, 336, 135628, 2020.

[71] Ogata, C., Kurogi, R., Hatakeyama, K., et al., All-graphene oxide device with tunable supercapacitor and battery behaviour by the working voltage. *Chem. Commun.*, 52, 3919, 2016.

[72] Sahoo, S., Sahoo, G., Jeong, S.M., et al., A review on supercapacitors based on plasma enhanced chemical vapor deposited vertical graphene arrays. *J. Energy Storage*, 53, 105212, 2022.

[73] Ham, D.J., and Lee, J.S., Transition metal carbides and nitrides as electrode materials for low temperature fuel cells. *Energies*, 2, 873, 2009.

[74] Abdel Maksoud, M.I.A., Fahim, R.A., Shalan, A.E., et al., Advanced materials and technologies for supercapacitors used in energy conversion and storage: A review. *Environ. Chem. Lett.* 19, 2021.

[75] Serov, A., and Kwak, C., Review of non-platinum anode catalysts for DMFC and PEMFC application. *Appl. Catal. B: Environ.*, 90, 313, 2009.

[76] Gasteiger, H.A., Kocha, S.S., Sompalli, B., et al., Activity benchmarks and requirements for Pt, Pt-alloy, and non-Pt oxygen reduction catalysts for PEMFCs. *Appl. Catal. B: Environ.*, 56, 9, 2005.

[77] Nguyen, T. and Montemor, M.F., Redox active materials for metal compound based hybrid electrochemical energy storage: A perspective view. *Appl. Surf. Sci.*, 422, 492, 2017.

[78] Dekrafft, K.E., Wang, C., and Lin, W., Metal-organic framework templated synthesis of Fe 2 O_3/TiO_2 nanocomposite for hydrogen production. *Adv. Mater.*, 24, 2014, 2012.

[79] Hou, L., Lian, L., Zhang, L., et al., Self-sacrifice template fabrication of hierarchical mesoporous Bi-Component-Active $ZnO/ZnFe_2O_4$ sub-microcubes as superior anode towards high-performance lithium-ion battery. *Adv. Funct. Mater.*, 25, 238, 2015.

[80] Yang, X., Xue, H., Yang, Q., et al., Preparation of porous $ZnO/ZnFe_2O_4$ composite from metal organic frameworks and its applications for lithium ion batteries. *Chem. Eng. J.*, 308, 340, 2017.

[81] Cao, H., Zhu, S., Yang, C., et al., Metal-organic-framework-derived two-dimensional ultrathin mesoporous hetero-$ZnFe_2O_4/ZnO$ nanosheets with enhanced lithium storage properties for Li-ion batteries. *Nanotechnology*, 27, 465402, 2016.

[82] Li, G.C., Liu, P.F., Liu, R., et al., MOF-derived hierarchical double-shelled NiO/ZnO hollow spheres for high-performance supercapacitors. *Dalton. Trans.*, 45, 13311, 2016.

[83] Kaneti, Y.V., Zakaria, Q.M., Zhang, Z., et al., Solvothermal synthesis of ZnO-decorated α-Fe_2O_3 nanorods with highly enhanced gas-sensing performance toward n-butanol. *J. Mater. Chem. A*, 2, 13283, 2014.

[84] Zhou, X., Li, X., Sun, H., et al., Nanosheet-assembled $ZnFe_2O_4$ hollow microspheres for high-sensitive acetone sensor. *ACS Appl. Mater. Interfaces*, 7, 15414, 2015.

[85] Huang, G., Zhang, F., Zhang, L., et al., Hierarchical $NiFe_2O_4/Fe_2O_3$ nanotubes derived from metal organic frameworks for superior lithium ion battery anode. *J. Mater. Chem. A*, 2, 8048, 2014.

[86] Conway, B.E., and Liu, T.C., Characterization of electrocatalysis in the oxygen evolution reaction at platinum by evaluation of behavior of surface intermediate states at the oxide film. *Langmuir*, 6, 268, 1990.

[87] Zhang, L., Wu, H.B., Madhavi, S., et al., Formation of Fe_2O_3 microboxes with hierarchical shell structures from metal—organic frameworks and their lithium storage properties. *J. Am. Chem. Soc.*, 134, 17388, 2012.

[88] Kong, S., Dai, R., Li, H., et al., Microwave hydrothermal synthesis of Ni-based metal—organic frameworks and their derived yolk—shell NiO for Li-ion storage and supported ammonia borane for hydrogen desorption. *ACS Sustain. Chem. Eng.*, 3, 1830, 2015.

[89] Hu, H., Guan, B., Xia, B., et al., Designed formation of $Co_3O_4/NiCo_2O_4$ double-shelled nanocages with enhanced pseudocapacitive and electrocatalytic properties. *J. Am. Chem. Soc.*, 137, 5590, 2015.

[90] Bae, J., Song, M.K., Park, Y.J., et al., Fiber supercapacitors made of nanowire-fiber hybrid structures for wearable/flexible energy storage. *Angew. Chem. Int. Ed. Engl*, 50, 1683, 2011.

[91] Ghodbane, O., Pascal, J.L., Fraisse, B., et al., Structural in situ study of the thermal behavior of manganese dioxide materials: Toward selected electrode materials for supercapacitors. *ACS Appl. Mater. Interfaces*, 2, 3493, 2010.

[92] Wang, J.G., Kang, F., and Wei, B., Engineering of MnO_2-based nanocomposites for high-performance supercapacitors. *Prog. Mater. Sci.*, 74, 51, 2015.

[93] Halper, S.M., and Ellenbogem, C.J., *Supercapacitor: A brief overview*, 2012. Virginia, USA: MITRE Corporation.

[94] Pullanchiyodan, A., Manjakkal, L., Ntagios, M., et al., MnO x-electrodeposited fabric-based stretchable supercapacitors with intrinsic strain sensing. *ACS Appl. Mater. Interfaces*, 13, 47581, 2021.

[95] Al Kiey, S.A., and Hasanin, M.S., Green and facile synthesis of nickel oxide-porous carbon composite as improved electrochemical electrodes for supercapacitor application from banana peel waste. *Environ. Sci. Pollut. Res.*, 28, 66888, 2021.

[96] Yi, T.F., Wei, T.T., Mei, J., et al., Approaching high-performance supercapacitors via enhancing pseudocapacitive nickel oxide-based materials. *Adv. Sustain. Syst.*, 4, 1900137, 2020.

[97] Xu, L., Ding, Y.S., Chen, C.H., et al., 3D flowerlike α-nickel hydroxide with enhanced electrochemical activity synthesized by microwave-assisted hydrothermal method. *Chem. Mater.*, 20, 308, 2008.

[98] Wang, X., Sun, J., Zhao, J., et al., All-solid-state fiber-shaped asymmetric supercapacitors with ultrahigh energy density based on porous vanadium nitride nanowires and ultrathin Ni(OH)$_2$ nanosheet wrapped NiCo$_2$O$_4$ nanowires arrays electrode. *J. Phys. Chem. C.*, 123, 985, 2018.

[99] Deyab, M.A., Awadallah, A.E., Ahmed, H.A., et al., Progress study on nickel ferrite alloy-graphene nanosheets nanocomposites as supercapacitor electrodes. *J. Energy Storage*, 46, 103926, 2022.

[100] Youssry, S.M., El-Hallag, I.S., Kumar, R., et al., Synthesis of mesoporous Co(OH)$_2$ nanostructure film via electrochemical deposition using lyotropic liquid crystal template as improved electrode materials for supercapacitors application. *J. Electroanal. Chem.*, 857, 113728, 2020.

[101] Yuan, R., Li, H., Zhang, X.A., et al., Facile one-pot solvothermal synthesis of bifunctional chrysanthemum-like cobalt-manganese oxides for supercapacitor and degradation of pollutants. *J. Energy Storage*, 29, 101300, 2020.

[102] Shen, L., Du, L., Tan, S., et al., Flexible electrochromic supercapacitor hybrid electrodes based on tungsten oxide films and silver nanowires. *Chem. Commun.*, 52, 6296, 2016.

[103] Chen, J., Xu, J., Zhou, S., et al., Amorphous nanostructured FeOOH and Co—Ni double hydroxides for high-performance aqueous asymmetric supercapacitors. *Nano Energy*, 21, 145, 2016.

[104] Xiong, P., Zhu, J., and Wang, X., Recent advances on multi-component hybrid nanostructures for electrochemical capacitors. *J. Power Sources*, 294, 31, 2015.

[105] Zhao, Y., Yuan, M., Chen, Y., et al., Construction of molybdenum dioxide nanosheets coated on the surface of nickel ferrite nanocrystals with ultrahigh specific capacity for hybrid supercapacitor. *Electrochim. Acta*, 260, 439, 2018.

[106] Anwar, S., Muthu, K.S., Ganesh, V., et al., A comparative study of electrochemical capacitive behavior of NiFe$_2$O$_4$ synthesized by different routes. *J. Electrochem. Soc.*, 158, A976, 2011.

[107] Li, M., Xiong, Y., Liu, X., et al., Facile synthesis of electrospun Mfe$_2$O$_4$ (M= Co, Ni, Cu, Mn) spinel nanofibers with excellent electrocatalytic properties for oxygen evolution and hydrogen peroxide reduction. *Nanoscale*, 7, 8920, 2015.

[108] Zhu, Z., Zhang, Z., Zhuang, Q., et al., Growth of MnCo2O4 hollow nano-spheres on activated carbon cloth for flexible asymmetric supercapacitors. *J. Power Sources*, 492, 229669, 2021.

[109] Fu, M., Zhu, Z., Zhuang, Q., et al., In situ growth of manganese ferrite nanorods on graphene for supercapacitor. *Ceram. Int.*, 46, 28200, 2020.

[110] El-Khodary, S.A., El-Enany, G.M., El-Okr, M., et al., Modified iron doped polyaniline/sulfonated carbon nanotubes for all symmetric solid-state supercapacitor. *Synth. Met.*, 233, 41, 2017.

[111] He, X., Zhao, Y., Chen, R., et al., Hierarchical FeCo$_2$O$_4$@polypyrrole core/shell nanowires on carbon cloth for high-performance flexible all-solid-state asymmetric supercapacitors. *ACS Sustain. Chem. Eng.*, 6, 14945, 2018.

[112] Qi, W., Li, X., Wu, Y., et al., Flexible electrodes of MnO$_2$/CNTs composite for enhanced performance on supercapacitors. *Surf. Coat. Technol.*, 320, 624, 2017.

[113] Wang, S., Xiao, Z., Zhai, S., et al., Construction of Sn—Mo bimetallic oxide nanoparticle-encapsulated P-doped 3D hierarchical porous carbon through an in-situ reduction and competitive cross-linking strategy for efficient pseudocapacitive energy storage. *Electrochim. Acta*, 343, 136106, 2020.

[114] Wang, T., Le, Q., Guo, X., et al., Preparation of porous graphene@Mn_3O_4 and its application in the oxygen reduction reaction and supercapacitor. *ACS Sustain. Chem. Eng.*, 7, 831, 2018.

[115] Hekmat, F., Unalan, H.E., and Shahrokhian, S., Biomass-derived wearable energy storage systems based on poplar tree-cotton fibers coupled with binary nickel—cobalt nanostructures. *Sustain. Energy Fuels*, 4, 643, 2020.

[116] Yang, G., and Park, S.J., Nanoflower-like $NiCo_2O_4$ grown on biomass carbon coated nickel foam for asymmetric supercapacitor. *J. Alloys Compd.*, 835, 155270, 2020.

[117] Kaverlavani, S.K., Moosavifard, S.E., and Bakouei, A., Designing graphene-wrapped nanoporous $CuCo_2O_4$ hollow spheres electrodes for high-performance asymmetric supercapacitors. *J. Mater. Chem. A*, 5, 14301, 2017.

[118] Pourfarzad, H., Shabani-Nooshabadi, M., Ganjali, M.R., et al., Synthesis of Ni—Co-Fe layered double hydroxide and Fe_2O_3Graphene nanocomposites as actively materials for high electrochemical performance supercapacitors. *Electrochim. Acta*, 317, 83, 2019.

7 Carbon-Based Heterostructures
A Circular Economy Energy Application towards Electrochemical Sensors

Ragu Sasikumar and Byungki Kim*

ABBREVIATIONS

AA	Ascorbic acid
AAS	Atomic absorption spectrometry
AFS	Atomic fluorescence spectrometry
ALA	Alpha lipoic acid
AS	Aspirin
CA	Caffeic acid
CE	Counter electrode
CHEM	Carbon-based heterostructure energy materials
CNT	Carbon nanotubes
CPD	Carbonized polymer dot
CQD	Carbon quantum dots
CS	Chitosan
CV	Cyclic voltammetry
DA	Dopamine
DPV	Differential pulse voltammetry
DSSC	Dye-sensitized solar cells
EIS	Electrochemical impedance spectroscopy
g-C_3N_4	Graphitic carbon nitride
GCE	Glassy carbon electrode
GO	Graphene oxide
GR	Graphene
***h*-BN**	Hexagonal boron nitride
HER	Hydrogen evolution reaction
HNC	Hybrid nanocomposite
ICP-MS	Inductively coupled plasma mass spectrometry

* byungki.kim@koreatech.ac.kr

DOI: 10.1201/9781003269779-7

ICP-OES	Inductively coupled plasma optical emission spectrometry
LOD	Limit of detection
LR	Linear range
LSV	Linear sweep voltammetry
N-CQD	Nitrogen-doped carbon quantum dots
OER	Oxygen evolution reaction
OPD	*ortho*-Phenylenediamine
PA	Propionic acid
PSSS	Poly(sodium-4-styrene sulfonate)
RE	Reference electrode
RF	Riboflavin
rGO	Reduced graphene oxide
SPCE	Screen-printed carbon electrode
Sr	Strontium
SWASV	Square wave anodic stripping voltammetry
SWV	Square wave voltammetry
TCN	Triclosan
TPEA	N-(2-aminoethyl)-N,N',N'-tris(pyridine-2-yl-methyl) ethane-1,2-diamine
t-RA	*trans*-Resveratrol
UV	Ultraviolet spectrophotometry
WE	Working electrode
XFS	X-fluorescence spectrometry

7.1 INTRODUCTION

In recent years, carbon-based heterostructure energy materials (CHEMs) are gaining significant attention in different research fields such as medicine, electronics, cosmetics, drug designing, electrochemical sensors, energy storage devices, supercapacitors, water splitting, and photocatalytic applications. Due to the high number of active sites, CHEMs and CHEM-based composites are used for the construction of electrochemical sensors for environmental pollution, biomarkers, organic chemicals, antibiotics, heavy metal ions, and others. CHEMs perform unique novel properties when compared with their bulk material. Additionally, CHEM-based nanocomposites possess low volume, which helps in the miniaturization of electrochemical sensor devices. Because of their chemical stability and high electronic transportation properties, CHEMs play a dual role in the research field and they can be easily fabricated/modified [1–4].

Different analytical techniques/methods have been reported for the detection of biomarkers, harmful metal ions, and organic/environmental pollutants including atomic absorption spectrometry (AAS), ultraviolet spectrophotometry (UV), electrochemical methods, atomic fluorescence spectrometry (AFS), X-fluorescence spectrometry (XFS), inductively coupled plasma optical emission spectrometry (ICP-OES), and mass spectrometry (ICP-MS) [5, 6]. Among them, the electrochemical techniques/methods are very convenient method due to their accumulation time, cost-effectiveness, fast response, portability, high sensitivity, and selectivity. Different types of polymers, proteins, metal nanoparticles, enzymes, small organic molecules, DNA, and metal oxide nanoparticles are used to fabricate/modify CHEMs. Figure 7.1 shows the total number of publications

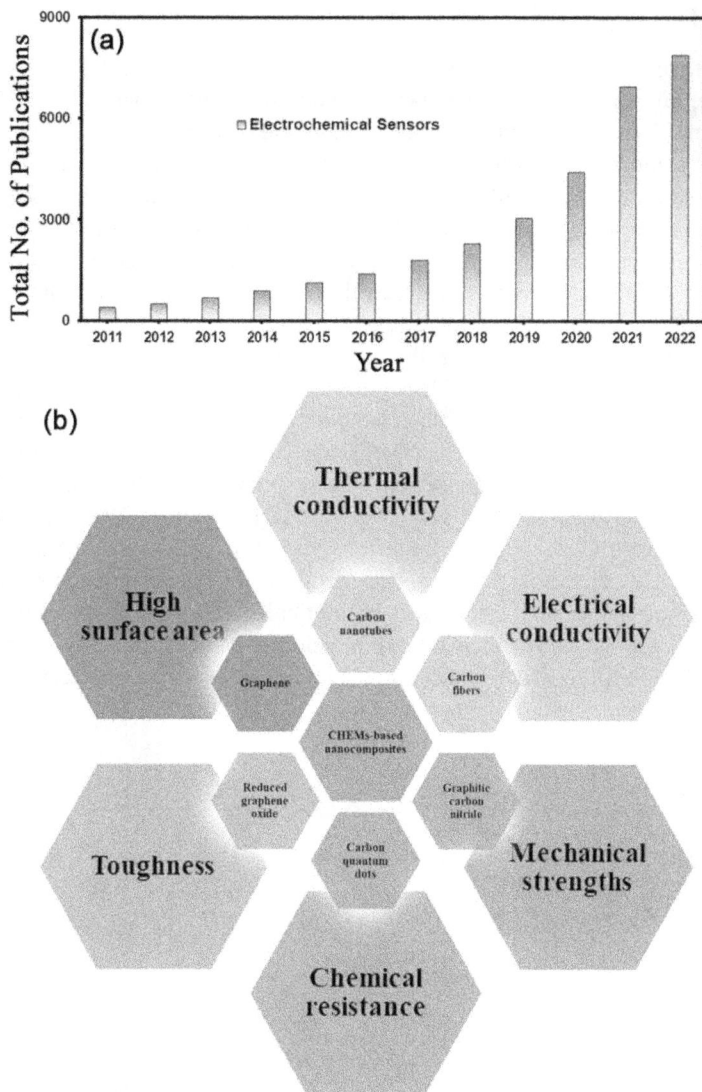

FIGURE 7.1 (a) The total quantity of publications in the field of electrochemical sensors using CHEM-based composites since 2011 and (b) schematic illustration of different allotropes of carbon materials and their properties.

Source: (a) Google Scholar.

in the field of electrochemical sensors using CHEM-based composites since 2011 and schematic illustrations of different allotropes of carbon materials and their properties.

In this chapter, we aim to briefly show the CHEM and their composite including GR, CQDs, CNTs, and g-C$_3$N$_4$ applications to electrochemical sensors. This review covers from 2012 until we reach the present, in which the main focus of this chapter is to know the current states of the CHEM and their composite to electrochemical sensors.

7.2 ELECTROCHEMICAL TECHNIQUES

Electrochemical sensors are a significant method with excellent properties of fast response, high accuracy, good reliability, and cost-effectiveness [7, 8]. When we design and develop high-quality electrochemical sensors, we should understand the basic principles. Figure 7.2a shows that first without any further process, an analyte (sample) can be added into the electrochemical cell. Then, depending upon applying voltage, the designed electrochemical sensor converts the physicochemical response of the analyte into detectable electrochemical signals. By analyzing the signal response between the detected analyte and matching electronic signals, the chemicals in samples can be identified. As shown in Figure 7.2b, a conventional three-electrode system contains a reference electrode (RE; Ag/AgCl), a counter (auxiliary) electrode (CE; Pt rod, carbon rod), and a working (active) electrode (WE; carbon-based nanocomposites, metal oxides, hybrid nanocomposites, polymer composites, 2D/3D materials, etc.). Recently, carbon-based heterostructures such as graphene, carbon nanotubes (SWCNTs and MWCNTs), carbon quantum dots (CQDs), and graphitic carbon nitride (g-C_3N_4) with their composites were used as active materials for the WE, because of their extraordinary stability, electron transportation, reproducibility, surface area, and durability. Carbon-based heterostructures dramatically improved the electrochemical properties such as the linearity, stability, selectivity, sensitivity, dynamic range, lower detection limit, wider linear range, and fast response of the electrochemical sensors [9–12].

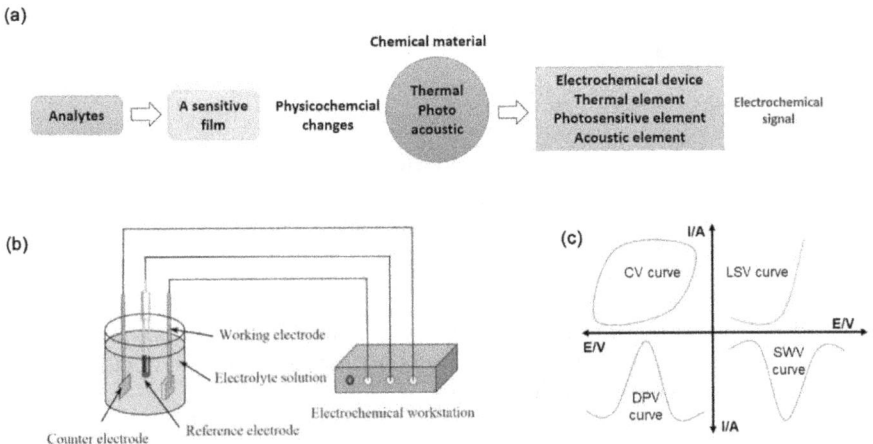

FIGURE 7.2 (a) Electrochemical detection principle. (b) Schematic diagram of an electrochemical sensor. (c) Different electrochemical techniques: cyclic voltammetry (CV), linear sweep voltammetry (LSV), differential pulse voltammetry (DPV), and square wave voltammetry (SWV) curves.

Sources: (a) Reproduced from Ref. [12] copyright (2022), with permission from Elsevier; (b) Reproduced from Ref. [11] copyright (2022), with permission from MDPI.

In 2018, Manikandan et al. reported various electrochemical techniques for the detection of samples. The electrochemical method used to measure the potential (voltage) and/or current (ampere) in an electrochemical cell contains the analyte (sample). The potentiometric technique is used to measure the electrochemical potential (voltage) of the metallic structure in a constant environment, and it can be controlled. An amperometric (i-t) method is used to measure the current (ampere) under a fixed applied voltage. Electrochemical impedance spectroscopy (EIS) is used to measure the electrode/electrolyte interface impedance. The next part of this chapter will discuss the properties and application to electrochemical sensors until now based on CHEM nanocomposites.

7.3 CARBON-BASED HETEROSTRUCTURES PROPERTIES AND APPLICATION TO ELECTROCHEMICAL SENSORS

In the chemistry world, carbon is one of the significant elements, and recently CHEMs have been continuously studied and employed in different research fields such as energy storage devices, supercapacitors, dye-sensitized solar cells (DSSCs), electrochemical sensors, photocatalytic sensors, piezoelectric devices, oxygen evolution reactions (OER), and hydrogen evolution reactions (HER). Usually, dimension aspects of carbon-based materials are further classified into zero (0D; CQDs), one (1D; carbon nanofibers and CNTs), and two (2D; GO and graphene) dimensional materials as shown in Figure 7.3a. All directions are in the order of nanometers called 0D carbon-based materials. Two directions are in the order of nanometers called 1D carbon-based materials. Only one direction is in the order of nanometers called 2D carbon-based materials [13–15] as shown in Figure 7.3b–e. Graphite and diamond were reported by sp^2 and sp^3 hybridized carbon atoms in the 18th century. Also, from

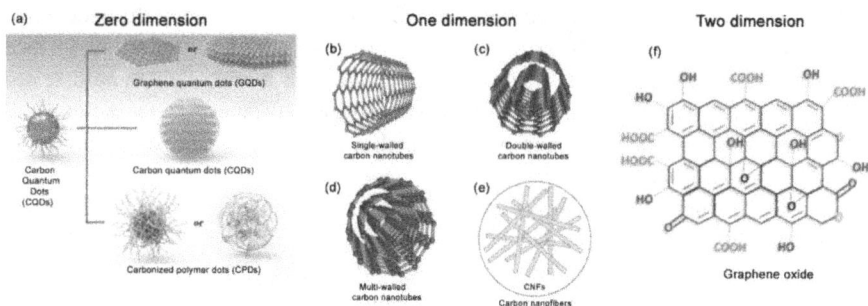

FIGURE 7.3 Different classifications of CHEMs are based on their dimensions. (a) 0D (GQDs, CQDs, and CPDs); (b–d) 1D (single-walled carbon nanotubes, double-walled carbon nanotubes, and multi-walled carbon nanotubes); (e) carbon nanofibers; and (f) 2D graphene oxide.

Sources: (a) Reproduced from Ref. [13] copyright (2020), with permission from American Chemical Society; (b–d) reproduced from Ref. [14] copyright (2019), with permission from Oat; (e) reproduced from Ref. [15] copyright (2019), with permission from MDPI; and (f) reproduced from Ref. [16] copyright (2012), with permission from MDPI.

1985 to 1991, CNTs, C60, and C70 were reported [16] as shown in Figure 7.3f. These reports indicate not only extended the possibility of the carbon material family but also notice the start of new research fields of CHEMs. Due to the wide potential window, excellent electron transportation, chemical modifiability, and high electrocatalytic activity, in 2004, graphene-based materials have been recognized as one of the important base materials for the preparation of electrochemical sensors [17] as shown in Figure 7.3g. In the following section, we reviewed the investigation and application of typical CHEMs, including graphene, carbon quantum dots (CQDs), carbon nanotubes (SWCNTs and MWCNTs), and graphitic carbon nitride (g-C_3N_4) with their composites, as electrochemical sensors because of their easy construction, good electron transportation, and cost-effective methods.

CHEM-based materials showed great attention in various research fields due to their excellent properties such as thermal conductivity, high surface area, more active sites, and electrochemical energy storage. Due to the strong sp^2 bonds CHEM-based materials exhibited high thermal conductivity and it has been reported for SWCNT and MWCNT 2980 and 200 W mK^{-1}, respectively making them a suitable candidate [18, 19]. A similar trend was observed in GR, for instance, Wang et al. reported a novel composite based on 3D-reduced graphene oxide/epoxy (3D-rGO/epoxy) and it exhibited high thermal conductivity owing to the 3D network within the composite [20]. In terms of electrical conductivity, Bourdo et al. reported increasing conductivity 10^{-1} S cm^{-1} when introducing 5-wt% foliated graphite due to the high concentration of foliated graphite [21].

7.3.1 GRAPHENE (GR)

Graphene and graphene families such as graphene oxide (GO), reduced graphene oxide (rGO), and hexagonal boron nitride (*h*-BN), are significant attention in the electrochemical sensor field. Electrochemically/chemically modified GR can produce graphene-based materials/composites [22]. Because of excellent physicochemical properties (similar to CNTs), such as high electron transportation, high surface area, and high electron transfer acceleration, graphene is an excellent candidate for electrochemical simultaneous detection of various biomolecules, heavy metals, and other biological analytes. GO can be dispersed in water due to hydroxyl groups on the edge and basal plane [23].

In 2017, Thiruppathi et al. reported the electrochemical detection of heavy metal ions such as Pb^{2+}, Cd^{2+}, Hg^{2+}, and Cu^{2+} ions by using fluorine doped on the surface of GO sheets with square wave anodic stripping voltammetry (SWASV) technique [24] as shown in Figure 7.4a. Graphene@nafion composite modified glassy carbon electrode (GCE) was utilized by Li and his colleagues to electrochemically detect heavy metal ions, such as Pb^{2+} and Cd^{2+}. Graphene nanosheets (NSs) combined with Nafion enhanced the synergistic effect, selectivity, and sensitivity of the electrochemical detection of metal ions. GO/MWCNTs composite is highly sensitive and selective for electrochemical detection of ascorbic acid (AA) and nitride.

Propionic acid (PA)-functionalized GO-modified (*f*-GO) GCE was reported for the electrochemical detection of the yeast using amperometric detection in a buffer solution, and the linear range (LR) is about 10–107 CFU mL^{-1}. Due to the number of

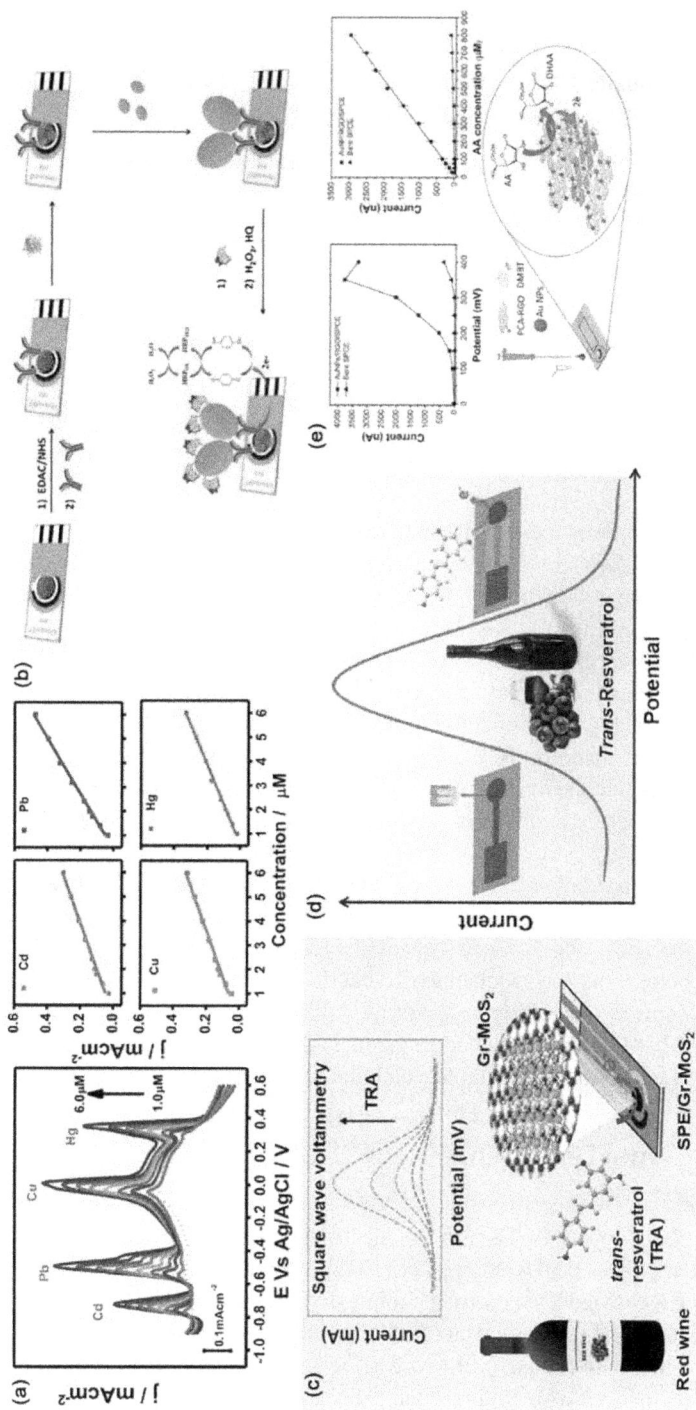

FIGURE 7.4 (a) SWASVs of *f*-GO for the simultaneous electrochemical sensing and calibration plots of Pb²⁺, Cd²⁺, Hg²⁺, Cu²⁺; (b) schematic illustration of the steps involved in the preparation and performance of the immunosensor; (c) electrochemical sensor for the determination of *t*-RA; (d) schematic illustration of the electrochemical *t*-RA sensor real-time analysis in grape skins and red wines; and (e) schematic illustration of the electrochemical AA sensor with AuNPs/rGO/SPCE-modified GCE and real-time analysis in milk.

Sources: (a) Reproduced from Ref. [24] copyright (2017), with permission from Elsevier; (b) reproduced from Ref. [25] copyright (2020), with permission from MDPI; (c) reproduced from Ref. [26] copyright (2020), with permission from Elsevier; (d) reproduced from Ref. [27] copyright (2020), with permission from Elsevier; and (e) reproduced from Ref. [28] copyright (2020), with permission from Elsevier.

TABLE 7.1

Graphene-Based Composite Electrochemical Sensors with Different Analytes, Techniques, LOD, LR, and Real-Sample

Analytes	Materials	Techniques	LR	LOD (nM)	Real-Sample	Ref.
Cd^{2+}	Fluorine-GO	SWASV	0.6–5.0 μM	10	–	[24]
Pb^{2+}			0.3–5.0 μM	10		
Saccharomyces cerevisiae	PA-GO/SPCE	CA	10–107 CFU mL^{-1}	ND	White wine	[25]
t-RV	GR-MoS$_2$	DPV	1.0–200 μM	450	Red wine	[26]
t-RV	LPG	DPV	0.2–50 μM	160	Red wine	[27]
Vitamin C	AuNPs/PCA-rGO	CV	50–500 μM	17000	Foods	[28]
L-Met	GO/SPCE	DPV	2–96 μM	184	–	[29]

active sensing sites, the PA-functionalized GO-modified GCE successfully detected the *Saccharomyces cerevisiae* with high selectivity, good reproducibility, a limit of detection (LOD), and storage stability [25] as shown in Figure 7.4b. In 2020, Zhang et al. reported a *trans*-resveratrol (*t*-RA) electrochemical sensor with LR (0.2–50 μM), LOD (0.16 μM), good repeatability, and reproducibility. Moreover, the modified *t*-RA sensor was performed in real-time analysis in grape skins and red wines with good recovery results [26] as shown in Figure 7.4c.

Graphene-molybdenum disulfide (GR-MoS$_2$) modified GCE was reported as a highly selective and sensitive electrochemical sensor for *t*-RA, as shown in Figure 7.4d. Due to the high synergistic effect of MoS$_2$ and GR, the *t*-RA electrochemical sensor showed LR from 1.0 to 200 μM with a LOD of about 0.45 μM [27]. As shown in Figure 7.4e, the oxidation of vitamin C, with the rGO surface functionalized 1-pyrenecarboxyl acid (PCA) decorated by gold nanoparticles (Au NPs) modified GCE showed good LR is about 50–500 μM and a LOD of 17 μM [28]. In 2017, Sasikumar et al. reported the electrochemical detection of the sulfur-containing amino acid (L-Methionine; L-Met) with GO-modified GCE using the DPV technique [29]. Graphene-based composite electrochemical sensors with different analytes, techniques, LOD, LR, and real-sample as tabulated in Table 7.1.

7.3.2 CARBON QUANTUM DOTS (CQDs)

Shao et al. [30] prepared *N*-(2-aminoethyl)-*N*,*N'*,*N'*-tris(pyridine-2-yl-methyl)ethane-1,2-diamine (TPEA)-doped CQDs for the electrochemical detection of Cu^{2+} metal ion. In this composite, CQDs showed excellent electron transportation and conductivity, and TPEA showed a very strong complexing agent with Cu^{2+} metal ion. Due to the good properties of the composite, they applied real-time electrochemical detection of Cu^{2+} metal ion in mouse brain with good results using plasma-atomic emission spectrometry (ICP-AES) (Figure 7.5a). Muthusankar et al. [31] prepared nitrogen-doped CQDs (N-CQDs) decorated copper oxide (Cu$_2$O) for the electrochemical detection of aspirin (AS). The reported composite showed an excellent synergistic effect by combining N-CQDs with Cu$_2$O. Muthusankar et al. applied this

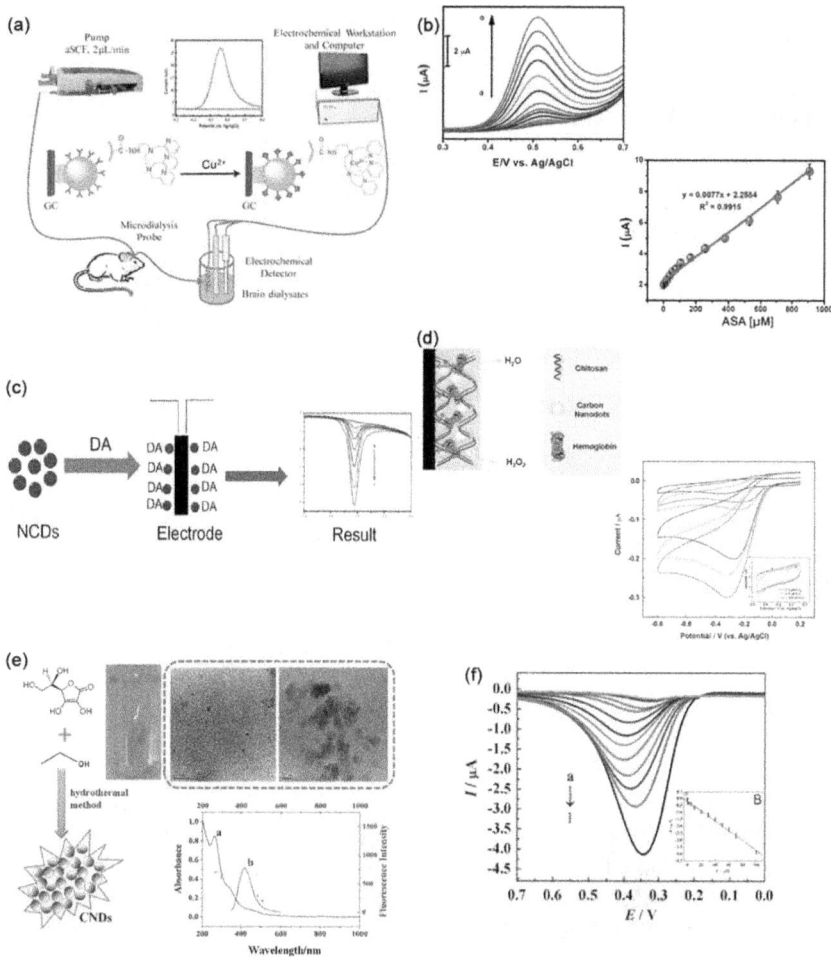

FIGURE 7.5 Schematic illustrations of (a) Cu²⁺ sensor using TPEA-doped CQDs; (b) DPVs of N-CQD/Cu₂O/GCE for electrochemical detection of AS; (c) DA sensor with N-CQDs; (d) H₂O₂ sensor with CQDs-CS/Hb; (e) TCN sensor using CQDs-CS/GCE; and (f) DA sensor with Au@CQDs-CS/GCE.

Sources: (a) Reproduced from Ref. [30] copyright (2015), with permission from Elsevier; (b) reproduced from Ref. [31] copyright (2013), with permission from Royal Society of Chemistry; (c) reproduced from Ref. [32] copyright (2012), with permission from Elsevier; (d) reproduced from Ref. [33] copyright (2014), with permission from Elsevier; (e) reproduced from Ref. [34] copyright (2012), with permission from American Chemical Society; and (f) reproduced from Ref. [35] copyright (2018), with permission from Elsevier.

composite to the real-time electrochemical analysis of AS in berries such as blackberry, raspberry, and cranberry using the DPV method (Figure 7.5b).

Jiang et al. [32] (Figure 7.5c) reported N-CQDs using a one-step microwave irradiation method. Compared to other methods, the product prepared by this method showed highly selective and sensitive electrochemical detection of dopamine (DA), with good LOD of 1.2 nM and LR 0.05–8.0 μM. As shown in Figure 7.5d, Sheng et al.

TABLE 7.2

CQD-Based Composite Electrochemical Sensors with Different Analytes, Techniques, LOD, LR, and Real-Sample

Analytes	Materials	Techniques	LR	LOD (nM)	Real-Sample	Ref.
Cu^{2+} ions	TPEA-doped CQDs	DPASV	1.0–60 μM	100	Spiked sample	[30]
AS	N-CQD/Cu_2O/GCE	DPV	1.0–907 μM	2	Berries	[31]
DA	N-CQDs	DPV	0.05–8 μM	1.2	Serum	[32]
H_2O_2	CQDs-CS/Hb	CV	1–118 μM	270	Toothpaste	[33]
TCN	CQDs-CS/GCE	CV	10 nM–1.0 μM	9.2	Water	[34]
DA	Au@CQDs-CS/GCE	DPV	0.01–100 μM	0.1	Spiked sample	[35]

[33] reported a new electrochemical sensor for H_2O_2 with CQDs-chitosan/hemoglobin (CQDs-CS/hemoglobin) nanocomposite with good selectivity and sensitivity. The LOD and LR for the electrochemical H_2O_2 sensor are about 0.27 μM and 1–118 μM with a signal-to-noise ratio of 3. The results revealed that the doping metal nanoparticles (M NPs) have enhanced the selectivity and sensitivity of the whole electrochemical sensor. CQDs have significant attention, to detect special chemicals, heavy metals, and environmentally hazardous chemicals due to the rich hydroxyl (–OH) and amino (–NH$_2$) functional groups, and easy modification with other materials.

Dai and co-workers [34] reported that CDs-chitosan modified GCE with simple construction and the current response was higher than that of bare GCE. Also, the modified GCE showed a higher sensitivity for the electrochemical detection of triclosan (TCN) and explored satisfactory recovery results with gargling daily water and toothpaste. The CDs-chitosan/GCE showed good LOD and LR is about 0.92 nM and 10.0–1.0 mM (Figure 7.5e). Huang et al. [35] reported an electrochemical DA sensor using Au@CDs-CS/GCE (Figure 7.5f), which is highly selective and sensitive and it can suppress the background currents of interference from UA and AA. Huang et al. reported good LOD and LR is about 0.1 nM and 0.01–100.0 μM using Au@CDs-CS modified GCE. CQDs-based composite electrochemical sensors with different analytes, techniques, LOD, LR, and real-sample as tabulated in Table 7.2.

7.3.3 CARBON NANOTUBES (CNTS)

Sasikumar et al. [36] (Figure 7.6a) reported Fe_3O_4@f-MWCNTs composite with an easy preparation method and it showed a higher current response towards o-phenylenediamine (OPD) with the electrochemical detection method. They reported satisfactory results in LOD and LR is about 50 μM and 0.6–80 μM. Sasikumar et al. [37] (Figure 7.6b) prepared a novel hybrid nanocomposite and reported nanomolar level electrochemical detection of α-Lipoic acid (ALA) and they applied real-time analysis using tomato and broccoli with f-MWCNTs-PIN/Ti_2O_3/GCE. Due to the fast electron transportation between f-MWCNTs and PIN, it showed a good electrochemical activity response as compared to other electrodes. The f-MWCNTs-PIN/Ti_2O_3/GCE

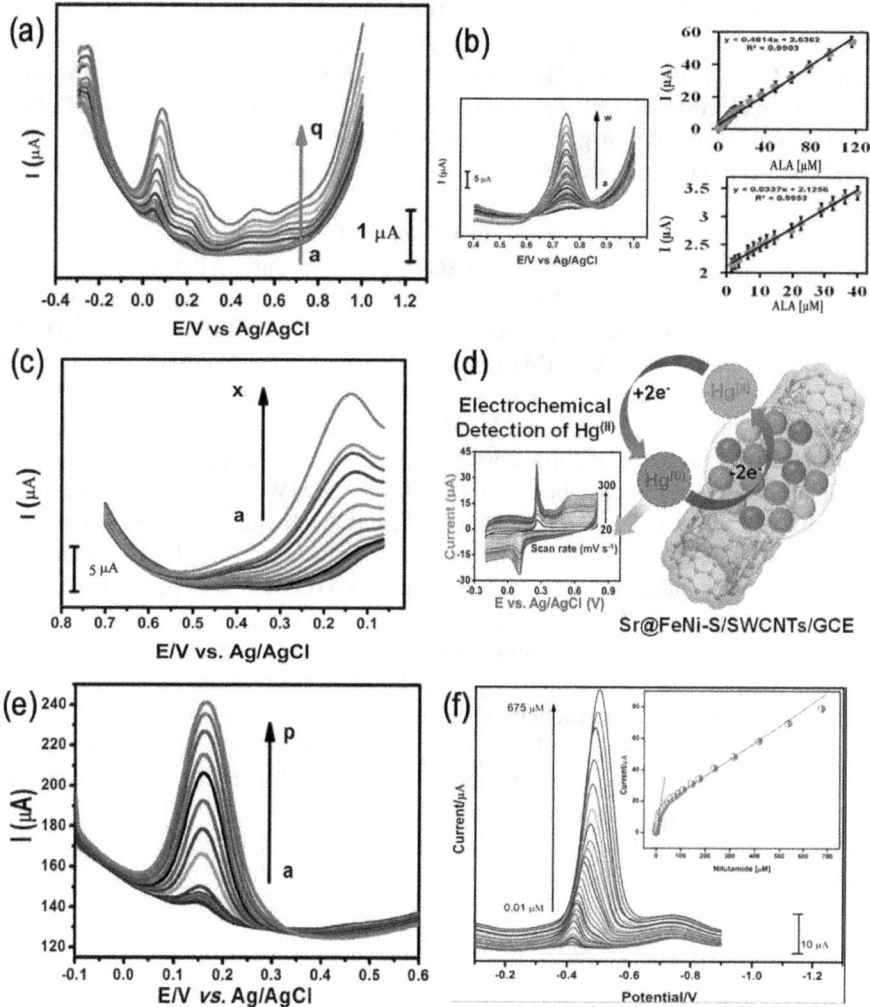

FIGURE 7.6 (a) DPV curves of OPD electrochemical sensor using $Fe_3O_4@f$-MWCNTs composite; (b) DPV curves of ALA electrochemical sensor using f-MWCNTs-PIN/Ti_2O_3/GCE; (c) DPV curves of CA electrochemical sensor using PSSS@f-MWCNTs composite; (d) the schematic illustration and different sweeping rates for Hg^{2+} ion electrochemical sensor using Sr-doped FeNi-S/SWCNTs composite; (e) DPV curves of CA electrochemical sensor using N-CQD/HP-Cu_2O/MWCNTs/GCE; and (f) DPV curves of nilutamide electrochemical sensor using f-MWCNTs/GCE.

Sources: (a) Reproduced from Ref. [36] copyright (2017), with permission from Elsevier; (b) reproduced from Ref. [37] copyright (2017), with permission from Elsevier; (c) reproduced from Ref. [38] under CCA license; (d) reproduced from Ref. [39] copyright (2020), with permission from American Chemical Society; (e) reproduced from Ref. [40] copyright (2019), with permission from Elsevier; and (f) reproduced from Ref. [41] copyright (2016), with permission from Elsevier.

showed LOD and LR is about 12 nM and 0.39–110 μM. Chen and his research group [38] (Figure 7.6c) prepared a novel composite based on poly(sodium-4-styrene sulfonate) coated with f-MWCNTs (PSSS@f-MWCNTs). The as-prepared composite applied the electrochemical sensor towards 3,4-dihydroxy-trans-cinnamate (Caffeic acid; CA) and reported a good LOD of 0.035 μM and an LR range is 0.4–173.9 μM. PSSS polymer enhanced the mechanical, electrical, and sensitivity property of f-MWCNTs.

Mariyappan et al. [39] (Figure 7.6d) developed Sr-doped FeNi-S/SWCNTs and applied them for the electrochemical sensor of Hg^{2+} ions. Alkaline-earth metal doped with bimetallic FeNi-S catalyst enhanced electronic conductivity, surface area, and electrochemical performance. Sr-doped FeNi-S/SWCNTs modified GCE showed excellent LOD (0.52 nM) and LR (0.05–279 μM). Moreover, the modified Hg^{2+} sensor was performed in real-time analysis in lake and river water with good recovery results. Muthusankar et al. [40] (Figure 7.6e) reported a novel composite with N-doped carbon quantum dots@hexagonal porous copper oxide decorated multi-walled carbon nanotubes (N-CQD/HP-Cu_2O/MWCNTs) and applied them for the electrochemical detection of caffeic acid (CA). In this composite, MWCNTs are used as an electrically conductive substrate because of their high surface area and electrical conductivity. N-CQD/HP-Cu_2O/MWCNTs/GCE showed excellent LOD (4 nM) and LR (0.05–43 μM). Moreover, the modified CA electrochemical sensor was performed in real-time analysis in the red wine with good recovery results. Chen and his research group [41] (Figure 7.6f) prepared f-MWCNTs and applied them for the electrochemical detection and determination of nilutamide with good LOD (0.2 nM) and LR (0.01–21 μM and 28–535 μM). Finally, they applied the real-time electrochemical sensor in the nilutamide tablet and human serum with excellent recovery results. CNTs-based composite electrochemical sensors with different analytes, techniques, LOD, LR, and real-sample as tabulated in Table 7.3.

TABLE 7.3

CNT-Based Composite Electrochemical Sensors with Different Analytes, Techniques, LOD, LR, and Real-Sample

Analytes	Materials	Techniques	LR	LOD (nM)	Real-Sample	Ref.
OPD	Fe_3O_4@f-MWCNTs	DPV	0.6–80 μM	50,000	–	[36]
ALA	f-MWCNTs-PIN/Ti_2O_3/ GCE	DPV	0.39–110 μM	12	Tomato, broccoli	[37]
CA	PSSS@f-MWCNTs	DPV	0.4–173.9 μM	35	–	[38]
Hg^{2+}	Sr-doped FeNi-S/SWCNTs	DPV	0.05–279 μM	0.52	Lake water, river water	[39]
CA	N-CQD/HP-Cu_2O/ MWCNTs/GCE	DPV	0.05–43 μM	4	Red wine	[40]
Nilutamide	f-MWCNTs/GCE	DPV	0.01–21 and 28–535 μM	0.2	Tablet, human serum	[41]

7.3.4 GRAPHITIC CARBON NITRIDE (G-C$_3$N$_4$)

Devi et al. (Figure 7.7a) achieved a nanomolar-level detection towards organophosphate pesticides in food samples using iron oxide@molybdenum carbide micro flowers decorated with graphitic-carbon nitride modified GCE (Fe$_3$O$_4$@MoC/g-CN/GCE) [42]. During the hierarchical nanostructure growth from the core level, it

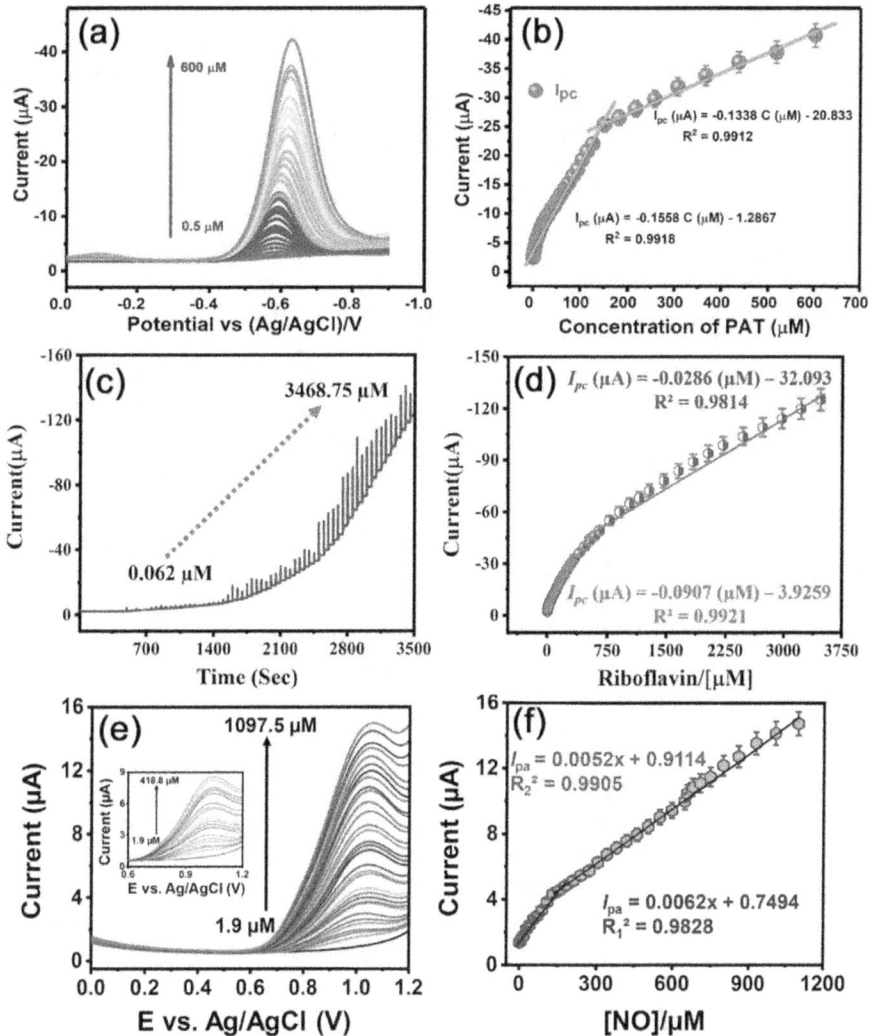

FIGURE 7.7 (a, b) DPV curves of MP electrochemical sensor using Fe$_3$O$_4$@MoC/g-CN/GCE; (c, d) Amperometric (i-t) current response of RF with R-CoP/GCN/RRDE; and (e, f) DPV curves of NO electrochemical sensor using Bi$_2$S$_3$/Zn-GCN modified GCE.

Sources: (a, b) Reproduced from Ref. [42] copyright (2022), with permission from Elsevier; (c, d) reproduced from Ref. [43] copyright (2022), with permission from Elsevier; and (e, f) reproduced from Ref. [44] copyright (2022), with permission from Elsevier.

TABLE 7.4

g-C₃N₄–Based Composite Electrochemical Sensors with Different Analytes, Techniques, LOD, LR, and Real-Sample

Analytes	Materials	Techniques	LR (µM)	LOD (nM)	Real-Sample	Ref.
MP	Fe₃O₄@MoC/ g-CN/GCE	DPV	0.5–600	7.8	Cabbage juice, tomato juice, onion	[42]
RF	R-CoP/GCN/RRDE	Amperometric (i-t)	0.062–3468.75	1.09	Human serum, urine, mushroom	[43]
NO	Bi₂S₃/Zn-GCN/GCE	DPV	1.9–1097.5	7.0	Blood serum, lake water, sausage	[44]

enhances the more active sites (easy diffusion of the sample) to detect the parathion (PAT). Fe_3O_4 NFs doped on the surface of MoC hydrothermally and further decorated with g-CN. This method further increases mechanical stability and electrical conductivity. The Fe_3O_4@MoC MFs/g-CN/GCE showed excellent LR (0.5–600 μM) and LOD (7.8 nM). Moreover, Fe_3O_4@MoC MFs/g-CN/GCE showed good recovery results towards PAT electrochemical real-time detection in the food samples (cabbage, tomato, and onion). Shanmugam et al. [43] (Figure 7.7b) reported a portable electrochemical sensor for vitamin B_2 (riboflavin [RF]) with ruthenium-doped cobalt phosphide embedded graphitic carbon nitride (R-CoP/GCN). The R-CoP embedded with GCN enhances the synergistic effect; due to this effect, it has a high surface area (69.43 m^2 g^{-1}) and low charge transfer resistance (R_{ct} = 367.89 Ω). With CV and i-t techniques, the as-prepared composite showed excellent LOD with about 1.09 nM and LR range from 0.062 to 3468.75 μM. Furthermore, R-CoP/GCN/GCE showed good recovery results towards RF electrochemical real-time detection in the environmental and biological samples (human serum, human urine, and mushroom).

Chen and his research group [44] (Figure 7.7c) reported a hybrid nanocomposite (HNC) based on zinc-doped graphitic carbon nitride (Bi_2S_3/Zn-GCN) using thermal and simple ultrasonication methods. The developed HNC was applied for the electrochemical detection of nitric oxide (NO). The as-prepared HNC showed unique properties such as high electrochemical active area, low charge transfer, selectivity, reproducibility, and stability. Due to the synergistic effect between Zn-GCN and Bi_2S_3, it showed a wide range of LR (1.9–1097.5 μM), LOD (7 nM), and low sensitivity (0.732 μA μM^{-1} cm^{-2}). The real-time analysis with good recovery results was performed with lake water, human blood serum, and sausage samples. g-C_3N_4-based composite electrochemical sensors with different analytes, techniques, LOD, LR, and real-sample as tabulated in Table 7.4.

7.4 CHALLENGES AND FUTURE PERSPECTIVES

CHEMs and CHEM-based nanocomposites have been developed as significant electrochemical-sensing materials and exhibited excellent performance due to their low toxicity, thermal stability, biocompatibility, higher surface area, more active sites, and good electrical conductivity properties. On the other hand, electrochemical devices for portable on-site detection are still facing many problems:

- Designing, developing, and modifying/functionalization of CHEMs and CHEM-based nanocomposites are still lacking. To resolve this issue, nowadays researchers are increasingly focused on coordinating the electrochemical sensors and analytes to study their behavior, functional groups, and other properties.
- CHEM-based nanocomposite-modified electrodes should be fabricated for wearable, portable sensors for electrochemical detection.
- More screen-printed and lab-on-chip fabrication works should be developed.
- Portable, wearable, and reusable screen-printed electrodes should be designed to take advantage of their user-friendliness, easy accessibility, and miniaturization.

We strongly believe that the above-mentioned challenges will be significantly resolved by upcoming researchers and continuous research development in materials science, electrochemical sensors, nanotechnology, and other research fields.

7.5 CONCLUSIONS

In this chapter, we discussed the developments in the field of electrochemical sensors utilizing CHEMs and CHEM-based nanocomposites. GR, CQDs, CNTs, and g-C_3N_4-based nanocomposites have garnered significant attention in electro-catalytic studies and electro-analytical research. Due to their physicochemical properties and structural morphologies, CHEMs and their hybrid/macro/nano-composites have been used to employed to fabricate electrodes for a wider range of applications in electrochemical sensors. These remarkable properties render them suitable for the sensitive and selective electrochemical determination and detection of various toxic metal ions, biomarkers, biological compounds (such as dopamine and ascorbic acid), food toxins, and drugs. These materials can yield excellent results not only in the electrochemical sensor field but also in other industrial applications.

7.6 ACKNOWLEDGMENTS

This work was supported by the Priority Research Program through the National Research Foundation of Korea (NRF) under the Ministry of Education (2018R1A6A1A03025526) and by Basic Science Research Program through the NRF funded by the Ministry of Science, ICT (2021R1A2C1004540).

REFERENCES

[1] M. Naveen, N.G. Gurudatt, Y.-B. Shim, Applications of conducting polymer composites to electrochemical sensors: A review, *Appl. Mater. Today*, 9, 419, 2017.

[2] A. Afkhami, F. Kafrashi, M. Ahmadi, T. Madrakian, A new chiral electrochemical sensor for the enantioselective recognition of naproxen enantiomers using L-cysteine self-assembled over gold nanoparticles on a gold electrode, *RSC Adv.*, 5, 58609, 2015.

[3] S. Ahmadzadeh, M. Rezayi, E. Faghih-Mirzaei, M. Yoosefian, A. Kassim, Highly selective detection of titanium (III) in industrial waste water samples using meso-octamethylcalix[4]pyrrole-doped PVC membrane ion-selective electrode, *Electrochim. Acta*, 178, 580, 2015.

[4] K. Anitha, S. Namsani, J.K. Singh, Removal of heavy metal ions using a functionalized single-walled carbon nanotube: A molecular dynamics study, *J. Phys. Chem.* 119, 8349, 2015.

[5] E. Archer, B. Petrie, B.K. Hordern, G.M. Wolfaardt, The fate of pharmaceuticals and personal care products (PPCPs), endocrine disrupting contaminants (EDCs), metabolites and illicit drugs in a WWTW and environmental waters, *Chemosphere*, 174, 437, 2017.

[6] F.J. Beltran, M. Checa, J. Rivas, J.F.G. Araya, Modeling the mineralization kinetics of visible led graphene oxide/titania photocatalytic ozonation of an urban wastewater containing pharmaceutical compounds, *Catalysts*, 10, 1, 2020.

[7] V.S. Manikandan, B. Adhikari, A. Chen, Nanomaterial based electrochemical sensors for the safety and quality control of food and beverages, *Analyst*, 143, 4537, 2018.

[8] K. Rajeshwar, J.G. Ibanez, G.M. Swain, Electrochemistry and the environment, *J. Appl. Electrochem.* 24, 1077, 1994.

[9] D. Grieshaber, R. MacKenzie, J. Voros, E. Reimhult, Electrochemical biosensors—sensor principles and architectures, *Sensors*, 8, 1400, 2008.

[10] D. Arrigan, Bioelectrochemistry. Fundamentals, experimental techniques and applications, *Chromatographia*, 72, 585, 2010.

[11] Z. Yang, X. Zhang, J. Guo, Functionalized carbon-based electrochemical sensors for food and alcoholic beverage safety, *Appl. Sci.* 12, 9082, 2022.

[12] Y. Wang, H. Liu, Z. Jia, B. Yang, L. He, The electrochemical performance of Al-Mg-Ga-Sn-xBi alloy used as the anodic material for al-air battery in KOH electrolytes, *Crystals* 12, 1785, 2022.

[13] J. Liu, R. Li, B. Yang, Carbon dots: A new type of carbon-based nanomaterial with wide applications, *ACS Cent. Sci.* 12, 2179, 2020.

[14] S.A. Harfoush, J. Nguyen, S. Heck, A. Mahdy, R. Bals, Q.T. Dinh, Nanoparticles and air pollutants as potential stimulators of asthmatic reaction, *Front. Nanotechnol.* 6, 5, 2019.

[15] Z. Wang, S. Wu, J. Wang, A. Yu, G. Wei, Carbon nanofiber-based functional nanomaterials for sensor applications, *Nanomaterials*, 9, 1045, 2019.

[16] A. Ricci, A. Cataldi, S. Zara, M. Gallorini, Graphene-oxide-enriched biomaterials: A focus on osteo and chondroinductive properties and immunomodulation, *Materials*, 15, 2229, 2022.

[17] Z. Wang, S. Wu, J. Wang, A. Yu, G. Wei, Carbon nanofiber-based functional nanomaterials for sensor applications, *Nanomaterials*, 9, 1045, 2019.

[18] D.J. Yang, Q. Zhang, G. Chen, S.F. Yoon, J. Ahn, S.G. Wang, Thermal conductivity of multiwalled carbon nanotubes, *Phys. Rev. B.* 66, 165440, 2002.

[19] J. Che, T. Cagin, W.A. Goddard, Thermal conductivity of carbon nanotubes, *Nanotechnology* 11, 65, 2000.

[20] R. Wang, C. Xie, B. Gou, H. Xu, J. Zhou, Significant thermal conductivity enhancement of polymer nanocomposites at low content via graphene aerogel, *Mater. Lett.* 305, 130771, 2021.

[21] K. Yang, F. Zhang, Y. Chen, H. Zhang, B. Xiong, H. Chen, Recent progress on carbon-based composites in multidimensional applications, *Compos. Part A* 157, 106906, 2022.

[22] Y. Ito, Y.H. Shen, D. Hojo, Y. Itagaki, T. Fujita, L.H. Chen, T. Aida, Z. Tang, T. Adschiri, M.W. Chen, Correlation between chemical dopants and topological defects in catalytically active nanoporous graphene, *Adv. Mater.* 28, 10644, 2016.

[23] X.F. Zhang, H.T. Liu, Y.N. Shi, J.Y. Han, Z.J. Yang, Y. Zhang, C. Long, J. Guo, Y.F. Zhu, X.Y. Qiu, Boosting CO_2 conversion with terminal alkynes by molecular architecture of graphene oxide-supported Ag nanoparticles, *Matter*, 3, 558, 2020.

[24] A.R. Thiruppathi, B. Sidhureddy, W. Keeler, A. Chen, Facile one-pot synthesis of fluorinated graphene oxide for electrochemical sensing of heavy metal ions, *Electrochem. Commun.* 76, 42, 2017.

[25] S. Campuzano, P.Y. Sedeno, J.M. Pingarron, Electrochemical affinity biosensors based on selected nanostructures for food and environmental monitoring, *Sensors*, 20, 5125, 2020.

[26] C. Zhang, J. Ping, Z. Ye, Y. Ying, Two-dimensional nanocomposite-based electrochemical sensor for rapid determination of trans-resveratrol, *Sci. Total Environ.* 742, 140351, 2020.

[27] C. Zhang, J. Ping, Y. Ying, Evaluation of trans-resveratrol level in grape wine using laser-induced porous graphene-based electrochemical sensor, *Sci. Total Environ.* 714, 136687, 2020.

[28] F. Bettazzi, C. Ingrosso, P. Sfragano, V. Pifferi, I. Palchetti, Gold nanoparticles modified graphene platforms for highly sensitive electrochemical detection of vitamin C in infant food and formulae, *Food Chem.* 344, 128692, 2020.

[29] R. Sasikumar, P. Ranganathan, S.-M. Chen, T. Kavitha, S.-Y. Lee, T.-W. Chen, W.-H. Chang, Electrochemical determination of sulfur-containing amino acid on screen-printed carbon electrode modified with graphene oxide, *Int. J. Electrochem. Sci.* 12, 4077, 2017.

[30] X. Shao, H. Gu, Z. Wang, X. Chai, Y. Tian, G. Shi, Highly selective electrochemical strategy for monitoring of cerebral Cu^{2+} based on a carbon Dot-TPEA hybridized surface, *Anal. Chem.* 85, 418, 2012.

[31] G. Muthusankar, R. Sasikumar, S.-M. Chen, G. Gopu, N. Sengottuvelan, S.-P. Rwei, Electrochemical synthesis of nitrogen-doped carbon quantum dots decorated copper oxide for the sensitive and selective detection of non-steroidal anti-inflammatory drug in berries, *J. Colloid Interface Sci.* 523, 191, 2018.

[32] Y. Jiang, B. Wang, F. Meng, Microwave-assisted preparation of N-doped carbon dots as a biosensor for electrochemical dopamine detection, *J. Colloid Interface Sci.* 452, 199, 2015.

[33] M. Sheng, G. Yue, J. Sun, G. Feng, Carbon nanodots—chitosan composite film: A plat form for protein immobilization, direct electrochemistry and bioelectrocatalysis, *Biosens. Bioelectron.* 58, 351, 2014.

[34] H. Dai, G. Xu, L. Gong, C. Yang, Y. Lin, Y. Tong, J. Chen, G. Chen, Electrochemical detection of triclosan at a glassy carbon electrode modifies with carbon nanodots and chitosan, *Electrochim. Acta* 80, 362, 2012.

[35] Q. Huang, H. Zhang, S. Hu, F. Li, W. Weng, J. Chen, Q. Wang, Y. He, W. Zhang, X. Bao, A sensitive and reliable dopamine biosensor was developed based on the Au carbon dots—chitosan composite film, *Biosens. Bioelectron.* 52, 277, 2013.

[36] R. Sasikumar, P. Ranganathan, S.-M. Chen, S.-P. Rwei, Muthukrishnan, Electro-oxidative determination of aromatic amine (o-phenylenediamine) using organic-inorganic hybrid composite, *J. Colloid Interface Sci.* 504, 149, 2017.

[37] R. Sasikumar, P. Ranganathan, S.-M. Chen, S.-P. Rwei, f-MWCNTs-PIN/Ti$_2$O$_3$ nano-composite: Preparation, characterization and nanomolar detection of α-Lipoic acid in vegetables, *Sens. Actuators B-Chem.* 255, 217, 2018.

[38] R. Sasikumar, T.-W. Chen, S.-M. Chen, Y.C. Chen, S.-P. Rwei, Ultrasonic synthesis of polysodium-4-styrene sulfonate coated functionalized MWCNTs for electrochemical detection of anti-oxidant drug in red wine, *Int. J. Electrochem. Sci.* 13, 9441, 2018.

[39] V. Mariyappan, S. Manavalan, S.-M. Chen, G. Jaysiva, P. Veerakumar, M. Keerthi, Sr@FeNi-S nanoparticle/carbon nanotube nanocomposite with superior electrocatalytic activity for electrochemical detection of toxic Mercury(II), *ACS Appl. Electron. Mater.* 2, 1943, 2020.

[40] G. Muthusankar, M. Sethupathi, S.-M. Chen, R.K. Devi, R. Vinoth, G. Gopu, N. Anandhan, N. Sengottuvelan, N-doped carbon quantum dots@hexagonal porous copper oxide decorated multiwall carbon nanotubes: A hybrid composite material for an efficient ultra-sensitive determination of caffeic acid, *Compos. Part B.* 174, 106973, 2019.

[41] R. Karthik, R. Sasikumar, S.-M. Chen, J. Vinothkumar, A. Elangovan, V. Muthuraj, P. Muthukrishnan, Fahad M.A. Al-Hemaid, M. Ajmal Ali, Mohamed S. Elshikh, A highly sensitive and selective electrochemical determination of non-steroidal prostate anti-cancer drug nilutamide based on f-MWCNT in tablet and human blood serum sample, *J. Colloid Interface Sci.* 487, 289, 2017.

[42] R.K. Devi, M. Ganesan, T.-W. Chen, S.-M. Chen, K.-Y. Lin, M. Akilarasan, W.A. Al-onazi, R.A. Rasheed, M.S. Elshikh, Tailored architecture of molybdenum carbide/iron oxide micro flowers with graphitic carbon nitride: An electrochemical platform for nano-level detection of organophosphate pesticide in food samples, *Food Chem.* 397, 133791, 2022.

[43] R. Shanmugam, C. Koventhan, S.-M. Chen, W. Hung, A portable Ru-decorated cobalt phosphide on graphitic carbon nitride sensor: An effective electrochemical evaluation method for vitamin B2 in the environment and biological samples, *J. Chem. Eng.* 446, 136909, 2022.

[44] J. Ganesamurthi, R. Shanmugam, S.-M. Chen, P. Veerakumar, Bismuth sulfide/zinc-doped graphitic carbon nitride nanocomposite for electrochemical detection of hazardous nitric oxide, *J. Electroanal. Chem.* 910, 116174, 2022.

8 Technologies for Conversion of Biomass to Valuable Chemicals and Fuels

*Uplabdhi Tyagi and Neeru Anand**

8.1 INTRODUCTION

Limited dependency on conventional resources and minimizing the generation of solid biowaste are two major challenges that confront contemporary societies. Biowaste including forestry wastes, agricultural residues, and municipal wastes are the most promising renewable and sustainable sources due to their easy availability, abundance, and sustainability. These wastes are projected to increase by 2030 which motivates the industries and societies to recycle or utilize them more effectively [1]. The world's transition to a post-carbon fossil social system may be facilitated by improvements and breakthroughs in the production and processing of cleaner biomass. Therefore, scientists have concentrated on finding cost-effective ways to turn industrial and agricultural waste into bioproducts and bioenergy. In this regard, a circular economy (CE) must be established in order to achieve sustainability, low carbon emissions, and resource efficiency. The circular economy employs asset reuse, reprocessing, repair, and refurbishment to increase asset effectiveness in terms of both usage and production. Importance of biomass feedstocks has been increased with the help of several policies that promote renewable energy and bio-based products which ultimately enable a circular economy. The creation of a circular modern economy has biological products as a vocal stakeholder as they are the biggest producers and consumers of natural resources. The creation of biological goods has favorable consequences on both the economy and the environment, such as the replacement of biomaterials generated using chemicals or fossil fuels. The circular economy is based on coordinated biorefineries that produce energy, vitality, and chemicals from the entire biomass waste stream. Pomace, peels, corn stover, seeds, and other byproducts are biomass-dense and contain a variety of essential components, such as proteins, lactase, extractives, carbohydrates, and minerals [71]. These waste streams are both solid and liquid processed waste streams [40–45]. These wastes could be used to produce enzymes, furfurals, pigments, bio-plastics, bioactive compounds, and biofuels

* neeruanand@ipu.ac.in

DOI: 10.1201/9781003269779-8

with a wide range of applications in cosmetics, plastics and therapeutic industries [2]. They could also serve as great, cost-effective raw materials for the production of these products. Biobased treatment is receiving a lot of attention in this respect and is contributing significantly to the expansion of the bioproducts refinery sector. This increases the amount of organic waste materials that can be recovered and increases the yields of bioproducts dramatically.

8.1.1 POTENTIAL OF BIO-BASED PRODUCTS IN TERMS OF CIRCULAR ECONOMY

The circular economy of biomass-based energy and products is an important concept in sustainable development. The basic paradigm for the sustainability of bioenergy and bioproducts is summarized in Table 8.1 together with contemporary circular economy concepts. Reducing our reliance on finite resources and addressing climate change are two ways that the circular economy of bioenergy can help create a more resilient and sustainable future. There are various factors responsible for the effective utilization of biomass including biomass feedstocks, conversion technologies, energy use, waste management, and circular supply chains. Figure 8.1 shows schematic representation of the interactions between lignin and cellulose that result from the pre-treatment of biomass. In addition, circular economy of bioenergy and bioproducts offers several benefits including reduced greenhouse gas emissions, resource efficiency, job creation, energy security, and increased

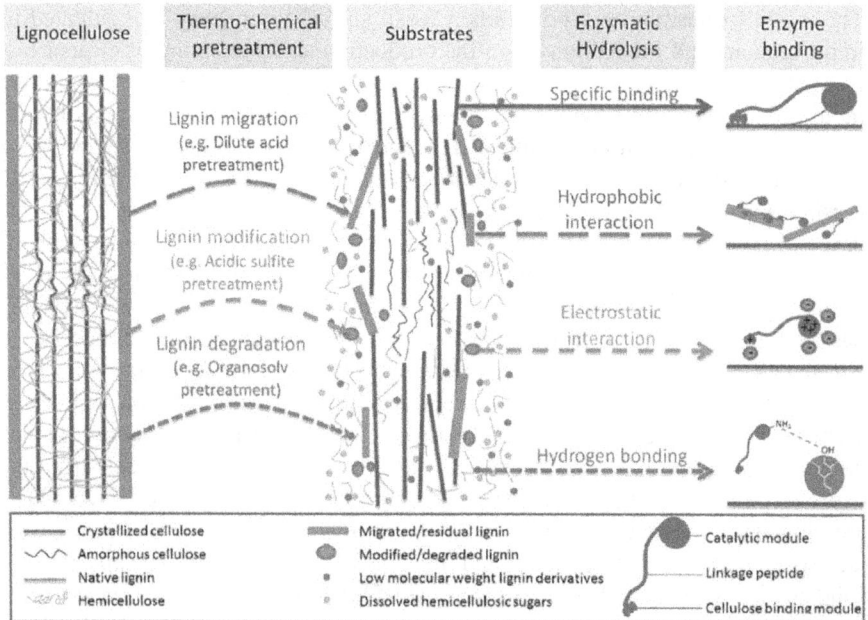

FIGURE 8.1   Schematic representation of the interactions between lignin and cellulose resulting from the pre-treatment of biomass.

TABLE 8.1

Contemporary Circular Economy Methods and Primary Paradigm for the Sustainability of Biomass-Based Energy and Products

S. No	Country	Feedstock	Strategy	Circular Economy Model	Advantages
1	Brazil	Solid waste	• Content and SWOT analysis for the management of organic waste	• Potential for recycling waste streams into useful resources • Enhancing the environment, lowering greenhouse gas emissions, and reducing expenses • Stimulation of collaborative initiatives and the manufacture of biobased chemicals, energy, new employment possibilities, and investment prospects because of the formation of a new business model • Developed an organic waste-based new value chain	• Focused on changing existing business structures, laws, and taxation while focusing on developing value chains
2	Belgium	Urban waste	• The trash flow study showed how much rubbish had been collected. • Percentage provided by each sector • Analyzed the site of treatment and the material composition	• Metropolitan settings may limit how local garbage flows are valued. • Examined how cities might contribute most to a circular economy • Completion of material cycles on national and international level	• They were created for the region so that they could be used to examine the metabolism of urban garbage in terms of waste. • Evaluation of the effectiveness of waste treatment and waste production intensity
3	Portugal	Biowaste	• Encourage the use of a circular economy for trash management and waste prevention • Getting ready for disposal, alternate recovery methods, and reuse	• In a circular economy framework, a waste hierarchy index (WHI) was proposed and applied to municipal wastes.	• According to Eurostat, recycling and preparation for reuse are viewed as beneficial elements of the circular economy. • Considered landfill and incineration as harmful factors

(Continued)

TABLE 8.1 (*Continued*)
Contemporary Circular Economy Methods and Primary Paradigm for the Sustainability of Biomass-Based Energy and Products

S. No	Country	Feedstock	Strategy	Circular Economy Model	Advantages
4	United Kingdom	Biomass	• A controversy about whether woody-biofuels cause deforestation has been sparked by biofuels made from forests.	• A true transition exists with a low-carbon, environment friendly, but the methods of accounting used to measure it are skewed and inconsistent.	• Disputes have arisen about how to allocate environmental consequences for companies and goods in highly interconnected systems. • Resources that were formerly regarded as trash are becoming more valued.
5	Denmark	Biowaste	• Concepts for mixed-biowaste refineries • Sustainable and economical alternative of biorefinery • Utilization of wastes to produce valuable products	• The creation of biorefineries using the organic MSW fraction as raw material offers a promising opportunity for technological advancements.	• MSW residue as raw material
6	Italy	Municipal solid waste	• Many suggestions to reduce errors and increase the accuracy of the information	• The circular economy's guiding principles reduce landfill's current function. • Construction of treatment facilities with specified capabilities is mandated by the circular economy. • Appropriate for addressing precisely what is not source segregated • From the perspective of authorization, direct landfilling is no longer practical.	• Hence, a novel characterization model is suggested. • Appropriate for waste management planning based on the principles of circular economy

economic value [46–49]. The transition of a circular economy has gained a lot of support both locally and internationally. The main challenges are (1) the significant negative effects of landfilling on the environment and society; (2) the significant reliance of national economy on resource recovery and extractive industries; and (3) the rapid creation of urban business models that are competitive with traditional recycling firms.

8.1.2 TRANSITION IN CIRCULAR ECONOMY BASED ON BIOMASS-DERIVED ENERGY AND PRODUCTS SUSTAINABILITY

A circular economy is a type of economic growth that focuses on avoiding waste and harm to the environment while maximizing the use of resources. This concept can be applied to the bioenergy and bioproducts sectors to achieve greater sustainability. Bioenergy is a term used to describe energy produced from biomass derived materials such as wood, plants, and agricultural waste, or biomass. Contrarily, materials and chemicals produced using sustainable biomass sources are known as bioproducts. In a circular economy for bioenergy and bioproducts, the focus is on using renewable resources efficiently and sustainably, minimizing waste and emissions, and creating a closed loop of material flows [3, 50, 51]. Several key elements of a circular economy for bioenergy and bioproducts sustainability are resource efficiency, closed-loop material flows, life-cycle thinking, collaboration, and innovation. Biomass is an essential and developing option to minimize economic dependence on oil and petrochemicals, in addition to contributing in the creation of a low-carbon and circular economy. The utilization of these materials as industrial feedstock as well as their recycling and processing hold great promise for decreasing environmental risks like greenhouse gas emissions and halting future climate change. The required and forward-thinking steps that have been made in this respect are summarized in Table 8.2. According to the literature, SWOT analysis for the management of waste is used to determine criteria based on circular economy. The goal of the circular economy is to minimize the consumption of raw materials and the production of biowaste by completing the economical and biological cycles of resources. According to

TABLE 8.2
Developed Measures for Low-Carbon and Circular Economy

S. No	Knowledge Centre	Remarks

1	Knowledge Centre for Bioeconomy	1. It aims to set the bar for better knowledge management in relation to the creation of policies for the biobased economy.
		2. Supports policymaking by—
		  2.1 Determining, categorizing, and organizing pertinent information, then making it accessible.
		  2.2 Evaluating, combining, and concisely presenting the analysis of the available evidence.

(Continued)

TABLE 8.2   (*Continued*)
Developed Measures for Low-Carbon and Circular Economy

S. No	Knowledge Centre	Remarks
2	Bio-Monitor project	1. Consists of a group of statisticians, academics, standardizers, consultants, and experts in data modelling.
		2. It will produce a comprehensive framework that is sustainable that diverse stakeholders may use to track and evaluate the bioeconomy and its varied effects.
		3. Using Bio-Monitor will make the work of statistics and customs officials easier.
		4. Politicians will be able to create bioeconomic plans that are more successful.
		5. With efficient and open communication of bio-based products, bio-based enterprises may encourage evidence-based business strategy.
3	BioEconomy Regional Strategy Toolkit project	1. Examining the possibilities for future bio-economics in various European Union areas
		2. Closing the potential gap between the existing and future bio-economies in the various European Union areas
		3. Creating clever initiatives for the regional bioeconomy of the European Union
4	Systems Analysis Tools Framework for the European Union Bio-Based Economy Strategy	1. The aim of the project is to build a framework that can monitor any change in the biobased economy as well as its impacts on the socioeconomic system and the environment.
		2. Outcomes:
		2.1 Structuring the bio-economics strategy principles.
		2.2 Identify the core components of a system and monitor the application of a bio-economics approach.
		2.3 To offer the theoretical foundation for a system that might be used to the bioeconomy agenda.
5	Biomass Policies project	1. It develops the integrated strategies that can mobilize the value chain of bioenergy and help to achieve the 2020 and 2030 bioenergy goals.
		2. Outcomes:
		2.1 Selecting the effective raw materials for each area and nation
		2.2 Comparing and contrasting various policy options to determine which are more effective in terms of resource usage, market impact, and other factors
		2.3 Government involvement to provide ideas to biomass value chains and national policy

certain research, more work must be made to guarantee that the essential data for MSW optimization is gathered with certainty [4, 52, 53]. Yet others found that the biogas economy depends on factors including how well biowaste can be utilized and handled logistically, how productively the preparation process works, and what the finished product looks like. By linking the biowaste and producing areas, the development of bioproduct refineries employing natural MSW fractions as feedstock offers an interesting opportunity to reorganize the biowaste hierarchy. Also, it was found that urban environment might restrict the extensive valorization of biowaste streams.

8.2 PROCESS ADVANCEMENT AND TECHNOLOGIES FOR BIOMASS PRETREATMENT

8.2.1 PHYSIO-CHEMICAL CONVERSION

8.2.1.1 Reduction in Size

Size reduction is a type of pretreatment technique used to reduce the size and increase the surface area of biomass particles. The aim of size reduction is to make the biomass more accessible to downstream processing. Further, to improve the digestibility of biomass, a variety of mechanical size reduction procedures have been applied, including milling, chopping, shredding, grinding, and extrusion. The size of biomass from logs is reduced by harvesting and preconditioning to coarse sizes of around 15–55 mm, chipping (15–35 mm), and grinding and milling to particles as small as 0.25–2.5 mm [4, 54, 55]. Nevertheless, further biomass particle size reduction below 40 or 36 mesh has very little impact on the biomass hydrolysis yields and conversion rates. The increase in specific surface area, degree of polymerization, porosity, and cellulose crystallinity significantly varies depending on milling type and time as well as the type of biomass. While, spruce and aspen chip cellulose crystallinity is reduced more efficiently by vibratory ball milling than by traditional ball milling. Compared to hammer milling, which creates finer bundles, disk milling enhances cellulose hydrolysis more. The amount of energy required to mechanically reduce the particle size of lignocellulosic biomass depends on the characteristics of the biomass and the ultimate particle size that is desired. Compared to agricultural leftovers, hardwoods take more energy to grow. Also, studies shows that milling treatment enhances the yield of biogas, bioethanol, and biohydrogen [5, 55, 56]. Milling is still unlikely economically feasible due to high energy requirements at large scale. Moreover, milling can be employed pre- and post-chemical pretreatment hence offers several benefits including low energy consumption, low solid liquid separation cost, lowers the production of fermentation inhibitors, demands a low liquid to solid ratio, and lessens the energy-intensive mixing of pretreatment slurries.

8.2.1.2 Steam Pretreatment

The processing of biomass using steam has gained wide attention and is the most frequently used technique. It is also called as steam explosion/autohydrolysis due to explosive action on fibers that leads to easily hydrolysis. This method uses high pressure saturated steam to treat the physically processed biomass in the temperature between 150°C and 230°C and pressure between 0.65 and 5.5 mPa [29, 57, 58]. To encourage hemicellulose hydrolysis, the pressure is sustained for a short period of time—a few seconds to a few minutes—and then released. Due to such pretreatment cellulose is more accessible and thus increases the digestibility of lignocellulosic biomass. Utilization of catalysts such as metal chlorides, organic and inorganic acids improve the efficacy of steam pretreatment process. The benefits of steam pretreatment include that it is a desirable process since it uses few chemicals, does not dilute the sugars produced excessively, requires little energy input, and has no negative environmental effects.

8.2.1.3 Liquid Hot Water Pretreatment

Liquid hot water pretreatment (LHW) involves heating the biomass in hot water under high pressure, which causes the hemicellulose to dissolve and the lignin to become more accessible to downstream processing. This results in higher yields of fermentable sugars, which can then be used to produce biofuels or other products. Depending on the biomass feedstock, hot water and biomass are combined during the LHW pretreatment process at temperatures between 160°C and 250°C and pressures between 10 and 25 mPa [6, 59]. The reaction time can range from 10 to 60 min. The resulting liquid contains dissolved hemicellulose and lignin. The hemicellulose can be recovered by evaporating the liquid and separating the solids from the resulting syrup. It is possible to recover the lignin as a liquid or solid byproduct. LHW pretreatment has several advantages over other pretreatment methods. It is a relatively mild process that uses only water and does not require the use of harsh chemicals. It is also a fast process that can be completed in less than an hour. However, some challenges associated with the LHW pretreatment are the high energy requirements associated with heating the water to high temperatures and pressures. These inhibitors can be removed by various methods, including adsorption or detoxification. LHW pretreatment is still in the process of being commercialized. While there has been significant research on the effectiveness of LHW pretreatment, there are currently only a few commercial-scale plants using this technology. Several companies that are actively working on commercializing LHW pretreatment are Renmatix, a biotechnology company based in the United States, which has developed a proprietary LHW pretreatment technology called Plantrose that can convert a wide range of biomass feedstocks into cellulosic sugars. In Europe, the Horizon 2020–funded project ValChem aims to develop a novel LHW pretreatment process that can be used to convert forestry residues and agricultural waste into biofuels and other high-value products. The project involves a consortium of companies and research institutions and aims to demonstrate the feasibility and scalability of the technology. Figure 8.2 shows biomass conversion techniques.

8.2.1.4 Dilute Acid Pretreatment

Dilute acid pretreatment is a method of pretreating lignocellulosic biomass, such as agricultural residues, forestry waste, and energy crops, to make them more amenable to downstream processing. The procedure involves the hydrolysis of hemicellulose and removal of lignin using dilute acids, usually sulfuric acid or hydrochloric acid, to make cellulose more accessible for enzymatic hydrolysis. During dilute acid pretreatment process biomass is subjected to an acid solution, typically 0.5%–2.0% sulfuric acid or hydrochloric acid, at high temperatures (100–220°C) and pressures (1–10 atm) for a specified time (usually 5–60 min) [7]. The acid solution breaks down the hemicellulose component of the biomass into its individual sugar monomers and

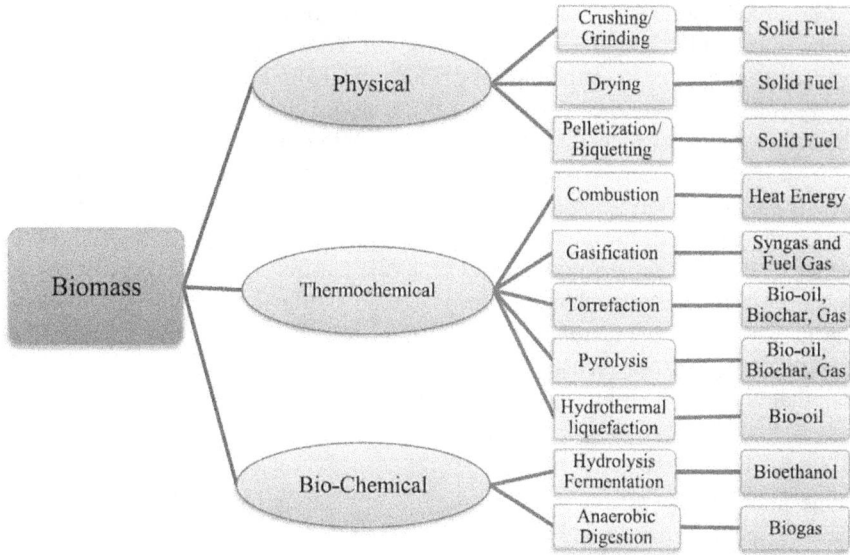

FIGURE 8.2   Biomass conversion techniques.

removes lignin by breaking its cross-linkages. The acid solution also acts as a cata-lyst for the hydrolysis of cellulose to glucose. After the acid treatment, the biomass is neutralized by adding a base, such as lime, to raise the pH and remove any remain-ing acid [31, 60, 61]. The remaining acid and contaminants are then flushed from the neutralized biomass with water. The cellulose component of the pretreated biomass is next exposed to enzymatic hydrolysis to turn it into glucose, which may then be used to make biofuels and other high-value chemicals. The advantages of dilute acid pretreatment include its effectiveness in removing lignin and hydrolyzing hemicel-lulose, as well as its compatibility with a wide range of lignocellulosic feedstocks [8, 62, 63]. However, there are some challenges associated with dilute acid pretreat-ment. The high temperatures and pressures required for the process can be energy-intensive, leading to high operating costs. The acid used in the process can also be corrosive, requiring specialized equipment and materials. Dilute acid pretreat-ment has been successfully demonstrated at the pilot and commercial scale. Some of the largest commercial-scale biorefineries that use dilute acid pretreatment are the POET-DSM Project Liberty plant in Emmetsburg, Iowa; the Beta Renewables plant in Crescentino, Italy; and Arundo donax by Rennovia Inc. company, California.

8.2.1.5   Ammonia Fiber or Freeze Explosion

The biomass is treated with liquid anhydrous ammonia during the ammonia fiber pretreatment (AFEX) process, which alters the structure of the biomass and makes it more suitable for subsequent processing. The process also reduces the formation of inhibitory compounds that can interfere with downstream processing and fer-mentation. Normally, the AFEX procedure is completed in two steps. The biomass is heated to between 75°C and 95°C in the first step before being subjected to liquid ammonia at a high pressure for 15 to 30 min [9]. The second step involves the rapid release of ammonia from the biomass, which results in swelling and fracturing of the biomass and additional exposure of the cellulose structure. Main advantages of the AFEX process are that it can be integrated with other pretreatment and conversion technologies, such as enzymatic hydrolysis, fermentation, and gasification [10, 64]. The commercialization of the technology is still limited due to the high cost of ammonia and the challenges associated with scaling up the process to commercial scale. One of the largest commercial-scale demonstrations of the AFEX process is the Raizen cellulosic ethanol plant in Brazil. The facility, which started operating in 2014, can manufacture 40 million liters of cellulosic ethanol annually using sugarcane bagasse as a feedstock. Praj Industries can manufacture 1 million liters of bioethanol annually using sugarcane bagasse as a feedstock.

8.2.1.6   Lime Pretreatment

Lime pretreatment process involves the use of calcium hydroxide, also known as lime, to modify the structure of the biomass by changing its pH and solubilizing some of the hemicellulose components. The lime pretreatment process can be divided into three stages: pre-soaking, lime impregnation, and post-soaking [11, 65, 66]. The first step of lime pretreatment is to pre-soak the biomass in water to remove any dirt or impurities that may be present. The pre-soaking stage also helps to hydrate the biomass and open up the pores, making it more susceptible to the lime impregnation step. While in lime impregnation, the pre-soaked biomass is mixed with a lime solution to impregnate the material with calcium ions. The calcium ions cause a disruption of the lignocellulosic structure, breaking down the bonds between the hemicellulose and cellulose components. The lime solution also increases the pH of the biomass, creating an alkaline environment that promotes the solubilization of hemicellulose. The lime impregnation stage can be performed in different ways, including batch, semi-continuous, or continuous processes [12, 67]. The effectiveness of the lime pretreatment process depends on various factors, including biomass type, concentration, and operating conditions. Lime pretreatment has been commercialized for various applications, including biofuels, pulp and paper, and biorefinery industries. Several companies are currently using lime pretreatment as part of their commercial processes. Companies such as Fiberight, Renmatix, LanzaTech, GranBio, and Praj Industries are using lime pretreatment to produce biofuels from various feedstocks.

8.2.1.7   Organosolv Pretreatment

This treatment involves the use of organic solvents to convert the lignocellulosic structure leaving behind a more accessible cellulose material for conversion into valuable products. The organosolv pretreatment process is categorized into three major stages: pre-treatment, extraction, and recovery. First, the biomass is pretreated by grinding it into small particles and mixing it with the organic solvent. The pre-treatment stage helps to increase the biomass surface area and allows better penetration of the solvent. During extraction the pre-treated biomass is mixed with an organic solvent, such as ethanol, ethylene glycol, methanol, or acetone, and heated to convert its lignocellulosic structure [13]. The solvent dissolves the hemicellulose and lignin components, leaving behind a more accessible cellulose material. The extraction can be performed under acidic or alkaline conditions, depending on the desired product. After the extraction stage, the solvent and extracted components are separated from the cellulose material. The extracted components can be further processed to produce biobased products such as sugars, lignin, and chemicals, while the cellulose material can be converted into biofuels, paper, or other products. Organosolv pretreatment has several advantages over other pretreatment methods, including high selectivity, low toxicity, and the ability to produce high-quality lignin [32]. The overall cost can be decreased by recovering and reusing the solvents that were utilized. However, one of the main challenges of organosolv pretreatment is the high cost of the organic solvents used in the process. Additionally, the recovery of the extracted components can also pose technical and economic challenges. Several companies have adopted this technology including Borregaard, Stl Nordic Oy, Renmatix, Domtar Corporation, Resolute Forest Products, and Sweetwater Energy are using organosolv pretreatment to produce biofuels from various feedstocks.

8.2.1.8   Carbon Dioxide Explosion Pretreatment

A relatively recent pretreatment technique for lignocellulosic biomass uses carbon dioxide (CO_2) to produce an explosion-like event in the presence of water which helps to break down the biomass structure and improve its accessibility for further processing. The process involves the use of high-pressure carbon dioxide, which is injected into a vessel containing water and the biomass to be treated. The pressure and temperature are increased rapidly, causing the water to turn into steam, and creating a sudden increase in volume and pressure within the vessel [30]. Carbon dioxide explosion pretreatment has several advantages over other pretreatment methods. For instance, it does not require the use of chemicals or enzymes, making it a more environmentally friendly process. Additionally, it is a rapid process that can be completed in minutes, and the high-pressure carbon dioxide used in the process can be recovered and reused [14]. The significant energy consumption needed to maintain high pressure and temperature conditions, however, is one of the key difficulties with carbon dioxide explosion preparation. There are currently only a few companies that are actively pursuing the commercialization of carbon dioxide explosion pretreatment including Finnish company KaiCell Fibers, AFYREN, and ENGIE Pvt Ltd.

8.2.2   THERMO-CHEMICAL CONVERSION

This conversion is a process that involves the conversion of biomass into energy or fuel by applying heat and chemical reactions. Figure 8.3 shows different thermo-chemical techniques for biomass conversion. It can transform a variety of biomass feedstocks, such as forestry and agricultural waste, seeds, municipal solid waste, and energy crops, into energy or fuel. These wastes could be a potential raw material for a variety of products, including biofuels, chemicals, and electricity. Thermo-chemical

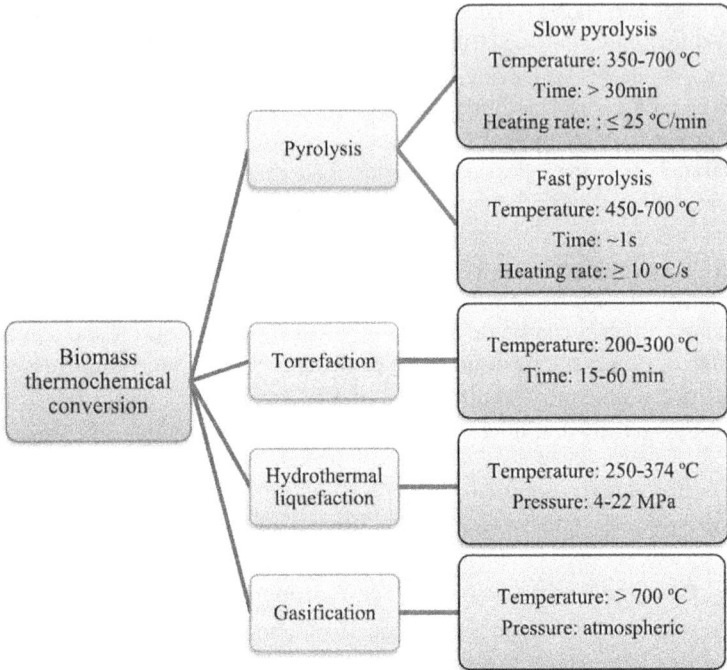

FIGURE 8.3 Biomass thermochemical conversion.

conversion can also be used to produce energy and fuel in a decentralized manner [36], which can reduce the dependence on centralized energy systems [15, 68]. However, there are also several challenges associated with thermo-chemical conversion. For instance, the process requires significant amounts of energy, which can make it less efficient compared to other methods of biomass conversion.

8.2.2.1 Combustion

The chemical process called combustion releases heat and light when a fuel is rapidly oxidized in the presence of oxygen. This process is commonly used to generate energy and is the basis for many power generation technologies, including steam turbines, internal combustion engines, and gas turbines. The combustion process can be divided into four stages: ignition, flame propagation, combustion, and extinction. Ignition stage occurs when the fuel is heated to its ignition temperature, causing it to react with oxygen in the air and release energy in the form of heat and light. The ignition temperature is different for each fuel and is the minimum temperature required to initiate combustion [16, 69]. While flame propagation stage occurs when the heat and light generated by the initial ignition cause the fuel and oxygen to react and produce a flame. The flame front moves through the fuel and oxygen mixture, causing the combustion to spread. Combustion stage occurs when the fuel and oxygen react to produce heat, light, and various combustion products, including water vapor and trace amounts of gases such as methane and carbon dioxide [33]. The amount and type of combustion products produced depend on the fuel being burned and the conditions of combustion, such as temperature and pressure. Last, the extinction stage occurs when the fuel and oxygen are no longer available to sustain the combustion process, causing the flame to extinguish. The combustion process produces various products depending on the type of fuel burned and the conditions of combustion, including temperature and oxygen availability. The primary products of combustion are heat, light, and various products including gases and particulate matter.

The combustion process has been commercialized for many decades and is used extensively in a wide range of industries for various applications such as power generation, transportation, and heating [17]. The technology has been refined over time, resulting in more efficient and cleaner-burning combustion systems. In the power generation sector, combustion-based systems such as coal-fired power plants and gas turbines have been widely deployed for many years. There has been a shift in recent years toward cleaner energy production including the use of combustion-based systems that continue to be crucial for producing steady electricity. In the transportation sector, combustion engines powered by fossil fuels such as gasoline and diesel are still the dominant technology, but there has been a growing shift towards electric vehicles powered by batteries or fuel cells [34, 70]. However, combustion engines will play an important role in transportation for many years to come, particularly in heavy-duty applications such as shipping and aviation. In the heating sector, combustion-based systems such as furnaces and boilers are widely used for residential, commercial, and industrial applications. Alternative heating methods, such as heat pumps and district heating systems that utilize renewable energy sources, are gaining popularity.

8.2.2.2 Gasification

By exposing fuels to high temperatures and a controlled amount of oxygen or steam, gasification is the process of turning solid or liquid fuels into gas. The gasification process comprises four main stages including drying, combustion, pyrolysis, and reduction. In the drying stage, the fuel is heated to remove any moisture that may be present. In the pyrolysis stage, the fuel is heated to between 600°C and 800°C in an oxygen-free environment, which causes it to break down into char, tar, and gas [12, 18, 24]. While the tar is a liquid that can be utilized as a feedstock for the chemical industry, the char is a solid residue that can be used as fuel or a source of carbon for other processes. In the combustion stage, oxygen is introduced to the gasification reactor, and the gas produced in the pyrolysis stage is burned to generate heat. This heat is used to maintain the temperature in the gasification reactor and to drive the gasification process. Finally, in the reduction stage, the remaining carbon dioxide and water vapor in the gas are removed, leaving a clean syngas that can be used for a variety of applications [35, 55]. Gasification exhibits several advantages such as it can be easily transported and stored, making it suitable for use in remote locations and produces low-emission fuel. Also, it can be combined with other technologies such as co-firing with natural gas, which can help increase the efficiency of power generation and reduce costs. However, gasification also has some drawbacks including the requirement of high energy, generation of complex mixture of gases and formation of toxic byproducts such as ash and tar.

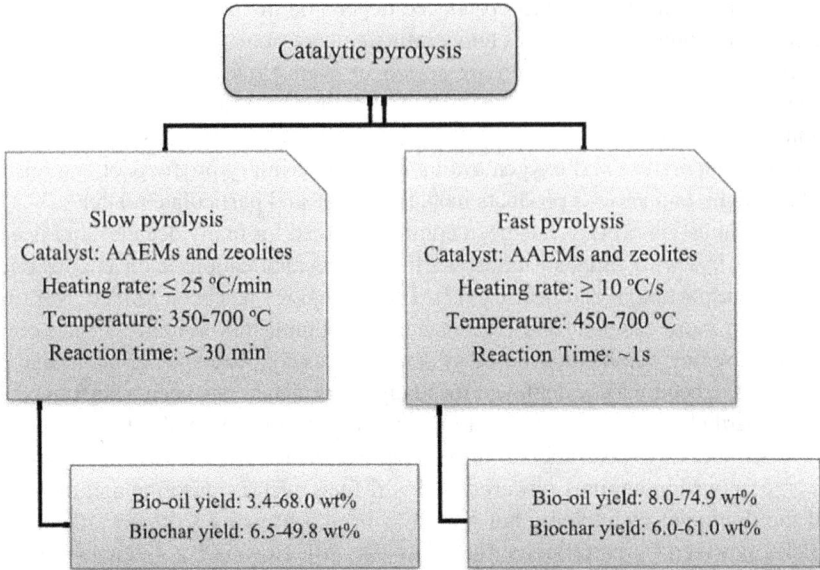

FIGURE 8.4 Catalytic pyrolysis process of biomass.

8.2.2.3 Pyrolysis

Pyrolysis is a type of thermal decomposition in which organic molecules are heated without oxygen to convert them into more basic chemical compounds. The obtained products include gases, liquids, and solids that can be used for a variety of applications. Pyrolysis is often used to convert biomass and waste materials into biofuels, such as bio-oil, biochar, and syngas. There are three main types of pyrolysis processes: fast pyrolysis, slow pyrolysis, and intermediate pyrolysis as shown in Figure 8.4. In fast pyrolysis, the biomass is heated for a brief period of time—usually less than 2 seconds—to temperatures between 500°C and 600°C [19, 39]. This process produces a high yield of bio-oil, with smaller amounts of char and gas. In slow pyrolysis, the material is heated to lower temperatures of around 300–500°C for a longer period of time, typically several hours [37, 56]. This process produces a higher yield of char and a lower yield of bio-oil and gas. Intermediate pyrolysis is a combination of fast and slow pyrolysis, with the material being heated to temperatures of around 350–450°C for several minutes [20]. This process produces a more balanced yield of bio-oil, char, and gas.

8.3 ADVANCEMENTS IN BIOWASTE REFINING PROCESS AND TECHNIQUES FOR THE DEVELOPMENT OF BIOENERGY AND BIOPRODUCTS

Biowaste was employed in the production of a wide variety of industrial items throughout the 20th century. Biowaste can be converted into biofuels using a variety of thermal and biobased processes. Recently, several biowaste-consuming microorganisms were activated to produce biobutanol and bioethanol with increased yields [21, 47]. A few research have shown the enhanced value of biopolymers such as polyhydroxy butyrates, polyurethane, and polyhydroxy alkonates from agricultural waste using maturation and biocatalysis procedures. For higher-quality bioproducts, which are necessary for companies operating in the bioeconomy to be financially sustainable, many studies have advocated creative economic strategies. The conversion of lignocellulosic biomass into various industrially significant bioproducts through thermochemical and biological processes is depicted in Figure 8.5. There are several ways to get organisms and their thermophilic proteins to work together dynamically, including genetic alteration and enzyme saccharification, solid fermentations, fed batch fermentation, and thermotolerant enzyme saccharification [22, 53]. Moreover, shrimp shells, a leftover from seafood restaurants, were traditionally used to make chitosan. Many procedures that are economically feasible exist, but they are not necessarily compatible with certain types of biowaste, such as hydrotreated herbal oil, ethanol fermentation, fatty acid methyl esters (FAME) biodiesel, or biomethane from anaerobic digestion (AD). Table 8.3 summarizes the current state of biowaste feedstock development for bioenergy and bioproducts.

8.3.1 ANAEROBIC DIGESTION AND CO-DIGESTION
FOR THE GENERATION OF BIOENERGY

Global interest has been generated in the production of sustainable biogas by the anaerobic fermentation, which also produces biohydrogen and biomethane. Yard waste (YW) and biowaste can be effectively processed to create bioenergy using solid-state anaerobic digestion. A large amount of methane (of 431 mL/gV) was produced after the co-digestion of yard waste with microwave pretreatment. Targeted bioproducts include hydrolytic compounds (derived from the autochthonous microbiome's cellulases and proteases), biosurfactants (produced from *Starmerella bombicola*'s sophorolipids), and biopesticides. It was shown that the addition of co-biomass greatly enhances the production of biogas (12%) and CH_4, which may be a suitable strategy for a sustainable plant [23, 38]. Several studies have developed a strategy for the installation of micro-scale and decentralized anaerobic digestion systems in urban and peri-urban settings. Additionally, of all the trace metals, cobalt chloride emits the fewest greenhouse gases when evaluated in CO_2-equivalent terms, with a concentration of 1.89 g/m^3. The attapulgite expansion affected the energy required to produce methane and increased methane production by 9.13% to 38.25%. Attapulgite produces a higher methane delivery of 221.56 mL/g volatile solids when paired with a 11.10 g/L expansion stacking [24]. Attapulgite enhanced acetogenesis, hydrolysis, and methanogenesis as shown by increases in beta-glucosidase, protease, coenzyme F420, and dehydrogenase activities. Additionally, there are much more methanogenic small-scale organisms (Clostridiales, Syntrophobacterales, and Fibrobacterales) as well as acetogenic and hydrolytic bacteria [25, 26, 59, 63].

TABLE 8.3

Current State of Biowaste Feedstock Development for Bioenergy and Bioproducts

S. No	Country	Strategies	Features	Raw Material	Bioproducts	Advantages
1	Barcelona, Spain	Solid state fermentation	The maximum spore generation by B. thuringiensis was 9.23 ± 0.05 (109) CFU g−1 DM and 3.81 ± 0.22 (107) CFU g−1 DM.	Biowaste	Hydrolytic enzymes, biosurfactants, and biopesticides	Contribute to the replacement of pesticides made with chemicals
2	Spain	Solid-state fermentation	Production of Bioethanol was 0.32 to 0.53 g/g	Biowaste	Bioethanol	For the development of a new SSF process, choosing the right bioreactor configuration is crucial.
3	France	Anaerobic digestion, thermochemistry, and biological therapy	Algal biomass, fatty waste, and the entire organic part of municipal solid waste	Organic solid sludge	Biogas	Supplements with plenty of nutrients for agriculture
4	China	Fed batch fermentation, Strain development and genome lumbering	Alter biowaste to bioproducts and improves the yield of cellulose	Lignocellulosic biomass	Bioproducts	Provides high bioproducts yield
5	Lodz, Poland	Chemical pre-treatments like ammonia pressurization, organosolv, lime, acid/base	Obtaining and preserving the beet, Beets are fluffed, cleaned, and sieved into thin slices before being heated in a hot water bath.	Sugar Beet Pulp waste	Pure sugar	Generate pure sugar from sugar beets for the cheapest possible price.
6	Italy	Anaerobic digestion	Utilized molten carbonate fuel cell, continuous stirred tank reactor, and biological hydrogen potential in an internal combustion engine.	Food waste	Bio-degradable substrates	Produced energy-recovering biodegradable substrates
7	Poland	Combustion, gasification and fermentation	Produce high-value biopolymers from biowaste, such as biopharmaceuticals, cosmetics, nutrition, chemicals, fertilizers, and materials	Biowaste	Biorefinery, Bioproduct and Bioenergy	Greater awareness of the biggest technology gaps, preventative long-term economics, and practical use of high-quality bioproducts

8.3.2 Composting or Co-Composting Techniques

There are number of benefits of composting biowaste including valuation, sterilization, stability, and reduction of waste biomasses. Vermin composting of biowaste and natural matter biodegradation was found to increase the physicochemical properties of the compost mixture while reducing nitrogen loss and other emissions. The crumbling of olive scraps during composting is supposed to produce better interfacial covering materials. Controlled fluid expansion into solid manure compost brought it significantly closer to the desired level of moisture, effectively managed the thermophilic stage, and decreased leachate production. Co-composting with biochar resulted in a more complex substance than compost having more neutral pH and better moisture retention. Irrespective of the co-substrate or mixing ratio, it is shown that a single rotation per week yields the lowest product quality. Composting can be done successfully in a lab setting in an air pack bioreactor with an oxygen concentration of 15.75%–23.74% and a carbon dioxide content of 0.15%–8.94% [5, 9, 69]. Ammonia recycling in the air bag bioreactor is effective, increasing the nitrate concentration from 73.48 to 1081.69 mg/kg [27]. Nevertheless, the ammonia concentration fluctuates owing to unpredictable pumping and waste gas fatigue. Growing attention is being shown in the black soldier fly, *Hermetiaillucens*, as a potential method for transforming biowaste into protein- and fat-induced biomass suited for animal feeding [28, 38]. Moreover, larvae may be regulated and used in unconventional animal feed in place of fishmeal, and the leftover material can be composted for soil improvement. Synergistic effect of fly ash with biomass and kitchen garbage in a co-composting process was investigated and found that maximum organic degradation occurs with a rate of 0.550 d^1. Additionally, it was investigated whether fly ash might be added to organic wastes to increase their microbial and enzymatic activity during in-vessel composting. Also, few studies demonstrated the existence of several chemical changes to plant leftovers during composting process which minimizes the soil mineralization. Figure 8.5 shows the conversion of lignocellulosic biomass through thermochemical and biological to various industrially useful bioproducts.

8.4 IMPLICATION AND CHALLENGES

The lack of upgrade, variations in measurable strategies, rigid framework boundaries, changes in product usability, product type selectivity, precise information accessibility, and local conditions are some of the limitations and challenges faced by circular bioeconomy strategies. Analyzing the circular bioeconomy for various types of biowaste is difficult since it depends on a number of variables, including the amount and type of biowaste produced, energy inputs, and carbon emissions. The socio-economic difficulties that Europe and the rest of the world are grappling with, such as population increase, unpredictability surrounding food security, and harmful effects of human activity on climate change are driving the development of the bioeconomy. Table 8.4 lists the effects of several biorefining processes along with suggestions for preserving a circular bio-economy. It is important to notice

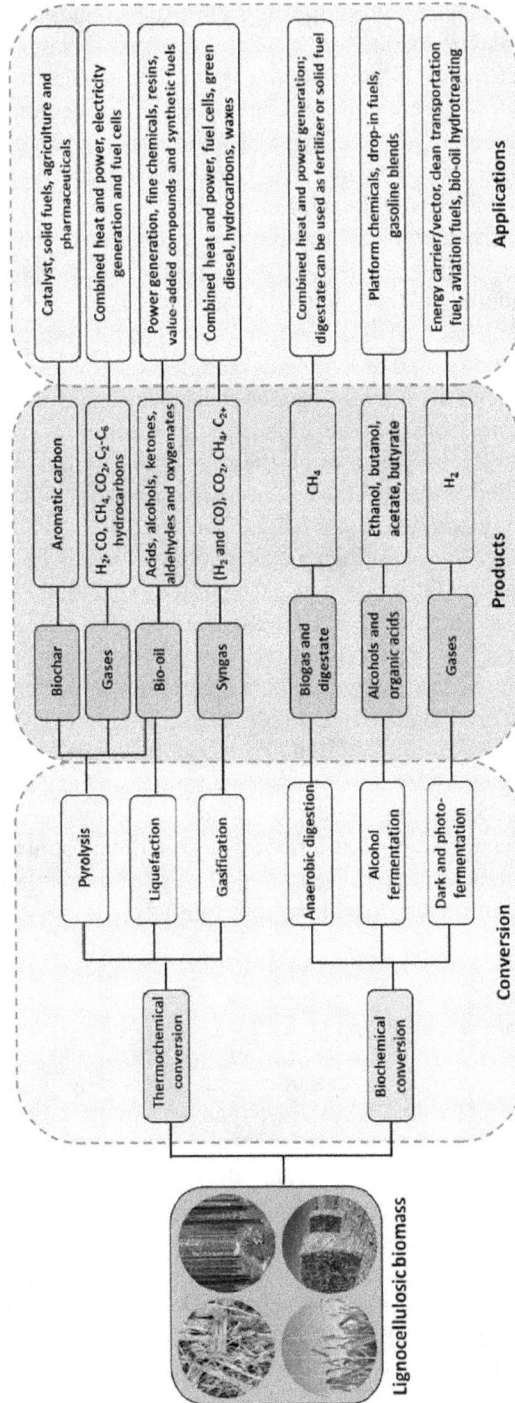

FIGURE 8.5   Conversion of lignocellulosic biomass through thermochemical and biological to various industrially useful bioproducts.

TABLE 8.4

Effects of Several Biorefining Processes along with Suggestions for Preserving a Circular Bio-Economy

S. No	Biomass	Impact Category	Evaluation Methods	Remarks
1	Bagasse transformed via gasification and saccharification	Net present value (NPV), economic parameter uncertainty during the plant's 25-year life span	Techno-economic analysis	Production of lactic acid has a larger NPV and is more successful while production of alcohols is profitable if government offers a 39% subsidy.
2	Sugarcane bagasse transformed via saccharification	Abiotic depletion potential, ozone layer depletion potential, and fresh water ecotoxicity	Life cycle assessment	Benefits related to economic and environment are enhanced with co-combustion of coal.
3	Olive tree pruning transformed via saccharification	Ratio of end usage energy	Energy analysis	Energy matrix significantly increases with increase in environmental regime.
4	Olive tree pruning transformed via saccharification	NPV and discounted cash flow	Economic analysis	Diminishing deficit of bioethanol production with monetizing the contribution of biofuel production
5	Sugarcane bagasse transformed via fermentation	Abiotic depletion potential and fresh water ecotoxicity	Life cycle assessment	Generation of high revenues that mitigate the effect of biorefineries in monetary terms
6	Wood residues transformed via oxidation	Terrestrial ecotoxicity potential, marine eutrophication, and acidification potential	Life cycle assessment	Assesses the process or product
7	Rice husk transformed via pyrolysis	Production cost, fixed operating cost, market price, market selling price	Techno-economic analysis	Economic and environmental benefit promotes circular bioeconomy.
8	Sugar beet pulp transformed via hydrolysis	Abiotic depletion potential, ozone layer depletion potential, and fresh water ecotoxicity	Life cycle assessment	Transformation of biomass presents significant economic and environmental challenges.
9	Dry wood transformed via hydrolysis	Climate change and marine aquatic ecotoxicity potential	Life cycle assessment	Sustainability of the environment and the bioeconomy are positively impacted by the utilization of alternative resources.
10	Vine shoots transformed via hydrolysis	Terrestrial ecotoxicity potential, terrestrial acidification, marine eutrophication, and acidification potential	Life cycle assessment	By raising the revenues and environmental profile, the closing-the-loop technique contributes to waste reduction.

how closely the source nexus notion is related to the circular economy and the bio-economy. Yet, there are still a few areas that require immediate attention if it is to become more feasible.

8.5 FUTURE IMPLICATIONS

Reducing the quantity of landfilled biomass is the main goal of sustainable biowaste management. Several nations also dump substantial volumes of unclassified munici-pal, forestry, and agricultural trash. As a result, biowaste handling is responsible for the greatest share of greenhouse gas emissions. Utilization is one of the major fac-tors that affects the creative development of biomass-related products and several commercialization activities. While operations for biomass conversion are organized by the supply chain. Low collection costs and poor rubbish cleaning continue to be hallmarks of plans for solid waste biorefineries in middle- and low-income countries. Around 18% of all anthropogenic methane emissions worldwide, or nearly 90% of all biowaste area outflows, are caused by landfills that also house trash. Biowaste is a resource for generating material energy or recovering materials that might assist rural communities in achieving sustainable rural development. Domestic composting is more rational and financially possible if biowaste is segregated at the source and the fundamental requirements are successfully controlled. It is essential to optimize conversion procedures for biowaste treatment facilities to generate energy. One of the fundamental components of advanced synergetic intelligence is the mixed propor-tion that is most productive for development, particularly in the creation of methane and digestate value. Anaerobic co-digestion optimized for vitality recovery improves natural performance while reducing carbon footprint. By establishing a connection between trash creation and plans for future circular economy consumption in inner cities, this technique can shut the loop. Two workable frameworks for the circular economy system were developed as a consequence of the application study of bio-waste biorefineries. These frameworks involve supporting niche markets, expanding the bioeconomy, and lowering transportation costs related to distributing biomass feedstocks. These biomass biorefinery designs might provide a remedy by establish-ing specialized markets and expediting the transportation of biowaste materials for sustainable production.

- To encourage the conversion to higher-quality bioproducts, thereby rein-forcing the commercialization function.
- To transform linear economic prototype models into a circular bioeconomic system.
- In order to facilitate movement and preserve stream order under the varied effect of bioresources, production variables connected with the bioresource assembly location must be developed.

Genetic bioengineering techniques may assist boost the output of biomass mixes through faster production rates and customized biochemical structures. The regen-erative bioindustry creates complex bioproducts including collagen, emulsifiers, sur-factants, thickeners, colors, and anticancer medications. To make these biocosmetics, biomass, algal biomass, or microbials may be employed. The best sustainability evaluations are those that assist innovative producers and businesses in steadily improving ecological, social, and economic conditions. Even with centrality, the classic three-pillar model's superior administration is jeopardized when it is enlarged, necessitating its implementation in sustainability assessment systems. The bioeconomy of today requires ongoing technique innovation to handle the growing variety of biomass-derived goods.

8.6 CONCLUSION

The biowaste biorefinery concept has garnered a lot of interest as a viable alternative for biomass-based products and energy. It is feasible to increase energy security and lower greenhouse gas emissions by turning readily available lignocellulosic biomass into biofuels for use as transportation fuels. Numerous physicochemical, structural, and compositional variables reduce the digestibility of the cellulose and hemicellulose found in lignocellulosic biomass. The chapter examines anaerobic digestion/co-digestion, hydrothermal, composting/co-composting, and thermochemical technologies that are based on a carbon neutrality concept with the integration of life cycle assessments. The biorefineries that are now being built use biowaste as a feedstock and have a strong chance of climbing up the biowaste hierarchy in comparison to coupling waste in the production sector.

REFERENCES

[1] Chaudhary, G., Chaudhary, N., Saini, S., et al. Assessment of pretreatment strategies for valorization of lignocellulosic biomass: Path forwarding towards lignocellulosic biorefinery. *Waste Biomass Valori.* 1–36, 2023.

[2] Saravanan, A., Kumar, P. S., Badawi, M., et al. Valorization of micro-algae biomass for the development of green biorefinery: Perspectives on techno-economic analysis and the way towards sustainability. *J. Chem. Eng.* 139754, 2022.

[3] Amjith, L. R., Bavanish, B. A review on biomass and wind as renewable energy for sustainable environment. *Chemosphere.* 293, 133579, 2022.

[4] Saravanan, A., Kumar, P. S., Jeevanantham, S., et al. Recent advances and sustainable development of biofuels production from lignocellulosic biomass. *Bioresour. Technol.* 344, 126203, 2022.

[5] Irfan, M., Elavarasan, R. M., Ahmad, M., et al. Prioritizing and overcoming biomass energy barriers: Application of AHP and G-TOPSIS approaches. *Technol. Forecast. Soc. Change.* 177, 121524, 2022.

[6] Li, T., Zhi, D. D., Guo, Z. H., et al. 3D porous biomass-derived carbon materials: Biomass sources, controllable transformation and microwave absorption application. *Green Chem.* 24(2), 647–674, 2022.

[7] Tezer, Ö., Karabağ, N., Öngen, A., et al. Biomass gasification for sustainable energy production: A review. *Int. J. Hydrog. Energy.* 27, 2022.

[8] Basak, B., Kumar, R., Bharadwaj, A. S., et al. Advances in physicochemical pretreatment strategies for lignocellulose biomass and their effectiveness in bioconversion for biofuel production. *Bioresour. Technol.* 369, 128413, 2023.

[9] Song, H., Yang, G., Xue, P., et al. Recent development of biomass gasification for H2 rich gas production. *Appl. Energy Combust. Sci.* 10, 100059, 2022.

[10] Ifeanyi-Nze, F. O., Omiyale, C. O. Insights into the recent advances in the pretreatment of biomass for sustainable bioenergy and bio-products synthesis: Challenges and future directions. *Eur. J. Sustain. Dev.* 7(1), 2022.

[11] Banerjee, N. Biomass to energy—an analysis of current technologies, prospects, and challenges. *BioEnergy Res.* 16(2), 683–716, 2023.

[12] Mariyam, S., Shahbaz, M., Al-Ansari, T., et al. A critical review on co-gasification and co-pyrolysis for gas production. *Renew. Sust. Energ. Rev.* 161, 112349, 2022.

[13] Velvizhi, G., Jacqueline, P. J., Shetti, N. P., et al. Emerging trends and advances in valorization of lignocellulosic biomass to biofuels. *J. Environ. Manage.* 345, 118527, 2023.

[14] Marsh, A. T., Velenturf, A. P., Bernal, S. A. Circular economy strategies for concrete: Implementation and integration. *J. Clean. Prod.* 132486, 2022.

[15] Liu, Q., Trevisan, A. H., Yang, M. A framework of digital technologies for the circular economy: Digital functions and mechanisms. *Bus Strategy Environ.* 31(5), 2171–2192, 2022.

[16] Neves, S. A., Marques, A. C. Drivers and barriers in the transition from a linear economy to a circular economy. *J. Clean. Prod.* 341, 130865, 2022.

[17] Bhushan, S., Jayakrishnan, U., Shree, B., et al. Biological pretreatment for algal biomass feedstock for biofuel production. *J. Environ. Chem. Eng.* 11(3), 109870, 2023.

[18] Awan, U., Sroufe, R. Sustainability in the circular economy: Insights and dynamics of designing circular business models. *Appl. Sci.* 12(3), 1521, 2023.

[19] Ganewatta, M. S., Lokupitiya, H. N., Tang, C., 2019. Lignin biopolymers in the age of controlled polymerization. *Polymers.* 11(7), 1176, 2022.

[20] Chauhan, C., Parida, V., Dhir, A. Linking circular economy and digitalisation technologies: A systematic literature review of past achievements and future promises. *Technol. Forecast. Soc. Change.* 177, 121508, 2022.

[21] Yu, Z., Khan, S. A. R., Umar, M. Circular economy practices and industry 4.0 technologies: A strategic move of automobile industry. *Bus Strategy Environ.* 31(3), 796–809, 2022.

[22] Yang, M., Chen, L., Wang, J., et al. Circular economy strategies for combating climate change and other environmental issues. *Environ. Chem. Lett.* 1–26, 2022.

[23] Kumar, J. A., Sathish, S., Prabu, D., et al. Agricultural waste biomass for sustainable bioenergy production: Feedstock, characterization and pre-treatment methodologies. *Chemosphere.* 331, 138680, 2023.

[24] Jain, A., Sarsaiya, S., Awasthi, M. K., et al. Bioenergy and bio-products from bio-waste and its associated modern circular economy: Current research trends, challenges, and future outlooks. *Fuel.* 307, 121859, 2023.

[25] Saravanan, A., Yaashikaa, P. R., Kumar, P. S., et al. A comprehensive review on techno-economic analysis of biomass valorization and conversional technologies of lignocellulosic residues. *Ind. Crops Prod.* 200, 116822, 2023.

[26] Lehmann, C., Cruz-Jesus, F., Oliveira, T., et al. Leveraging the circular economy: Investment and innovation as drivers. *J. Clean. Prod.* 360, 132146, 2022.

[27] Sehnem, S., De Queiroz, A. A. F. S., Pereira, S. C. F., et al. Circular economy and innovation: A look from the perspective of organizational capabilities. *Bus. Strategy Environ.* 31(1), 236–250, 2022.

[28] Rusch, M., Schöggl, J. P., Baumgartner, R. J. Application of digital technologies for sustainable product management in a circular economy: A review. *Bus. Strategy Environ.* 78, 2022.

[29] Ogunmakinde, O. E., Egbelakin, T., Sher, W. Contributions of the circular economy to the UN sustainable development goals through sustainable construction. *Resour. Conserv. Recycl.* 178, 106023, 2022.

[30] Hoang, A. T., Nguyen, X. P., Duong, X. Q., et al. Steam explosion as sustainable biomass pretreatment technique for biofuel production: Characteristics and challenges. *Bioresour. Technol.* 129398, 2023.

[31] Prasad, B. R., Padhi, R. K., Ghosh, G. A review on key pretreatment approaches for lignocellulosic biomass to produce biofuel and value-added products. *Int. J. Environ. Sci. Technol.* 20(6), 6929–6944, 2023.

[32] Li, J., Liu, B., Liu, L., et al. Pretreatment of poplar with eco-friendly levulinic acid to achieve efficient utilization of biomass. *Bioresour. Technol.* 376, 128855, 2023.

[33] Mohanakrishna, G., Modestra, J. A. Value addition through biohydrogen production and integrated processes from hydrothermal pretreatment of lignocellulosic biomass. *Bioresour. Technol.* 369, 128386, 2023.

[34] Valladares-Diestra, K. K., De Souza Vandenberghe, L. P. The potential of imidazole as a new solvent in the pretreatment of agro-industrial lignocellulosic biomass. *Bioresour. Technol.* 372, 128666, 2023.

[35] Basak, B., Kumar, R., Bharadwaj, A. S., et al. Advances in physicochemical pretreatment strategies for lignocellulose biomass and their effectiveness in bioconversion for biofuel production. *Bioresour. Technol.* 369, 128413, 2023.

[36] Lee, J., Kim, S., You, S. Bioenergy generation from thermochemical conversion of lignocellulosic biomass-based integrated renewable energy systems. *Renew. Sust. Energ. Rev.* 178, 113240, 2023.

[37] Wang, Q., Zhang, X., Cui, D., et al. Advances in supercritical water gasification of lignocellulosic biomass for hydrogen production. *J. Anal. Appl. Pyrolysis.* 170, 105934, 2023.

[38] Awasthi, M. K., Sar, T., Gowd, S. C., et al. A comprehensive review on thermochemical, and biochemical conversion methods of lignocellulosic biomass into valuable end product. *Fuel.* 342, 127790, 2023.

[39] Osman, A. I., Farghali, M., Ihara, I., et al. Materials, fuels, upgrading, economy, and life cycle assessment of the pyrolysis of algal and lignocellulosic biomass: A review. *Environ. Chem. Lett.* 21(3), 1419–1476, 2023.

[40] Mohanakrishna, G., Modestra, J. A. Value addition through biohydrogen production and integrated processes from hydrothermal pretreatment of lignocellulosic biomass. *Bioresour. Technol.* 369, 128386, 2023.

[41] Woiciechowski, A. L., Neto, C. J. D., De Souza Vandenberghe, et al. Lignocellulosic biomass: Acid and alkaline pretreatments and their effects on biomass recalcitrance—Conventional processing and recent advances. *Bioresour. Technol.* 304, 122848, 2020.

[42] Ullah, A., Zhang, Y., Liu, C. Process intensification strategies for green solvent mediated biomass pretreatment. *Bioresour. Technol.* 128394, 2022.

[43] Zhang, Y., Ding, Z., Hossain, M. S., et al. Recent advances in lignocellulosic and algal biomass pretreatment and its biorefinery approaches for biochemicals and bioenergy conversion. *Bioresour. Technol.* 128281, 2022.

[44] Sidana, A., Yadav, S. K. Recent developments in lignocellulosic biomass pretreatment with a focus on eco-friendly, non-conventional methods. *J. Clean. Prod.* 335, 130286, 2022.

[45] Yeganeh, F., Chiewchan, N., Chonkaew, W. Hydrothermal pretreatment of biomass-waste-garlic skins in the cellulose nanofiber production process. *Cellulose.* 29(4), 2333–2349, 2022.

[46] Del Mar Contreras-Gamez, M., Galan-Martin, A., Seixas, N., da Costa Lopes, A. M., et al. Deep eutectic solvents for improved biomass pretreatment: Current status and future prospective towards sustainable processes. *Bioresour. Technol.* 128396, 2022.

[47] Varjani, S., Sivashanmugam, P., Tyagi, V. K., et al. Breakthrough in hydrolysis of waste biomass by physico-chemical pretreatment processes for efficient anaerobic digestion. *Chemosphere.* 294, 133617, 2022.

[48] Mohanakrishna, G., Modestra, J. A. Value addition through biohydrogen production and integrated processes from hydrothermal pretreatment of lignocellulosic biomass. *Bioresour. Technol.* 369, 128386, 2023.

[49] Scapini, T., Dos Santos, M. S., Bonatto, C., et al. Hydrothermal pretreatment of lignocellulosic biomass for hemicellulose recovery. *Bioresour. Technol.* 342, 126033, 2021.

[50] Ruiz, H. A., Sganzerla, W. G., Larnaudie, V. Advances in process design, techno-economic assessment and environmental aspects for hydrothermal pretreatment in the fractionation of biomass under biorefinery concept. *Bioresour. Technol.* 128469, 2022.

[51] Selig, M. J., Tucker, M. P., Sykes, R. W. Lignocellulose recalcitrance screening by integrated high-throughput hydrothermal pretreatment and enzymatic saccharification. *Ind. Biotechnol.* 6(2), 104–111, 2022.

[52] Charnnok, B., Laosiripojana, N. Integrative process for rubberwood waste digestibility improvement and levulinic acid production by hydrothermal pretreatment with acid wastewater conversion process. *Bioresour. Technol.* 360, 127522, 2022.

[53] Sun, S. F., Yang, H. Y., Yang, J., et al. Structural characterization of poplar lignin based on the microwave-assisted hydrothermal pretreatment. *Int. J. Biol. Macromol.* 190, 360–367, 2021.

[54] Tezer, Ö., Karabağ, N., Öngen, A., et al. Biomass gasification for sustainable energy production: A review. *Int. J. Hydrog. Energy.* 47(34), 15419–15433, 2022.

[55] Amjith, L. R., Bavanish, B. A review on biomass and wind as renewable energy for sustainable environment. *Chemosphere.* 293, 133579, 2022.

[56] Su, Z., Yang, Y., Huang, Q., et al. Designed biomass materials for "green" electronics: A review of materials, fabrications, devices, and perspectives. *Prog. Mater. Sci.* 125, 100917, 2022.

[57] Buffi, M., Prussi, M., Scarlat, N. Energy and environmental assessment of hydrogen from biomass sources: Challenges and perspectives. *Biomass Bioenergy.* 165, 106556, 2022.

[58] Deng, W., Feng, Y., Fu, J. Catalytic conversion of lignocellulosic biomass into chemicals and fuels. *Green Energy Environ.* 8(1), 10–114, 2023.

[59] Thengane, S. K., Kung, K. S., Gomez-Barea, A., et al. Advances in biomass torrefaction: Parameters, models, reactors, applications, deployment, and market. *Prog. Energy Combust. Sci.* 93, 101040, 2022.

[60] Pata, U. K., Kartal, M. T., Adebayo, T. S., et al. Enhancing environmental quality in the United States by linking biomass energy consumption and load capacity factor. *Geosci. Front.* 14(3), 101531, 2023.

[61] Saravanan, A., Kumar, P. S., Jeevanantham, S., et al. Recent advances and sustainable development of biofuels production from lignocellulosic biomass. *Bioresour. Technol.* 344, 126203, 2022.

[62] Zhu, J. Y., Pan, X. Efficient sugar production from plant biomass: Current status, challenges, and future directions. *Renew. Sust. Energ. Rev.* 164, 112583, 2022.

[63] Ramos, A., Monteiro, E., Rouboa, A. Biomass pre-treatment techniques for the production of biofuels using thermal conversion methods—A review. *Energy Convers. Manag.* 270, 116271, 2022.

[64] Sun, Y., Shi, X. L., Yang, Y. L., et al. Biomass-derived carbon for high-performance batteries: From structure to properties. *Adv. Funct. Mater.* 32(24), 2201584, 2022.

[65] Qamar, O. A., Jamil, F., Hussain, M., et al. Feasibility-to-applications of value-added products from biomass: Current trends, challenges, and prospects. *J. Chem. Eng.* 454, 140240, 2023.

[66] Calijuri, M. L., Silva, T. A., Magalhães, I. B., et al. Bioproducts from microalgae biomass: Technology, sustainability, challenges and opportunities. *Chemosphere.* 305, 135508, 2022.

[67] Siddique, I. J., Salema, A. A., Antunes, E., et al. Technical challenges in scaling up the microwave technology for biomass processing. *Renew. Sust. Energ. Rev.* 153, 111767, 2022.

[68] Pal, D. B., Singh, A., Bhatnagar, A. A review on biomass-based hydrogen production technologies. *Int. J. Hydrog. Energy.* 47(3), 1461–1480, 2022.

[69] Ni, J., Qian, L., Wang, Y., et al. A review on fast hydrothermal liquefaction of biomass. *Fuel.* 327, 125135, 2022.

[70] Vuppaladadiyam, A. K., Vuppaladadiyam, S. S. V., Awasthi, A., et al. Biomass pyrolysis: A review on recent advancements and green hydrogen production. *Bioresour. Technol.* 128087, 2022.

[71] Singh, A. D., Gajera, B., Sarma, A. K. Appraising the availability of biomass residues in India and their bioenergy potential. *Waste Manag.* 152, 38–47, 2022.

9 Microgrid Structure for Tapping Renewable Energy

Manjeet Singh*, Kulbir Singh, Surinder Singh, Arashdeep Singh, and Harjot Gill

9.1 INTRODUCTION

In the current system, earth provides us with resources that are turned into products and then eventually discarded as waste. The circular economy model is going to be feasible and economical if and only if waste generation is avoided. The circular economy is founded on three design-driven principles: reduce waste and pollution, move goods and resources, and restore the natural world. To reduce waste and pollution renewable energy resources are to be exploited to their full extent. The circular economy in context is the need of sustainable developments in rural and urban areas with continuous improvement in the use of available renewable energy resources. With the available renewable energy resources, the sustainable rural areas with sound economy can be achievable. Also, the emphasis is on making the rural areas self-reliant in terms of energy, with minimum carbon emission and waste pollution. Accordingly, rural microgrids are proposed for the safe circular economy with minimum pollution and distribution losses. In circular economy the microgrid structure setup will not only reduce pollution (carbon emission) along with minimum distribution losses but will also make the rural areas self-sufficient in energy. Microgrids, with their distributed and localized energy generation and management, align well with the principles of the circular economy in several ways, as follows:

- **Local energy generation and consumption:** Microgrids often rely on renewable energy sources like solar, wind, and biomass, which are abundant and locally available. By generating energy locally from renewable sources, microgrids reduce reliance on fossil fuels and minimize the environmental impact associated with long-distance energy transmission.
- **Energy efficiency and demand response:** Microgrids can implement advanced energy management systems, demand response strategies, and load-balancing techniques to optimize energy consumption and reduce waste. Energy efficiency measures in microgrid structures ensure that

* jeet.mtech@gmail.com

DOI: 10.1201/9781003269779-9

resources are used more efficiently, aligning with the circular economy's goal of maximizing resource utilization.

- **Integration of energy storage and grid flexibility:** Energy storage systems (e.g., batteries) in microgrids facilitate energy storage and retrieval, allowing excess energy to be stored and utilized during periods of high demand or low renewable generation. This integration promotes a more flexible and responsive energy system, reducing the need for energy curtailment and enabling better resource utilization.
- **Closed-loop systems for waste and heat recovery:** Microgrids can integrate waste-to-energy systems, where waste materials are converted into energy through processes like anaerobic digestion or biomass conversion. Heat recovery systems can also be implemented to capture waste heat from industrial processes or energy generation and repurpose it for other applications, promoting circularity.
- **Decentralization and local resilience:** The decentralized nature of microgrids enhances local resilience in the face of external disruptions or disasters, reducing the dependence on a single centralized energy source. During grid outages, microgrids can continue to operate in islanded mode, utilizing local renewable resources, energy storage, and distributed generation to meet local energy needs.
- **Sharing economy and peer-to-peer trading:** Microgrids can facilitate peer-to-peer energy trading, where energy surplus from one prosumer (a consumer who also produces energy) can be directly shared with another in the microgrid. This enables the efficient utilization of excess energy, promoting a sharing economy and fostering a more circular energy system.

By incorporating circular economy principles into microgrid structures, communities and industries can transition towards more sustainable and resilient energy systems. Microgrids offer an opportunity to reduce waste, optimize resource use, and promote a regenerative energy model that aligns with the broader goals of a circular economy. For the proper zero carbon emission implementation with circular economy perspective the microgrid structures are to be selected and designed as followed by the preceding structure selection.

9.2 MICROGRID STRUCTURE SELECTION

The circular economy, a component of the present energy cycle, is dependent on non-renewable energy sources in order to prevent social unrest brought on by the never-ending depletion of our planet's natural resources with wastage and pollution, in this moment the transition to renewable energy sources is must. In the current Indian renewable energy scenario, the NITI Aayog (2015) decided that there is a need to establish the concept/utilization of more renewable energy from DERs and utilize public finance to achieve 175 GW of renewable energy by 2022 [1]. As per the committee recommendations, India must achieve target of renewable energy as source by 2022 as follows:

1. 100 GW from solar energy
2. 60 GW from wind energy
3. 5 GW from small hydro power plants
4. 10 GW from bioenergy.

Microgrid structure selection for circular economy is to be implemented with local energy generation and consumption with energy efficiency and demand response so that transmission and distribution losses can be minimized as no transmission and distribution lines are involved (no I^2R/heat losses from the line) in rural microgrid structure with generation and consumption at the load end. With more and more exploitation of photovoltaics (PV) and wind energy, more and more microgrid can be installed at the demand points with no lines involved, whereas existing lines will be used during no output from PV and wind sources. The proposed rural microgrid whether it is PV-based, wind, or hybrid microgrid must fulfil the NITI Aayog (2015) commitment by the year 2022, [1]. This need cannot be fulfilled without the hybrid microgrid structures as PV and wind operating together are economical and reliable. Table 9.1 shows the statewise renewable energy need for PV and wind hybrid generation [1]. It is clear that in India there is good scope for microgrid based on PV DGs, wind DGs, and combined PV and Wind DGs (hybrid). To reduce the conventional coal import, the states may play the role with a pure macro-economic perspective for helping India reaching 175 GW renewable energy commitment by 2022. To achieve this target hybrid, microgrids are the best perspective as they together harness maximum of the available renewable energy resource potential in given states. Use of renewable energy has environmental benefits (reduction in emissions of harmful gases—environment pollution), increase in employment opportunities, and increase in investment. But, at same time there is need of the necessary capital. So, India

TABLE 9.1
Renewable Energy Potential from DERs in India by 2022

State	Solar Power (MW)	Wind (MW)	Small Hydro (MW)	Biomass Power (MW)
Tamil Nadu	8884	11,900	75	649
Maharashtra	11,926	7600	50	2469
Andhra Pradesh	9834	8100	0	543
Gujarat	8020	8800	25	288
Rajasthan	5762	8600	0	0
Karnataka	5697	6200	1500	1420
Madhya Pradesh	5675	6200	25	118
Uttar Pradesh	10,697	0	25	3499
Punjab	4772	0	50	244

must be able in managing the uncertainty and variability of DERs generation with incremental decrease of investment in the existing conventional fossil fuel–based large power plants. This new infrastructure development is under the 12th Five Year Plan. As per the Planning Commission of India estimates, it would require more than a trillion US dollars. Financial capability is certain to be a challenge for new technologies based on DERs. So, loans are to be provided by centre or states at lower interest rates. It should be taken into count that with DERs set there is no fuel costs for 25–30 years.

9.2.1 MICROGRID

When the DGs are integrated to existing power system, where they can feed loads individually depending on their capability to share load that structure involving DGs and load they are feeding is a simple microgrid [2,3]. DGs that are normally used now a days are mostly PV and wind based. Integration of renewable energy and grid flexibility with decentralization and local resilience is not only going to utilize the existing structures but also with local generation dependency on existing will reduce the carbon emissions. Decentralized DGs with local resilience involves the DGs that are be used in microgrids are [4,5]:

1. Microturbine
2. Based on geothermal energy
3. Based on biomass energy
4. Based on ocean energy
5. Based on energy storage.

These microgrid structures can operated in two modes that are classified as either DGs are connected to grid or maybe not, depending upon their connection with grid, two modes of operation are defined as:

a) Grid connected
b) Islanded (not connected to grid).

Main structure of microgrid includes number of controllable elements in microgrid which are given as:

1. Intermittent DERs
2. Dispatchable micro sources
3. Energy storage units (ESUs)
4. Demand-side integration.

9.3 COMPONENTS OF MICROGRID STRUCTURE

The microgrid structure is shown in Figure 9.1. In general, microgrid structure is defined in three parts as distribution generation part, centralized control part, and local control [4,5]. The decentralized generation is ready for consumption minimizing the heat losses in the lines while using the existing feeders at minimal cost.

1. The distribution generation part microgrid control is regulated by the distribution network. The microgrid is kept safe and economic by coordinating and dispatching.
2. Centralized control is a microgrid central controller (MGCC), and it is the core of the microgrid control system. DGs, ESUs, and loads are monitored, managed, and controlled by MGCC. It ensures smooth transfer between different modes of operation that are grid connected and islanded.
3. The local control part of the microgrid has local protection and controller. Voltage of DGs and the frequency are controlled locally with local protection taking care for fault protection of the microgrid.

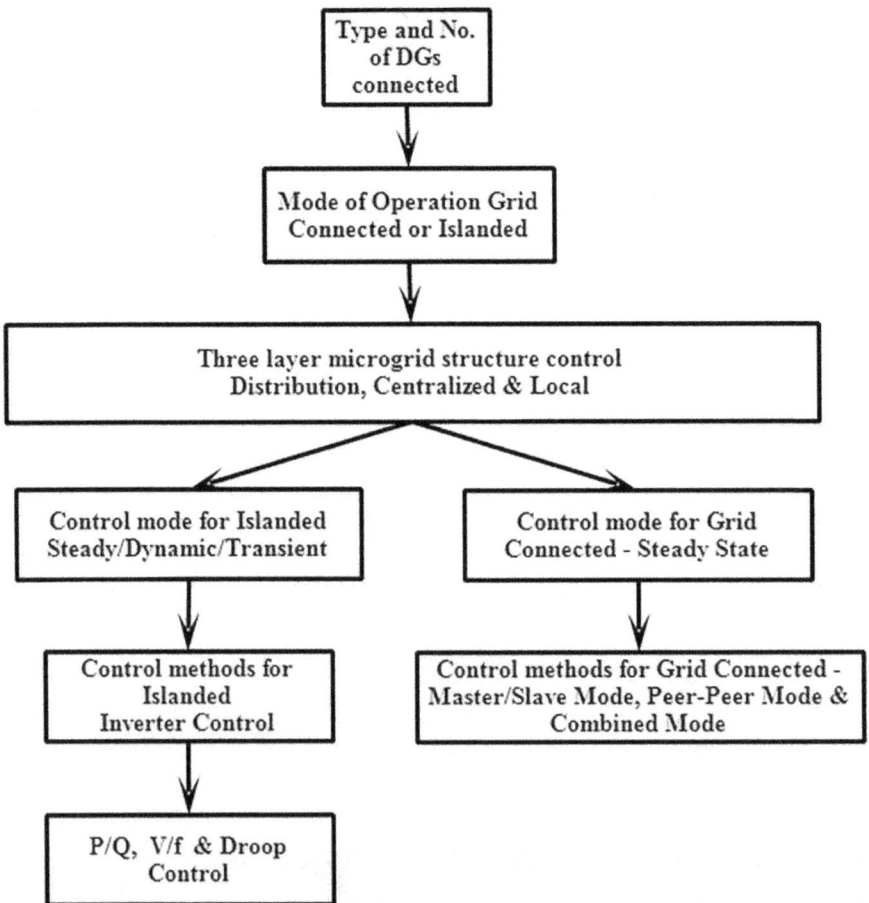

FIGURE 9.1   Block diagram of microgrid structure.

9.3.1 CONTROL METHODS OF MICROGRID STRUCTURES

Different control methods that are involved in control of above three layers of microgrid operating in different modes [4,5]. Based on whether it is islanded or grid connected control methods. For grid connected mode of operation, control method approach is given as:

Steady state control: Involves steady state constant frequency and voltage control.

For islanded mode of operation control method approaches are given as:

1. Steady state control—It involves steady state constant frequency and voltage control.
2. Dynamic control—It involves tripping of generators and shedding of loads.
3. Transient control—It involves quick removal of fault after fault occurrence.

9.3.2 CONTROL MODES FOR MICROGRID OPERATION IN GRID-CONNECTED MODE

Three control modes are used in microgrid operation for grid connected modes are as follows:

1. Master/slave mode—One or more number of DGs can either act as the master or slave. DGs connected are under P/Q control. DGs that are acting as master also manage load fluctuation that means power output is controllable to an extent, and the slave DGs are under P/Q control.
2. Peer-peer mode—In this mode, all the DGs of the microgrid are working at same level with no master or slave. Based on droop control active and reactive powers are regulated by all DGs in a pre-set control mode. Voltage and frequency stability of the system is maintained.
3. Combined mode—Based on different types of DGs with variation in generation capability, their ability to get easily stabilized and controlled, operation requirements cannot be satisfied. Master/slave control and peer-peer control are considered together.

9.3.3 CONTROL MODES FOR MICROGRID OPERATION IN ISLANDED MODE

Inverter control—In islanded mode the rated voltage and frequency of microgrid are maintained by DGs.

P/Q control—Inverters have ability to maintain active and reactive power as per reference power which is the pre-requisite for power management and control.

V/f control—Inverters ensures constant voltage-frequency for continual operation of sensitive loads and slave DGs.

Droop control—Depending up-on variation in output power the droop control is implemented from droop characteristics of generator (grid) and by controlling the voltage source inverter for output voltage and frequency.

9.4   CONCERNS OF MICROGRID STRUCTURE

A circular economy approach in microgrid implementation involves efficient resource management. It means using locally available renewable resources, such as solar, wind, or biomass, to generate electricity. By utilizing these renewable sources, the microgrid can reduce its dependence on fossil fuels, thus lowering emissions and promoting sustainability. The main concern while designing the efficient microgrid is the availability of DERs in that area with proximity of load end as most of the microgrids are low and medium voltage DGs. So, to avoid transmission losses and capital cost, DGs are set nearest to load end. Now depending on type of load requirement microgrid could be AC or DC, further in case whether load is single phase or three-phase, DGs feeding in microgrid could be single-phase or three-phase. With integration of DGs to power system, from power system reliability perspective still more work on standards and procedures for basic requirements is required [6]. The basic requirement of microgrid design is to cover [6]:

1. The voltage regulation—reactive power capability: Voltage regulation ensures that the voltage at various points in the microgrid is maintained within acceptable limits to prevent overvoltage or undervoltage conditions. This helps avoid potential equipment damage and ensures consistent performance. Renewable sources like solar and wind have limited or no inherent capability to provide reactive power support. As a result, a microgrid must have sufficient reactive power capability to ensure voltage stability.
2. High/low voltage ride-through: LVRT capability allows the microgrid to maintain its connection to the main grid during these low voltage events, avoiding unnecessary disconnections and preserving the continuity of power supply to critical loads. HVRT capability ensures that the microgrid can tolerate these high voltage conditions without being disconnected from the main grid, avoiding unnecessary interruptions and protecting sensitive equipment from potential damage.
3. Inertia response: Response mimics the stabilizing effect provided by traditional rotating generators connected to the grid, which have inherent kinetic energy and contribute to grid stability during disturbances. In microgrids with a high share of non-synchronous renewable sources like solar and wind, the lack of inertia can lead to grid stability challenges, such as frequency fluctuations, during sudden changes in power supply or demand.
4. Controlling MW ramp rates: Play a critical role in ensuring grid stability, preventing excessive stress on equipment, and complying with grid interconnection requirements. Microgrid controllers can be designed to regulate the MW ramp rates of different energy sources and loads to meet operational and grid stability objectives.
5. Frequency control: In a microgrid, where multiple distributed energy resources (DERs) operate in parallel, frequency control becomes more complex compared to a traditional centralized power grid. The frequency needs to be maintained within narrow bounds (typically around 50 or 60 Hz, depending on the region) to prevent damage to equipment and ensure proper functioning of electrical devices.

TABLE 9.2
Standards Used in Microgrid Design

Standard	Area
IEEE1547	Microgrid Connection and Integration
IEEE2030.8	**Smartgrid (Microgrid Controller)**
IEEE1459	**Power Measurement**
IEEE1588	**Precision Time Protocol in Power System Relaying**
IEEE242–2001	**Protection and Coordination**
IEC61850	**Communication Architecture in Power System**
IEC61400	**Wind Energy Generation Systems**
IEC62738	**Design for Ground Mounted PV Power Plants**
IEC62790	**PV Module—Safety Requirements and Tests**
IEC60904	Photovoltaic Devices
IEC-TS62257	**Hybrid Systems for Rural Electrification**
IEC60909	**Short Circuit Calculations**

Bold items in table refer to wind based power grids and its important components. While unbold refer to other categories.

9.5 STANDARDS FOR MICROGRID STRUCTURES

Microgrid standards play a significant role in advancing the circular economy within the energy sector. By guiding the development and operation of microgrids in a sustainable and resource-efficient manner, these standards contribute to a more resilient, low-carbon, and environmentally friendly energy future. Adoption of microgrid standards can receive regulatory support and incentives, further promoting their implementation and encouraging stakeholders to incorporate circular economy principles into their microgrid projects. In the microgrid design architecture a number of standards and protocols are considered because of involvement of different types of DERs, control and protection models. Standards mostly used in microgrid design are mentioned in Table 9.1. Basic standards used in microgrid design structures are as follows (see Table 9.2) [1–5].

9.6 PARAMETERS IN MICROGRID DESIGN

Design the microgrid to prioritize the integration of renewable energy sources, such as solar, wind, hydro, and biomass. By relying on locally available and renewable resources, the microgrid can reduce its dependence on fossil fuels and minimize carbon emissions. Implement smart grid technologies to enable demand response capabilities. This allows the microgrid to adjust energy consumption based on demand patterns, maximizing energy efficiency and avoiding unnecessary energy wastage. Emphasize the use of locally available resources for energy generation and storage. This reduces transportation-related emissions and supports local economic development. Design the microgrid in a modular and scalable manner to allow for easy upgrades, expansions, or replacements of components. This approach enhances the microgrid's flexibility and extends its life span. In respect of different standards used in microgrid design various parameters are being identified that generally decide the different architectures involved in reliable and economical design of microgrid structure. They are identified as follows [7,8]:

1. Nature of supply in microgrid (AC or DC)
2. Nature of DERs used (PV, wind, diesel generator, fuel cell etc.)
3. Voltage level of DERs and load (low or medium)
4. Distance of distribution lines connecting DERs to load end (few kilometers)
5. Feeder connection (radial, ring, radial-ring)
6. Nature of loads connected (R, L, RL)
7. Protective relaying
8. Grounding at DERs and load end
9. Power quality
10. Reliability of DERs
11. Static switch at PCC
12. Voltage level and frequency at PCC
13. Fault current at PCC
14. Central controller and methods of control for DERs.

9.7   GUIDELINES FOR DISTRIBUTION OF ELECTRICITY IN INDIA

At present, Indian electrical energy system is mostly based on thermal and hydro energy and at some places DERs solely based on PV DGs also feeding distribution system. Considering global trend towards onshore and offshore wind DGs, Indian DERs usage could be equally distributed among wind and PV DGs. This study is carried out for proper selection of DGs and the microgrid structure from Indian rural and domestic perspective. Punjab State Electricity Regulatory Commission (PSERC) in compliance to Indian Electricity Act 2003 has issued guidelines under "Conditions of Supply" to licensee (supplier) for giving supply to consumers, the following should be considered while distributing electrical energy [10].

1. Low tension (LT) AC supply is defined at 50 cycles, 230–440 V.
2. For a single-phase supply, voltage between phase and neutral connections should be 230 V, for general load not exceeding 7 KW, and motive load not exceeding 2 brake horsepower (BHP—induction motors).
3. Three-phase 400 volts between phases (line to line) connections for motive load more than 2 brake horsepower (BHP—induction motors) with general load exceeding 7 kW but not exceeding 100 kW. Supply given to loads above 7 kW is three-phase with neutral.
4. Motive load supply for agricultural sector with load up to 100 kW and supply to street lighting is from three-phase LT.

9.8 COST OF RENEWABLE ENERGY

The cost of renewable energy in the circular economy can vary depending on several factors, including the specific type of renewable energy technology, geographic location, economies of scale, government policies, and advancements in technology. Generally, as the circular economy principles are applied, the cost of renewable energy can be influenced by ways resource efficiency, technological advancements, life cycle cost considerations, policy and incentives, and external cost savings. Resource efficiency, technological advancements, life cycle cost, incentives, and external cost savings total cost is in the form of number of DGs in grid connected and islanded mode for specific areas with other design options, is easily evaluated using HOMER software. Cost of DG is calculated as per following equations [9]:

$$\text{Present Cost} = \frac{\text{Total annualized cost}}{\text{Capital recover factor}\left(\text{Interest Rate}(\%), \text{Project lifetime}(\text{Years})\right)}$$

$$\text{Cost of Energy, COE} = \frac{\text{Total annualized cost of system}}{\text{AC primary load served} + \text{Deferrable load served} + \text{Total grid sales}}$$

where present cost is in Rs, annualized cost is in Rs/yr, cost of energy is the average cost per kWh, and loads served with total grid sales are in kWh/yr.

9.9 PROPOSED MICROGRID MODELS

As per the guidelines given in "Conditions of Supply," it has been decided that connection above 7 kW load for domestic consumers should be fed three-phase AC supply [10]. From "The supply voltage and classification of consumers" it is clear for DGs to be reliable and economical, generation should be three-phase with minimum voltage level 230 and maximum voltage level be 440 at frequency of 50 Hz. This is considered because India being agriculture-dominated nation maximum of the load in rural India is three-phase motive load. It comes under the low-tension (LT) supply operating as AC of 50 cycles with voltage limits from 220 to 440 V.

Three phase supply is considered because of its saving of conductors, three-phase distribution system is economical, better efficiency due to fewer line losses, possibility of self-revolving field, efficient and reliable as compared to single phase system. During further studies in the chapter LL, LLG, and LLL faults are simulated at DG side nodes, grid side nodes, and Load1–2 side nodes.

9.9.1 MICROGRID BASED ON LOW X/R RATIO

Incorporating low X/R ratio DG in the circular economy can offer several benefits such as improved power quality: low X/R ratio DG systems generally provide better power quality due to reduced voltage drop and improved voltage regulation. This is especially beneficial in sensitive industrial processes and facilities that require stable and reliable power. Low X/R ratio DG also enhanced grid stability by providing fast-reacting power support, voltage support, and frequency regulation. This helps in integrating renewable energy sources more smoothly and maintaining grid reliability. Efficient energy distribution: Low X/R ratio DG can help reduce energy losses during transmission and distribution. By locating DG closer to load centers, the energy doesn't need to travel long distances, resulting in lower transmission losses and improved energy efficiency. Based on above microgrid design and control structures a microgrid is designed for Indian rural areas. The suggested microgrid model is based on a typical small rural Indian community, with a range of 400 kW for its inductive load and 415 V for its feeders. For rural areas of India with home and motive load, the positions of the DGs and the load with regard to the utility approximate a microgrid. The suggested microgrid model's schematic design is displayed in Figure 9.2. In the microgrid system designed the X/R ratio of utility grid system is kept 7 and X/R ratio of synchronous generator acting as DG is considered with low ratio (X/R) of 5.

Before the point of common coupling (PCC), a three-phase AC system with a step-down transformer of 1 MVA, 6 kV/440 V, running at 50 Hz and 0.8 lagging power

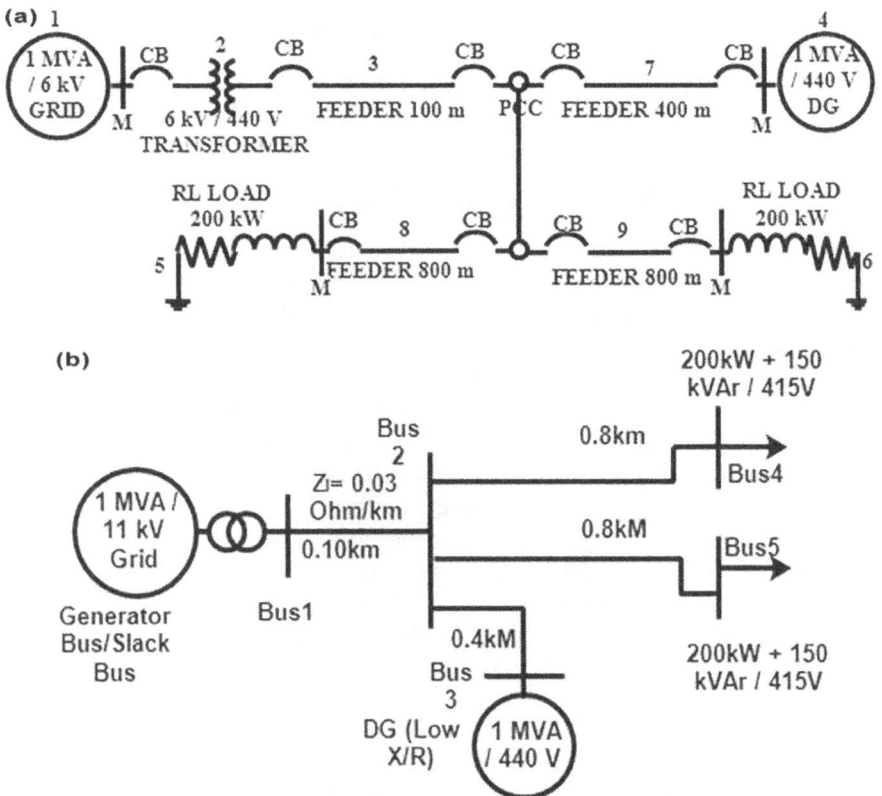

FIGURE 9.2   (a) Block diagram of microgrid system. (b) Single-line diagram of microgrid system.

factor is attached. So, taking into account that low X/R ratio's effects, a synchronous generator with a low X/R ratio is classified as a DER when it is linked to a PCC through a feeder of a 500 m distribution line and is able to serve an 800 kW load at a 0.8 lagging power factor. The distance between PCC and DG is 500 m, and two loads of 200 kW each are linked to DER at a distance of 1.2 km. Using MATLAB/Simulink, the aforementioned system is represented as a four-bus system with one slack bus, one generator bus, and two load buses. Its layout and single line diagram are shown below.

9.9.2 MICROGRID BASED ON PV DG

Solar panels convert sunlight directly into electricity without producing greenhouse gas emissions, supporting the circular economy's goal of transitioning away from fossil fuels and reducing carbon footprints. By using solar energy, the microgrid maximizes resource efficiency. It utilizes an abundant and freely available resource, reducing the need for resource extraction and minimizing environmental impacts associated with energy production. As the circular economy encourages local production and consumption. A PV DG-based microgrid generates electricity locally, reducing the dependence on centralized power plants and long-distance transmission lines. This decentralized approach can enhance energy resilience and contribute to local economic development. Circular economy principles advocate for considering the entire life cycle of products, including their end-of-life. In a PV DG-based microgrid, this involves planning for responsible recycling or repurposing of solar panels and batteries when they reach the end of their useful life to close the material loop and minimize waste. PV DG-based microgrids often incorporate energy efficiency measures to optimize energy consumption. Energy-efficient appliances, lighting, and building design contribute to reducing energy waste and increasing the overall efficiency of the system.

Different possible microgrid structures are designed and simulated as per the rural India requirements. Here, PV-based microgrid is designed and simulated. Figure 9.3 displays the suggested microgrid model's schematic diagram. Before the point of common coupling (PCC), a three-phase, three-bus AC system with a stepdown transformer of 1 MVA, 6 kV/440 V working at 50 Hz and a 0.8 lagging power factor is attached. A PV-based distributed generator with a 500 m feeder linked to the PCC and a 400 kW load capacity at a 0.8 lagging power factor. The distance between PCC and DG is 500 m, and two loads of 200 kW each are linked to DER at a distance of 1.2 km. As illustrated in Figure 9.3, the aforementioned system is simulated using MATLAB/Simulink as a four-bus system with one slack bus, one generator bus, and two load buses.

9.9.3 MICROGRID BASED ON WIND DG

Wind energy is a sustainable and abundant resource. By using wind turbines to generate electricity, the microgrid optimizes resource efficiency and minimizes environmental impact compared to conventional power plants that rely on finite fossil fuels. Wind DG microgrids enable local energy production, bringing the generation closer to the point of consumption. This localization reduces the need for long-distance transmission, leading to lower energy losses and enhancing energy resilience. In a wind DG microgrid, this

(a)

(b)

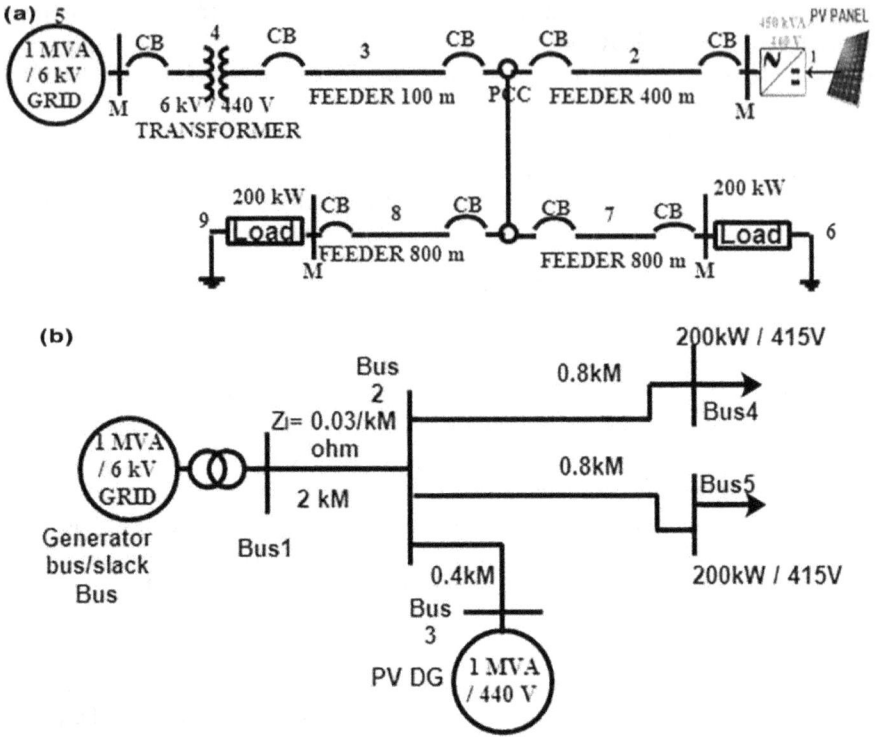

FIGURE 9.3 (a) Block diagram of PV microgrid system. (b) Single-line diagram of PV microgrid system.

involves planning for responsible decommissioning and recycling of wind turbines and other components at the end of their useful life to promote resource reuse and mini-mize waste. Keeping in mind the wind energy potential of coastal India, a wind-based microgrid system is also designed. Figure 9.4 displays the suggested microgrid model's schematic diagram. It depicts the design of a three-phase AC microgrid system with a 450 kVA/440 V photovoltaic (PV) system and a 1 MVA/440 V wind energy (WE) sys-tem that is connected to a 1 MVA and 6 kV utility grid via a step-down transformer of 1 MVA, 6 kV/440 V, 50 Hz at the point of common coupling (PCC).

The entire system has resistive load of 400 kW and reactive load of 300 kVAR. As mentioned above, hybrid microgrid system is shown in Figure 9.4 and described as three-phase AC system of 10 MVA and 6 kV utility with step-down transformer of 10 MVA, 6 kV/440 V operating at 50 Hz, connected before point of common coupling (PCC).

9.9.4 HYBRID MICROGRID BASED ON PV AND WIND DGS

Figure 9.5 illustrates single-fed and double-fed induction generators (types I, III, and IV) based on wind energy with ratings of 1 MVA/440 V and low X/R ratio. A 400 m

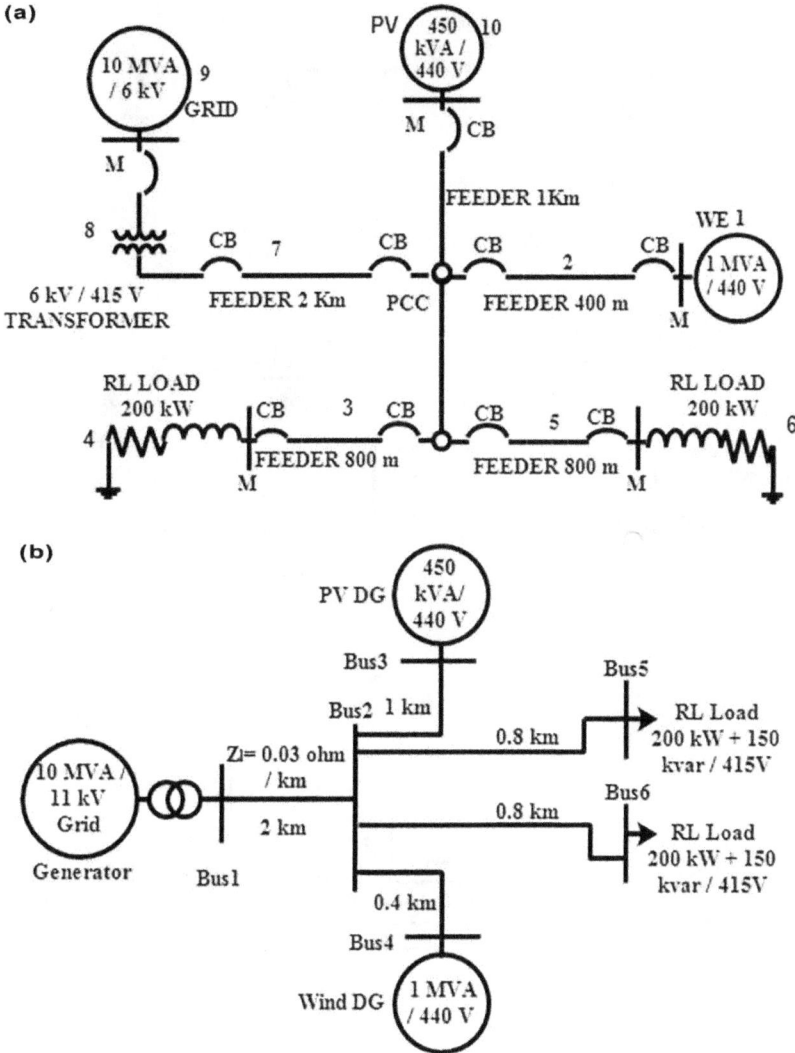

(a)

PV 450 kVA / 440 V 10

M CB

10 MVA / 6 kV 9 GRID

M

FEEDER 1Km

WE 1

8

CB 7   CB   CB   CB   1 MVA / 440 V

6 kV / 415 V   FEEDER 2 Km   PCC   FEEDER 400 m   M
TRANSFORMER

RL LOAD
200 kW   CB   3   CB   CB   5   CB   RL LOAD
200 kW   6

4   FEEDER 800 m   FEEDER 800 m
M   M

(b)

450
PV DG   kVA/
440 V

Bus3   Bus5

Bus2 1 km   RL Load
0.8 km   200 kW + 150
Zl= 0.03 ohm   kvar / 415V
10 MVA /   / km
11 kV   Bus6
Grid   2 km   0.8 km
Generator   Bus1   RL Load
200 kW + 150
0.4 km   kvar / 415V

Bus4

Wind DG   1 MVA / 440 V

FIGURE 9.4 (a) Block diagram of wind-PV microgrid system. (b) Single-line diagram of wind-PV microgrid system.

feeder is connected to the PCC, and a 450 kVA/440 V PV-based distributed generator additionally feeds the PCC through a 0.4 Km feeder. The distance between the PCC and the wind-based DG is 400 m, and two loads of 200 kW each are linked to the wind-based DER at a distance of 1.2 km. Similar to a hybrid microgrid, distributed generators (DGs) based on a wind-photovoltaic (hybrid system), load, and utilities are interconnected. Regardless of the availability of the utility grids, the proposed hybrid microgrid is self-sufficient to feed the loads linked to it. The locations of the intelligent electronic devices (IEDs) (indicated as IED 1 to IED 7), seven current

(a)

(b)

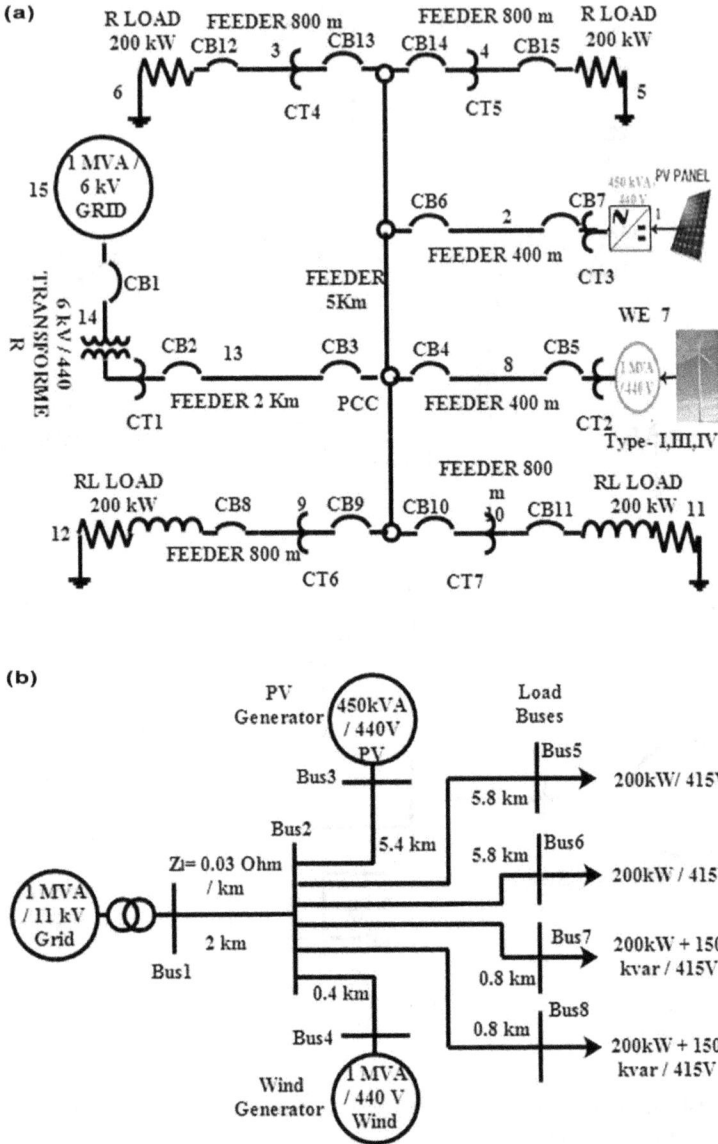

FIGURE 9.5 (a) Block diagram of wind-PV hybrid microgrid. (b) Single-line diagram of wind-PV hybrid microgrid.

transformers (CTs), and 15 circuit breakers (CBs; identified as CB1 to CB15) are also shown. A three-phase AC microgrid system with 450 kVA/440 V photovoltaic (PV) and 1 MVA/440 V wind energy (WE) systems is shown in Figure 9.5. These systems are connected to a utility grid with 1 MVA and 6 kV through a step-down transformer with 1 MVA, 6 kV/440 V, 50 Hz at the point of common coupling (PCC).

The entire system has resistive load of 800 kW and reactive load of 300 kVAR. For protection studies faults are simulated at node 9 near RL load (node 12), at node 2 near PV DG, at node 8 near wind DG, at node 3 near R load (node 6) and at node 15.

9.10 MICROGRID BEHAVIOR CHARACTERISTICS

Resource monitoring for efficient and optimized utilization of renewable energy microgrid characteristics of voltage, current, and frequency are of significant importance. Voltage-current (VI) characteristics enable real-time resource monitoring and optimization allow for more efficient energy use, reducing waste and supporting the circular economy's resource efficiency objectives. Steady state behavior of designed microgrid system is studied in terms of dq (power flow based energy storage system) components for to check the feasibility of different modes of microgrid operation. The following study is carried out on microgrid system shown in Figure 9.6(a).

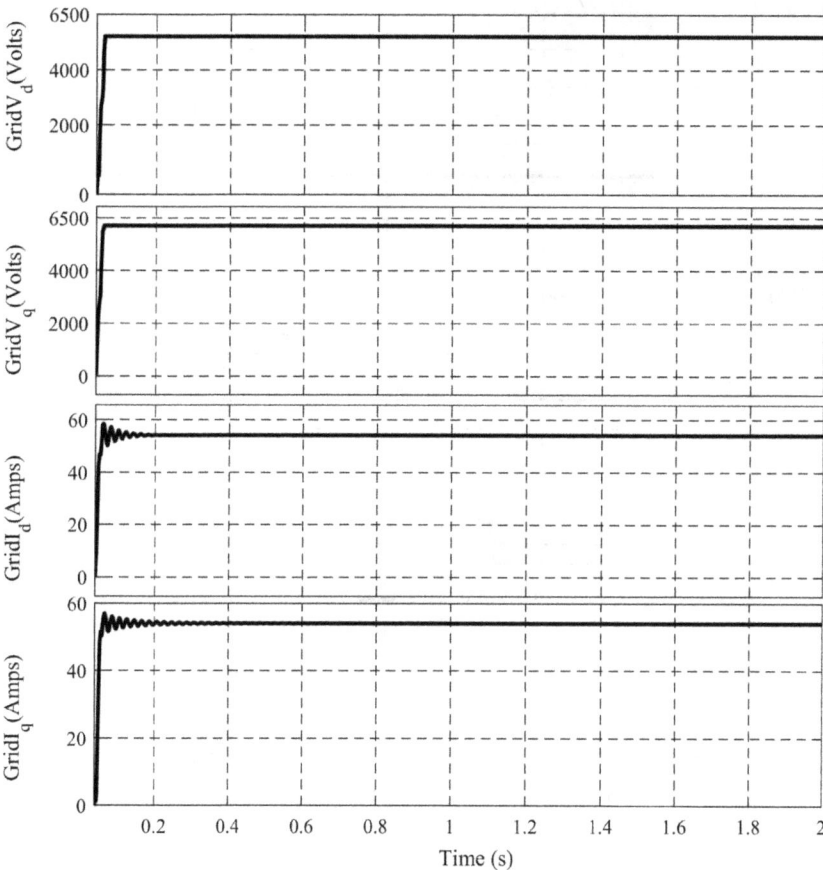

FIGURE 9.6 Grid side voltages and current in terms of their dq components.

9.10.1 VOLTAGE AND CURRENT CHARACTERISTICS—GRID-CONNECTED MODE

The results are shown for grid connected mode of operation with utility grid of high X/R ratio and synchronous generator as DG of low X/R ratio are together feeding load as shown in Figures 9.6. to 9.9. This study is carried out by simulating microgrid using MATLAB/Simulink.

Figure 9.6 shows voltages and currents of utility grid side in terms of dq components. Grid is shown operating at 6 kV and 55 A. In Figure 9.7(a) the voltage and currents of low voltage DG in terms of dq components are shown. Voltage generated by DG is 440 V and current is 780. Figure 9.7(b) shows the Load1 and Load2 voltages and currents in terms of dq components. Load1 and Load2 are of same wattage and connected in parallel. They are fed at 400 V and 340 A each.

Figure 9.8 shows the frequency at point of common coupling. During integration of DG having low X/R ratio there is small drop in frequency of the range of

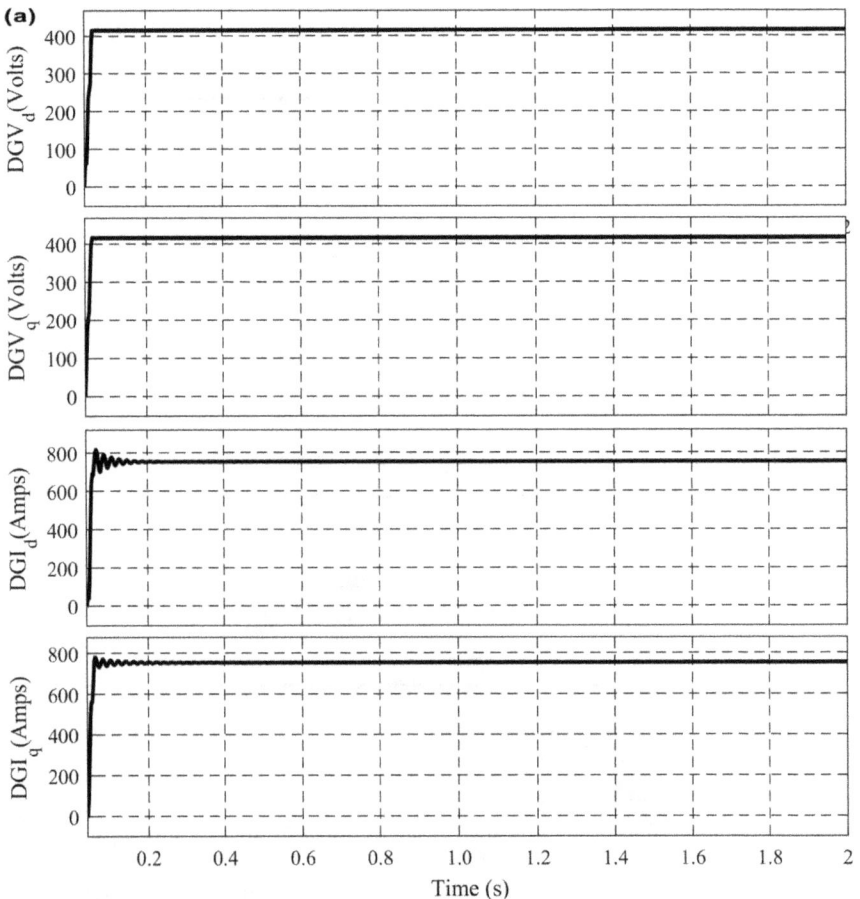

FIGURE 9.7   (a) DG side voltages and current in terms of their dq components. (b) Load1–2 side voltages and current in terms of their dq components.

(b)

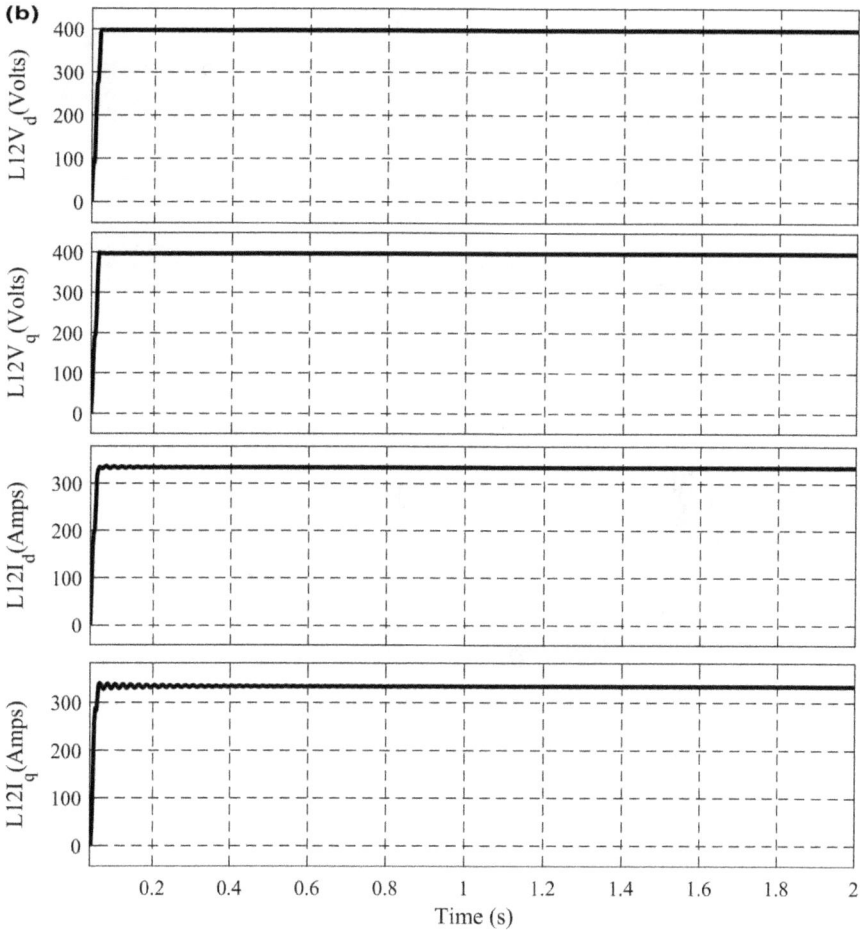

FIGURE 9.7   (Continued)

0.35 Hz. It is important to check whether the designed microgrids have ability to operate in different modes of microgrid operation that are grid connected and islanded. This study is simulated as detection of availability utility side grid and DG to feed the load which is based on comparison of current infeeds at PCC. If infeed from utility side grid is small and but DG is giving most of the infeed at PCC it is grid connected mode. DG gives maximum infeed at PCC for the implementation of plug and play model of microgrid, DG is sufficient to take care of load individually. When only DG is giving infeed at PCC this condition is considered islanded mode of operation.

Figure 9.9(a) shows the availability of DG and grid in grid connected and islanded mode of operation. When grid is available, its status is 1 and during the interval of 1.5 to 1.7 s it is not available and its status falls to zero during this interval of 0.2 seconds. When DG is available, its status is set to 2 and during

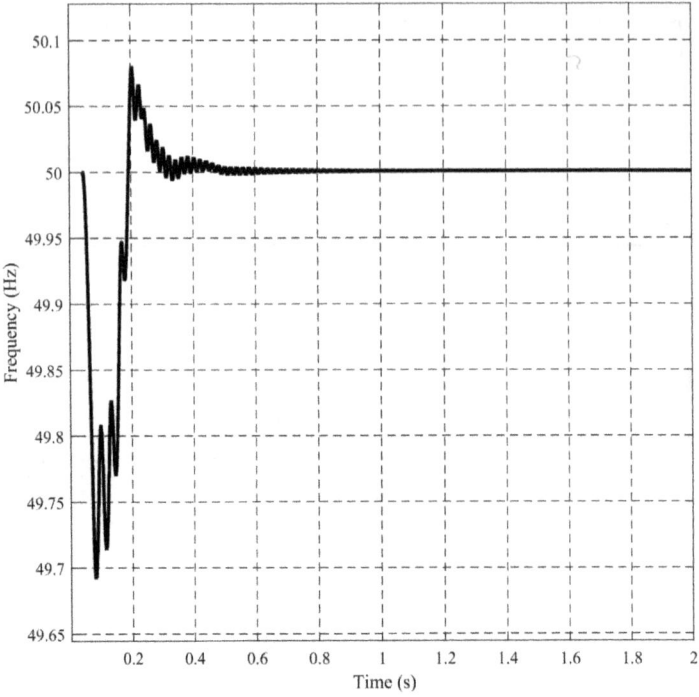

FIGURE 9.8 Frequency measured at point of common coupling.

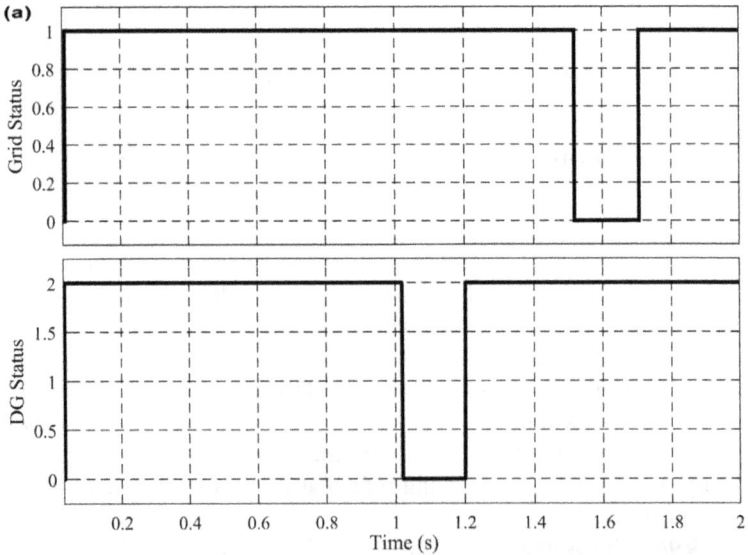

FIGURE 9.9 (a) Status of grid and DG feeding Load1 and Load2 using dq components of current. (b) Relay settings of grid, DG and Load1 and Load2 using dq components of current during normal operation.

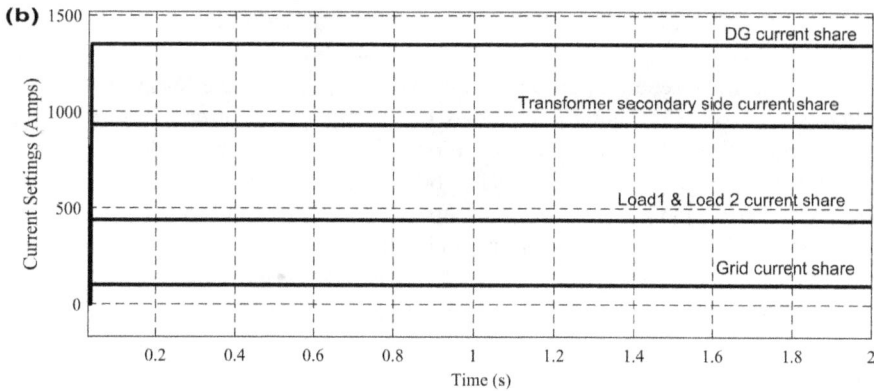

FIGURE 9.9   (Continued)

the interval of 1.0 to 1.2 s it is not available and its status falls to zero during this interval of 0.2 seconds.

Grid range 80 A; transformer range 950 A; main feeder range 950 A; DG range 1400 A; DG feeder range 1400 A; Load1 range 480 A; and Load2 range 480 A.

Figure 9.9(b) shows the relay settings of conventional relays which are shown as pickup current range, for different conventional considering maximum 15%–25% overloading and above these current values steady state of utility side grid, DG and Load1 and Load2 are lost. In this case there is no change in mode of operation, grid and DG remains in the system. During grid connected mode of operation intelligent relays present at different locations as at utility grid side, DG side, Load1 side and at Load2 side sense no change in available currents. Due to this setting of all the relays present at different locations remain same as like that of conventional relays. But if during microgrid operation somehow either grid or DG are disconnected and fault occurs, this may lead to maloperation of relays present at all the locations. As for only grid feeding the load different magnitude of current is present in the system. And if grid and DG both are present in the system than different current magnitudes are present.

9.10.2   VOLTAGE AND CURRENT CHARACTERISTICS—ISLANDED MODE

In this case only DG is feeding the load at PCC. Only grid feeding the load is not considered as it the conventional existing method. During islanded mode of operation only DG is feeding the Load1 and Load2. Voltages and currents of utility side grid and transformer are not considered as grid not feeding the load only DG is feeding. So, during islanded mode of operation no feed from grid side only DG and load sides are studied. As discussed above in islanded mode of operation throughout the operation time only DG remains present and feeds connected loads. Steady state behavior of designed microgrid system with only DG feeding the loads in terms of dq components is studied to check the feasibility of islanded mode of microgrid operation.

The following study is carried out on microgrid system shown in Figure 9.2.(a) but without utility side grid. The results are shown for islanded mode of operation with synchronous generator as DG of low X/R ratio feeding the loads as shown in Figures 9.10 to 9.12. Figure 9.10. shows DG voltages and currents in terms of dq components. The voltage and currents of low voltage DG in terms of dq components are shown. Voltage generated by DG is 600 V and current is 680 A. Figure 9.11 shows the Load1 and 2 voltage and currents in terms of dq components. Load1 and 2 are of same wattage and connected in parallel they are fed at 400 V and 340 A each. The results as shown in Figures 9.10 and 9.11 validate the feasibility and availability of proposed microgrid in any mode of microgrid operation.

Figure 9.12(a) shows the frequency at point of common coupling. During islanded mode of DG having low X/R ratio there is small increase in frequency as speed of DG is

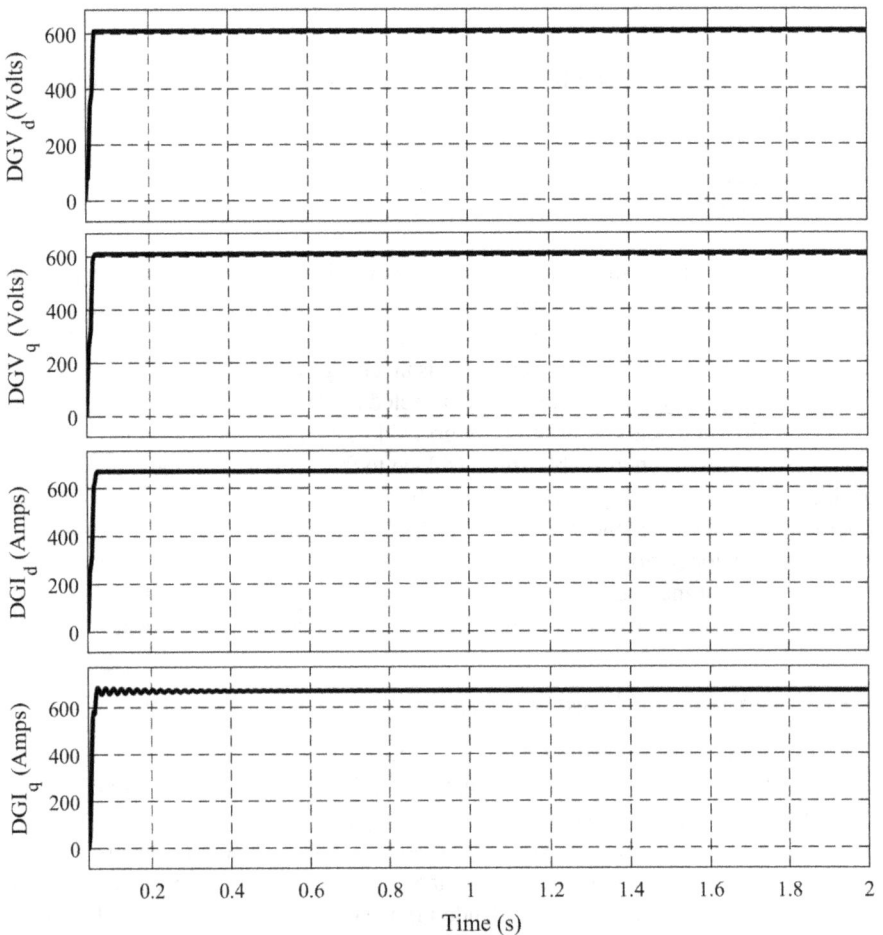

FIGURE 9.10 DG side voltages and current in terms of their dq components during islanded mode.

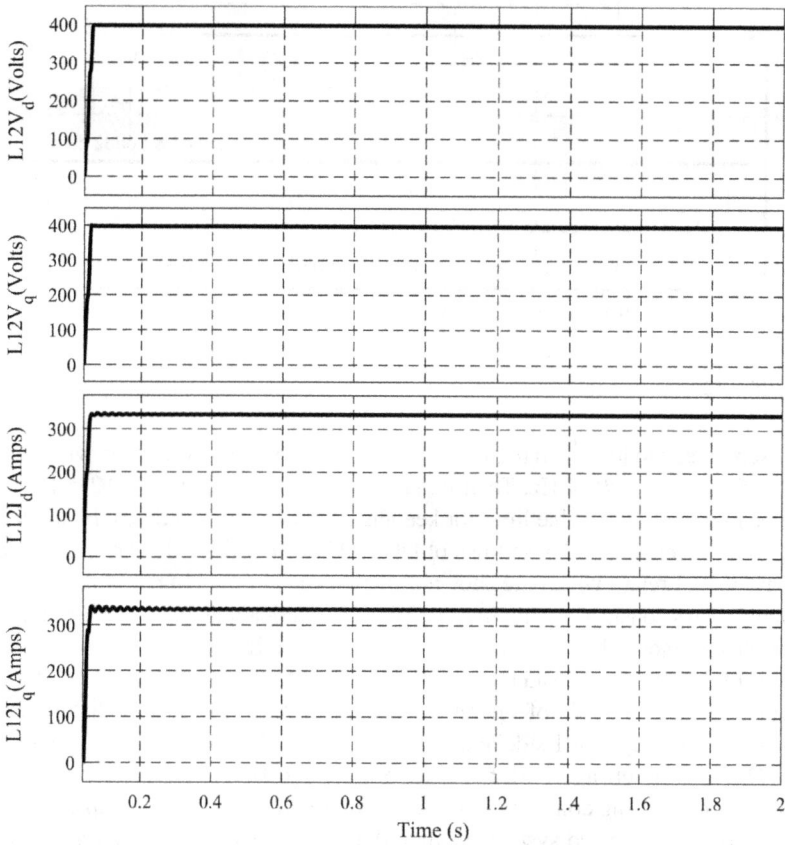

FIGURE 9.11 Load1 and Load2 side voltages and current in terms of their dq components during islanded mode.

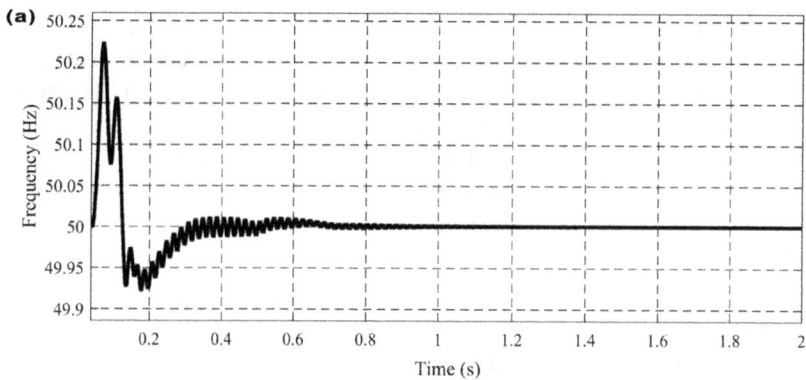

FIGURE 9.12 (a) Frequency measured at point of common coupling during islanded mode. (b) Relay settings of grid, DG and Load1 and Load2 using dq components of current during islanded mode.

(b)

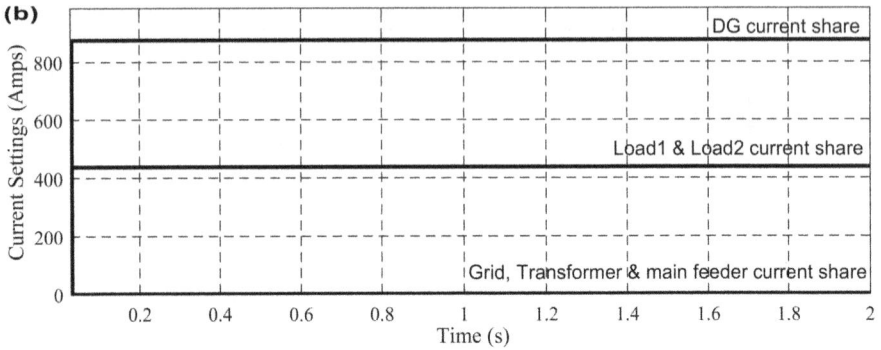

FIGURE 9.12   (Continued)

increased to feed to load with higher voltage generation. It is of the range with increase of 50.235 to drop of 49.94 Hz. Total variation in frequency at PCC is 0.295 Hz which is in the maximum allowable limit for keeping power quality standards. Figure 9.12(b) shows the conventional relay settings of DG and Load1 and Load2. The relay settings of conventional relays which are shown as the pickup current range, for different conventional relays considering maximum 15%–25% overloading and above these current values steady state of DG and Load1 and Load2 is lost. In this case there is no change in mode of operation, only DG remains in the system for feeding the Load1 and Load2.

During islanded mode of operation intelligent relays present at different locations as at DG side, Load1 side and at Load2 side sense no change in available currents. Due to this setting of all the relays present at different locations remain same as like that of existing conventional relays. Figure 9.12(b) shows pickup currents set as per existing microgrid system (islanded). Pickup current ranges are as follows:

DG range 880 A, DG feeder range 880 A, Load1 range 440 A; and Load2
   range 440 A.

Comparing relay settings for grid connected and islanded mode of microgrid operation there is drastic change in pickup currents of same DG and loads connected in system. In grid connected mode DG pickup current range is 1400 A and in islanded mode it is 880 A, for loads it remains nearly same. But this will lead to maloperation of certain conventional relays present in the system during grid connection and islanding. If in any circumstance either grid or DG fails and fault occurs than relay may lead to maloperation. This necessitates the need of adaptive relaying for efficient, reliable and economical operation of microgrid structures [11–14]. An adaptive relay is that intelligent relay which changes the relay settings as per change in microgrid modes of operation.

9.10.3   VOLTAGE AND CURRENT CHARACTERISTICS—
   UTILITY SIDE GRID AND ISLANDED MODE

Voltage and current characteristics of only grid connected throughout the microgrid operation and only islanded mode throughout the microgrid operation are studied earlier. It is also of utmost importance to study together the grid connected and islanded modes of microgrid operation.

9.10.3.1 Voltage and Current Characteristics of Grid and DG Sides at PCC

Further a new study is also carried out on newly designed microgrid to compare the voltage and current of utility side grid, DG and Load1 and Load2 during intentional islanded, only grid and grid connected mode. Grid connected mode goes from 0.0 to 2 seconds, only grid mode is simulated at 2 to 2.2 seconds and islanded modes is simulated at 3.5 to 3.7 seconds. Comparison among voltages, currents and frequency during different modes of operation is done as shown in Figures 9.13 to Figure 9.15. From Figure 9.13, during the duration of 0.0 to 2 seconds during grid

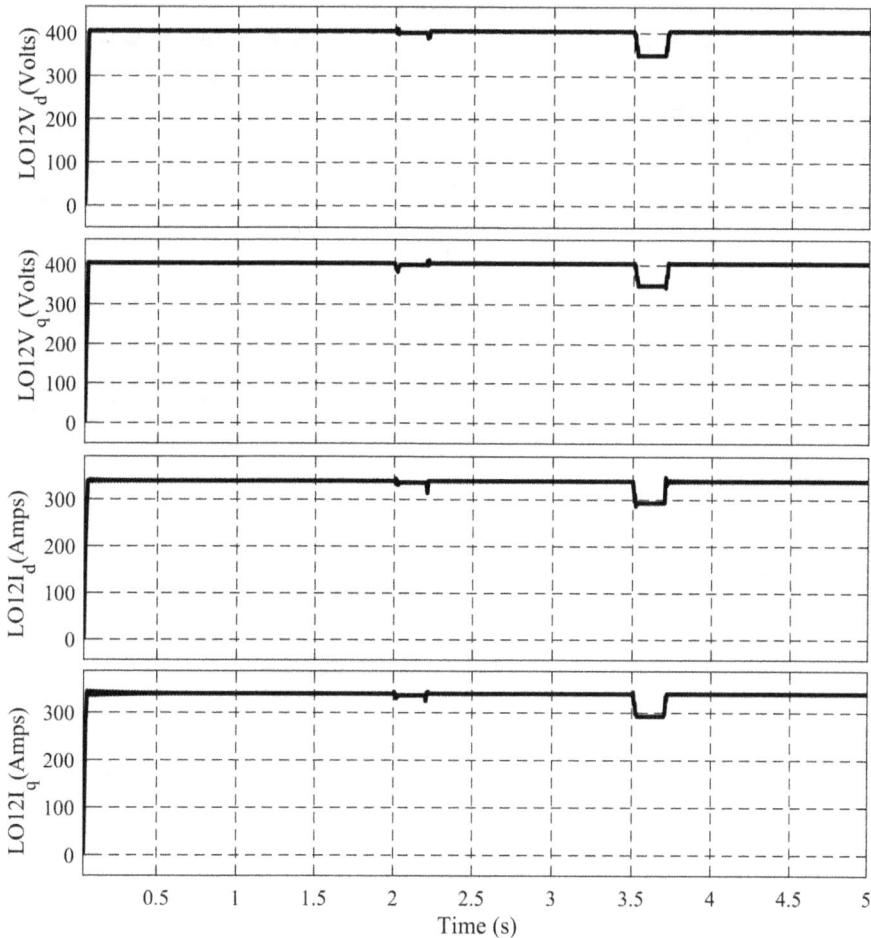

FIGURE 9.13 Voltages of grid and DG in terms of their dq components during different modes of operation.

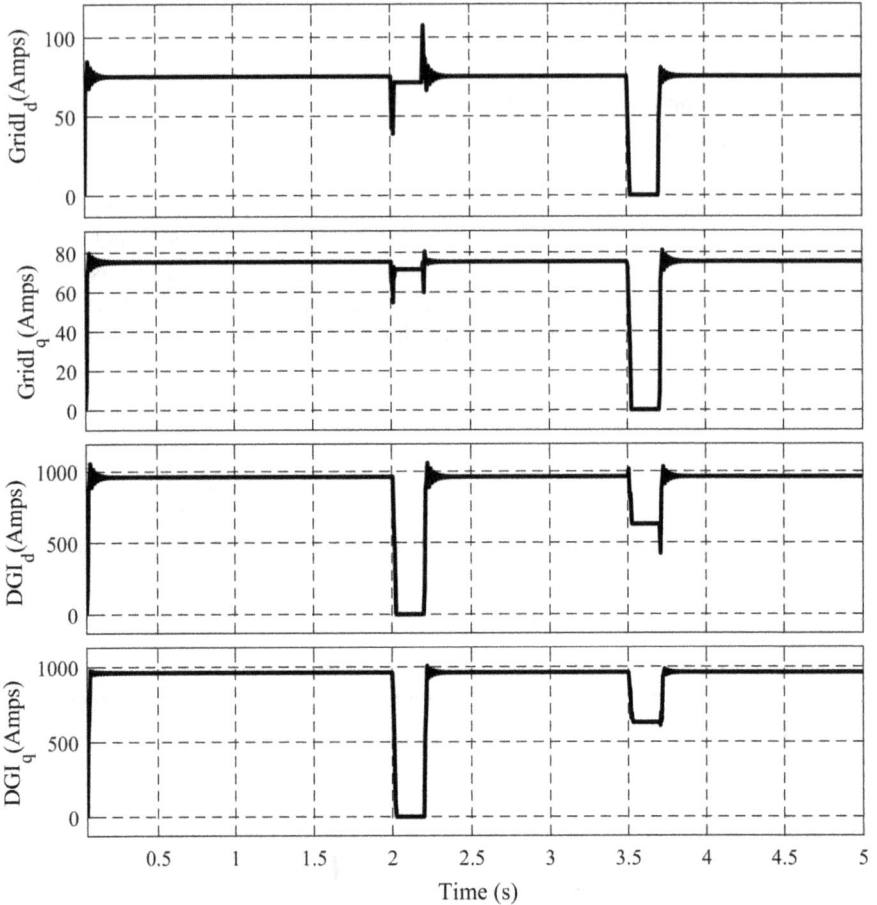

FIGURE 9.14 Currents of grid and DG in terms of their dq components during different modes of operation.

connected mode utility side grid voltage near PCC in d and q components of voltage is near to 6 kV and for same duration DG side voltage in d and q components of voltage is near to 580 V.

For the duration of 2.0 to 2.2 seconds during only grid connected mode utility side grid voltage near PCC in d and q components of voltage is near to 6 kV with drop of 80 V and for same duration DG side voltage near PCC in d and q components of voltage is near to 620 V with increase in potential of 40 V due to presence of grid. For the duration of 2.2 to 3.5 seconds again grid connected mode is present with changes as discussed for 0.0 to 2.0 seconds. For the duration of 3.5 to 3.7 seconds during islanded mode utility side grid voltage near PCC in d and q components of voltage is has a drop of 1500 volts and for same duration DG side voltage near PCC in d and q components of voltage are near to 570 V with drop in potential of 10 V. For the

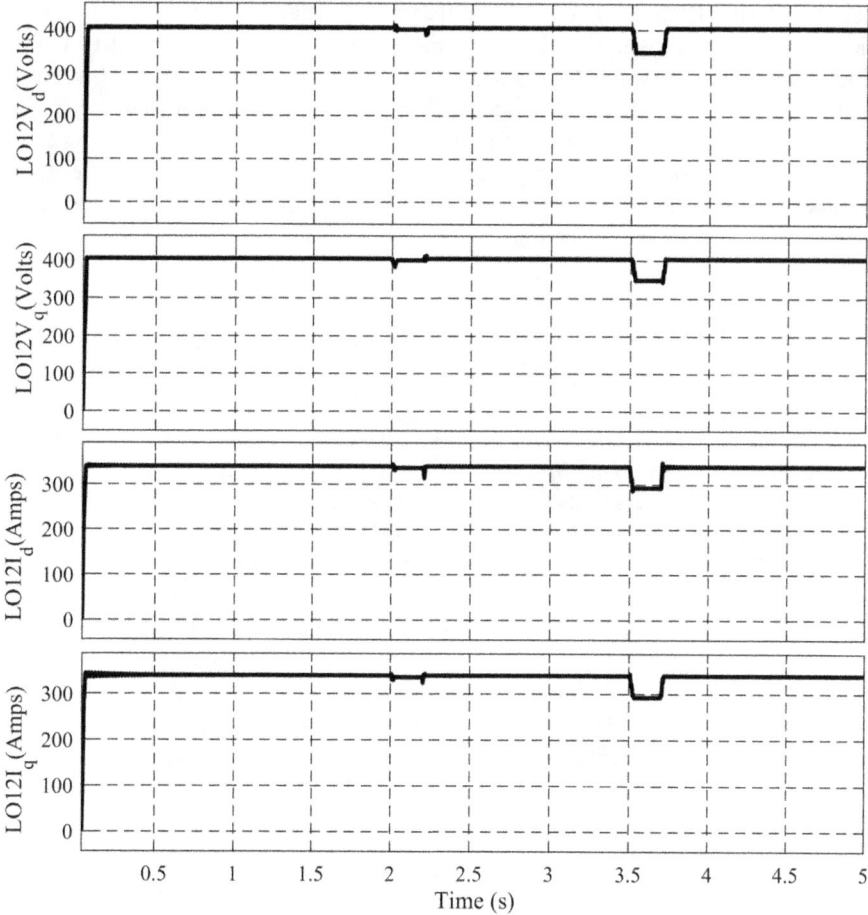

FIGURE 9.15 Voltages and currents of Load1 and Load2 in terms of their dq components during different modes of operation.

duration of 3.7 to 5.0 seconds again grid connected mode is present with changes as discussed for 0.0 to 2.0 seconds.

From the results as shown in Figure 9.14, for the duration of 0.0 to 2 seconds during grid connected mode utility side grid current at PCC in d and q components of current is near to 80 A and for same duration DG side current at PCC in d and q components of current is near to 900 A. For the duration of 2.0 to 2.2 seconds during only grid connected mode utility side grid current at PCC in d and q components of current is near to 80 A but after sudden drop in current 40 A and for same duration DG side current at PCC in d and q components of current falls to 0 A. During reconnection of DG the sudden increase in grid side current 40 A is felt at 2.2 seconds.

For the duration of 2.2 to 3.5 seconds again grid connected mode is present with changes in current as discussed for 0.0 to 2.0 seconds. For the duration of 3.5 to 3.7

seconds during islanded mode utility side grid current at PCC in d and q components of current drops to 0 A and for same duration DG side current at PCC in d and q components of current falls to 600 A. For the duration of 3.7 to 5.0 seconds again grid connected mode is present with changes in current as discussed for 0.0 to 2.0 seconds. With the above comparison of voltages and currents in terms of dq components for both utility side grid and DG, it has been made clear that d and q components of current and voltage show nearly same changed but while changing from steady state to transient state more change is felt in q components for both voltage and current.

9.10.3.2 Voltage and Current Characteristics of Load1 and Load2 Sides at PCC

During voltage comparison for the duration of 0.0 to 2 seconds during grid connected mode Load1 and Load2 side voltage near PCC in d and q components of voltage is near to 400 V as shown in Figure 9.15. For the duration of 2.0 to 2.2 seconds during only grid connected mode Load1 and Load2 side voltage near PCC in d and q components of voltage is near to 400 V with drop of nearly 20 V each for both loads. For the duration of 2.2 to 3.5 seconds again grid connected mode is present with changes as discussed for 0.0 to 2.0 seconds. For the duration of 3.5 to 3.7 seconds during islanded mode Load1 and Load2 side voltage near PCC in d and q components of voltage is 360 V and has a drop of 40 volts each for both loads. For the duration of 3.7 to 5.0 seconds again grid connected mode is present with changes as discussed for 0.0 to 2.0 or 2.2 to 3.5 seconds.

During current comparison for the duration of 0.0 to 2 seconds during grid connected mode Load 1 and Load2 side current at PCC in d and q components of current is near to 350 A. For the duration of 2.0 to 2.2 seconds during only grid connected mode Load1 and Load2 side current at PCC in d and q components of current is near to 320 A with drop of nearly 30 A each for both loads. For the duration of 2.2 to 3.5 seconds again grid connected mode is present with changes as discussed for 0.0 to 2.0 seconds. For the duration of 3.5 to 3.7 seconds during islanded mode Load1 and Load2 side current at PCC in d and q components of current is 295 A and has a drop of 55 A each for both loads. For the duration of 3.7 to 5.0 seconds again grid connected mode is present with changes as discussed for 0.0 to 2.0 or 2.2 to 3.5 seconds.

9.10.4 FREQUENCY AT PCC

During frequency comparison as shown in Figure 9.16, for the duration of 0.0 to 2 seconds during grid connected mode frequency at PCC is near to 50 Hz. Initial frequency fluctuations are due to integration of different X/R ratio generators. For the duration of 2.0 to 2.2 seconds during only grid connected mode frequency at PCC is near to 50 Hz with initial drop of nearly 0.2 Hz and rise of 0.2 Hz. For the duration of 2.2 to 3.5 seconds again grid connected mode is present with changes as discussed for 0.0 to 2.0 seconds. For the duration of 3.5 to 3.7 seconds during islanded mode frequency at PCC is near to 50 Hz with rise of 0.25 Hz and drop of 0.2 Hz. For the duration of 3.7 to 5.0 seconds again grid connected mode is present with changes as discussed for 0.0 to 2.0 or 2.2 to 3.5 seconds.

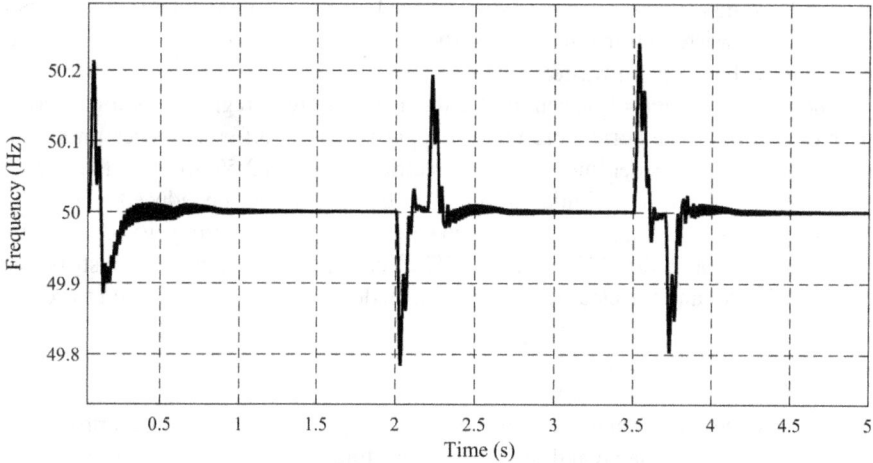

FIGURE 9.16 Frequency at PCC during different modes of operation.

With the discussion and comparison among results for voltages and currents in dq components it has been made clear that there are observable changes in d and q components. For the change in microgrid modes of operation voltage and current levels change. With the change in current magnitudes at different locations, relays must also change their pickup currents as per the change in current during change in microgrid mode of operation for proper relaying [11–14]. This necessitates the importance and need of adaptive relaying for the efficient protection system. By optimizing energy use and reducing waste in energy distribution, it could contribute to resource efficiency and overall sustainability [15–18]. From circular economy perspective India has to utilize the available renewable energy resources in such a manner that the existing wastage of electrical energy can be minimized [19–25]. Microgrid development in rural areas is not only going to reduce the carbon emission but will also reduce the dependency on fossil fuels. Zero waste microgrid architecture is going to reduce the electrical power wastage by 11% in power transfer with utilization of existing refurbished insulators and conductors [26–32]. Microgrid architecture is also going to reduce the dependency on conventional grid architecture which in turn is going to minimize the wastage of heat across the steam boilers and insulation loss of generators.

The sourcing value and environmental value are to be achieved in the current zero waste microgrid structure [11]. Carbon footprints are to be reduced by deploying the PV DG, wind DG, and Biomass-based DG depending on residual value and customer needs. The rural microgrid structure provides a feasible service to optimize the use of their renewable energy resources and biomass (waste) to reduce transmission-distribution waste (loss) [33–40]. It also involves controlling the negative externalities like use of insulators on transmission and distribution lines is avoided due to non-usage of transmission and distribution line. In phase 2, the utilization of reconditioned parts from the dismantled transmission and distribution lines are utilized in the microgrid that covers the acquisition, re-processing, and re-marketing.

So, while going for rural microgrid structure not only carbon footprint is reduced but also the zero waste in transmission and distribution will help to achieve the circular economy with minimum waste.

The region division of agricultural circular economy is regarded as a clustering problem. In India the agriculture power supply is fed via different feeder lines with respect to domestic feeder lines that adds electricity waste of 2.5% to 6% to the electrical power transfer with rural microgrid structure this waste can be reduced or avoided. While going for zero waste rural microgrid structure the 2% electricity wasted in transmission lines (400/220/132 kV), [41–43], 5% electricity wasted in primary distribution lines (66/33 kV) and 6% electricity wasted in feeder lines (11 kV) is reduced to zero.

9.11   CONCLUSIONS

Microgrids and the circular economy share a strong synergy, presenting a promising path towards a sustainable and resilient energy future. Microgrids, as decentralized energy systems, embody the core principles of the circular economy, which focuses on resource efficiency, waste reduction, and sustainable practices throughout the entire energy life cycle. By integrating renewable energy sources like solar, wind, and biomass, microgrids reduce reliance on finite fossil fuels and mitigate greenhouse gas emissions, aligning perfectly with circular economy goals of environmental protection and sustainability. Microgrids enable local energy production and consumption, promoting community engagement, ownership, and economic development. They offer flexibility, scalability, and modularity, facilitating product life extension through upgrades and replacements, contributing to circular economy principles of responsible resource management. Through smart grid technologies, energy storage integration, and demand response capabilities, microgrids optimize energy utilization, minimizing wastage and aligning energy supply with demand. Moreover, circular economy principles guide microgrid design, promoting the use of recyclable materials, and planning for responsible end-of-life management, fostering a closed-loop system that reduces waste and encourages resource recovery. Incorporating microgrids into the circular economy paves the way for resilient, efficient, and environmentally conscious energy systems. By embracing these interconnected concepts, societies can make significant strides towards a low-carbon future, combat climate change, and ensure a more sustainable world for future generations. Emphasizing circularity in microgrid development serves as a powerful solution for addressing the global energy challenge while creating economic opportunities and fostering a healthier planet.

REFERENCES

[1] National Institution for Transforming India, *Government of India: Report of the Expert Group on 175 GW RE by 2022*, 2015, https://library.niti.gov.in/cgi-bin/koha/opac-detail.pl?biblionumber=79492&shelfbrowse_itemnumber=90633
[2] Lasseter, R. H., Microgrid, *Proc. IEEE Power and Energy Society Winter Meeting*, Vol. 1, New York: IEEE, pp. 305–308, 2002
[3] Nikkhajoei, H. and Lasseter, R. H., Microgrid protection, *Proc. IEEE Power Engineering Society General Meeting*, New York: IEEE, pp. 1–6, 2007

[4] Oudalov, A., et al., *Microgrids: Architectures and Control*, New Jersey: John Wiley & Sons, Ltd., 2014

[5] Fusheng, L., et al., *Microgrid Technology and Engineering Application*, Cambridge, MA: Academic Press, pp. 47–67, 2016

[6] Beheshtaein, S., Cuzner, R., Savaghebi, M., et al., Review on microgrids protection, *IET Generation, Transmission & Distribution*, Vol. 13, No. 6, pp. 743–759, 2019

[7] Memon, A.A. and Kauhaniemi, K., A critical review of ac microgrid protection issues and available solutions, *Elsevier, Electric Power Systems Research*, Vol. 129, pp. 23–31, 2015

[8] Basak, P., et al., A literature review on integration of distributed energy resources in the perspective of control, protection and stability of microgrid, *Elsevier Renewable and Sustainable Energy Reviews*, Vol. 16, pp. 5545–5556, 2012

[9] Mohanty, P., Bhuvaneswari, G. and Balasubramanian, R., Optimal planning and design of distributed generation based micro-grids, *Proc. IEEE New York, 7th International Conference on Industrial and Information Systems (ICIIS)*, pp. 1–6, 2012

[10] Punjab State Electricity Regulatory Commission. Conditions of Supply, *Punjab State Power Corporation Limited (PSPCL)*, P.S.E.R.C. Chandigarh, India, pp. 5–8, 2013

[11] Singh, M. and Basak, P., Fractionalization of microgrid protection system through detection of zero sequence component of fault current, *Proc. 7th India International Conference on Power Electronics (IICPE)*, Patiala, India, New York: IEEE, pp. 1–5, 2016

[12] Sheta, A. N., Abdulsalam, G. M., Sedhom, B. E., et al. Comparative framework for AC-microgrid protection schemes: challenges, solutions, real applications, and future trends. *Prot Control Mod Power Syst* 8, 24 (2023). https://doi.org/10.1186/s41601-023-00296-9

[13] Singh, M. and Singh, K., Adaptive protection of microgrid through abrupt change analysis and fractional Fourier transform, *Proc. IET Conference on Developments in Power System Protection Conference (DPSP)*, New Castle, UK, London: IET, pp. 1–6, 2022

[14] Singh, M., Ganguli, S. and Gill, A., Optimum adaptive relaying using fault current quadrature (q) component, *Proc IEEE 10th Power India International Conference (PIICON)*, IEEE, New Delhi, India, pp. 1–6, 2022

[15] Krikke, H., et al., Circular economic surplus asset management: A game changer in life sciences, *IEEE Engineering Management Review*, Vol. 50, No. 2, 2022, pp. 117–126

[16] Meng, X., et al., Fuzzy min-max neural network with fuzzy lattice inclusion measure for agricultural circular economy region division in Heilongjiang Province in China, *IEEE Access*, Vol. 8, 2020, pp. 36120–36130

[17] Dewick, P., et al., Hand in glove? Processes of formalization and the circular economy post-COVID-19, *IEEE Engineering Management Review*, Vol. 48, No. 3, 2020, pp. 176–183

[18] Aceleanu, M. I., et al., The management of municipal waste through circular economy in the context of smart cities development, *IEEE Access*, Vol. 7, 2019, pp. 133602–133614

[19] Arruda, E. H., et al, Circular economy: A brief literature review (2015–2020), *Sustainable Operations and Computers*, Vol. 2, pp. 79–86, 2021

[20] Andersen, A.D. and Geels, F. W., Multi-system dynamics and the speed of net-zero transitions: Identifying causal processes related to technologies, actors, and institutions, *Energy Research & Social Science*, Vol. 102, p. 103178, 2023

[21] Geissdoerfer, M., et al., The circular economy—A new sustainability paradigm?, *Journal of Cleaner Production*, Vol. 143, pp. 757–768, 2017

[22] Kirchherr, J., Urbinati, A. and Hartley, K., Circular economy: A new research field?, *Journal of Industrial Ecology*, pp. 1–13, 2023. https://doi.org/10.1111/jiec.13426

[23] Heshmati, A., A review of the circular economy and its implementation, *International Journal of Green Economics*, Vol. 11, No. 3, pp. 251–288, 2017

[24] Heshmati, A., A review of the circular economy and its implementation, *International Journal of Green Economics*, Vol. 11, No. 3, pp. 251–288, 2017

[25] Vinante, C., et al., Circular economy metrics: Literature review and company-level classification framework, *Journal of Cleaner Production*, Vol. 288, p. 125090, 2021

[26] Hartley, K., Van Santen, R. and Kirchherr, J., Policies for transitioning towards a circular economy: Expectations from the European Union (EU), *Resources, Conservation and Recycling*, Vol. 155, p. 104634, 2020

[27] Centobelli, P., et al., Determinants of the transition towards circular economy in SMEs: A sustainable supply chain management perspective, *International Journal of Production Economics*, Vol. 242, p. 108297, 2021

[28] Panchal, R., Singh, A. and Diwan, H., Does circular economy performance lead to sustainable development? A systematic literature review, *Journal of Environmental Management*, Vol. 293, p. 112811, 2021

[29] Khan, K., et al., Is technological innovation a driver of renewable energy?, *Technology in Society*, Vol. 70, p. 102044, 2022

[30] Yildizbasi, A., Blockchain and renewable energy: Integration challenges in circular economy era, *Renewable Energy*, Vol. 176, pp. 183–197, 2021

[31] Seetharam, D.P., Khadilkar, H. and Ganu, T., Circular economy enabled by community microgrids, In *An Introduction to Circular Economy*, Springer, Singapore, pp. 179–199, 2020

[32] Sanjeev, P., Padhy, N. and Agarwal, P., Peak energy management using renewable integrated dc microgrid, *IEEE Transactions Smart Grid*, Vol. 9, No. 5, pp. 4906–4917, 2018

[33] Thomas, A., A green energy circular system with carbon capturing and waste minimization in a smart grid power management, *Energy Report*, Vol. 8, pp. 14102–14123, 2022

[34] Olabi, A.G., Circular economy and renewable energy, *Energy*, Vol. 181, pp. 450–454, 2019

[35] Prakash, K., et al., Planning battery energy storage system in line with grid support parameters enables circular economy aligned ancillary services in low voltage networks, *Renewable Energy*, Vol. 201, pp. 802–820, 2022

[36] Yu, G., et al., Research on the investment decisions of PV micro-grid enterprises under carbon trading mechanisms, *Energy Science & Engineering*, Vol. 10, No. 8, pp. 3075–3090, 2022

[37] Liaros, S., A network of circular economy villages: Design guidelines for 21st century Garden Cities, *Built Environment Project and Asset Management*, Vol. 12, No. 3, pp. 349–364, 2022

[38] Yu, G., et al., Research on the investment decisions of PV micro-grid enterprises under carbon trading mechanisms, *Energy Science & Engineering*, Vol. 10, No. 8, pp. 3075–3090, 2022

[39] Khadilkar, H., Seetharam, D.P. and Ganu, T., A quantitative analysis of energy sharing in community microgrids, *Materials Circular Economy*, Vol. 2, No. 1, p. 3, 2020

[40] Kiehbadroudinezhad, M., Merabet, A. and Hosseinzadeh-Bandbafha, H., A life cycle assessment perspective on biodiesel production from fish wastes for green microgrids in a circular bioeconomy, *Bioresource Technology Reports*, Vol. 21, p. 101303, 2023

[41] Singh, M. and Basak, P., Q and frequency component-based fault and nature detection for hybrid microgrid, *Sustainable Energy, Grids and Networks*, Vol. 28, p. 100552, 2021

[42] Singh, M. and Basak, P., Behavior of fault current in microgrid systems, *IEEE, 7th India International Conference on Power Electronics (IICPE)*, IEEE, Patiala, India, pp. 1–6, 2016

[43] Singh, M. and Basak, P., Fractionalization of microgrid protection system through detection of zero sequence component of fault current, *IEEE, 7th India International Conference on Power Electronics (IICPE)*, IEEE, Patiala, India, pp. 1–6, 2016

10 Plastic Waste to Value-Added Products via Recycling and Upcycling

Navneet Kaur Bhullar and Ashok Prabhakar*

10.1 INTRODUCTION

10.1.1 Plastic Pollution Overview

Plastic plays a significant role an everyday life as it is available at every place of our need. The role of plastic came into play from the early 1870s and it contributed to the making life comfortable in many aspects. It has diversified features ranging from its elasticity, hardness transparency, and durability to customized shaping with ease. The inclusive characteristics of plastic helped to raise the production from 0.5 to 550 (million tons) from 1940 to 2018, in the time span of 78 years [1]. The durability of plastic poses the negative implications in terms of it disposal as the consumption rate around the globe is increasing with rapid pace. The resulting implications are associated with the marine living beings and the reduction in crop yield of soil.

Environmental Impacts of Plastics However, the durability of plastics has some drawbacks because of the increasing rate of plastic material consumption globally and the lower degradation rate that leads to its accumulation in pelagic and benthic biota in coastal and marine sediments, as well as in **coastal boundaries** at all latitudes [2].

Plastic pieces are divided into "main" and "secondary" categories based on their sources and creation procedures. A new synthesis technology for producing plastic material results in dispersion of micro size material into the atmosphere causing pollution.

This definition identifies the sources of primary plastics such as polymers that are purposefully created and utilized as such; personal care products, industrial or commercial goods, and others containing polymeric micro-beads fall under this category. Other items include naturally occurring byproducts of other industrial processes or plastic produced as an unintentional or intentional spillage, such as pellets lost during manufacturing and shipping. The production of smaller plastic pieces results from the fragmentation of larger plastic items during plastic degradation, which starts with photo-oxidative degradation and is followed by thermal and/or chemical degradation. In contrast, primary plastics are linked to primary pollution sources.

* ashok04che@gmail.com

DOI: 10.1201/9781003269779-10

233

Presently people are addicted to the single-use plastics, which is having a bad impact on nature, society, economy and health. The global consumption of plastic bottles reaches to 1 million in every minute, while approximately five trillion bags are being circulated every year globally. The generated wastes from plastic were manageable during the period 1950–1970, as the production and consumption was very low as compared to the present date. The production increase of plastic products tripled the waste plastic generation during the period 1970–1990. Presently, the global production of plastic per year has reached 400 million tonnes. Based on the historical growth data, the forecasted production of plastic is 1100 tonnes in the year of 2050, which itself create a huge amount of plastic waste. The packaging itself account for 36% of total plastic uses of which a majority percentage of plastics ended up with landfilling or nonregulated waste.

It is interesting to note that approximately 98% of single-use plastics are made from fossil fuel or virgin feedstock. Moreover, the production, use, and disposal of plastic will account for a 19% rise in the emission of greenhouse gases by 2040. The systematic and scientific handling of plastic waste can only safeguard the environment and living beings. The recycle percentage of plastic waste is nearly 10% only, while millions of tonnes are settled in the lap of nature. For the sorting and processing of plastic waste from the packaging segment alone, $80 billion to $120 billion is spent, and this is a huge loss to the economy.

Microplastics are small plastic particles that measure less than 5 mm in size. They are created through the breakdown of larger plastic items, such as bottles, bags, and other debris, or are intentionally manufactured for use in various products like cosmetics, cleaning agents, and industrial applications. Microplastics can also be derived from the degradation of larger plastic items in the environment due to factors like weathering, UV radiation, and mechanical abrasion [3]. There are two primary types of microplastics:

Primary Microplastics: These are intentionally manufactured small plastic particles, often used in products like exfoliating scrubs, toothpaste, and as abrasives in various industrial processes.

Secondary Microplastics: These are formed through the fragmentation and breakdown of larger plastic items that have been discarded into the environment. Over time, larger plastic debris undergoes physical and chemical degradation, breaking down into smaller and smaller particles, eventually becoming microplastics [4].

Microplastics have become a major environmental concern due to their prevalence and potential ecological impacts. They can contaminate various ecosystems, including oceans, rivers, lakes, soil, and even the air. Microplastics can be ingested by marine life and can also find their way into the human food chain through seafood consumption. The long-term effects of microplastic exposure on human health are still under investigation, but there is growing concern about their potential health impacts. Therefore, there is an urgent need for better waste management practices, reducing plastic consumption, and developing sustainable alternatives to plastic to mitigate the issue of microplastic pollution. Microplastics and the circular economy are closely connected as they both address the challenges of plastic pollution and strive for sustainable resource management.

The concept of the circular economy represents an economic model with the objective of reducing waste and enhancing the value of resources. This is achieved by advocating for the perpetual utilization, reutilization, repair, and recycling of products and materials. It encourages moving away from the traditional linear economy, where resources are extracted, used once, and then discarded as waste. Instead, the circular economy seeks to create a closed-loop system where materials are continuously cycled back into the production process, reducing the need for virgin resources and minimizing environmental impact. When it comes to microplastics, the circular economy approach can play a significant role in mitigating their presence in the environment. Here are some ways in which the circular economy can address the microplastics issue:

Reducing Plastic Consumption: One of the primary strategies in the circular economy is to reduce the consumption of single-use plastics and non-essential plastic products. By using less plastic overall, there will be fewer opportunities for plastic waste to break down into microplastics.

Designing for Recyclability: In a circular economy, products are designed with end-of-life recycling in mind. By making plastic products easier to recycle and encouraging the use of recycled plastic in new products, the circular economy can help prevent plastic waste from ending up as microplastics in the environment.

Improving Recycling Infrastructure: A robust recycling infrastructure is essential for effectively managing plastic waste. In a circular economy, investments in recycling facilities and technologies can increase the recycling rates, thereby reducing the amount of plastic that ends up as litter or waste in the environment.

Promoting Extended Producer Responsibility (EPR): In a circular economy, manufacturers are encouraged to take responsibility for their products throughout their entire life cycle, including managing the waste generated. Extended producer responsibility programs can help ensure that producers take measures to prevent the release of microplastics into the environment.

Encouraging Innovation: The circular economy fosters innovation in materials and product design, including the development of biodegradable and compostable alternatives to conventional plastics. Such innovations can help reduce the persistence of plastic waste and, in turn, the presence of microplastics in the environment.

By adopting circular economy principles, society can address the issue of plastic pollution, including microplastics, and move towards a more sustainable and environmentally friendly approach to resource management. However, it requires collaboration and commitment from governments, industries, consumers, and other stakeholders to make significant progress in tackling this complex environmental challenge.

10.1.2 Importance of Circular Economy for Plastics

The circular economy embraces the principles of reduction, reuse, and recycling, carefully balancing economic and environmental considerations to promote sustainability [5]. At the core of the circular economy lies the hierarchy of reduction, reuse, and recycling. The first step is reduction, where efforts are made to minimize the use of resources, including plastic, at the source. This involves adopting more efficient processes, designing products with fewer materials, and encouraging responsible consumption practices. By reducing our reliance on plastics, we can curb waste generation and its subsequent impact on the environment. Next comes reuse, which involves extending the life span of products and materials through refurbishment, repair, or repurposing. By encouraging reuse, we reduce the need for constant production of new items, conserving resources and energy. Additionally, promoting a culture of repairability and encouraging consumers to choose reusable alternatives over single-use plastics fosters a more sustainable consumption pattern. Recycling, the final step in the hierarchy, involves converting plastic waste into new products or materials. While recycling is essential in diverting waste from landfills and reducing demand for virgin resources, it is not without its challenges. Contamination, limited infrastructure, and high processing costs have hindered the efficiency of recycling systems. However, with advancements in technology and better waste management practices, the potential for recycling to close the loop and maintain the value of materials is steadily increasing. By harmonizing economic and environmental concerns, the circular economy optimizes resource use and minimizes waste. Companies can find new revenue streams by adopting circular business models that focus on product longevity, service-based offerings, and material recovery. Embracing a circular approach enhances resource efficiency, mitigates supply chain risks, and builds resilience in the face of resource scarcity. Moreover, environmental benefits are significant. The circular economy minimizes pollution and reduces the pressure on natural ecosystems caused by plastic waste. By keeping plastics in the economy and out of the environment, we can prevent them from breaking down into harmful microplastics and contaminating oceans and wildlife.

In conclusion, the circular economy offers a comprehensive and sustainable framework for managing resources, particularly plastics. By combining the principles of reduction, reuse, and recycling, and carefully considering both economic and environmental aspects, the circular economy fosters sustainability. This approach not only reduces plastic waste but also stimulates innovation, creates green jobs, and promotes responsible consumption and production practices. To achieve a truly sustainable future, collaboration among businesses, governments, and consumers is essential to embrace the circular economy and ensure the well-being of our planet for generations to come [6–10]. The same is true with plastic, a significant class of materials that serves numerous purposes in civilization. Plastic is utilized across a multitude of sectors, including office supplies, toys, footwear, civil construction, electronics, aerospace, food, medicine, textiles, packaging, paint, varnish, and the automotive industry. Its adaptability and attractive qualities make plastic an essential player, fulfilling a diverse array of roles within these various domains [11–12]. Not all plastics, however, go back into manufacture after being used. Just 9% of the plastics manufactured globally in 2015 were recycled; 12% were burned and the remainder were dumped in landfills [13]. According to Bernardo et al. (2016), recycling has a lower environmental impact [14]. After assessing recycling's impact on global warming and overall energy consumption, their analysis led to the deduction that, across the entirety of their life cycles, plastic materials typically surpass traditional materials in terms of both ecological and economic advantages, spanning from the extraction of raw materials to synthesis, transformation, transportation, utilization, recovery, and final disposition. In addition, Duval and Maclean (2007) discovered a decrease in the amount of energy needed for recycling [15,16]. To retain qualities, mechanical recycling requires the insertion of virgin or recycled material [17]. Chemical recycling involves physical processes, like remoulding, and the end result is a monomer or oligomer that can be employed in the synthesis of new goods. Mechanical recycling requires mental activities, such as rearranging [18–20]. The energy released when garbage is burned is recycled in this process [18].

When recycled material replaces virgin material within the same production cycle as the original product, this process is known as closed-loop recycling [21,22]. Open-loop recycling occurs when a recycled product is employed in a different stage of the manufacturing process, creating a new product that is distinct from the original [23,24]. Finding a new market for recycled goods and using more ecologically friendly technologies are essential to encouraging recycling. Recycled goods typically compete with virgin materials, which could make it more difficult for them to enter the market. Research on recycling should also aim to find recycled materials that are more commercially and technically viable. High-density polyethylene (HDPE), polypropylene (PP), and low-density polyethylene are the three most commonly used and produced polymers.

In the realm of the automotive industry, PP finds extensive application in crafting a variety of components. These encompass car trunk lids, battery trays and enclosures, heater boxes, toolboxes, seatbelt buckling enclosures, rearview mirror housings, electric junction enclosures, hubcaps, carpets, battery safeguards (providing defense against short circuits), steering wheel overlays, shock absorber casings, vacuum and air hoses, consoles, bumpers, and glove compartments. These varied items, all based on PP, assume pivotal roles in augmenting the performance and functionality of vehicles within the automotive sector [25–28]. In order to meet the primary technical requirements of vehicle manufacturers, PP compounds must have a good thermal resistance, a good balance between stiffness and tenacity, and use fewer imported raw ingredients, resulting in more competitive prices. PP exhibits strong process ability in addition to these characteristics [29,30]. The industry producing laminated materials is a crucial source of supplies for the automotive industry.

However, throughout the tempering and laminating procedures, industrial waste made of glass powder is produced and dumped in landfills without being put to any particular purpose [31]. Since car windows cannot be reused, they are discarded in addition to where glass powder comes from in the laminating and tempering processes [32]. In such circumstances, these components can be gathered and recovered before being subjected to grinding and polymer protection film separation operations. The created glass powder can be mixed with polymer components to create composites with a variety of features [33]. Studies on the creation of materials with various qualities have focused on the incorporation of mineral loading into PP [34,35]. The type of load, the size of the mineral load being utilized, and the degree of dispersion of these particles in the polymer matrix all affect how the final product's qualities might be improved. Talc and calcium carbonate are the two commercial mineral loading that are most frequently used [36,37]. In order to acquire reinforcement qualities and contrast them with those of traditional composites, this study explains the addition of glass powder to a PP matrix. The goal is to reuse a residue while acquiring new qualities in polypropylene composites (in this case, glass powder). Also, we evaluated how adding recycled polypropylene affected the final qualities.

Polypropylene's inclusion in a composite or combination as homo polymers, polymers are being used more frequently in a variety of applications as a replacement for conventional materials like metal and ceramic. They are also being blended with other materials to create mixtures and polymer composites, or they are used for their unique qualities like lightness, low cost of transformation, resistance to corrosion, ideal thermal and electrical insulation, and ease of conformation into complex shapes [38]. Due to their weaker resistance compared to metals and ceramics, polymers' mechanical characteristics are generally not acceptable in a variety of applications. Yet, in addition to the exciting possibilities of these materials as mixes or a composite matrix, the thermoplastic sector is expanding because of environmental concerns [39].

Because of the cost-effectiveness relationship, compound systems formed by combining two polymer materials (mixtures) or a polymer material and a filler (composites) hold significant technological intrigue. In both cases, the material comprises a continuous (matrix) phase and a dispersed phase, and the efficiency of their interactions dictates the material's behavior [39].

10.2   PROBLEM STATEMENT

10.2.1   PLASTIC POLLUTION

Plastic pollution refers to the widespread and detrimental presence of plastic waste in the environment, particularly in oceans, rivers, lakes, and terrestrial ecosystems. Plastic pollution has become a global environmental crisis due to the exponential increase in plastic production and consumption over the past few decades. Plastic is a versatile and durable material, making it indispensable in various industries and daily life. However, its durability is also a major problem when it comes to waste management. Plastic does not readily biodegrade, and as a result, large quantities of plastic waste end up accumulating in the environment. Many plastic items, especially single-use plastics like bags, bottles, and packaging, are discarded after a short period of use, adding to the accumulation of plastic pollution. Plastic pollution has severe impacts on the environment, wildlife, and human health. It poses a significant threat to marine life, as marine animals can mistake plastic debris for food or become entangled in plastic waste, leading to injury and death. Moreover, when plastic breaks down into smaller particles, it can enter the food chain and potentially affect human health through the consumption of contaminated seafood. Plastic pollution also contributes to the formation of microplastics, which are tiny plastic particles less than 5 mm in size. Microplastics can be found in water bodies, soil, and even the air, and their long-term effects on ecosystems and human health are still being studied.

Thermoplastics, elastomers, thermosets, and polymer compounds are the four major classes under which plastics are categorized. The class of every plastic material is determined by its macromolecular structure as well as by its physical characteristics.

Both thermosets and elastomers have hard and soft elasticity, but thermosetting plastics, unlike thermoplastics, cannot be melted for recycling due to their irreversible chemical structure. Thermosetting plastic, on the other hand, can be either amorphous or semi-crystalline. In contrast to semi-crystalline resins, which are embedded with crystalline phases, amorphous resins have statistically orientated macromolecules that are nearly ordered. Common semi-crystalline resins include polyamide (PA) and polypropylene, whereas typical amorphous resins include polycarbonate (PC), polystyrene (PS), and polyvinyl chloride (PVC) (PP). As PP falls under the semi-crystalline class, this group will be the subject of our attention [40]. Figure 10.1 is a basic flowchart outlining the main components, processes, and consequences of plastic pollution. Please note that this is a simplified representation of the complex issue of plastic pollution. The actual processes and consequences involved in plastic pollution are multifaceted and interconnected. The flowchart (Figure 10.1) aims to provide an overview of the main stages and impacts in a more straightforward format.

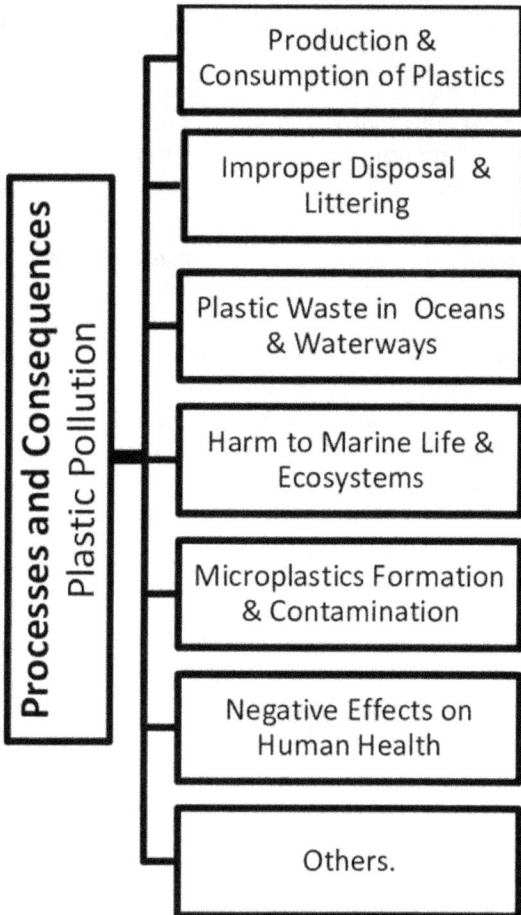

FIGURE 10.1 Processes and consequences of plastic pollution.

Explanation of the Flowchart:

- **Production & Consumption of Plastics:** This first step includes the manufacturing of plastic products and the widespread use of single-use plastics in various industries.
- **Improper Disposal & Littering:** Due to improper waste management practices and littering, a significant amount of plastic waste ends up in the environment, especially in oceans, rivers, and landfills.
- **Plastic Waste in Oceans & Waterways:** The plastic waste that is not properly disposed of finds its way into oceans and waterways through various means such as stormwater runoff and improper waste handling.
- **Harm to Marine Life & Ecosystems:** Plastic pollution has severe consequences on marine life and ecosystems. Marine animals can ingest or become entangled in plastic debris, leading to injury, suffocation, and death. This disrupts marine ecosystems and food chains.
- **Microplastics Formation & Contamination:** Over time, larger plastic items break down into smaller particles known as microplastics. These tiny particles are now found throughout the environment, including in water, soil, and even the air.
- **Negative Effects on Human Health:** Humans are exposed to plastic pollution through various sources like seafood consumption, drinking water, and inhalation of airborne microplastics. The potential impacts on human health are a growing concern.

Efforts to combat plastic pollution include waste management practices such as recycling, proper waste disposal, and reducing the use of single-use plastics. Various organizations and governments are implementing policies and initiatives to raise awareness about the issue and promote sustainable alternatives to plastic.

Addressing plastic pollution requires a collective effort from individuals, industries, governments, and organizations worldwide. It involves finding innovative solutions, investing in research and technology, and fostering a global commitment to reduce plastic waste and protect the environment for current and future generations.

Despite the significant potential of recycling and upcycling plastic waste, there are challenges that must be overcome. These include the lack of proper waste collection and sorting infrastructure, limited consumer awareness and participation, and the complexity of dealing with mixed plastics.

However, these challenges present opportunities for collaboration among governments, industries, and consumers. Investing in recycling infrastructure, raising awareness about the importance of recycling, and promoting eco-friendly products can accelerate the transition towards a circular economy for plastics.

10.2.2 Types of Plastic

Plastic, a synthetic polymer material, has become an integral part of modern life due to its versatility, durability, and cost-effectiveness. Plastics are used in a vast range of products and applications, from everyday household items to cutting-edge technologies. While plastics offer numerous benefits, they come in various types, each with distinct properties and characteristics.

Understanding the different types of plastic is crucial for proper waste management, recycling, and making informed choices about the products we use. In this chapter we will explore the main types of plastic commonly found in the market, their characteristics, applications, and environmental considerations.

By delving into the diverse world of plastics, we can gain valuable insights into their strengths, limitations, and potential for sustainability in our rapidly evolving world. Let's explore the fascinating realm of plastics and their impact on our lives and the environment. Some of these examples are mentioned in Table 10.1.

TABLE 10.1

Types of Plastics, Their Properties, Uses, and Recycling

Plastic Type	Properties	Common Uses	Recycling Information
PET (Polyethylene Terephthalate)	—Clear, lightweight, strong—Resistant to moisture and chemicals—Good barrier properties	—Beverage bottles (water, soda)—Food packaging (salad containers, condiment bottles)—Polyester fibers (clothing, carpets)	—Recyclable into fibers, fabrics, and new bottles—#1 symbol for identification
HDPE (High-Density Polyethylene)	—Rigid, tough, and durable—Resistant to chemicals and UV light	—Milk jugs—Cleaning product bottles—Trash bags	—Recyclable into containers, pipes, and plastic lumber—#2 symbol for identification
PVC (Polyvinyl Chloride)	—Rigid or flexible— Excellent resistance to chemicals	—Pipes and fittings—Vinyl flooring—Window frames	—Recyclable but difficult due to additives—#3 symbol for identification
LDPE (Low-Density Polyethylene)	—Flexible and tough—Resistant to chemicals	—Plastic bags—Squeezable bottles (e.g., shampoo)— Six-pack rings	—Recyclable into trash can liners, plastic lumber—#4 symbol for identification
PP (Polypropylene)	—High melting point— Resistant to heat and chemicals—Lightweight	—Food containers (e.g., yogurt cups)—Bottle caps—Auto parts	—Recyclable into battery cases, brooms, etc.—#5 symbol for identification
PS (Polystyrene)	—Rigid and lightweight—Good insulating properties	—Foam cups and plates (EPS)—Packaging peanuts—CD cases	—Recyclable but not widely accepted—#6 symbol for identification
Other (includes PC, ABS, etc.)	—Diverse properties depending on the type—May be rigid or flexible—Used in various applications	—Safety helmets (ABS)—Compact discs (PC)—Toys (various)	—Recycling options vary based on the specific plastic—#7 symbol for identification (others)

10.2.2.1 Polyethylene Terephthalate

Polyethylene terephthalate (PET or PETE) is a widely used thermoplastic polymer. It is known for its clarity, strength, and ability to hold carbonated beverages, making it a popular choice for food and beverage packaging, particularly for bottles and containers. Despite its advantages, PET also comes with environmental considerations that need to be addressed.

Environmental impact:

Recycling: PET is one of the most recycled plastics globally, and recycling PET bottles and containers helps reduce the amount of plastic waste in landfills and oceans. However, recycling rates vary widely between countries, and inadequate recycling infrastructure can limit its full potential for waste reduction.

Litter: Improper disposal of PET products, such as bottles and packaging, can lead to litter and contribute to plastic pollution. PET is lightweight and can easily be carried by wind and water, ending up in rivers, lakes, and oceans, where it can harm marine life and ecosystems.

Microplastics: Over time, PET products can break down into smaller fragments known as microplastics. These tiny particles can enter the environment and food chain, potentially posing risks to aquatic life and human health.

Energy consumption: The production of PET requires significant energy resources, and its manufacturing process contributes to greenhouse gas emissions. Reducing the environmental impact of PET involves improving energy efficiency and exploring sustainable alternatives.

Biodegradability: PET is not biodegradable under typical environmental conditions, which means it persists in the environment for a long time. Efforts are being made to develop more environmentally friendly alternatives, such as biodegradable plastics, to address this issue.

Mitigating environmental impact:

Recycling: Encouraging and supporting PET recycling initiatives can help divert plastic waste from landfills and reduce the demand for virgin plastic production.

Waste management: Implementing proper waste management practices, such as recycling, collection, and disposal, is crucial to prevent PET from entering the environment and causing pollution.

Sustainable alternatives: Promoting the use of sustainable packaging materials, such as biodegradable or compostable plastics and other eco-friendly alternatives, can reduce the environmental impact of PET.

Public awareness: Raising awareness among consumers about the importance of responsible plastic use and recycling can encourage more environmentally friendly practices.

While PET offers numerous benefits as a versatile and widely used plastic, its environmental impact requires careful consideration and responsible management. By promoting recycling, exploring sustainable alternatives, and improving waste management practices, we can mitigate the negative effects of PET on the environment and work towards a more sustainable future for plastic use [41].

10.2.2.2 High-Density Polyethylene

High-density polyethylene (HDPE) is a widely used thermoplastic known for its strength, rigidity, and versatility. It is commonly found in various applications, such as bottles, containers, pipes, and packaging materials. While HDPE offers many advantages, its environmental effects should be considered to understand its sustainability and potential impact on the environment.

Environmental effects:

Recycling: HDPE is one of the most commonly recycled plastics, and recycling it helps reduce the amount of plastic waste in landfills and the environment. HDPE's recyclability makes it a valuable material in a circular economy, where it can be reused to create new products.

Litter: Improper disposal of HDPE products, particularly single-use items like bottles and bags, can lead to litter and contribute to plastic pollution. Littered HDPE can find its way into water bodies, harming marine life and ecosystems.

Non-biodegradable: HDPE is non-biodegradable, meaning it does not naturally break down in the environment over time. It can persist for hundreds of years, accumulating in the environment and posing long-term challenges for waste management.

Landfill space: The disposal of HDPE in landfills takes up valuable space, leading to increased demand for landfill sites and potential land use issues.

Greenhouse gas emissions: HDPE production contributes to greenhouse gas emissions, primarily through the extraction and processing of fossil fuels used as raw materials. Efforts to reduce these emissions involve improving energy efficiency and transitioning to more sustainable energy sources.

Mitigating environmental effects:

Recycling: Encouraging the recycling of HDPE materials helps divert plastic waste from landfills and reduce the demand for virgin HDPE production.

Waste management: Proper waste management practices, such as recycling, collection, and disposal, are crucial to prevent HDPE from entering the environment and causing pollution.

Plastic alternatives: Promoting the use of alternative materials or more sustainable forms of packaging can help reduce HDPE consumption and its environmental impact.

Public awareness: Educating consumers about responsible plastic use, recycling, and the importance of reducing single-use plastic items can lead to more eco-friendly practices.

HDPE is a valuable and versatile plastic material with numerous applications, but it is essential to understand its environmental effects. By focusing on recycling, proper waste management, and exploring sustainable alternatives, we can mitigate the negative environmental impact of HDPE and work towards a more sustainable and responsible use of this plastic in our modern world [42]

10.2.2.3 Polyvinyl Chloride

Polyvinyl chloride (PVC), commonly known as vinyl, is a widely used synthetic thermoplastic polymer. It is known for its versatility, durability, and cost-effectiveness, making it a popular choice in various applications, including pipes, window frames, flooring, electrical cables, and packaging. While PVC offers numerous benefits, it also comes with environmental considerations that need to be addressed.

Environmental effects:

Toxic chemicals: The production of PVC involves the use of chlorine and other chemicals, which can release toxic pollutants, including dioxins, during manufacturing and disposal. Dioxins are harmful substances that can persist in the environment and have adverse effects on human health and ecosystems.

Phthalates: PVC products, especially flexible vinyl, may contain phthalates, which are chemical additives used to increase flexibility. Some phthalates are known to be endocrine disruptors and can have potential health impacts, especially in children and pregnant women.

Non-biodegradable: PVC is a non-biodegradable material, meaning it does not break down naturally in the environment. Improper disposal of PVC waste can lead to its accumulation in landfills and marine environments, contributing to plastic pollution.

Greenhouse gas emissions: The production of PVC involves the use of fossil fuels, leading to greenhouse gas emissions. These emissions contribute to climate change and its associated environmental impacts.

Recycling challenges: PVC recycling is more challenging compared to other plastics, mainly due to the complexity of separating PVC from other plastics and its potential for contamination. As a result, PVC recycling rates are relatively low.

Mitigating environmental effects:

Phasing out harmful additives: Efforts are being made to phase out the use of harmful additives, such as certain phthalates, in PVC products to reduce potential health and environmental risks.

Improved waste management: Proper waste management practices, including recycling and responsible disposal, can help prevent PVC waste from entering the environment and causing pollution.

Sustainable alternatives: Promoting the use of alternative materials, such as biodegradable plastics or other eco-friendly alternatives, can reduce the environmental impact of PVC.

Green chemistry and manufacturing: Implementing green chemistry principles in PVC production can help reduce the use of hazardous chemicals and minimize environmental impacts.

Product design and extended producer responsibility: Encouraging product design for recyclability and implementing extended producer responsibility programs can increase PVC recycling rates and reduce overall environmental impact.

Polyvinyl chloride (PVC) is a versatile and widely used plastic with various applications. While it offers many advantages, it is essential to address its environmental effects and take measures to minimize potential risks. By adopting sustainable practices, promoting recycling, and exploring safer alternatives, we can work towards a more responsible and environmentally friendly use of PVC in our society [43].

10.2.2.4 Low-Density Polyethylene

Low-density polyethylene (LDPE) is a plastic variant extensively utilized across diverse applications, primarily owing to its adaptability and comparatively affordable production. While LDPE has many benefits, its environmental impact can be a cause for concern. Here are some of the key environmental effects associated with LDPE:

Non-biodegradability: LDPE is a non-biodegradable material, meaning it does not readily break down into natural elements when disposed of. This attribute can contribute to the buildup of plastic waste within the environment, resulting in pollution and the potential endangerment of wildlife [44].

Pollution: Improper disposal of LDPE products, such as plastic bags, packaging materials, and single-use items, can lead to littering. These items can end up in water bodies, where they contribute to marine pollution and pose a threat to marine life. LDPE debris can also clog drainage systems, leading to localized flooding [45–47].

Microplastics: Over time, LDPE products can degrade into smaller plastic particles known as microplastics. These tiny particles can be found in soil, water, and even the air, and they have been detected in various ecosystems worldwide. Microplastics can enter the food chain, potentially causing health issues for both animals and humans [48–51].

Greenhouse gas emissions: The production of LDPE involves the use of fossil fuels, primarily natural gas, as a feedstock. The extraction, transportation, and processing of these fossil fuels contribute to greenhouse gas emissions, which contribute to climate change and its associated environmental impacts [52].

Energy consumption: The production of LDPE requires significant amounts of energy. High energy consumption not only contributes to greenhouse gas emissions but also puts additional stress on energy resources and can lead to increased carbon footprints.

Recycling challenges: While LDPE is recyclable, the recycling rate for this type of plastic is relatively low compared to other plastics. The recycling process for LDPE can be more complex and less economically viable than for some other plastics. As a result, a considerable amount of LDPE ends up in landfills or incineration facilities [53,60–64].

Mitigating the environmental repercussions of LDPE requires a multifaceted approach encompassing several strategies. These strategies include diminishing plastic consumption, enhancing waste management and recycling facilities, advocating for the adoption of more environmentally friendly materials, and fostering the innovation of biodegradable or compostable substitutes to conventional plastics [65–69].

Governments, industries, and individuals all play crucial roles in adopting responsible practices to minimize the negative effects of LDPE and plastic waste on the environment. Awareness, education, and sustainable policies are essential in the efforts to mitigate plastic pollution and protect our ecosystems.

In the early 1980s in Japan, approximately 45% of municipal plastic waste was disposed of in landfills as a proportion of municipal solid waste (MSW), while around 50% was subjected to incineration, and the remaining 5% was sorted and recycled, according to the Plastic Waste Management Institute's report in 1985. By 1990, the United States had only managed to recycle about 1% of post-consumer plastics, even though over 15% of all MSW was incinerated [44–47].

According to recent data from the Central Pollution Control Board in New Delhi, India, it has been estimated that a significant 8 million tonnes of plastic products are used on an annual basis in India alone. A study conducted on plastic waste generation across 60 major Indian cities revealed that the country generates a substantial 15,340 tonnes of plastic waste per day [54].

10.2.2.5 Polypropylene

Polypropylene (PP) is another thermoplastic polymer that garners extensive use and bears a substantial influence on the environment. Similar to other plastics, PP has both positive and negative environmental effects. Here are some key aspects to consider:

Positive environmental aspects of PP:

Lightweight: PP is a lightweight material, which means it requires less energy to transport compared to heavier materials. This can potentially reduce greenhouse gas emissions during transportation.

Recyclability: PP is recyclable, and the recycling process can help reduce the demand for virgin PP production. Recycling PP products also helps to divert waste from landfills, reducing pollution.

Versatility: PP's versatility allows it to be used in a wide range of applications, including packaging, automotive parts, and textiles. Its durability and resistance to various chemicals can lead to longer product life spans, reducing the need for frequent replacements.

Negative environmental aspects of PP:

Non-biodegradability: Like other conventional plastics, PP is non-biodegradable, leading to persistent plastic pollution in the environment.

Recycling challenges: While PP is recyclable, its recycling rate is still relatively low compared to other plastics. The recycling process for PP can be more complicated due to the different grades and types of PP, as well as potential contamination from food residues or other substances.

Energy consumption: The production of PP involves the use of fossil fuels, primarily natural gas, as a feedstock. The extraction, transportation, and processing of these fossil fuels contribute to greenhouse gas emissions and energy consumption.

Pollution: Improper disposal of PP products can lead to littering, and like other plastics, PP can break down into microplastics that can pollute ecosystems and harm marine life.

Environmental efforts related to PP:

Recycling initiatives: Promoting and endorsing recycling initiatives aimed at PP and other plastics holds the potential to curtail their environmental footprint. Governments, corporations, and local communities can allocate resources toward enhancing recycling facilities and disseminating awareness about the significance of recycling.

Biodegradable alternatives: Researchers and industries are exploring the development of biodegradable or compostable alternatives to traditional PP. These materials have the potential to break down more quickly in the environment, reducing long-term pollution.

Sustainable sourcing: Initiatives directed at utilizing renewable and sustainable feedstock for PP manufacturing can play a pivotal role in lessening dependence on fossil fuels and diminishing the carbon footprint associated with the plastic industry [14].

Plastic waste reduction: Reducing overall plastic consumption and promoting the use of reusable or less harmful materials can significantly decrease the environmental impact of PP and other plastics.

To address the environmental effects of PP, a combination of responsible consumption, improved waste management, innovation in material development, and policy changes is necessary. The shift towards a more circular economy, where products are designed with recycling and environmental impact in mind, is a crucial step in mitigating the negative effects of plastic materials like PP on the environment.

10.2.2.6 Polystyrene

Polystyrene (PS), commonly known by the brand name Styrofoam, is a type of plastic that has significant environmental effects. It is widely used in various industries for packaging, food containers, disposable cups, and insulation, among other applications. However, its environmental impact is a cause for concern due to the following factors:

Non-biodegradability: Like many other plastics, polystyrene is non-biodegradable. It can persist in the environment for hundreds of years, contributing to plastic pollution.

Litter and marine pollution: Polystyrene products, such as foam cups and takeout containers, are often used once and then discarded. Improper disposal of these items can lead to littering, and they are particularly prone to breaking down into small pieces and becoming microplastics. These microplastics can enter water bodies, posing a threat to marine life and potentially entering the food chain.

Recycling challenges: Recycling polystyrene can be challenging due to its low density and volume, which makes it uneconomical to transport and process. Moreover, polystyrene foam often contains additives and contaminants, making recycling more complicated and less viable.

Environmental toxins: Polystyrene is derived from styrene, a compound that the International Agency for Research on Cancer (IARC) has classified as a potential human carcinogen. While direct exposure to styrene is a concern for workers in the manufacturing process, the leaching of styrene and other chemicals from polystyrene products can also be a concern, especially when used for hot liquids or acidic foods.

Energy consumption: The production of polystyrene requires significant amounts of energy, contributing to greenhouse gas emissions and environmental degradation associated with fossil fuel consumption.

Waste management challenges: Because of its lightweight and bulky nature, polystyrene takes up a lot of space in landfills. Its resistance to decomposition means that polystyrene waste continues to accumulate, exacerbating waste management issues.

Environmental efforts related to polystyrene:

Bans and restrictions: Many regions have implemented bans or restrictions on the use of polystyrene foam products to reduce its environmental impact. These measures aim to promote the use of more sustainable alternatives, such as compostable or recyclable materials.

Recycling and reuse: Some communities have established recycling programs specifically for polystyrene, attempting to divert it from landfills. Additionally, efforts are made to recycle and repurpose polystyrene waste into other products like picture frames or building materials.

Biodegradable alternatives: Research and development are ongoing to create biodegradable alternatives to traditional polystyrene foam. These materials can break down more quickly in the environment, reducing long-term pollution.

Education and awareness: Raising public awareness about the environmental impact of polystyrene and promoting responsible consumption and waste management practices can play a significant role in reducing its negative effects.

To mitigate the environmental effects of polystyrene, a combination of legislative actions, industry initiatives, and individual responsibility is necessary. Transitioning towards more sustainable materials and improving waste management practices are crucial steps in reducing the environmental impact of polystyrene and other single-use plastics. Plastics have become a significant environmental issue due to their widespread use, improper disposal, and slow decomposition in the environment.

However, the concept of the circular economy offers potential solutions to mitigate the environmental impact of plastics. Here are some plastic types related to environmental issues and their relationship to the circular economy:

Single-use plastics: Single-use plastics, such as plastic bags, straws, and cutlery, have become a symbol of plastic pollution. These items are often used briefly and then discarded, leading to significant littering and marine pollution. In a circular economy, efforts are made to reduce the consumption of single-use plastics by promoting reusable alternatives and encouraging responsible consumption.

Polystyrene (PS) or Styrofoam: As mentioned earlier, polystyrene is a type of plastic with significant environmental effects. Due to its non-biodegradability and difficulties in recycling, polystyrene has been targeted for bans or restrictions in many places. In a circular economy, the focus is on finding alternatives to polystyrene or improving recycling systems to close the loop on its life cycle.

Polyethylene terephthalate (PET): PET is commonly used for beverage bottles and food containers. While PET is highly recyclable, the recycling rates for PET bottles can still be improved in many regions. In a circular economy, efforts are made to optimize PET recycling systems to ensure that more bottles are collected, recycled, and turned into new products.

High-density polyethylene (HDPE): HDPE is used in various applications, including milk jugs, detergent bottles, and plastic pipes. It is one of the more easily recyclable plastics. In a circular economy, increasing the demand for recycled HDPE and creating a market for products made from recycled HDPE is essential.

Low-density polyethylene (LDPE): LDPE is commonly used in plastic bags, cling wraps, and various packaging materials. Similar to other plastics, LDPE poses environmental challenges due to its non-biodegradability and low recycling rates. In a circular economy, promoting the recycling of LDPE and supporting the development of biodegradable alternatives are essential strategies.

Polyvinyl chloride (PVC): PVC is used in pipes, vinyl flooring, and various consumer products. Its production can lead to the release of toxic chemicals, and it is not easily recyclable. In a circular economy, efforts are made to phase out or replace PVC with more environmentally friendly materials and promote safer recycling processes.

In a circular economy, the focus is on designing products with recycling and end-of-life considerations in mind, maximizing the use of recycled content, and ensuring that plastics are recovered, recycled, and reintroduced into the production cycle. This approach aims to minimize waste, reduce the consumption of finite resources, and limit the environmental impact of plastics on ecosystems and human health. It requires collaboration among industries, governments, and consumers to create a more sustainable and circular system for plastics and other materials.

10.3   CIRCULAR ECONOMY FOR PLASTICS

The circular economy for plastics is a model that aims to create a more sustainable and resource-efficient approach to the production, use, and disposal of plastic materials. In a traditional linear economy, products are manufactured, used, and then discarded as waste. In contrast, the circular economy seeks to close the loop, keeping materials in use for as long as possible, minimizing waste, and promoting recycling and reuse.

10.3.1   KEY PRINCIPLES OF THE CIRCULAR ECONOMY FOR PLASTICS

Reduce and rethink: The first step is to reduce the overall consumption of plastics, especially single-use plastics. This involves rethinking product design, packaging, and consumption patterns to minimize plastic waste generation.

Design for circularity: Products and packaging should be designed with recycling and reuse in mind. Using recyclable materials and designing for easy disassembly can facilitate the recycling process and promote the creation of new products from recycled plastics.

Extended producer responsibility: Manufacturers and producers take responsibility for the entire life cycle of their products, including recycling and proper disposal. This encourages them to design products that are easier to recycle and to use more recycled materials in their production.

Recycling and recovery: Enhancing recycling infrastructure and technologies is crucial to transform plastic waste into valuable resources. This involves investing in sorting and processing facilities to increase the recycling rates for plastic materials.

Recycled content: Promote the use of recycled plastic in new products to reduce the demand for virgin plastic production. Encouraging industries to use recycled plastic in their manufacturing processes helps to close the loop and reduce the reliance on fossil fuel-based feedstock.

Circular business models: Adopting circular business models, such as product-as-a-service or leasing arrangements, encourages companies to retain ownership of products and packaging, incentivizing longer product life and better materials management.

Consumer awareness and behavior change: Educating consumers about the benefits of the circular economy and responsible plastic consumption can drive changes in behavior, such as choosing products with less packaging or opting for reusable alternatives.

10.3.2   BENEFITS OF THE CIRCULAR ECONOMY FOR PLASTICS

Reduced plastic waste: By promoting recycling and reuse, the circular economy reduces the amount of plastic waste entering landfills and oceans, mitigating environmental pollution.

Resource conservation: The circular economy conserves natural resources by using recycled plastics instead of relying solely on new fossil fuel-based materials.

Lower carbon emissions: Recycling plastics and using recycled content requires less energy compared to producing plastics from raw materials, leading to reduced greenhouse gas emissions.

Economic opportunities: The circular economy can create new business opportunities and jobs in recycling, remanufacturing, and eco-friendly product design.

10.3.3 CHALLENGES OF IMPLEMENTING THE CIRCULAR ECONOMY FOR PLASTICS

Infrastructure and technology: Building and upgrading recycling infrastructure and technologies to handle various types of plastics can be challenging and costly.

Consumer behavior: Encouraging consumers to adopt circular practices, such as recycling correctly and choosing sustainable products, requires concerted efforts in education and awareness.

Policy and regulation: Governments need to implement supportive policies and regulations to incentivize circular economy practices and discourage harmful plastic consumption and waste.

The circular economy for plastics is a vital approach to address the environmental challenges associated with plastic waste. It requires collaboration among governments, industries, consumers, and other stakeholders to create a more sustainable and circular system for plastics, reducing their impact on the environment and contributing to a more resource-efficient future.

10.3.3.1 Recycling Method

Recycling is a fundamental pillar of the circular economy, as it enables the reuse of materials and reduces the need for extracting new resources. Various recycling methods play a crucial role in the circular economy for different materials, including plastics. Here are some common recycling methods used in the circular economy [55]:

- **Mechanical recycling:** Mechanical recycling is one of the most common methods used for recycling plastics. It involves sorting, cleaning, and processing plastic waste into small flakes or pellets, which can then be melted and molded into new products. Mechanical recycling is suitable for clean and homogenous plastics, such as PET and HDPE [56–59].
- **Chemical recycling:** Chemical recycling, also known as feedstock recycling or advanced recycling, involves breaking down plastic polymers into their chemical constituents using various processes such as pyrolysis, depolymerization, or gasification. These processes convert plastic waste into feedstock or building blocks that can be used to produce new plastics or other valuable chemicals and materials [13, 56–58].
- **Biological recycling:** Biological recycling, also known as biodegradation or composting, is a method used for some biodegradable plastics. These

plastics can be broken down naturally by microorganisms, turning them into harmless substances like water, carbon dioxide, and compost.

- **Feedstock recycling:** As mentioned earlier, certain chemical recycling methods convert plastic waste into raw materials or feedstock that can be used in the production of new plastics or other materials.

Upcycling: Upcycling involves transforming plastic waste into higher-value products with improved properties or functionality. For example, turning plastic bottles into fashionable clothing or creating decorative items from discarded plastic containers.

- **Closed-loop recycling:** Closed-loop recycling refers to recycling processes where products are recycled back into the same type of product they came from. For example, recycling PET bottles to produce new PET bottles.
- **Open-loop recycling:** Open-loop recycling involves recycling materials into different types of products. For instance, recycling PET bottles to create polyester fibers for textiles.
- **Design for recycling:** In the circular economy, design for recycling is a crucial concept where products are designed with recycling in mind. This means using easily recyclable materials and ensuring that products can be efficiently disassembled and processed at the end of their life [70].

Recycling plays a pivotal role in the circular economy by reducing the environmental impact of waste, conserving resources, and minimizing the need for virgin materials. However, to maximize the effectiveness of recycling in the circular economy, it is essential to improve recycling infrastructure, increase recycling rates, enhance waste collection systems, and promote consumer awareness and participation in recycling initiatives. Collaborative efforts among governments, industries, and consumers are necessary to make recycling a central component of the circular economy and achieve a more sustainable and resource-efficient future. Figure 10.2 shows the circular economy sequence.

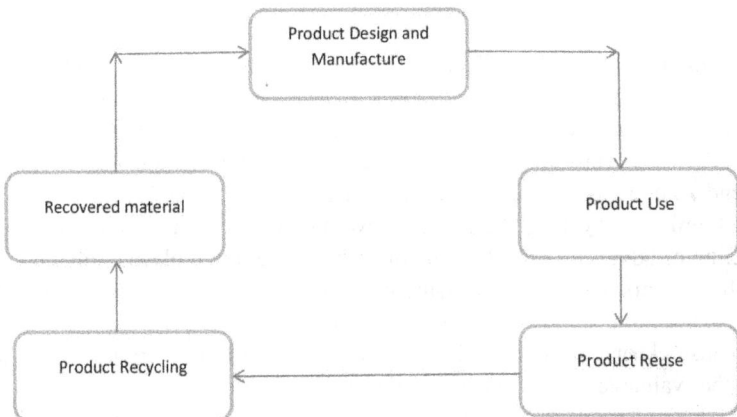

FIGURE 10.2 Circular economy sequence.

10.3.3.2 Upcycling Methods for Plastic Utilization to Value-Added Products

Upcycling methods for plastic utilization involve transforming plastic waste into value-added products with improved properties or functionality. These methods aim to divert plastic waste from landfills or the environment while creating new, valuable items. Here are some common upcycling methods for plastic:

Plastic extrusion: Plastic extrusion involves melting plastic waste and forcing it through a mold to create new shapes or profiles. This method is commonly used to create plastic lumber, which can be used as an alternative to wood in construction and landscaping.

3D printing: 3D printing, also known as additive manufacturing, can utilize plastic waste as the raw material to create custom-designed products, prototypes, and small-batch manufacturing.

Plastic art and crafts: Plastic waste can be creatively transformed into various art pieces, sculptures, and crafts. Artists and designers often use discarded plastic materials to create innovative and visually appealing artworks.

Plastic filament for 3D printing: Plastic waste can be processed and converted into filaments suitable for 3D printing. This process helps close the loop by using plastic waste to produce new products through additive manufacturing.

Plastic molding: Plastic waste can be melted and molded into various shapes using traditional molding techniques. This approach can create new products with unique designs and functionality.

Plastic fabrication: Plastic waste can be cut, shaped, and fabricated into new items, such as household items, storage containers, or decorative pieces.

Plastic composite materials: Plastic waste can be combined with other materials, such as wood fibers, to create composite materials with enhanced properties. These materials can be used in construction, furniture, and other applications.

Plastic bottle planters: Plastic bottles can be upcycled into planters for gardening. By cutting the top section and adding drainage holes, plastic bottles can become functional containers for growing plants.

Plastic jewelry: Discarded plastic materials, such as bottle caps or pieces of plastic, can be transformed into unique and stylish jewelry items.

Plastic bags and wrappers: Plastic bags and wrappers can be creatively transformed into fashion accessories, such as handbags, wallets, or even raincoats.

Implementing upcycling techniques for plastic provides a wide range of benefits, including reducing waste, conserving resources, and fostering creativity and innovation. These methods contribute to a more sustainable and circular approach to plastic utilization, tackling the global concern of plastic pollution while generating value-added goods. Upcycling stands as an effective means to transform discarded plastic into valuable items, significantly mitigating the environmental repercussions of plastic production and consumption. This chapter delves into recent achievements and future prospects within the realm of plastic waste upcycling technology. The focus is primarily on the production of high-value merchandise through processes such as pyrolysis, gasification, photoreforming, and mechanical reprocessing.

10.3.3.2.1 Conventional Pyrolysis

Conventional pyrolysis of plastics is a thermal degradation process that involves heating plastic waste in the absence of oxygen to produce useful products like liquid oil, gas, and solid residue. This process helps to recycle and convert plastic waste into valuable resources, contributing to waste management and reducing environmental pollution. Here is an overview of the steps and products involved in conventional pyrolysis of plastics:

Feedstock preparation: Plastic waste, such as mixed plastics, plastic films, or plastic containers, is collected and sorted. Contaminants and non-plastic materials are removed to ensure a consistent feedstock.

Heating and pyrolysis: The sorted plastic waste is fed into a pyrolysis reactor, where it is heated to high temperatures (typically around 300–500°C) in an oxygen-free environment (anaerobic conditions). This lack of oxygen prevents complete combustion and leads to thermal degradation of the plastic.

Vaporization: As the temperature rises, the plastic waste undergoes depolymerization, breaking down large polymer chains into smaller molecules, including various hydrocarbons.

Condensation: The vapors produced during pyrolysis are cooled rapidly, causing them to condense into different phases: liquid, gas, and solid.

Product collection: The condensed products are collected and separated. The main products obtained from the conventional pyrolysis of plastics include:

1. **Pyrolysis oil (pyrolysis liquid):** Also known as plastic oil or pyrolysis liquid fuel, this dark brown to black liquid is a complex mixture of hydrocarbons. It can be further refined and used as a substitute for conventional fuels like diesel, gasoline, or fuel oil.

2. **Pyrolysis gas:** The gas phase includes light hydrocarbons and non-condensable gases, such as methane, ethane, propane, and carbon monoxide. This gas can be used for heat or energy generation.

3. **Char (pyrolysis solid):** The solid residue left behind after pyrolysis is called char or carbon black. It consists of carbon and ash, and its composition depends on the plastic feedstock and pyrolysis conditions.

Post-treatment: The obtained pyrolysis oil may undergo further refining processes, such as hydrotreating or distillation, to improve its quality and remove impurities.

It's important to note that the quality and composition of the products obtained from conventional pyrolysis can vary depending on the plastic feedstock, pyrolysis conditions, and process design. Additionally, conventional pyrolysis methods may face challenges related to emissions, energy consumption, and the need for post-treatment processes to obtain higher-quality products. As a result, ongoing research and advancements in pyrolysis technology aim to improve the efficiency and environmental performance of this recycling method for plastic waste.

10.3.3.2.1.1 Temperature Conditions for Conventional Pyrolysis The temperature conditions for conventional pyrolysis of plastics typically range between 300°C and 500°C. Within this temperature range, the thermal degradation of plastic waste occurs, leading to the breakdown of large polymer chains into smaller hydrocarbons and other compounds [71].

The specific temperature used in conventional pyrolysis can vary depending on the type of plastic feedstock, the desired product yield and quality, and the pyrolysis reactor design. Different plastics have varying thermal degradation points, and the temperature range is chosen to ensure efficient decomposition while avoiding complete combustion (which requires higher temperatures and oxygen).

Lower temperatures (around 300–400°C) are often used for softer plastics with lower melting points, while higher temperatures (around 400–500°C) are applied to process harder plastics with higher melting points. Operating at higher temperatures generally results in faster pyrolysis rates, but it also requires more energy input.

Controlling the temperature in the pyrolysis process is critical to achieving the desired product composition and quality. Precise temperature control helps optimize the yield of valuable products, such as pyrolysis oil, gas, and char, while minimizing the formation of undesirable byproducts.

It's worth noting that the temperature range mentioned here is specific to conventional pyrolysis, where plastics are heated in a batch or continuous reactor without additional catalysts. Advanced pyrolysis techniques, such as catalytic pyrolysis or hydrothermal liquefaction, may involve different temperature conditions to enhance the yield and properties of the produced products.

The product yield of waste material via pyrolysis can vary depending on the type of waste being processed, the pyrolysis conditions, and the specific equipment used. Pyrolysis is a process that involves heating organic materials in the absence of oxygen, leading to the decomposition of the materials into various products. Table 10.2 is a general representation of the product yield from typical waste materials via pyrolysis.

TABLE 10.2

Representation of the Product Yield from Typical Waste Materials via Pyrolysis

Waste Material	Product Yield	Products Produced
Plastic (Polyolefins)	50%–75% liquid, 15%–40% gas 5%–15% char	Pyrolysis oil, pyrolysis gas, char
Tire (Rubber)	45%–50% liquid, 35%–45% gas 10%–15% char	Pyrolysis oil, pyrolysis gas, char
Biomass (Wood, Agricultural Residue)	60%–75% liquid, 10%–25% gas 10%–25% biochar	Bio-oil, biochar, syngas
Sewage Sludge	30%–60% liquid, 20%–30% gas 10%–30% biochar	Pyrolysis oil, pyrolysis gas, biochar
Food Waste	40%–60% liquid, 30%–40% gas 5%–20% biochar	Pyrolysis oil, pyrolysis gas, biochar

Please note that these product yields are approximate ranges and can vary based on the specific waste feedstock and pyrolysis process conditions. Additionally, the composition and quality of the products can vary depending on the waste material and the pyrolysis equipment used. Some pyrolysis processes may also be optimized to maximize the yield of a particular product, such as bio-oil for biofuel production.

The pyrolysis process has the potential to convert various waste materials into valuable products, such as biofuels, chemicals, and biochar, contributing to waste management and resource recovery efforts. However, the economic feasibility and environmental impact of pyrolysis depend on various factors, including feedstock availability, technology efficiency, and market demand for the end products.

Many studies have concentrated on the pyrolysis of plastic trash to produce high-quality bio-oil. The performance of pyrolysis is influenced by reaction parameters such as reaction temperature, feedstock, and reaction time etc. pyrolysis of mixed plastic waste like styrene butadiene and polyester, produce high-quality bio-oil with characteristics resembling those of diesel fuel at different temperatures (700°C and 900°C) [72]. Bio-oil received at 900°C showed high brake thermal efficiency than that received at 700°C and which reduced oxides of nitrogen, carbon monoxide, and carbon dioxide [72].

Combining biomass and plastic waste, co-pyrolysis is a promising step towards effective recycling of plastic since there may be a synergistic impact that enhances the yield and properties of the bio-oil [73]. The biomass pyrolysis towards the quality of the bio-oil benefits from co-feeding with plastic. It is well-known that plastic garbage contains a lot of hydrogen. The carbon/hydrogen ratio of the mixture can be adjusted by adding biomass, which will also improve the selectivity of the process and overall worth for products conversion [73].

Compared to cracking without a catalyst, the heat reduction of the molecular weight of molecules by using a catalyst can lower the upcycling plastic waste energy requirement. Various zeolites and metal oxides have lately been investigated and created to enhance the performance of pyrolysis. For instance, a two-stage fixed bed reactor system was used to study the impact of bimetallic catalysts (Ni-Fe) on plastic waste pyrolysis [74].

More carbon and hydrogen were deposited with the catalyst that had more iron. Its greater cracking capacity and comparatively less interaction between the support and active sites were cited as the reasons for this. The catalyst's Ni surface increased both the thermal stability and the degree of graphitization of the generated carbons [68]. In a further work, Ni-Co-Al catalysts were utilized to convert plastic waste using a two-stage reactor; the first stage involved pyrolyzing the plastic, while the second stage involved dry reforming [75]. The catalyst feedstock interaction mode and other factors have also been investigated to gauge the effectiveness of the catalyst. In catalytic pyrolysis in situ, the catalyst and feedstock are combined in a single reactor. In catalytic pyrolysis ex situ, the catalyst is placed in a separate reactor. In both processes (in situ and ex situ), the catalytic pyrolysis of plastic wastes was shown to occur via various reaction pathways [76]. In the catalytic pyrolysis of PS, the heat breakdown of PS was mostly caused by the free radical mechanism [76]. In the pyrolysis of PET, CO_2 provides a novel approach to improve energy recovery and fewer acidic byproducts (such as benzoic acid) produced [77]. Techno-economic analyses have been carried out by several researchers in order to enhance the pyrolysis process in the future. For instance, Fivga and Dimitriou investigated the generation of heavy fuel substitutes using the pyrolysis of plastic trash. According to their calculations, the facility would need to operate for about a year before making back the capital investment in the 10,000 kg/h plant case. When the cost of plastic garbage declines, the cost of producing fuel can be decreased [78].

10.3.3.2.2 Microwave-Assisted Pyrolysis

Microwave-assisted pyrolysis is an advanced pyrolysis technique that uses microwave energy to heat the feedstock, including plastics, in the absence of oxygen [79]. This method offers several advantages over conventional pyrolysis, such as faster heating rates, more uniform heating, and potentially better control over product yields and qualities. The temperature conditions for microwave-assisted pyrolysis of plastics can vary depending on the specific process and desired outcomes.

The temperature range for microwave-assisted pyrolysis of plastics is generally similar to that of conventional pyrolysis, which is around 300°C to 500°C. However, in microwave-assisted pyrolysis, the heating process can be more rapid and selective, allowing for localized heating of the feedstock and achieving higher temperatures in specific regions [80].

The advantage of using microwave energy lies in its ability to selectively heat polar and high-dielectric materials, like water and some plastics, more effectively. This can lead to faster and more efficient thermal degradation of the plastic feedstock, resulting in increased product yields and potentially improved product qualities.

The use of microwaves can also help reduce energy consumption and processing time compared to conventional pyrolysis methods. Additionally, the localized heating offered by microwave-assisted pyrolysis can improve the control of temperature gradients, reducing the risk of hot spots and thermal stress on the reactor [69].

However, microwave-assisted pyrolysis has its own set of challenges, including the potential for non-uniform heating and reactor design complexities. Optimization of parameters such as feedstock composition, particle size, and microwave power level is essential to achieve the desired outcomes.

In summary, the temperature conditions for microwave-assisted pyrolysis of plastics are similar to conventional pyrolysis, but the use of microwave energy offers advantages in terms of faster heating rates and potentially better control over product yields and qualities. Researchers and engineers continue to explore and develop this technology to make it more efficient and viable for large-scale plastic waste recycling applications.

10.3.3.2.3 Conventional Gasification

Conventional gasification is a thermochemical process that converts carbonaceous materials, including plastics, into a gas mixture known as synthesis gas or syngas [81]. The process occurs in a high-temperature, low-oxygen environment, enabling the conversion of organic materials into valuable gases, such as carbon monoxide, hydrogen, and methane. The temperature conditions for conventional gasification can vary depending on the type of feedstock and desired syngas composition [82].

Temperature Conditions for Conventional Gasification: The temperature range for conventional gasification typically falls between 700°C and 1200°C. Operating at high temperatures is essential to achieve the desired reactions, including the gasification of carbonaceous [83] materials and the reforming of hydrocarbons into syngas components. The actual temperature used depends on factors such as feed-stock composition, reactor design, and desired syngas quality.

10.3.3.2.3.1 Applications of Conventional Gasification

Syngas production: The primary application of conventional gasification is the production of syngas, which is a versatile fuel and feedstock. Syngas can be used as a substitute for natural gas in heating and power generation or further processed to produce hydrogen and other valuable chemicals [84].

Energy generation: The syngas produced from gasification can be used in gas turbines, internal combustion engines, or fuel cells to generate electricity and heat.

Chemical production: Syngas serves as a vital feedstock in the chemical industry. It can be converted into various chemicals, such as ammonia, methanol, and synthetic fuels.

Integrated gasification combined cycle (IGCC): In IGCC power plants, the syngas generated from gasification is used in a combined cycle system, which improves overall energy efficiency and reduces emissions compared to conventional coal-fired power plants.

Waste-to-energy (WtE): Gasification can be applied in WtE facilities, where it helps convert organic waste, including certain plastics, into syngas. This process reduces landfill volume and harnesses energy from waste materials [84].

Biomass gasification: Conventional gasification is also widely used for converting biomass, such as agricultural residues, wood chips, and energy crops, into syngas. This contributes to renewable energy generation and sustainable waste management.

Conventional gasification has several advantages, including the ability to handle a wide range of feedstocks and produce valuable syngas [85]. However, it also faces challenges, such as high capital costs, process complexities, and handling of potential contaminants in the feedstock. As technology advances and environmental concerns grow, efforts continue to improve gasification efficiency and make it a more sustainable and viable option for waste utilization and clean energy production.

10.3.3.2.3.2 Gasification Aided by Plasma and Supercritical Water
Plasma-assisted and supercritical water–assisted gasification are advanced gasification techniques that offer unique advantages for the upcycling of plastic waste. These methods involve subjecting plastic waste to extreme conditions, such as high temperatures and pressures, to facilitate the conversion of plastics into valuable products [86]. Here's an overview of both plasma-assisted and supercritical water–assisted gasification for plastic waste upcycling:

10.3.3.2.4 Plasma-Assisted Gasification
Plasma gasification involves using plasma, an ionized gas with high energy, to generate extremely high temperatures (up to several thousand degrees Celsius) [87].

This high-temperature environment allows for the efficient decomposition of complex organic materials like plastic waste [88]. In the case of plastic waste, plasma gasification breaks down the long polymer chains into smaller hydrocarbon fragments and other compounds [89].

Advantages:

Rapid and efficient thermal degradation of plastics, even for mixed or contaminated feedstock.

Enhanced control over gasification reactions, leading to a more selective and optimized product output.

Destruction of hazardous or toxic components in the feedstock.

Syngas produced can be used for energy generation or as a feedstock for chemical production.

Challenges:

High capital and operating costs due to the complexity of plasma generation equipment.

Challenges in the proper handling and treatment of byproducts, such as the solid ash residue.

Scalability and integration with existing waste management infrastructure [83].

Supercritical Water-Assisted Gasification:

Supercritical water-assisted gasification involves using water at supercritical conditions (above its critical point, where temperature and pressure are high enough to remove the distinction between gas and liquid phases) as a medium for the gasification. Under these conditions, water becomes an excellent solvent for organic materials like plastics, promoting the rapid and efficient decomposition of plastic waste.

Advantages:

Uses water as the reaction medium, eliminating the need for additional gasifying agents.

Efficient breakdown of complex polymers in plastic waste into syngas components.

Minimal emissions of greenhouse gases.

Potential to operate at lower temperatures compared to other gasification methods.

Challenges:

High energy requirements for maintaining supercritical water conditions. Challenges in handling and processing the generated syngas, as it may contain dissolved organic compounds. Corrosion and erosion issues due to the aggressive nature of supercritical water.

Both plasma-assisted and supercritical water-assisted gasification have shown promise in upcycling plastic waste into valuable products like syngas, which can be used for energy generation or as feedstock for the chemical industry process [90]. However, these advanced gasification techniques are still in the research and development phase, and further advancements are needed to make them economically and environmentally viable on a larger scale [91].

10.3.3.2.5 Emerging Conversion Technologies

As the global focus on sustainable waste management and plastic recycling intensifies, several emerging conversion technologies are being developed to tackle the challenges of plastic waste. These technologies aim to convert plastic waste into valuable products, reduce environmental pollution, and contribute to a more circular economy. Here are some of the prominent emerging conversion technologies for plastic recycling:

10.4 CHEMICAL RECYCLING (ADVANCED RECYCLING OR FEEDSTOCK RECYCLING)

Chemical recycling involves breaking down plastic waste into its chemical building blocks or converting it back into raw materials that can be used to produce new plastics or other valuable chemicals. Different methods, such as pyrolysis, depolymerization, and gasification, fall under chemical recycling. These processes offer the potential to recycle mixed or contaminated plastics that are challenging to process through mechanical recycling.

- **Enzymatic recycling:** Enzymatic recycling uses enzymes to break down polymers into their monomers, facilitating the recycling of certain types of plastics that are difficult to recycle by traditional methods. This technology holds promise for dealing with complex plastics and multilayer materials.
- **Solvent-based recycling:** Solvent-based recycling involves dissolving plastic waste in a specific solvent, separating it from contaminants, and then recovering the solvent for reuse. This method can be effective in recycling certain plastics that are not easily recyclable by mechanical methods.
- **Hydrothermal liquefaction:** Hydrothermal liquefaction uses high-temperature and high-pressure water to convert plastic waste into a bio-oil-like substance. This technology offers the potential to convert a wide range of plastic waste, including mixed plastics and post-consumer plastics, into valuable fuels and chemicals.
- **Biodegradable plastics and composting:** The development of biodegradable plastics that can break down naturally in composting conditions is gaining traction. These biodegradable plastics can reduce plastic pollution in certain applications, especially in organic waste streams.
- **Upcycling and additive manufacturing:** Advanced recycling technologies enable the upcycling of plastic waste into high-value products. Additionally, additive manufacturing (3D printing) using recycled plastics is being explored as a way to create custom-designed products and prototypes.

- **Chemical upcycling:** Chemical upcycling involves transforming plastic waste into chemicals that can be used as building blocks for various industries, such as the production of specialty chemicals and polymers.
- **Monomer recovery:** Technologies are being developed to recover valuable monomers, such as ethylene and propylene, from mixed plastic waste. These monomers can then be used as feedstock for new plastics.
- **Waste-to-hydrogen:** Some technologies are focused on converting plastic waste into hydrogen gas, which can be used as a clean energy source for various applications.

It's important to note that while these emerging technologies show promise, many are still in the early stages of development or limited to pilot-scale operations. Scaling up these technologies and addressing technical, economic, and regulatory challenges are crucial steps toward realizing their potential for large-scale plastic recycling and waste management.

10.5   ASSOCIATED CHALLENGES AND FUTURE PERSPECTIVES

The primary challenges hindering the extensive adoption of plastic waste conversion are elevated expenses (stemming from factors like energy consumption and the need for pre-sorting plastics), time constraints, and limited resources. Additionally, the prevalence of contaminants and non-polymeric substances in most waste streams, along with the considerable diversity in plastic compositions (especially in the case of mixed plastics), underscores the necessity for pre-sorting.

- To further increase the conversion efficiency and selectivity of plastic waste, it is necessary to investigate and develop affordable, sophisticated catalysts with higher activity. Due to metal sintering and coke production, conventional catalysts employed in thermal conversions frequently deactivate.
- To further increase conversion efficiencies, systematic approach is needed for development in experimental process and condition adjustment are required. The mild circumstances in photoreforming as well as the heating rate in pyrolysis and gasification need to be optimized.
- The use of thermosets for upcycling requires more study. Thermoplastics are the main focus of the great majority of existing conversion technologies. Reduced global plastics demand and increased awareness of plastic reuse are two important implementation strategies to reduce plastic waste.

Transforming plastic waste into value-added products through recycling and upcycling methods is essential for building a circular economy that reduces environmental impact and fosters sustainable resource management. By adopting a circular economy approach, we can minimize plastic waste, protect ecosystems and wildlife, and create a more resilient and responsible future for generations to come. To achieve this, a collective effort is required from all stakeholders to embrace innovative solutions, invest in sustainable practices, and prioritize the long-term well-being of the planet. Through these combined efforts, we can turn the plastic waste crisis into an opportunity for positive change and build a more sustainable and circular future.

10.6   CONCLUSIONS

Much study has been done on renewable alternatives as a result of rising environmental concerns about the buildup of plastic trash around us. Because to its green credentials, PLA (polylactic acid) has thus been one of the top contenders during the previous 20 years. As a result, extensive research has concentrated on creating ecologically safe and long-lasting catalyst replacements to Sn (Oct)2 for PLA manufacture, with a current emphasis on stereoselectivity. Comparative analysis, suggested for improvement in the research on the EOL (end of life) possibilities for PLA waste and plastic. In fact, traditional approaches to managing plastic trash are in line with theory, of a linear economy, which is a major cause of the waste crisis.

So, industries are looking for the transition towards, sustainable circular economy model to cut down plastic waste while also maintaining material value, which will necessitate the creation of other waste handling techniques. Mechanical recycling is one option; however, the technology is still constrained by eventual material degradation brought on by thermomechanical degradation, which forces material repurposing to less honourable applications. Chemical recycling offers a viable alternative to this process because it includes both PLA depolymerization and degradation, the latter of which makes it possible to obtain value-added compounds like lactate esters. In turn, this will add value to the PLA supply chain, fostering improved economic performance—a quality that industries find particularly appealing.

Plastic garbage upcycling has recently drawn more notice. The conversion of plastic trash into high-value goods like H_2, carbon nano material, monomers, and liquid fuel has enormous potential. There are uses and categories of plastic trash that each conversion technique now in use is best suited for. Several solutions should be combined and scaled up to meet the huge amount of plastic garbage produced. Several thermophysical factors affect the process's ability to transform waste plastic into a product of equal or higher value. For PS (polystyrene) pyrolysis, the favourable conditions for enhanced bio-oil production is 400–450°C, and for MSW (municipal solid waste) gasification the temperature ranges from 550°C to 850°C. The synergism of the processes, which produces a high quality plus increased product yield, and co-pyrolysis of plastic waste and biomass has been identified as a potential technology. Due to its large specific surface, pore volume, and metal content, a zeolite-based catalyst with promoter added has been reported to be the viable catalyst in the pyrolysis of plastic waste.

REFERENCES

[1] Plastics Europe. *Annual Review 2017–2018*. Available from: www.plasticseurope.org/en/resources/publications/498-plasticseuropeannual-review-2017-2018.

[2] Wright SL, Thompson RC, Galloway TS. The physical impacts of microplastics on marine organisms: A review. *Environmental Pollution*. 2013;178:483–492.

[3] McDevitt J, Criddle CS, Morse M, Hale RC, Bott C, Rochman C. Addressing the issue of microplastics in the wake of the Microbead-Free Waters Act—A new standard can facilitate improved policy. *Environmental Science and Technology.* 2017;51(12):6611–6617.

[4] Ward CP, Armstrong CJ, Walsh AN, Jackson JH, Reddy CM. Sunlight converts polystyrene to carbon dioxide and dissolved organic carbon. *Environmental Science & Technology Letters.* 2019;6:669–674. https://doi.org/10.1021/acs.estlett.9b00532.

[5] Kirchherr J, Reike D, Hekkert M. Conceptualizing the circular economy: An analysis of 114 definitions. *Resources, Conservation and Recycling.* 2017;127:221–232. https://doi.org/10.1016/j.resconrec.2017.09.005.

[6] Korhonen J, Honkasal JS. Circular economy: The concept and its limitations. *Ecological Economics.* 2018;143:37–46. https://doi.org/10.1016/j.ecolecon.2017.06.041.

[7] Michelini G, Moraes RN, Cunha RN, Costa JMH, Ometto AR. From linear to circular economy: PSS conducting the transition. *Procedia CIRP.* 2017;64:2–6. https://doi.org/10.1016/j.procir.2017.03.012.

[8] Ritzén S, Sandström GÖ. Barriers to the circular economy—Integration of perspectives and domains. *Procedia CIRP.* 2017;64:7–12. https://doi.org/10.1016/j.procir.2017.03.005.

[9] Pomponi F, Moncaster A. Circular economy for the built environment: A research framework. *Journal of Cleaner Production.* 2017;143:710–718. https://doi.org/10.1016/j.jclepro.2016.12.055.

[10] Ghisellini P, Cialani C, Ulgiati S. A review on circular economy: The expected transition to a balanced interplay of environmental and economic systems. *Journal of Cleaner Production.* 2016;114:11–32. https://doi.org/10.1016/j.jclepro.2015.09.007.

[11] Andrady AL, Neal MA. Applications and societal benefits of plastics. *Philosophical Transactions of the Royal Society of London. Series B, Biological Sciences.* 2009;364(1526):1977–1984. https://doi.org/10.1098/rstb.2008.0304.

[12] North EJ, Halden RU. Plastics and environmental health: The road ahead. *Reviews on Environmental Health.* 2013;28(1):1–8. https://doi.org/10.1515/reveh-2012-0030.

[13] Geyer R, Jambeck JR, Law KL. Production, use, and fate of all plastics ever made. *Science Advances.* 2017;3:1–5. https://doi.org/10.1126/sciadv.1700782.

[14] Bernardo CA, Simões C, Pinto LMC. Environmental and economic life cycle analysis of plastic waste management options. A review. In: *Proceedings of the Regional Conference Graz 2015—Polymer Processing Society PPS: Conference Papers*; 2016. https://doi.org/10.1063/1.4965581.

[15] Duval D, MacLean HL. The role of product information in automotive plastics recycling: A financial and life cycle assessment. *Journal of Cleaner Production.* 2007;15:1158–1168. https://doi.org/10.1016/j.jclepro.2006.05.030.

[16] Arena U, Mastellone ML, Perugini F. Life cycle assessment of a plastic packaging recycling system. *The International Journal of Life Cycle Assessment.* 2003;8(2):92–98. https://doi.org/10.1065/lca2003.02.106.

[17] Rajendran S, Hodzic A, Soutis C, Al-Maadeed MA. Review of life cycle assessment on polyolefins and related materials. *Plastics, Rubber and Composites.* 2012;41(4/5):159–168. https://doi.org/10.1179/1743289811Y.0000000051.

[18] Horodytska O, Valdés FJ, Fullana A. Plastic flexible films waste management—A state of art review. *Waste Management.* 2018;77:413–415. https://doi.org/10.1016/j.wasman.2018.04.023.

[19] Conceição RDP, Pereira C, Pessoa G, Pacheco EBAV. The post-consumer PET recycling chain and definition of their stages: A case study in Rio de Janeiro. *Revista Brasileira de Ciências Ambientais.* 2016;39:80–96. https://doi.org/10.5327/Z2176-9478201613514.

[20] Sinha V, Patel MR, Patel JV. PET waste management by chemical recycling: A review. *Journal of Polymers and the Environment.* 2010;18(1):8–25. https://doi.org/10.1007/s10924-008-0106-7.

[21] Huysman S, Debaveye S, Schaubroeck T, DeMeester S, Ardente F, Mathieux F, et al. The recyclability benefit rate of closed-loop and open-loop systems: A case study on plastic recycling in Flanders. *Resources, Conservation and Recycling.* 2015;101:53–60. https://doi.org/10.1016/j.resconrec.2015.05.014.

[22] Chilton T, Burnley S, Nesaratnam S. A life cycle assessment of the closed-loop recycling and thermal recovery of post-consumer PET. *Resources, Conservation and Recycling.* 2010;54(12):1241–1249. https://doi.org/10.1016/j.resconrec.2010.04.002.

[23] Williams TGJL, Heidrich O, Sallis P. A case study of the open-loop recycling of mixed plastic waste for use in a sports field drainage system. *Resources, Conservation and Recycling.* 2010;55(2):118–128. https://doi.org/10.1016/j.resconrec.2010.08.002.

[24] Shen L, Worrel E, Patel MK. Open-loop recycling: A LCA case study of PET bottle-to-fibre recycling. *Resources, Conservation and Recycling.* 2010;55(1):34–52. https://doi.org/10.1016/j.resconrec.2010.06.014.

[25] Fernandes BL, Domingues AJ. Mechanical characterization of recycled polypropylene for automotive industry. *Polímeros.* 2007;17:85–87. https://doi.org/10.1590/S0104-14282007000200005.

[26] Hemais CA. Polymers and the automobile industry. *Revista Polímeros: Ciência e Tecnologia.* 2003;13:107–114. https://doi.org/10.1590/S0104-14282003000200008.

[27] Maddah HA. Polypropylene as a promising plastic: A review. *American Journal of Polymer Science.* 2016;6(1):1–11. https://doi.org/10.5923/j.ajps.20160601.01.

[28] Moritomi S, Watanabe T, Kanzaki S. Polypropylene Compounds for Automotive Applications [Internet]; 2010. Translated from R&D Report, "SUMITOMO KAGAKU" Vol. 2010-I. Available from: www.sumitomo-chem.co.jp/english/rd/report/theses/docs/20100100_a2g.pdf [Accessed: 09–05–2018]

[29] Santos LS, Silva AHMFT, Pacheco EBAV, Silva ALN. Avaliação do efeito da adição de PP reciclado nas propriedades mecânicas e de escoamento de misturas PP/EPDM. *Revista Polímeros: Ciência e Tecnologia.* 2013;23(3):389–394. https://doi.org/10.4322/polimeros.2013.083.

[30] Da Silva Spinacé MA, De Paoli MA. The technology of polymer recycling. *Química Nova.* 2005;28:65–72. https://doi.org/10.1590/S0100-40422005000100014doi:10.1039/A809539F.

[31] Valera TS, Sakai ACV, Wiebeck H, Toffoli SM. Propriedades do compósito poliamida-6/vidroempó. *AnaisCongressoBrasileiro de Engenharia e Ciência dos Materiais*, São Pedro/SP; n°14. 2000, pp. 49401–49411. Available from: www.ipen.br/biblioteca/cd/cbecimat/2000/Docs/TC403-010.pdf [Accessed: 25–08–2018]

[32] Medina HV, Gomes DA. The automobile industry designing for recycling. In: *CETEM*, 2002. Available from: http://web resol.org/textos/ferro_reciclagem_automoveis_brasil.pdf [Accessed: 25–08–2018]

[33] Swain B, Ryang Park J, Yoon Shin D, Park KS, Hwan Hong M, Gi Lee C. Recycling of waste automotive laminated glass and valorization of polyvinyl butyral through mechanochemical separation. *Environmental Research.* 2015;142:615–623. https://doi.org/10.1016/j.envres.2015.08.017.

[34] Sheril RV, Mariatti M, Samayamutthirian P. Single and hybrid mineral fillers (talc/silica and talc/calcium carbonate)-filled polypropylene composites: Effects of filler loading and ratios. *Journal of Vinyl and Additive Technology.* 2014;20:160–167. https://doi.org/10.1002/vnl.21347.

[35] Yu M, Huang R, He C, Wu Q, Zhao XD, Hom N. Hybrid composites from wheat straw, inorganic filler, and recycled polypropylene: Morphology and mechanical and thermal expansion performance. *International Journal of Polymer Science.* 2016;2016:1–12. ID2520670. https://doi.org/10.1155/2016/2520670.

[36] Srivabut C, Ratanawilai T, Hiziroglu S. Effect of nanoclay, talcum, and calcium carbonate as filler on properties of composites manufactured from recycled polypropylene and rubberwood fiber. *Construction and Building Materials.* 2018;162:450–458. https://doi.org/10.1016/j.conbuildmat.2017.12.048.

[37] Rabello M. *Aditivação de Polímeros*, EditoraArtliber. São Paulo: ABPol; 2011.

[38] Fernandes BL, Domingues AJ. Mechanical characterization of recycled polypropylene for automotive industry. *Revista Polímeros: Ciência e Tecnologia.* 2007;7:85–87. https://doi.org/10.1590/S0104-14282007000200005.

[39] Ota WN. *Análise de compósitos de polipropileno e fibras de vidro utilizados pela indústria automotiva nacional* [Dissertation]. Curitiba: Universidade Federal do Paraná; 2004. 90 f. Available from: www.pipe.ufpr.br/portal/defesas/dissertacao/058. pdf [Accessed: 25–08–2018]

[40] Maddah HA. Polypropylene as a promising plastic: A review. *American Journal of Polymer Science.* 2016;6(1):1–11.

[41] Ferrotto MF, Asteris PG, Borg RP, Cavaler L. Strategies for waste recycling: The mechanical performance of concrete based on limestone and plastic waste. *Sustainability.* 2022;14:1706.

[42] Silviyati I, Zubaidah N, Amin JM, Supraptiah E, Utami RD, Ramadhan I. The effect of addition of high-density polyethylene (HDPE) as binder on Hebel Light Brick (Celcon). *Journal of Physics: Conference Series.* 2020;1500(1):012083. IOP Publishing.

[43] Comaniță ED, Ghinea C, Roșca M, Simion IM, Petraru M, Gavrilescu M. Environmental impacts of polyvinyl chloride (PVC) production process. In *2015 E-Health and Bioengineering Conference (EHB).* IEEE; 2015, pp. 1–4.

[44] Thompson RC, Moore CJ, vomSaal FS, Swan SH. Plastics, the environment and human health: Current consensus and future trends. *Philosophical Transactions of the Royal Society B.* 2009;364:2153–2166. https://doi.org/10.1098/rstb.2009.0053.

[45] Zhu Y, Romain C, Williams CK. Sustainable polymers from renewable resources. *Nature.* 2016;540:354–362. https://doi.org/10.1038/nature21001.

[46] Ellen MacArthur Foundation. *The New Plastics Economy: Catalysing Action*; 2016. Available from: www.ellenmacarthurfoundation.org/publications/new-plastics-economy-catalysing-action [Accessed 9–01–2019]

[47] Rabnawaz M, Wyman I, Auras R, Cheng S. A roadmap towards green packaging: The current status and future outlook for polyesters in the packaging industry, *Green Chemistry.* 2017;19:4737–4753. https://doi.org/10.1039/c7gc02521a.

[48] Okkerse C, van Bekkum H. From fossil to green. *Green Chemistry.* 1999;1:107–114.

[49] Jambeck JR, Geyer R, Wilcox C, Siegler TR, Perryman M, Andrady A, Narayan R, Law KL. Plastic waste inputs from land into the ocean. *Science.* 2015;347:768–771. https://doi.org/10.1126/science.1260352.

[50] Hopewell J, Dvorak R, Kosior E. Plastics recycling: Challenges and opportunities. *Philosophical Transactions of the Royal Society B.* 2009;364:2115–2126. https://doi. org/10.1098/rstb.2008.0311.

[51] Coszach P, Bogaert J-C, Wilocq J. *Chemical Recycling of PLA by Hydrolysis*, US 8431683 B2, 2013.

[52] Piemonte V, Sabatini S, Gironi F. Chemical recycling of PLA: A great opportunity towards the sustainable development? *Journal of Polymers and the Environment.* 2013;21:640–647. https://doi.org/10.1007/s10924-013-0608-9.

[53] Soroudi A, Jakubowicz I. Recycling of bioplastics, their blends and biocomposites: A review. *European Polymer Journal*. 2013;49:2839–2858. https://doi.org/10.1016/j.eurpolymj.2013.07.025.23.

[54] Song JH, Murphy RJ, Narayan R, Davies GBH. Biodegradable and compostable alternatives to conventional plastics. *Philosophical Transactions of the Royal Society B*. 2009;364:2127–2139. https://doi.org/10.1098/rstb.2008.0289.

[55] Clercq RD, Dusselier M, Sels BF. Heterogeneous catalysis for bio-based polyester monomers from cellulosic biomass: Advances, challenges and prospects. *Green Chemistry*. 2017;19:5012–5040. https://doi.org/10.1039/C7GC02040F.

[56] Lambert S, Wagner M. Environmental performance of bio-based and biodegradable plastics: The road ahead. *Chemical Society Reviews*. 2017;46:6855–6871. https://doi.org/10.1039/C7CS00149E.

[57] Auras R, Harte B, Selke S. An overview of polylactides as packaging materials. *Macromolecular Bioscience*. 2004;4:835–864. https://doi.org/10.1002/mabi.200400043.

[58] Datta R, Henry M. Lactic acid: Recent advances in products, processes and technologies—a review. *Journal of Chemical Technology & Biotechnology*. 2006;81:1119–1129. https://doi.org/10.1002/jctb.1486.

[59] Weber CJ, Haugaard V, Festersen R, Bertelsen G. Production and applications of bio-based packaging materials for the food industry. *Food Additives & Contaminants*. 2002;19:172–177. https://doi.org/10.1080/02652030110087483.

[60] Nampoothiri KM, Nair NJ, John RP. An overview of the recent developments in polylactide (PLA) research. *Bioresource Technology*. 2010;101:8493–8501. https://doi.org/10.1016/j.biortech.2010.05.092.

[61] Lasprilla AJR, Martinez GAR, Lunelli BH, Jardini AL, Filho RM. Poly-lactic acid synthesis for application in biomedical devices—a review. *Biotechnology Advances*. 2012;30:321–328. https://doi.org/10.1016/j.biotechadv.2011.06.019.

[62] Jones MD. Sustainable catalysis: With non-endangered metals—part 1. *Sustainable Catalysis: With Non-endangered Metals*. 2015:199–215.

[63] Garlotta D. A literature review of poly (Lactic Acid). *Journal of Polymers and the Environment*. 2001;9:63–84. https://doi.org/10.1023/a:1020200822435.

[64] Van Wouwe P, Dusselier M, Vanleeuw E, Sels B. Lactide synthesis and chirality control for polylactic acid production. *ChemSusChem*. 2016;9:907–921. https://doi.org/10.1002/cssc.201501695.

[65] Vink ETH, Rábago KR, Glassner DA, Springs B, O'Connor RP, Kolstad J, Gruber PR. The sustainability of NatureWorks polylactide polymers and Ingeo polylactidefibers: An update of the future. *Macromolecular Bioscience*. 2004;4:551–564. https://doi.org/10.1002/mabi.200400023.

[66] Vink ETH, Davies S. Life cycle inventory and impact assessment data for 2014 Ingeo TM polylactide production. *Industrial Biotechnology*. 2015;11:167–180. https://doi.org/10.1089/ind.2015.0003.

[67] Dusselier M, Van Wouwe P, Dewaele A, Jacobs PA, Sels BF. Shape-selective zeolite catalysis for bioplastics production. *Science*. 2015;349:78–80. https://doi.org/10.1126/science.aaa7169.

[68] Ishihara K, Ohara S, Yamamoto H. Direct condensation of carboxylic acids with alcohols catalyzed by Hafnium(IV) salts. *Science*. 2000;290:1140–1142. https://doi.org/10.1126/science.290.5494.1140.

[69] Zhao Y, Wang Y, Duan D, Ruan R, Fan L, Zhou Y, Dai L, Lv J, Liu Y. Fast microwave-assisted ex-catalytic co-pyrolysis of bamboo and polypropylene forbio-oil production. *Bioresource Technology*. 2018;249:69–75.

[70] Ellen Mac Arthur Foundation. *The New Plastics Economy: Rethinking the Future of Plastics*, 2016. Available from: www.ellenmacarthurfoundation.org/publications/the-new-plastics-economy-rethinkingthe-future-of-plastics [Accessed: 7–01–2019].

[71] Kaimal VK, Vijayabalan P. A study on synthesis of energy fuel from waste plastic and assessment of its potential as an alternative fuel for diesel engines. *Waste Management.* 2016;51:91–96.

[72] Kalargaris I, Tian G, Gu S. The utilisation of oils produced from plastic waste at different pyrolysis temperatures in a DI diesel engine. *Energy.* 2017;131:179–185.

[73] Ozsin G, Pütün AE. Insights into pyrolysis and co-pyrolysis of biomass and polystyrene: Thermochemical behaviors, kinetics and evolved gas analysis. *Energy Conversion and Management.* 2017;149:675–685.

[74] Yao D, Wu C, Yang H, Zhang Y, Nahil MA, Chen Y, Williams PT, Chen H. Co-production of hydrogen and carbon nanotubes from catalytic pyrolysis of waste plastics on Ni-Fe bimetallic catalyst. *Energy Conversion and Management.* 2017;148:692–700.

[75] Saad JM, Williams PT. Pyrolysis-catalytic-dry reforming of waste plastics and mixed waste plastics for syngas production. *Energy & Fuels.* 2016;30:3198–3204.

[76] Xue Y, Johnston P, Bai X. Effect of catalyst contact mode and gas atmosphere during catalytic pyrolysis of waste plastics. *Energy Conversion and Management.* 2017;142:441–451.

[77] Lee J, Lee T, Tsang YF, Oh J-I, Kwon EE. Enhanced energy recovery from polyethylene terephthalate via pyrolysis in CO2 atmosphere while suppressing acidic chemical species. *Energy Conversion and Management.* 2017;148:456–460.

[78] Fivga A, Dimitriou I. Pyrolysis of plastic waste for production of heavy fuel substitute: A techno-economic assessment. *Energy.* 2018;149:865–874.

[79] Fodah AEM, Ghosal MK, Behera D. Microwave-assisted pyrolysis of agricultural residues: Current scenario, challenges, and future direction. *International Journal of Environmental Science and Technology.* 2021:1–26.

[80] Suriapparao DV, Boruah B, Raja D, Vinu R. Microwave assisted co-pyrolysis of biomasses with polypropylene and polystyrene for high quality bio-oil production. *Fuel Processing Technology.* 2018;175:64–75.

[81] Couto N, Monteiro E, Silva V, Rouboa A. Hydrogen-rich gas from gasification of Portuguese municipal solid wastes. *International Journal of Hydrogen Energy.* 2016;41:10619–10630.

[82] Zhao X, Li K, Lamm ME, Celik S, Wei L, Ozcan S. Solid waste gasification: Comparison of single- and multi-staged reactors, in: Silva V (Ed.), *Gasification*, Intech Open; 2021.

[83] Materazzi M, Lettieri P, Taylor R, Chapman C. Performance analysis of RDF gasification in a two stage fluidized bed-plasma process. *Waste Management.* 2016;47:256–266.

[84] Bora RR, Wang R, You F. Waste polypropylene plastic recycling toward climate change mitigation and circular economy: Energy, environmental, and technoeconomic perspectives. *ACS Sustainable Chemistry & Engineering.* 2020;8:16350–16363.

[85] Ng KS, Phan AN. Evaluating the techno-economic potential of an integrated material recovery and waste-to-hydrogen system. *Resources, Conservation and Recycling.* 2021;167:105392.

[86] Mazzoni L, Janajreh I. Plasma gasification of municipal solid waste with variable content of plastic solid waste for enhanced energy recovery. *International Journal of Hydrogen Energy.* 2017;42:19446–19457.

[87] Ramos A, Berzosa J, Espí J, Clarens F, Rouboa A. Life cycle costing for plasma gasification of municipal solid waste: A socio-economic approach. *Energy Conversion and Management.* 2020;209:112508. https://doi.org/10.1016/j.enconman.2020.112508.

[88] Ma W, Chu C, Wang P, Guo Z, Liu B, Chen G. Characterization of tar evolution during DC thermal plasma steam gasification from biomass and plastic mixtures: Parametric optimization via response surface methodology. *Energy Conversion and Management.* 2020;225:113407. https://doi.org/10.1016/j.enconman.2020.113407.

[89] Yayalık I, Koyun A, Akgün M. Gasification of municipal solid wastes in plasma arc medium. *Plasma Chemistry and Plasma Processing.* 2020;40:1401–1416.

[90] Messerle VE, Mosse AL, Ustimenko AB. Processing of biomedical waste in plasma gasifier. Waste Management. 2018;79:791–799.

[91] Bai B, Liu Y, Zhang H, Zhou F, Han X, Wang Q, Jin H. Experimental investigation on gasification characteristics of polyethylene terephthalate (PET)microplastics in super-critical water. *Fuel.* 2020;262:116630.

11 Closing the Loop
Circular Economy for Food-Based Emulsions and Films for Energy Materials

Himanshi Bansal, Surinder Singh*,
Alex Ibhadon, and S.K. Mehta

ABBREVIATIONS

CEO	clove essential oil
CUR	curcumin
HIPE	high internal phase emulsion
LSPR	localised surface plasma resonance
NPs	nanoparticles
O/W	oil in water
O/W/O	oil in water in oil
W/O	water in oil
W/O/W	water in oil in water

11.1 INTRODUCTION

Worldwide plastic pollution, which occurs on land and in the oceans, is as a consequence of the continuous production of plastic solid waste. Accordingly, recycling plastics like polymers is a viable, long-term option for waste management. Researchers have created techniques to recycle polymers in a "closed loop," or without losing these qualities, in order to tackle this issue. As an outcome, it will make it possible to recycle enormous quantities of various kinds of polymer materials, reducing plastic waste and possibly finding a solution to the issues related to ecological deterioration and pollution caused by plastics.

The closed-loop recycling approach unquestionably "opens" new avenues for the effective and environmentally responsible recycling of polymer materials! These ideas can help us develop a more environmentally friendly and circular economy that is good for individuals, the natural world, and the economy. The fact that edible coatings are biodegradable and consumed alongside the product gives them a variety of advantages versus artificial packaging, not the least of which is that they do not contaminate the environment [1].

---

* ssbdcet@gmail.com

DOI: 10.1201/9781003269779-11

The European Commission has launched a circular economy–based policy with a focus on the growth of sustainable materials, identifying the proper handling of plastic waste as a top priority (European Commission, 2018). The creation of films and emulsions utilising food byproducts is regarded as a "greener" method of replacing traditional and non-biodegradable materials for packaging while also satisfying customer demands for sustainable and healthier food packaging.

Food and agricultural waste can be decreased and additionally converted into brand-new items or goods that encourage recycling, repair, reconstruction, and reuse which ensures circular economy practices. By providing a source of money and, over time, reducing environmental damage, this can strengthen area businesses [2]. The circular economy of plastics serves as a blueprint for a closed system that encourages the reuse of plastics-based products, creates value from trash, and keeps recyclable plastics out of landfills. Plastic trash is a useful resource that can be used to make novel substances based on polymers, make parts out of plastic, or even create electricity when recycling is not an option [3]. By implementing the three fundamental tenets of reduce, reuse, and recycle, the circular economy aims to optimise the material resources at our disposal. In this approach, waste is utilised, product life cycles are prolonged, and over time, a more effective and sustainable production model is formed. The goal of this chapter is to examine the fundamentals of biopolymer-based food-based emulsions and films, as well as their usage in energy material manufacture, energy conversion, storage, and their application for encapsulation purposes.

11.2 SIGNIFICANCE OF FOOD-BASED FILMS AND EMULSIONS IN SUSTAINABLE ENERGY AND FOOD SYSTEMS

The standard of life in human civilisation may be preserved whilst moving a single step forward to a truly sustainable society if materials are able to be readily break down into individual resources after use and recycled without losing their properties [4–9]. On the other hand, a lot of materials are currently burned or dumped in landfills, which results in resource scarcity and deterioration of the environment on an international level [10]. There aren't many readily available recyclable polymer materials, especially those found in polyethylene terephthalate (PET) bottles, which are able to be recycled without losing any of their original properties [11]. Therefore, fundamental study that would enable polymers to be recovered without losing their characteristics (also known as "closed-loop" recycling) is currently attracting a lot of interest.

Edible coatings are being employed to preserve the quality of various foods since the end of the 20th century in an effort to reduce pollution in the environment and in response to the global circular economy trends. To develop the next wave of emulsion-based products, various natural emulsifiers are being researched. Among these, biosurfactants (such as substances obtained from microbial fermentation) and organic-based solid particles (Pickering stabilisers) are being utilised or are beginning to attract interest from the food business. Novel technologies have been developed recently for drug delivery systems using emulsions. The use of herbal formulations for novel drug delivery systems using encapsulation strategy is more advantageous and has more benefits compared to others.

11.2.1 EMULSIONS: DEFINITION, CLASSIFICATION, AND APPLICATION

Emulsions are a common ingredient in food, and years of investigation have produced sophisticated, albeit frequently empirical, mechanisms to regulate their creation and operation. However, given the changes in the food industry that have evolved in recent years, it is necessary to review the conventional approaches for creating food emulsions. Oil and water are common immiscible components that are present in numerous food items as one phase scattered in the other as colloidal particles. These include water-in-oil (W/O) emulsions, such as butter and margarine, or oil-in-water (O/W) emulsions, which are present in a variety of beverages, including milk, infant formula, and various other dairy-based items [12]. Undoubtedly, this trend has had an impact on the chemical makeup and layout of food emulsions. When coupled directly to the dispersed oil phase, bioactive, health-promoting compounds like vitamins, polyunsaturated fatty acids, or phytochemicals [13,14] that are frequently lipophilic can be retained within emulsions. For the encapsulation of hydrophilic elements, sophisticated emulsion-based systems, like double emulsions, can also be designed.

11.2.1.1 Classification of Emulsions

Emulsions are broadly divided into two categories: (1) food-based emulsions and (2) droplet size. Various types of food-based emulsions are depicted in Figure 11.1.

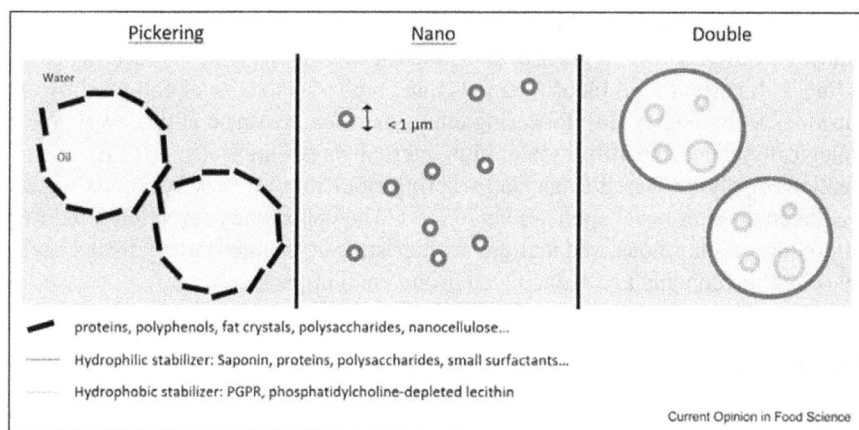

FIGURE 11.1   Illustrations of various food-based emulsions [15].

Source: Copyright 2023. Reproduced with permission from Elsevier.

TABLE 11.1

Some Characteristics of Microemulsions, Nanoemulsions, and Emulsions

Sr. No.	Criteria	Microemulsion	Nanoemulsion	Emulsion
1.	Method of formation	Spontaneous	High-pressure homogenisation and ultrasonic	Manual and mechanical stirring
2.	Stability	Thermodynamically stable	Kinetically stable	Kinetically stable
3.	Size of droplet (μm)	0.01–0.10	0.05–0.50	0.5–50
4.	Visual appearance	Transparent	Transparent/opaque	Opaque/cloudy
5.	Surface area ($m^2 g^{-1}$)	High (200)	Moderate (50–100)	Low (15)
6.	Tension at interface	Ultra-low	Low	Low
7.	System type	W/O, O/W and bi-continuous	W/O/O/W	W/O, O/W/W/O/W, and O/W/O
8.	Type of co-surfactant	Short-chain alcohol	Long-chain alcohol	None

In terms of droplet size, emulsions are divided into three major categories: micro-emulsion, nanoemulsion and emulsion. Some characteristics of emulsions, nano-emulsions, and microemulsions are given in Table 11.1.

11.2.1.2 Sustainable Pickering Emulsion Applications

Pickering emulsions are systems stabilised by solid colloidal particles adsorbed at the oil-water interface in an almost irreversible method. This results in a coating that surrounds the droplets, which can take the form of a single or multiple layer, forming an effective steric barrier and greater stability [16]. The quest for natural-based particles in the form of Pickering emulsions is nowadays a popular topic to meet market needs for fresh clean label products, i.e., in the absence of emulsifiers [17]. Pickering emulsions have received a great deal of interest recently as their characteristics are in line with current food industry trends that emphasise the use of eco-friendly and nutritious technologies [18]. Pickering emulsion research aimed at discovering new biological systems, the utilisation of high internal phase emulsions (HIPPE), and the creation of polymer-based films via Pickering emulsions are now hot topics regarding the creation of novel applications in food. The opportunity for valuing industry co-products is enormous, and it aligns with existing corporate strategy for a sustainable circular economy and reducing environmental impact.

11.2.2 Films and Materials Used for Film Formation

Films and coatings are linked to the class of food-safe packaging. These are termed as a thin covering of substance that might either get consumed with the food product or can get dissolved during their preparatory procedures. Polymers made from nature can be produced from a variety of materials, including plants, animals, including algae, that could be used to create environmentally friendly food packaging. Moreover, ancient cultures used a variety of organic substances for food packaging, such as shells, pottery, and portions of plants and animals [19]. Different types of packaging are being created in today's marketplace today in accordance with marketing criteria, technological specifications, simple opening their doors, attractiveness, and mobility. The natural polymers are classified into three major types based on their chemical properties: (1) polysaccharides, (2) proteins, and (3) polyesters. A few examples of polymers that occur naturally include proteins, DNA, starch, cellulose, starch, silk, and wool. Numerous industrial fields, including those involving food, papers, adhesives, textiles, wood, and pharmaceuticals, have made significant use of these polymers. Inorganic liquids that are favourable to cells, which include phosphate-buffered saline and cell culture media, can dissolve naturally occurring water-soluble polymers in order to create solutions or hydrogels. Gelatin, fibrinogen, alginate, and hyaluronic acid are some of these polymers [20].

11.2.2.1 Biopolymer-Based Materials

The food quality preservation relies on different variables like rancidity, temperature, and moisture content [21]. The subclass of bioplastics and biodegradable plastics known as biopolymer-based substances has mesmerised the food packaging sector in recent years [22]. Owing to their biological compatibility, protection, and rate of biodegradability, HPP (high-pressure processing), biopolymer-based packaging materials—which originate from proteins, lipids, aliphatic polyesters, or polysaccharides—are seen to be potential substitutes for conventional plastics [23–25]. These commodities are procured from farming, agriculture, or fisheries and hence are environmentally friendly in nature. Given the right conditions of temperature, moisture, and oxygen, microbes in nature breakdown polymer chains quickly and without inflicting any ecological harm [26]. Fibres including chitosan, cellulose, starch, pectin, alginate, gum, and pullulan are all included in polysaccharides [27]. Classification of bio-based polymers is given in Figure 11.2.

11.2.2.2 Nanocomposites Made Up of Biopolymers

Classical packaging technique makes use of non-biodegradable polymeric composites, which have a negative influence on soil fertility, the well-being of humans, and the environment [29]. Biopolymeric recyclable composites are used in the food business to solve this problem [30]. Such composites can be made either with or without the addition of additives or by combining multiple biopolymeric components. Both mechanical and functional characteristics are enhanced by the incorporation of bio polymers, which is advantageous in some applications, such as food packaging. As an example, Pantelic et al. (2021) remarked on the compostable behaviour of PVA/TPS. Strong water solubility in the PVA/TPS-based film contributed to its microbial breakdown properties [31].

Oil-water as well as air-water interfaces are frequently seen in food systems because they comprise elaborate combinations of different components which sometimes aren't adequately combined with each other. These interfaces can be stabilised employing the right emulsifying or surface stabilising chemicals. For this reason, it is crucial for researchers linked to the field of food to become familiar with the scientific and technological basis of emulsion systems since many natural and processed foods, including milk, soft beverages, and cakes, are created either entirely or partially of emulsion [32].

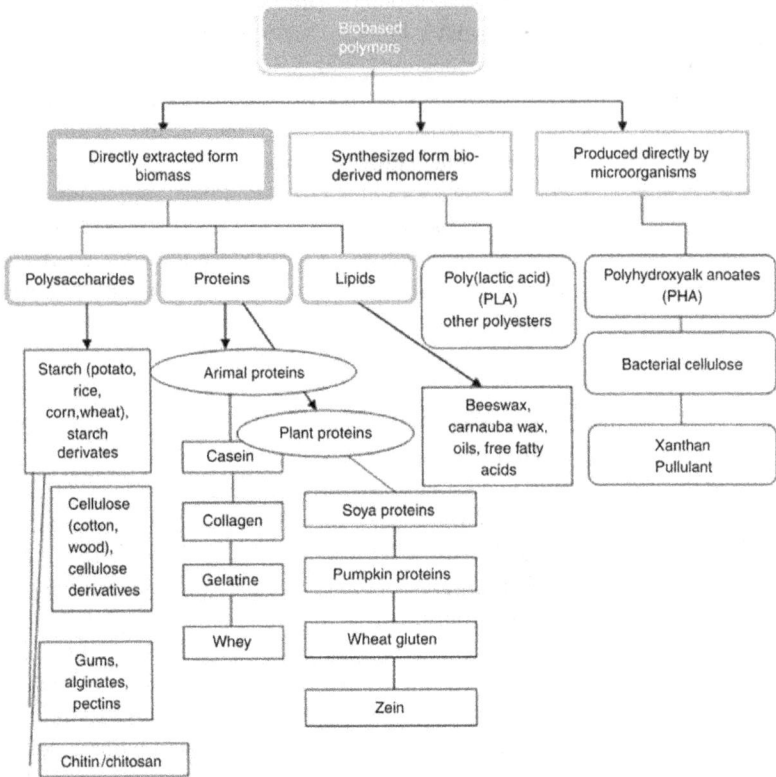

FIGURE 11.2   Schematic representation of the type and source of biopolymers [28].

Source: Copyright 2018. Reproduced with permission from Elsevier.

11.2.3   FOOD-BASED EMULSIONS AND FILMS CHARACTERIZATION TECHNIQUES

11.2.3.1   Emulsion Stability and Morphology

As per the surface-tension theory of emulsion formation, the emulsifiers/stabilisers lower the tension at interface between the two immiscible liquids, and hence reduce the repellent force between them and minimises the attractive forces between the particles of the same liquid [32]. To study the stability behaviour of emulsions, these two approaches are used: (a) zeta potential measurements for the particle surfaces and (b) ageing tests, which tells the study of the changes of emulsion properties with respect to time [33]. The feature of a food system's colour is arguably the one that is most frequently highlighted. Once an emulsion's microstructure is smaller than 100 μm, it cannot be seen with the naked eye [34]. In Table 11.2, a number of parameters that show how stable an emulsion is are presented.

11.2.3.2   Characterisation of Films and Coatings

Basic characterisation techniques for emulsions and films are given in Figure 11.3.

TABLE 11.2

Various Parameters Depicting the Stability and Morphology of Emulsion

Sr. No.	Parameter	Analysis	Reference
1.	Visual observation	Creaming and sedimentation are the two phenomena that make up gravitational separation. Whenever the dispersed phase, which is lesser in density than the continuous phase, rises upward, it causes creaming, which creates a thick separating layer. The droplets travel downwards as a result of sedimentation, which occurs when the dispersed phase has greater density compared to the continuous phase.	[32,35]
2.	Microscopy observation	Microscopy is employed to monitor particles that are indistinguishable by naked eyes particularly to investigate the variables which affect the long-term viability of the system of emulsions because visual inspection is insufficient for researching most unstable mechanisms and droplets less than 100 μm.	[36]
3.	Particle size analysis	Stokes' law states that the size of particles of the dispersed phase influences the stability of an emulsion system, which is made up of two immiscible liquids of different densities.	[37]
4.	Charge analysis	The ionisation feature of the absorbing emulsifier component on the surface of the dispersed emulsion particle determines the size and sign of the charge.	[32]
5.	Rheology	In accordance with Stokes' law, when viscosity rises, the emulsion system is more resistant to instability brought on by particulate flocculation.	[37]

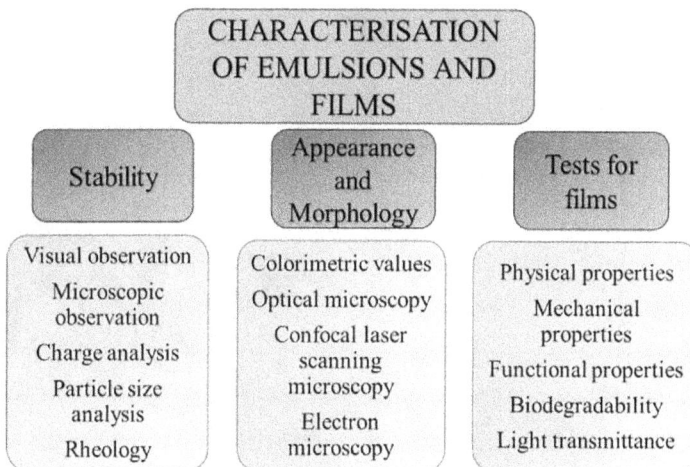

FIGURE 11.3 Basic characterisation techniques for emulsions and films.

11.3 FOOD-BASED EMULSIONS FOR ENERGY MATERIALS

11.3.1 EMULSIONS AS TEMPLATES FOR ENERGY MATERIAL SYNTHESIS

Particle surfactants can be used to create structures from composites, and emulsions offer a convenient template for creating a variety of frameworks, such as capsule-like structures, foams, and armoured particles. Macromolecular synthesis (stiff structures and porous frameworks with intrinsic microporosity), phase inversion, porogen inclusion, and templating are just a few strategies that can be employed to create porous polymers [38–41]. Block copolymer (BCP) templating, emulsion templating, breath figure templating, as well as solid particle templating—the subject of this chapter—are some of the accessible templating approaches [42,43]. This viewpoint offers a comprehensive examination of the current state of "emulsion templating" today, highlighting the theoretical innovations that transformed the technique from one with a constrained use to one with a steadily growing application. The initial objective of emulsion templating was to create "poly-HIPEs," (which are hydrophobic cross-linked porous polymer monoliths. A poly-HIPE was typically made by free radical polymerising monomers of a water-in-oil (w/o) in the external phase, HIPE, which is a strongly viscous, paste-like emulsion with an internal-phase content greater than 74% [42,44,45]. To get such large internal phase contents in emulsions, either polyhedrally deform monodisperse droplets or create a poly-disperse distribution of size of droplets [46]. A schematic representation of the preparation of NPC using HIPE template is given in Figure 11.4.

11.3.2 EMULSION-TEMPLATED POROUS MATERIALS

In this context, four major forms of emulsion-templated porous materials are identified. By modifying the materials after they have been synthesised, it is possible to develop and manufacture materials with size-selective pores/apertures, favourable geometry, the necessary chemistry of the surface, and good chemical-based and thermal stability.

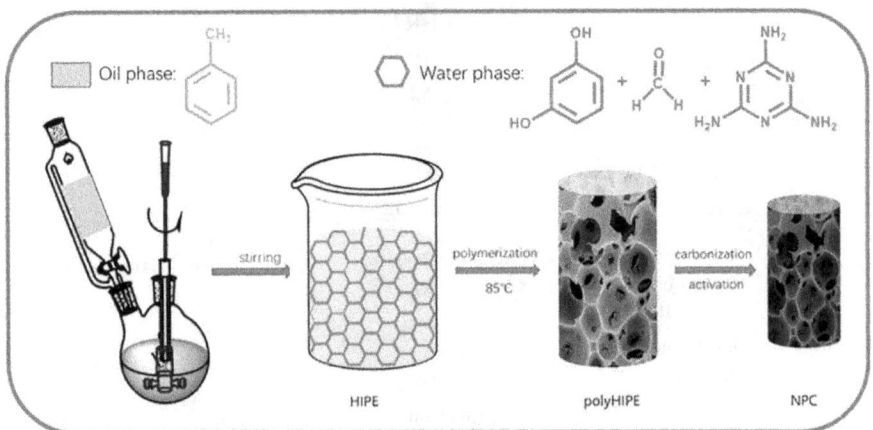

FIGURE 11.4 Representation of the synthesis of NPC using HIPE template [47].

Source: Copyright 2021. Reproduced with permission from Elsevier.

These porous polymers are thin materials with special mechanical, thermal, and acoustic capabilities linked to their distinguishing features. From an application standpoint, the size, shape (spherical or polyhedral), hierarchy, and interconnectivity of pores, in addition to the surface areas, design (shape and morphology), structural gradients, and size of porous polymeric materials are of utmost relevance [48].

The size and arrangement of pores determine how liquids and gases are transported through bulk materials. The activated transport behaviour of the microporous substances is accompanied by slowly mass transfer, mainly for large molecules. Knudsen and surface diffusion mechanisms control capillary transport in mesoporous materials, where molecules collide with pore walls frequently. Molecular diffusion and viscous flow and are made possible by macroporous materials, while mass transport is made possible by intermolecular collisions that control momentum and energy transfer with the least amount of diffusion [49]. With the least amount of diffusion and transport resistance, the hierarchy of materials based on porosity ensures effective (quick and widespread) distribution of liquid and gas throughout the bulk material. Since closed-cell structures are helpful for encapsulation operations and open-cell designs are beneficial for absorption, adsorption, controlled release, chemical supports and tissue engineering, the interconnectivity of pores is of more significance. Increased surface area results in sites with greater activity, which increases surface reactivity and enhances the efficiency of applications [49,50–52].

11.3.3 NANOSTRUCTURED EMULSIONS FOR ENERGY CONVERSION AND STORAGE

Aerogel applications for energy conversion and medicine administration are equally relevant in the state of microspheres/granules. However, it is highly challenging to produce microspherical particles by merely milling or grinding the monolithic form typically achieved by the sol-gel process. An emulsion-based technique was made by Alnaief et al. whereby by keeping the internal structural features of micro-spherical aerogels equal to those of monolithic aerogels, this approach is useful for producing them [53]. In order to create stable microspheres using this method, the dispersed phase must first be prepared using the sol-gel procedure, followed by emulsification in a continuous phase and a crosslinking process. As a cross-linking medium, the oil-water emulsion is used. Inorganic SiO_2 and polymeric (Starch and Alginate) aerogels' microspherical particles were created using this technique [54,55]. With a high surface area, customisable 3-D hierarchical structure, electrical conductivity, and chemical inertness network organisation, carbon aerogels are a special family of materials. Biener et al. [56] reports on the most current developments in carbon aerogel synthesis techniques and how they are used in hydrogen storage, supercapacitors, fuel cells, batteries and catalysis. Investigation of novel commercial meso-porous silica materials is depicted in Figure 11.5.

A significant problem in material chemistry is the development of materials having novel and enhanced capabilities for energy storage and conversion. However, this issue could be solved by the invention of composite materials, which combine the two well-known materials with extraordinary physical and chemical characteristics [58]. Composite materials are made of multiple components with distinct physical and chemical characteristics that, at the microscopic level, stay distinct from one another while remaining combined into a single physical substance [59,60]. The required

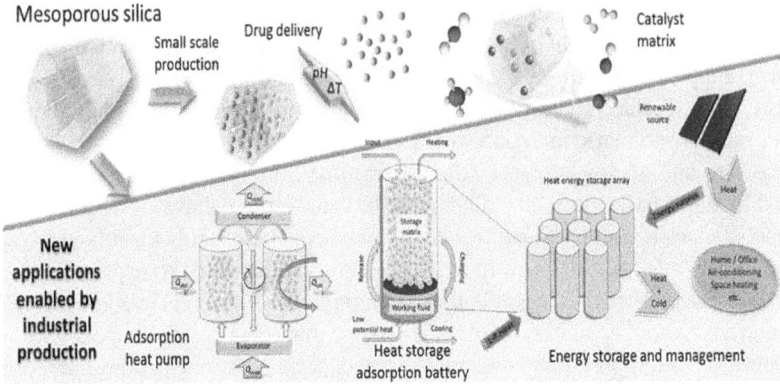

FIGURE 11.5   Novel commercial mesoporous silica materials [57].

Source: Copyright 2019. Reproduced with permission from Elsevier.

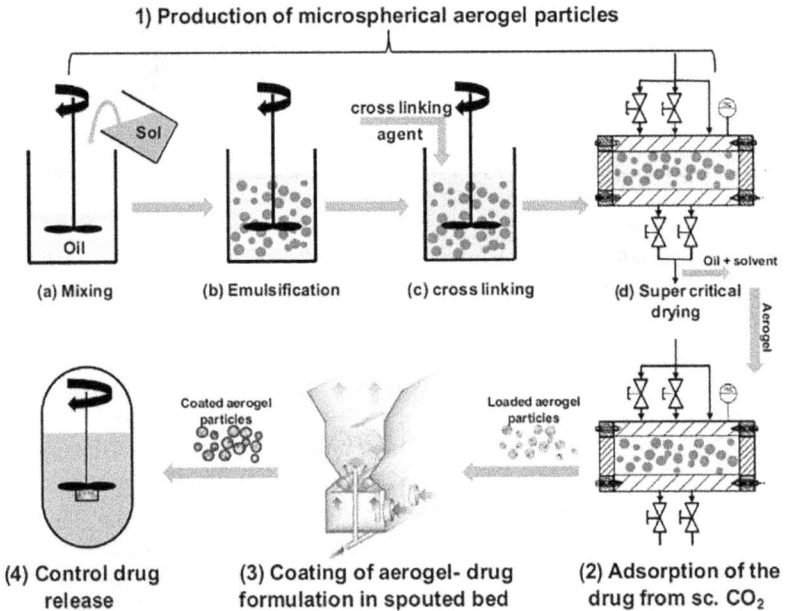

FIGURE 11.6   Coating of silica aerogel microspheres for controlled drug release applications [62].

Source: Copyright 2012. Reproduced with permission from Elsevier.

guest is added to the sol immediately before gelation in Anderson et al.'s approach for making composite aerogel. This keeps the intriguing qualities of both components from being completely encapsulated by the aerogel network. Consequently, sol-gel chemistry allows considerable versatility in the design of these materials for particular purposes [61]. Figure 11.6 illustrates the coating of silica aerogel microspheres for controlled drug release applications.

11.4 FOOD-BASED FILMS FOR SUSTAINABLE ENERGY APPLICATIONS

11.4.1 EDIBLE AND BIODEGRADABLE FILMS APPLICATIONS FOR ENCAPSULATION

Environmental factors (such as humidity and temperature), oxidation deterioration during transit, microbes and mechanical damage, storage as well as consumption processes result in the waste of around one-third of food raw materials and products worldwide each year [63,64]. Innovative packaging solutions, which are crucial for the preservation and commercialisation of either raw or processed items in the food sectors, have been actively developed to meet this difficulty [64,65]. Sources, types, and increased stabilisation by encapsulation and adsorption strategy of flavonoids, betalains, CUR, anthocyanins, and carotenoids are shown in Figure 11.7.

Foods can be packaged to prevent deteriorating impacts from the environment and keep their structural integrity for a longer duration of time, maintaining flavours and nutritional benefits [67]. Critical reviews and various studies on the creation of NPs for the delivery and release of bioactive substances are published [68–70]. Saifullah et al. [71], for instance, conducted a detailed review of the micro- or nanoencapsulation, preservation, and controlled discharge of flavour and aroma components.

11.4.2 ENERGY APPROACHES FOR THE DEVELOPMENT OF FILMS AND EMULSIONS

The two energy approaches of films and emulsions formation are listed below.

11.4.2.1 Low-Energy Approaches Applications

The Pickering emulsion in question was created by Hua et al. [72] and contained CEO-loaded NPs along with 1:2 (w/w) zein and sodium caseinate (NaCas). They proved that these NPs possess a greater encapsulation effectiveness (over 60%), smaller particle size,

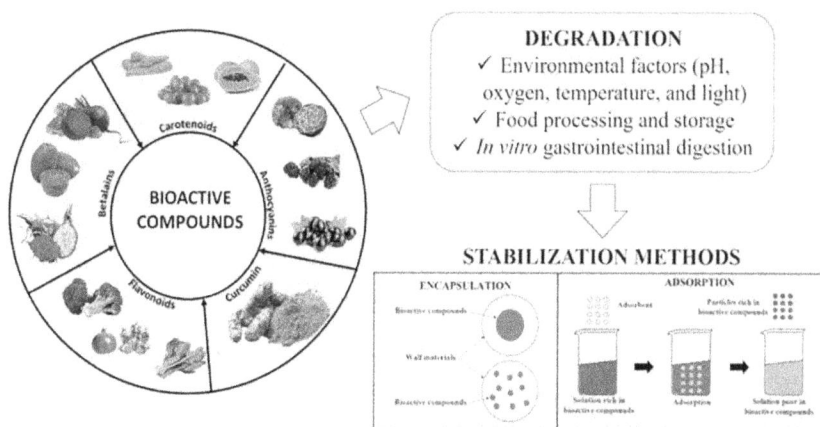

FIGURE 11.7 Sources, classification, and improved stabilisation by encapsulation and adsorption strategy [66].

Source: Copyright 2022. Reproduced with permission from Elsevier.

the right zeta potential, and a high polydispersity index [72]. The potential for their food packaging application is therefore very high. Additionally, CUR, a hydrophobic polyphenol that is brilliant yellow and can be extracted from the root of the turmeric plant, is a superb sensor of changes in dietary pH levels. While being prepared and stored, however, CUR is susceptible to losing its properties due to light, heat, and oxidation. So, in a piece of study by D. Liu et al., Pickering emulsion was used to encase CUR [73].

11.4.2.2   High-Energy Approaches Applications

A nanoemulsion using lemongrass essential oil (LEO) and sodium alginate (2% w/v)/ Tween 80 (1% v/v) in a microfluidiser at 150 mPa was created by Salvia-Trujillo et al. [74]. Additionally, compared to traditional methods, the ultrasound technique can create emulsions of nano size and with stronger kinetic stability, smaller polydispersity and smaller particle sizes. This technique can produce a high-intensity force, which is a sequence of events when the vapour bubble forms suddenly, grows quickly, and explodes, producing strong shock waves and shear stresses in the liquid medium which cause droplet fragmentation [75]. In this instance, Chu et al. created nanoemulsions of cinnamon essential oil (CNEO) using ultrasound. According to their findings, CEO nanoemulsion had the smallest droplet size of nearly 60 nm, as well as the maximum stability to heat, centrifugation, and storage [76]. As a result, such a nanoemulsion has the potential to be included as an active component to edible films made of pullulan. Low and high energy strategies for films and emulsions formation are given in Figure 11.8.

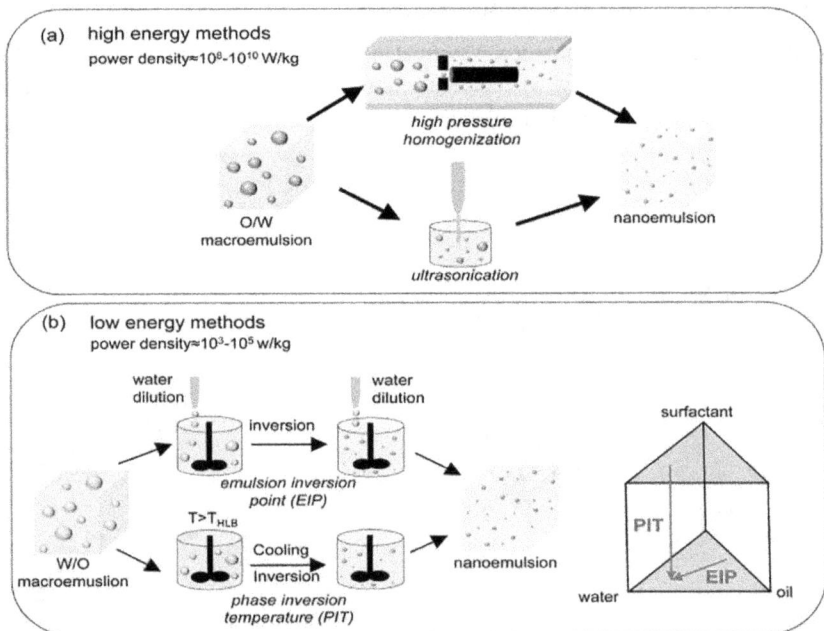

FIGURE 11.8   Low and high energy strategies for films and emulsions formation [77].

Source: Copyright 2016. Reproduced with permission from Royal Society of Chemistry.

An overview of the techniques used to create O/W nanoemulsions using high and low energy includes (a) high-energy processes that shatter macroemulsion drops to smaller droplets include high pressure homogenisation (HPH) and ultrasonication and (b) low-energy techniques begin with W/O macroemulsions and divide coarse emulsions into smaller droplets when they move through a low interfacial tension environment throughout phase inversion. While the phase inversion temperature (PIT) approach causes a phase inversion on cooling of the mixture, the emulsion inversion point (EIP) technique produces a phase reversal by aqueous dilution.

11.4.3 FOOD-BASED FILMS AS PROTECTIVE COATINGS FOR ENERGY DEVICES

The use of probiotics in films that are edible is another cutting-edge bio-protective method for preserving food. It is generally agreed that the application of edible films and coatings will alleviate the limitations linked to the restricted use of probiotics in fresh and low-processed foods [78]. The generation of certain compounds that can change the amount of water, pH, and nutritional kinetics in the target fruit and vegetable is thought to be responsible for their bio-protective function [79]. Therefore, bio-protection employing advantageous microbes may be a useful method to control the existence of harmful and rotting microorganisms, particularly in vegetables and fruits, which might be maintained fresher for an extended amount of time rather than using chemical-based antimicrobial therapies [80].

Colloidal metallic NPs possessing exceptional LSPR are widely used as probes in photothermal therapy, optical imaging, photoelectric devices, and sensing [81–85]. Since the shape, composition, size, particle number, and interparticle interaction of metal NPs are significantly connected with their LSPR, these factors can be controlled to alter the LSPR signal emission. Colorimetric antioxidant detection by the use of silver nanocage is given in Figure 11.9.

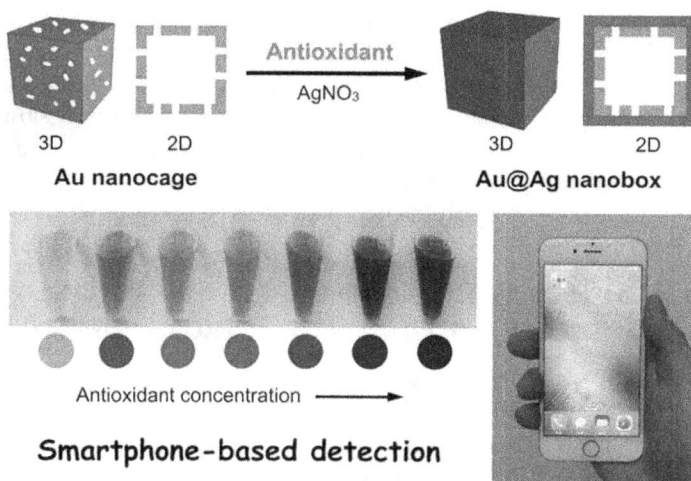

FIGURE 11.9 Colorimetric antioxidant detection [86].

Source: Copyright 2018. Reproduced with permission from Elsevier.

11.5 SUSTAINABILITY AND ENVIRONMENTAL CONSIDERATIONS IN FOOD-BASED EMULSIONS AND FILMS

The need to find alternatives to the use of plastics in the food packaging business is a crucial one to take into account as environmental protection concerns increase [87]. The existing environment uses plastics widely but seeks to replace them with new biomaterial-based food packaging that will take the place of the plastics now employed in a huge number of containers. From the perspective of food safety, it's important to keep an eye on these cutting-edge materials on the market for consumers and to ensure that no dangerous product can move in bulk that pose a threat to public health. Recent breakthroughs in nanotechnology have revolutionised the food business because of its many uses in food packaging and safety and its success in raising the nutritional value of supplements, extending their shelf lives, and reducing packaging waste [88,89]. Advances in nanobiotechnology have been significant in addressing issues with food safety, such as contamination by bacteria and improved toxin proof of identity, shelf life, and packaging methods. Additionally, biomaterials can be employed to create microbe quantifying biosensors as well as additional applications for food safety monitoring, such as metal oxide NPs, nanowires, nanotubes, and quantum dots [90–92].

A few investigations focused on the dangers of NPs transferred through active packaging to food-related items and their effect on consumer health [87,90,93]. It is crucial to make sure that the physiological chemical characteristics and dosage of biomaterials in foods as well as food packaging dictate their effects on human health and final fate. These factors need to be evaluated using various in vitro and in vivo research. Absorption, transmission, ultimate toxicity, digestion and excretion are all characteristics that can be evaluated and analysed for risk assessment along with physical interactions, chemical elements, pH, osmotic concentration, and biological parameters.

Even when a substance has been determined to be GRAS (generally considered as safe), more research is needed to examine the risk posed by its nano equivalents because their physio-chemical properties are entirely distinct from those of the macro state [94]. For example, NPs of silica utilised as anti-caking substances in human lung cells may be hazardous to the cells if exposed for an extended period of time [95]. In addition, regulatory bodies must create such standards for commercial goods in order to ensure the safety, quality, and compliance with environmental laws.

11.6 FUTURE PERSPECTIVES AND CHALLENGES IN FOOD-BASED EMULSIONS, FILMS, AND ENERGY MATERIALS

In the coming years, modern agrochemicals will be used to supplement existing agricultural and sophisticated food processing. Organic farming along with active food packaging should adopt more efficient green chemistry methods for the synthesis of nanoscale components for improved crop protection and more secure food packaging approaches. The majority of green techniques using biopolymers that have been reported on at the laboratory scale have reduced toxicity and provided great mechanical qualities for packaging ensuring circular economy. The next step is to

introduce these antimicrobial polymers into the commercial scale packaging industries. Although there's no specific regulation for these kinds of products, Regulation
(1935)/2004/CE protects any food contact materials (FCM), so it is crucial to learn
about the components that can be passed from the package to the product. In contrast
to the utilisation of plastic materials, the growing need for this type of material as a
more environmentally friendly and long-lasting substitute suggests their investigation to ensure safety [88].

11.7   CONCLUSION

It has been made apparent that research and development initiatives ought to be
scheduled to reduce plastic packaging and adopting biopolymer-based products,
thereby reducing waste to a minimum. A potential source of essential components
that is presently underutilised are non-edible agri-food wastes, including industrial
waste from processing fruits, vegetables, or grains. A circular economy model is
being investigated in order to repurpose these resources, which include structural
biopolymers (like polysaccharides and proteins), as well as a wide range of bioactive
molecules (like antioxidant carotenoids, anthocyanins, as well as antibacterial essential oils). It is also possible to take use of the incorporation of bioactive chemicals
as high-value ingredients in the creation of bioplastic films and emulsions to create
new capabilities like radical scavenging or antioxidant defence. Thus, the circular
economy model greatly benefits from the utilisation of biofilms and emulsions made
from agro-industrial byproducts that are consistent with zero waste. In this way, it
can reduce packaging waste and enhance packaging design, with for instance clear
labelling to encourage reuse and recycling; and push for the adoption of bio-based,
biodegradable, and compostable plastics.

REFERENCES

[1] Pashova, S., *Application of Plant Waxes in Food Technologies*, SAFO Publishing
House, Lovech, 193, 2011
[2] Lüdeke-Freund, F., Gold, S., & Bocken, N.M., A review and typology of circular economy business model patterns, *Journal of Industrial Ecology*, *23*(1), 36–61, 2019
[3] Plastics Europe. The circular economy for plastics—A European overview. *Technical
Report*, 2019. https://plasticseurope.org/knowledge-hub/the-circular-economy-for-
plastics-a-european-overview/
[4] Garcial, J., & Robertson, M., The future of plastics recycling chemical advances are
increasing the proportion of polymer waste that can be recycled, *Science*, *358*(6365),
870–872, 2017
[5] Fortman, D. J., Brutman, J. P., De Hoe, G. X., et al., Approaches to sustainable and continually recyclable cross-linked polymers, *ACS Sustainable Chemistry & Engineering*,
6(9), 11145–11159, 2018
[6] Vollmer, I., Jenks, M. J., Roelands, M. C., et al., Beyond mechanical recycling: Giving
new life to plastic waste, *Angewandte Chemie International Edition*, *59*(36), 15402–
15423, 2020
[7] Schyns, Z. O., & Shaver, M. P., Mechanical recycling of packaging plastics: A review,
Macromolecular Rapid Communications, *42*(3), 2000415, 2021

[8] Rosenboom, J. G., Langer, R., & Traverso, G., Bioplastics for a circular economy, *Nature Reviews Materials*, *7*(2), 117–137, 2022

[9] Zuin, V. G., & Kümmerer, K., Chemistry and materials science for a sustainable circular polymeric economy, *Nature Reviews Materials*, *7*(2), 76–78, 2022

[10] Geyer, R., Jambeck, J. R., & Law, K. L., Production, use, and fate of all plastics ever made, *Science Advances*, *3*(7), e1700782, 2017

[11] Ghosal, K., & Nayak, C., Recent advances in chemical recycling of polyethylene terephthalate waste into value added products for sustainable coating solutions—hope vs. Hype, *Materials Advances*, *3*(4), 1974–1992, 2022

[12] McClements, D. J., *Food Emulsions: Principles, Practices, and Techniques*, CRC Press; Taylor and Francis, Boca Raton, 2004

[13] Yeung, A. W. K., Mocan, A., & Atanasov, A. G., Let food be thy medicine and medicine be thy food: A bibliometric analysis of the most cited papers focusing on nutraceuticals and functional foods, *Food Chemistry*, *269*, 455–465, 2018

[14] McClements, D. J., Enhanced delivery of lipophilic bio actives using emulsions: A review of major factors affecting vitamin, nutraceutical, and lipid bioaccessibility, *Food & Function*, *9*(1), 22–41, 2018

[15] Gisle, Ø. Y. E., Simon, S., Rustad, T., & Kristofer, P. A. S. O., Trends in food emulsion technology: Pickering, nano and double emulsions, *Current Opinion in Food Science*, 101003, 2023

[16] Yan, X., Ma, C., Cui, F., McClements, D. J., Liu, X., & Liu, F., Protein-stabilized Pickering emulsions: Formation, stability, properties, and applications in foods, *Trends in Food Science and Technology*, *103*, 293–303, 2020

[17] Sharkawy, A., Barreiro, M. F., & Rodrigues, A. E., Chitosan-based Pickering emulsions and their applications: A review, *Carbohydrate Polymers*, *250*, 116885, 2020

[18] Tan, C., & McClements, D. J., Application of advanced emulsion technology in the food industry: A review and critical evaluation, *Foods*, *10*, 812, 2021

[19] Teixeira-Costa, B. E., & Andrade, C. T., Natural polymers used in edible food packaging—History, function and application trends as a sustainable alternative to synthetic plastic, *Polysaccharides*, *3*(1), 32–58, 2021

[20] Liu, F., Chen, Q., Liu, C., et al., Natural polymers for organ 3D bioprinting, *Polymers*, *10*(11), 1278, 2018

[21] Kumar, L., Ramakanth, D., Akhila, K., et al., Edible films and coatings for food packaging applications: A review, *Environmental Chemistry Letters*, 1–26, 2022

[22] Khalid, M. Y., Imran, R., Arif, Z. U., et al., Developments in chemical treatments, manufacturing techniques and potential applications of natural-fibers-based biodegradable composites, *Coatings*, *11*(3), 293, 2021

[23] Horue, M., Berti, I. R., Cacicedo, M. L., et al., Microbial production and recovery of hybrid biopolymers from wastes for industrial applications-a review, *Bioresource Technology*, *340*, 125671, 2021

[24] Horue, M., Berti, I. R., Cacicedo, M. L., et al., Microbial production and recovery of hybrid biopolymers from wastes for industrial applications-a review, *Bioresource Technology*, *340*, 125671, 2021

[25] Vinod, A., Sanjay, M. R., Suchart, S., et al., Renewable and sustainable biobased materials: An assessment on biofibers, biofilms, biopolymers and biocomposites, *Journal of Cleaner Production*, *258*, 120978, 2020

[26] Garrido-Romero, M., Aguado, R., Moral, A., et al., From traditional paper to nanocomposite films: Analysis of global research into cellulose for food packaging, *Food Packaging and Shelf Life*, *31*, 100788, 2022

[27] Meindrawan, B., Suyatma, N. E., Wardana, A. A., et al., Nanocomposite coating based on carrageenan and ZnO nanoparticles to maintain the storage quality of mango, *Food Packaging and Shelf Life*, *18*, 140–146, 2018

[28] Popović, S. Z., Lazić, V. L., Hromiš, N. M., et al., Biopolymer packaging materials for food shelf-life prolongation, In *Biopolymers for Food Design*, Academic Press; Elsevier, Amsterdam, Netherlands, 2018, 223–277

[29] Huang, K., Wang, Y., et al., Recent applications of regenerated cellulose films and hydrogels in food packaging, *Current Opinion in Food Science*, *43*, 7–17, 2022

[30] As'ad Mahpuz, A. S., Muhamad Sanusi, N. A. S., Jusoh, A. N. C., et al., Manifesting sustainable food packaging from biodegradable materials: A review, *Environmental Quality Management*, *32*(1), 379–396, 2022

[31] Pantelic, B., Ponjavic, M., Jankovic, V., et al., Upcycling biodegradable PVA/starch film to a bacterial biopigment and biopolymer, *Polymers*, *13*(21), 3692, 2021

[32] Kale, S. N., & Deore, S. L., Emulsion micro emulsion and nano emulsion: A review, *Systematic Reviews in Pharmacy*, *8*(1), 39, 2017

[33] Indo, K., Ratulowski, J., Dindoruk, B., et al., Asphaltene nanoaggregates measured in a live crude oil by centrifugation, *Energy & Fuels*, *23*(9), 4460–4469, 2009

[34] Aguilera, J. M., & Stanley, D. W., *Microstructural Principles of Food Processing and Engineering*, Springer Science & Business Media, 1999

[35] Friberg, S., Larsson, K., & Sjoblom, J., editors. *Food Emulsions*, 4th edition, Marcel Dekker, Inc., New York, NY U.S.A., 2004

[36] Russ, J. C., *Image Analysis of Food Microstructure*, Boca Raton: CRC Press, 2004, publisher location is: Boca Raton DOI: https://doi.org/10.1201/9781420038996

[37] Hunter, R. J., White, L. R., & Chan, D. Y., *Foundations of Colloid Science*, Oxford: Clarendon Press, 1987.

[38] Wu, D., Xu, F., Sun, B., et al., Design and preparation of porous polymers, *Chemical Reviews*, *112*(7), 3959–4015, 2012. https://doi.org/10.1021/cr200440z

[39] Gokmen, M. T., & Du Prez, F. E., Porous polymer particles—A comprehensive guide to synthesis, characterization, functionalization and applications, *Progress in Polymer Science*, *37*(3), 365–405, 2012. https://doi.org/10.1016/j.progpolymsci.2011.07.006

[40] Silverstein, M. S., Cameron, N. R., & Hillmyer, M. A., *Porous Polymers*, John Wiley & Sons, New York, 2011.

[41] Qiu, S., & Ben, T., *Porous Polymers: Design, Synthesis and Applications*, Royal Society of Chemistry, Cambridge, 2016.

[42] Cameron, N. R., Krajnc, P., & Silverstein, M. S., Colloidal templating, In *Porous Polymers*, Silverstein, M. S., Cameron, N. R., Hillmyer, M. A., Eds., Wiley, New York, 2011, 119–172

[43] Todd, E. M., & Hillmyer, M. A., Porous polymers from self-assembled structures, In *Porous Polymers*, Silverstein, M. S., Cameron, N. R., Hillmyer, M. A., Eds., John Wiley & Sons, New York, 2011, 31–78

[44] Cameron, N. R., & Sherrington, D. C., High internal phase emulsions (HIPEs)— Structure, properties and use in polymer preparation, *Advances in Polymer Science*, *126*, 163–214, 1996. https://doi.org/10.1007/3-540-60484-7_4

[45] Silverstein, M. S., & Cameron, N. R., PolyHIPEs—Porous polymers from high internal phase emulsions, In *Encyclopedia of Polymer Science and Technology*, John Wiley & Sons, Inc., New York, 2010.

[46] Lissant, K. J., Geometry of emulsions. *Journal of the Society of Cosmetic Chemists*, *21*(3), 141–154, 1970

[47] Zhao, Y., Zhao, Z., Zhu, Z., et al., Preparation of N-doped porous carbons via high internal phase emulsion template, *Progress in Natural Science: Materials International*, *31*(2), 270–278, 2021

[48] Khin, M. M., Nair, A. S., Babu, V. J., et al., A review on nanomaterials for environmental remediation, *Energy & Environmental Science*, *5*(8), 8075–8109, 2012

[49] Mudassir, M. A., Hussain, S. Z., Asma, S. T., at al., Fabrication of emulsion-templated poly (vinylsulfonic acid)—Ag nanocomposite beads with hierarchical multimodal porosity for water cleanup, *Langmuir*, *35*(40), 13165–13173, 2019

[50] Yang, X. Y., Chen, L. H., Li, Y., Rooke, J. C., et al., Hierarchically porous materials: Synthesis strategies and structure design, *Chemical Society Reviews*, *46*(2), 481–558, 2021.

[51] Zhang, T., Sanguramath, R. A., Israel, S., et al., Emulsion templating: Porous polymers and beyond, *Macromolecules*, *52*(15), 5445–5479, 2019

[52] Schwieger, W., Machoke, A. G., Weissenberger, T., et al., Hierarchy concepts: Classification and preparation strategies for zeolite containing materials with hierarchical porosity, *Chemical Society Reviews*, *45*(12), 3353–3376, 2016

[53] Alnaief, M., & Smirnova, I., In situ production of spherical aerogel microparticles, *The Journal of Supercritical Fluids*, *55*(3), 1118–1123, 2021

[54] Alnaief, M., Alzaitoun, M. A., García-González, C. A., & Smirnova, I., Preparation of biodegradable nanoporous microspherical aerogel based on alginate, *Carbohydrate Polymers*, *84*(3), 1011–1018, 2011

[55] García-González, C. A., Uy, J. J., Alnaief, M., et al., Preparation of tailor-made starch-based aerogel microspheres by the emulsion-gelation method, *Carbohydrate Polymers*, *88*(4), 1378–1386, 2012

[56] Biener, J., Stadermann, M., Suss, M., et al., Advanced carbon aerogels for energy applications, *Energy & Environmental Science*, *4*(3), 656–667, 2011

[57] Mikšík, F., Miyazaki, T., & Inada, M., Detailed investigation on properties of novel commercial mesoporous silica materials, *Microporous and Mesoporous Materials*, *289*, 109644, 2019

[58] Heiligtag, F. J., Cheng, W., de Mendonca, V. R., et al., Self-assembly of metal and metal oxide nanoparticles and nanowires into a macroscopic ternary aerogel monolith with tailored photocatalytic properties, *Chemistry of Materials*, *26*(19), 5576–5584, 2014

[59] Guo, Z., Chen, Y., & Lu, N. L. (Eds.), *Multifunctional Nanocomposites for Energy and Environmental Applications*, New Jersey, USA: John Wiley & Sons, 2018

[60] Alwin, S., Bhat, S. D., Sahu, A. K., et al., Modified-pore-filled-PVDF-membrane electrolytes for direct methanol fuel cells, *Journal of the Electrochemical Society*, *158*(2), B91, 2010

[61] Anderson, M. L., Stroud, R. M., Morris, C. A., et al., Tailoring advanced nanoscale materials through synthesis of composite aerogel architectures, *Advanced Engineering Materials*, *2*(8), 481–488, 2000

[62] Alnaief, M., Antonyuk, S., Hentzschel, C. M., et al., A novel process for coating of silica aerogel microspheres for controlled drug release applications, *Microporous and Mesoporous Materials*, *160*, 167–173, 2012

[63] Hoseinnejad, M., Jafari, S. M., & Katouzian, I., Inorganic and metal nanoparticles and their antimicrobial activity in food packaging applications, *Critical Reviews in Microbiology*, *44*(2), 161–181, 2018

[64] Rangaraj, V. M., Rambabu, K., Banat, F., et al., Natural antioxidants-based edible active food packaging: An overview of current advancements, *Food Bioscience*, *43*, 101251, 2021

[65] Kumar, S., Mukherjee, A., & Dutta, J., Chitosan based nanocomposite films and coatings: Emerging antimicrobial food packaging alternatives, *Trends in Food Science & Technology*, *97*, 196–209, 2020

[66] Koop, B. L., da Silva, M. N., da Silva, F. D., et al., Flavonoids, anthocyanins, betalains, curcumin, and carotenoids: Sources, classification and enhanced stabilization by encapsulation and adsorption, *Food Research International*, *153*, 110929, 2022

[67] Jiang, W., Zhao, P., Song, W., et al., Electrospun zein/polyoxyethylene core-sheath ultrathin fibers and their antibacterial food packaging applications, *Biomolecules*, *12*(8), 1110, 2022

[68] Delshadi, R., Bahrami, A., Tafti, A. G., et al., Micro and nano-encapsulation of vegetable and essential oils to develop functional food products with improved nutritional profiles, *Trends in Food Science & Technology*, *104*, 72–83, 2020

[69] Tavares, L., Santos, L., & Noreña, C. P. Z., Bioactive compounds of garlic: A comprehensive review of encapsulation technologies, characterization of the encapsulated garlic compounds and their industrial applicability, *Trends in Food Science & Technology*, *114*, 232–244, 2021

[70] Zhang, Q., Zhou, Y., Yue, W., et al., Nanostructures of protein-polysaccharide complexes or conjugates for encapsulation of bioactive compounds, *Trends in Food Science & Technology*, *109*, 169–196, 2021

[71] Saifullah, M., Shishir, M. R. I., Ferdowsi, R., et al., Micro and nano encapsulation, retention and controlled release of flavor and aroma compounds: A critical review, *Trends in Food Science & Technology*, *86*, 230–251, 2019

[72] Hua, L., Deng, J., Wang, Z., et al., Improving the functionality of chitosan-based packaging films by crosslinking with nano encapsulated clove essential oil, *International Journal of Biological Macromolecules*, *192*, 627–634, 2021

[73] Liu, D., Dang, S., Zhang, L., et al., Corn starch/polyvinyl alcohol-based films incorporated with curcumin-loaded Pickering emulsion for application in intelligent packaging, *International Journal of Biological Macromolecules*, *188*, 974–982, 2021

[74] Salvia-Trujillo, L., Rojas-Graü, M. A., Soliva-Fortuny, R., et al., Use of antimicrobial nano emulsions as edible coatings: Impact on safety and quality attributes of fresh-cut Fuji apples, *Postharvest Biology and Technology*, *105*, 8–16, 2015

[75] Rodrigues, D. C., Cunha, A. P., Brito, E. S., et al., Mesquite seed gum and palm fruit oil emulsion edible films: Influence of oil content and sonication, *Food Hydrocolloids*, *56*, 227–235, 2016

[76] Chu, Y., Cheng, W., Feng, X., et al., Fabrication, structure and properties of pullulan-based active films incorporated with ultrasound-assisted cinnamon essential oil nano emulsions, *Food Packaging and Shelf Life*, *25*, 100547, 2020

[77] Gupta, A., Eral, H. B., Hatton, T. A., et al., Nano emulsions: Formation, properties and applications, *Soft Matter*, *12*(11), 2826–2841, 2016

[78] Guimaraes, A., Abrunhosa, L., Pastrana, L. M., et al., Edible films and coatings as carriers of living microorganisms: A new strategy towards biopreservation and healthier foods, *Comprehensive Reviews in Food Science and Food Safety*, *17*(3), 594–614, 2018

[79] Al-Tayyar, N. A., Youssef, A. M., & Al-Hindi, R., Antimicrobial food packaging based on sustainable Bio-based materials for reducing foodborne Pathogens: A review, *Food Chemistry*, *310*, 125915, 2020

[80] de Oliveira, K. Á. R., Fernandes, K. F. D., & de Souza, E. L., Current advances on the development and application of probiotic-loaded edible films and coatings for the bio protection of fresh and minimally processed fruit and vegetables, *Foods*, *10*(9), 2207, 2021

[81] Fang, C., Jia, H., Chang, S., et al., (Gold core)/(titania shell) nanostructures for plasmon-enhanced photon harvesting and generation of reactive oxygen species, *Energy & Environmental Science*, *7*(10), 3431–3438, 2014

[82] Dreaden, E. C., Mackey, M. A., Huang, X., et al., Beating cancer in multiple ways using nanogold, *Chemical Society Reviews*, *40*(7), 3391–3404, 2011

[83] Rycenga, M., Cobley, C. M., Zeng, J., et al., Controlling the synthesis and assembly of silver nanostructures for plasmonic applications, *Chemical Reviews*, *111*(6), 3669–3712, 2011.

[84] Wang, Y., Zou, H. Y., & Huang, C. Z., Real-time monitoring of oxidative etching on single Ag nanocubes via light-scattering dark-field microscopy imaging, *Nanoscale*, 7(37), 15209–15213, 2015

[85] Xiong, Y., Long, R., Liu, D., et al. Solar energy conversion with tunable plasmonic nanostructures for thermoelectric devices, *Nanoscale*, 4(15), 4416–4420, 2012

[86] Wang, Y., Zhang, P., Fu, W., et al., Morphological control of nanoprobe for colorimetric antioxidant detection, *Biosensors and Bioelectronics*, 122, 183–188, 2018

[87] Duncan, T. V., Applications of nanotechnology in food packaging and food safety: Barrier materials, antimicrobials and sensors, *Journal of Colloid and Interface Science*, 363(1), 1–24, 2011

[88] Asensio, E., Montañés, L., & Nerín, C., Migration of volatile compounds from natural biomaterials and their safety evaluation as food contact materials, *Food and Chemical Toxicology*, 142, 111457, 2020

[89] Pastor, C., Sánchez-González, L., Marcilla, A., et al., Quality and safety of table grapes coated with hydroxypropylmethylcellulose edible coatings containing propolis extract, *Postharvest Biology and Technology*, 60(1), 64–70, 2011

[90] Ahmed, T., Shahid, M., Azeem, F., et al., Biodegradation of plastics: Current scenario and future prospects for environmental safety, *Environmental Science and Pollution Research*, 25, 7287–7298, 2018

[91] Poverenov, E., Danino, S., Horev, B., et al., Layer-by-layer electrostatic deposition of edible coating on fresh cut melon model: Anticipated and unexpected effects of alginate—chitosan combination, *Food and Bioprocess Technology*, 7, 1424–1432, 2014

[92] Ramos, Ó. L., Silva, S. I., Soares, J. C., et al., Features and performance of edible films, obtained from whey protein isolate formulated with antimicrobial compounds, *Food Research International*, 45(1), 351–361, 2012

[93] Ashori, A., Wood—plastic composites as promising green-composites for automotive industries!, *Bioresource Technology*, 99(11), 4661–4667, 2008

[94] Savolainen, K., Pylkkänen, L., Norppa, H., et al., Nanotechnologies, engineered nanomaterials and occupational health and safety—A review, *Safety Science*, 48(8), 957–963, 2010

[95] Athinarayanan, J., Periasamy, V. S., Alsaif, M. A., et al., Presence of nanosilica (E551) in commercial food products: TNF-mediated oxidative stress and altered cell cycle progression in human lung fibroblast cells, *Cell Biology and Toxicology*, 30, 89–100, 2014

12 MOF-Based Nanocomposites for Energy Storage and Supercapacitor Applications

Rekha Gaba and Ramesh Kataria*

12.1 INTRODUCTION

The necessity to switch to electronic devices to generate power in the coming era to combat the threat posed by the continued use of fossil fuels, such as changes in climatic conditions and environmental pollution, is no longer debatable [1–4]. Although it is well recognized that electricity is a versatile kind of energy, its energy-storing capacity can have negative effects. Large amounts of energy may be stored by batteries, although they take a while to charge. However, only a tiny amount of energy can be stored effectively in regular capacitors, which may be charged fairly quickly. There is a huge need for environmentally friendly, sustainable, and high-efficiency energy assets, as well as for powerful and effective energy storage and new conversion technologies, in particular with the fast-expanding market for portable electronic devices [5–8].

To create clean energy production and storage to lower carbon footprints, the engineering or fabrication of materials with pores demands a significant level of research. A new class of crystalline material named metal-organic frameworks (MOFs) has resulted from the interaction of metal ions/clusters and organic bridging ligands. MOFs are used in a variety of processes, such as sensing [9–11], catalysis [12–15], gas storage and separation [16,17], water harvesting [18–21], and purification of water [22], as well as in batteries [23–28] and supercapacitors [29–33]. Even though MOFs are either poor conductors or insulators, which limits their natural conductivity [34–35], they can work incredibly well in energy storage devices like supercapacitors and batteries and in catalytic applications. In the context of the circular economy, MOFs offer several advantages. First, their high porosity allows for efficient adsorption and storage of gases, such as carbon dioxide (CO_2). CO_2 capture is a crucial step in mitigating greenhouse gas emissions and addressing climate

* rkatariapu@gmail.com

DOI: 10.1201/9781003269779-12

change. MOFs have shown promise in selectively capturing CO_2 from flue gas emissions and industrial processes. The evaluation of MOFs for CO_2 capture involves considering parameters such as economic cost, stability, and tolerance to impurities. By optimizing the performance of MOFs in CO_2 capture, they can contribute to the circular economy by reducing emissions and enabling the utilization of captured CO_2 for other purposes, such as carbon capture and utilization (CCU) or carbon storage.

Due to its prolonged cycle life as well as high power density [36], supercapacitors are a form of energy storage device that has attracted a lot of attention recently. Unlike batteries, which store energy through chemical reactions, supercapacitors store energy through the separation of charges at the interface of electrode-electrolyte [37]. It also allows for rapid charge and discharge rates, making supercapacitors suitable for applications that require high-power delivery [36].

The electrode materials used in supercapacitors play a crucial role in determining their performance. Carbon-based materials, such as microporous carbons, are commonly used as electrodes due to their high conductivity and enlarged surface area [38]. Other materials, such as conducting polymers and mixed metal oxides, have also been investigated [37]. The choice of electrode material affects the power density as well as the energy density of the supercapacitor [38]. Efforts have been made to improve the energy density of supercapacitors. Nanostructuring, nano-/micro-combination, hybridization, pore-structure control, surface modification, and composition optimization are some of the approaches employed for enhancing the performance of supercapacitor materials [36]. These approaches aim to enhance the surface area and boost the charge storage capacity of the electrodes. In addition to energy storage, researchers have also explored the integration of additional functionalities into supercapacitors. For example, developed a smart supercapacitor that not only stores energy but also communicates the level of stored energy through multiple-stage pattern indications integrated into the device [39]. This demonstrates the potential for supercapacitors to be used in novel applications beyond traditional energy storage. Supercapacitors are among the advanced energy storage materials that are being developed. Researchers are investigating new electrode materials, electrolytes, current collectors, and separators to improve the performance of supercapacitors. The fabrication techniques for supercapacitor electrodes are also being explored, with a focus on developing flexible and thin devices for wearable applications.

12.2   CLASSIFICATION OF SUPERCAPACITORS

Supercapacitors can be divided into electric double-layer capacitors (EDLC), hybrid supercapacitors, and pseudocapacitors based on the storing mechanisms (Figure 12.1). Pseudocapacitors store energy using the faradaic redox process, whereas EDLC stores energy electrostatically. The two mechanisms are shared by the third one, which is a hybrid supercapacitor. The chosen storage technique and, thus, the supercapacitor's performance are significantly influenced by the electrode materials.

12.2.1   ELECTRIC DOUBLE-LAYER CAPACITORS

The EDLC is often referred to as a super or ultra-capacitor. An electrochemical capacitor known as an EDLC uses conducting polymers as electrodes. The EDLC only has a storage capacity of about 10 Wh/kg but allows for huge power effects per weight with a goal of up to 10 kW/kg. The storage time is brief, usually between 30 and 60 seconds. EDLCs are energy storage devices that rely on the electrical double layer (EDL) formed at the interface between the electrolyte and the electrode. The EDL is a complex system where ions are distributed near the surface, resulting in an exponential decrease in electrical potential with increasing distance from the surface. Due to its geometry and structure, which are similar to an electric capacitor, the EDL is also referred to as an EDLC. Numerous studies have explored the EDL and its electrokinetic phenomena, such as electro-osmosis, electrophoresis, and streaming current. There have been attempts to generate electrical energy in microfluidic channels using the streaming current approach. EDLCs have advantages over other capacitors, including higher energy density and the ability to design more compact systems with the same capacity. They have been implemented in peak-cut energy storage systems and utility-interactive energy storage systems. Recent research has focused on developing EDLCs with higher energy density and investigating the properties of the EDL formed near carbon-based electrodes. Additionally, there have been efforts to explore the potential of mechanically modulating the EDL to generate AC electric current, which could be useful for constructing microfluidic power generation systems. Overall, EDLCs offer promising potential for energy storage and power generation applications.

12.2.2 PSEUDOCAPACITOR

Pseudocapacitors are a type of electrochemical capacitor that rely on charge storage involving fast surface redox reactions. Unlike EDLCs, which store charge through ion adsorption, pseudocapacitors utilize reversible redox reactions at the electrode-electrolyte interface to store and release energy. This mechanism allows pseudocapacitors to achieve higher energy storage capacities compared to EDLCs, although still lower than those of secondary batteries. Pseudocapacitive materials, such as transition metal oxides or conducting polymers, exhibit faradaic behaviour and can undergo rapid and reversible redox reactions, enabling efficient charge storage. The incorporation of pseudocapacitive materials in supercapacitors enhances their energy density and power density, making them attractive for various applications, including energy storage systems, portable electronics, and electric vehicles. Ongoing research in the field of pseudocapacitors focuses on developing new materials with improved redox properties, optimizing electrode structures, and exploring advanced characterization techniques to better understand their electrochemical behaviour. Overall, pseudocapacitors offer a promising avenue for advancing energy storage technologies and addressing the growing demand for highly efficient energy storage devices.

12.2.3 HYBRID SUPERCAPACITORS

Hybrid supercapacitors are energy-storing devices that combine the advantages of both EDLCs and pseudocapacitors. They typically consist of a combination of carbon-based materials, such as graphene or carbon nanotubes, and transition metal oxides or hydroxides, such as MnO_2 or $Ni(OH)_2$. The incorporation of

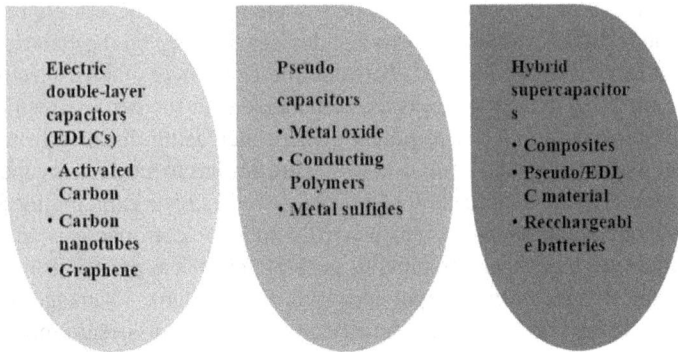

FIGURE 12.1 Classification of supercapacitors and the material used for their fabrication.

pseudocapacitive materials in hybrid supercapacitors allows for higher specific capacitance values compared to EDLCs alone. This is because pseudocapacitive materials can undergo faradaic reactions, which involve reversible redox reactions at the electrode-electrolyte interface, leading to additional charge storage mechanisms. The combination of both electric double-layer capacitance and pseudocapacitance in hybrid supercapacitors results in improved energy storage performance, including higher energy density and power density. Furthermore, hybrid supercapacitors have shown promise in applications requiring high-rate capability, as they can maintain their performance even at high scan rates. The development of hybrid supercapacitors has attracted significant attention due to their potential applications in portable electronics, hybrid electric vehicles, and smart electricity grids. Ongoing research in this field aims to further enhance the performance of hybrid supercapacitors by exploring new materials and optimizing their electrode structures. Overall, hybrid supercapacitors offer a promising solution for efficient energy storage and have the potential to revolutionize various industries.

12.3 STRUCTURE OF MOFS

The functions of MOFs are supported by their crystalline structures. To create MOFs, metals or metal clusters, sometimes referred to as secondary building units (SBUs), are connected with organic linkers, usually carboxylic acid or nitrogen comprehending ligands [40,41]. MOFs can be tuned, which makes them differ from other spongy materials like carbons and zeolites. Since the morphology and geometry of SBUs and organic ligands, respectively, determine the pore size, geometry, and functionality of MOFs, these structures can be intentionally tailored by selecting SBUs and linkers. Because MOFs have unique structures, researchers can examine the relationship between structure and property, which is essential for the cogent design of innovative MOFs for particular applications [42–44]. Due to the use of only one type of SBU and organic ligand, the heterogeneity and complexity of the majority of reported MOFs remain limited despite major advancements in MOF chemistry. The careful selection of MOF constituents allows for the formation of crystals with ultrahigh porosity and high thermal and chemical stability. One of the key advantages of MOFs is their ability to undergo precise chemical modification without changing the underlying topology. This flexibility in chemical composition and shape of building units within a particular structure offers the potential for materials with a synergistic combination of properties. MOFs can expand their metrics, allowing for the creation of materials with tailored properties. The chemistry of coordination polymers, from which MOFs are derived, has advanced significantly in recent years. This includes the use of crystal engineering to construct porous frameworks, characterize and catalogue the porous properties, and explore the next generation of porous functions based on dynamic crystal transformations. The design and synthesis of MOFs with exceptional stability and high porosity have been a focus of research. The aim is to create MOFs that can withstand harsh conditions and exhibit high adsorption capacities. By understanding the chemistry and physics of the micropores within MOFs, researchers can further optimize their properties and explore new applications.

12.3.1 METHOD OF SYNTHESIS OF MOF

MOFs can be synthesized using various methods, each offering unique benefits and control over the resulting properties based on their structure. The following methods are commonly employed in the formation of MOFs (Figure 12.2):

1. *Solvothermal method*: This method involves the reaction of metal ions or clusters with different organic linkers in a high-temperature solvent under autogenous pressure. The solvothermal method allows for the formation of well-defined crystalline MOFs with high porosity and surface area.
2. *Hydrothermal method*: Similar to the solvothermal method, the hydrothermal method involves the reaction of organic linkers with different metal ions, but at lower temperatures and pressures. This method is advantageous for the synthesis of MOFs that are sensitive to high temperatures or solvents.
3. *Microwave-assisted method*: Microwave irradiation can be used to accelerate the synthesis of MOFs by providing rapid and efficient heating. This method offers the advantage of reduced reaction times and improved control over the crystal size and morphology of the resulting MOFs.
4. *Mechanochemical method*: The mechanochemical method involves the grinding or milling of solid reactants in the presence of a solvent or grinding agent. This technique enables the synthesis of MOFs without the need for high temperatures or solvents, making it a more environmentally friendly approach.
5. *Layer-by-layer method*: The layer-by-layer method involves the sequential deposition of metal ions or clusters and organic linkers onto a substrate. This method allows for the controlled growth of MOF thin films with precise control over the film thickness and composition.
6. *Post-synthetic modification*: Post-synthetic modification involves the modification of pre-formed MOFs by introducing new functional groups or replacing existing linkers. This method allows for the fine-tuning of MOF properties and the introduction of additional functionalities.

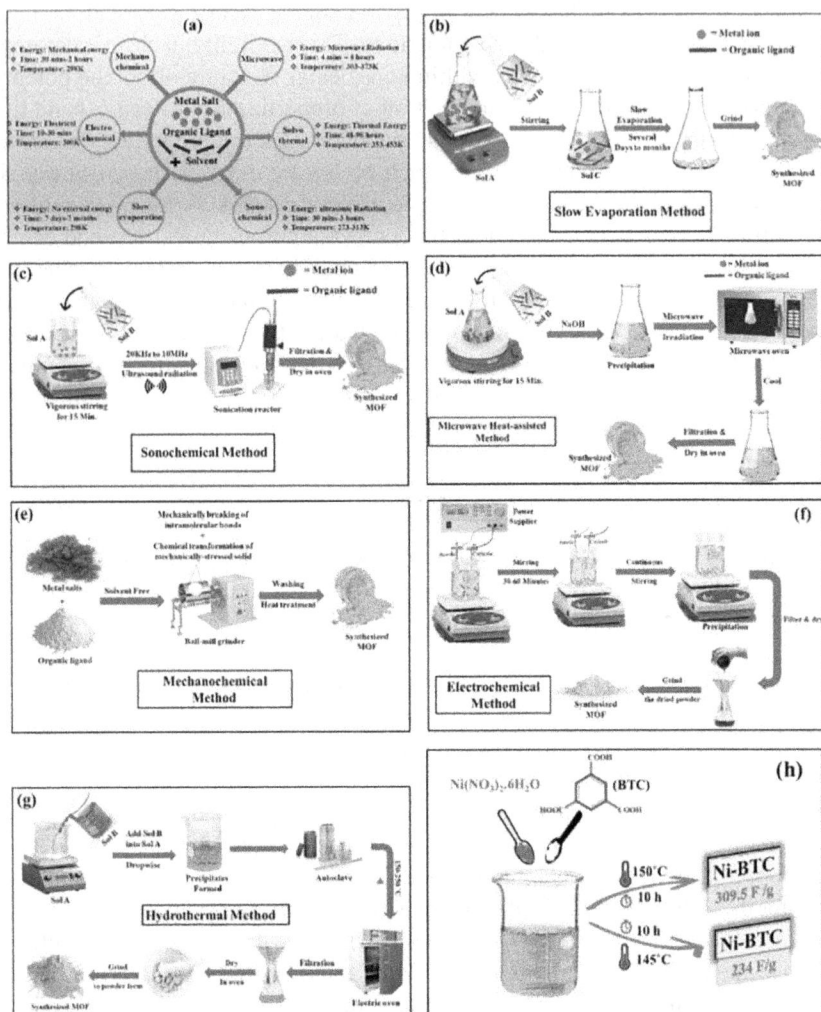

FIGURE 12.2 (a) Description of various methods used for the synthesis of required material. (b) Schematic diagram of slow evaporation method. (c) Schematic diagram of sonochemical method. (d) Schematic diagram of microwave heat-assisted method. (e) Schematic diagram of mechanochemical method. (f) Schematic diagram of electrochemical method. (g) Schematic diagram of hydrothermal method. (h) Depiction of the influence of hydrothermal temperature on specific capacitance.

Source: Reproduced with permission from [45]. Copyright 2022 Elsevier.

These methods offer a range of options for synthesizing MOFs with tailored properties and structures. The choice of method depends on factors such as the desired MOF composition, porosity, and application requirements. By utilizing these methods, researchers can continue to explore the vast potential of MOFs in different fields, like gas separation, catalysis, and energy storage.

12.3.2 MOF-Derived Metal Oxide/Carbon Nanocomposites for Supercapacitors

Synthesis of MOF-derived metal oxide/carbon nanocomposites has gained significant attention in recent years due to their potential applications in various fields [45]. Several methods have been employed to fabricate these nanocomposites. One common approach is the use of MOF/graphene oxide (GO) composites as sacrificial templates. In this method, MOFs are uniformly distributed on the surface of GO, and subsequent pyrolysis or thermal treatment leads to the formation of metal oxide/porous carbon nanocomposites uniformly dispersed on reduced graphene oxide (rGO). This strategy allows for the precise control of the composition and structure of the resulting nanocomposites. Another method involves the immobilization of metal nanoparticles (NPs) on MOF-derived nanocomposites. For example, ultrafine PdAg NPs can be anchored on zirconia/porous carbon/reduced graphene oxide ($ZrO_2/C/rGO$) nanocomposites derived from MOF/GO. This immobilization not only enhances the dispersion of the metal NPs but also optimizes the catalytic performance of the resulting catalysts. Furthermore, MOF-derived metal oxide/carbon nanocomposites can be prepared through hydrothermal or solvothermal methods. These methods involve the reaction of metal ions or clusters with organic linkers in high-temperature or high-pressure solvents, resulting in the formation of well-defined crystalline MOFs. Subsequent thermal treatment or carbonization leads to the formation of metal oxide/carbon nanocomposites. In addition, the impregnation of MOF-derived structures with active components has been explored. For instance, the skeleton structure of bimetallic Ni-Zr MOF-derived nickel-zirconium oxide (Ni-Zr-O) can be impregnated with silicotungstic acid (HSiW) to prepare a nanocomposite catalyst.

Metal oxide/carbon nanocomposites made from MOFs have sparked a lot of interest in research due to their enormous application potential in electrochemical energy storage. Although several MOFs have been transformed into different active materials via thermolysis, a great number of in situ changes in chemical composition, phase(s), and shape need careful control over heating settings. For example, by using a novel two-stage method, involving annealing and heating at different temperatures $Mn_3O_4@C$ has been prepared. First, Mn-MIL-100 is annealed at a high temperature of about 700°C and converted into MnO@C under N_2 flow; and second, the obtained product is converted into $Mn_3O_4@C$ by heating to 200°C in air while maintaining a large surface area. With careful management of the heating period, the proper retention of carbon content for $Mn_3O_4@C$ is likewise simple to achieve. In contrast, higher temperature thermolysis of MnO@C results in manganese oxides with significantly reduced surface area and zero carbon content. The optimized $Mn_3O_4@C$-2 h, which was produced by heating MnO@C to 200°C for two hours, displayed a capacitance that was significantly higher than that of MnO@C and other derivatives. It displayed a capacitance of 328.4 F cm^3 when coupled with GO nanosheets (NSs) to create a bendable $Mn_3O_4@C/rGO$ paper electrode. The asymmetric supercapacitor constructed using $Mn_3O_4@C/rGO$ also performs well. To maximize the electrochemical performance of active materials formed from MOFs used as electrode materials, this work illustrates the great controllability provided by the novel two-stage thermolysis method [46]. A simple method combining the MOF templating process and electrode-position is used to generate Co_3O_4@MnO_2 NS hybrid nanostructures on carbon fabric has been reported [47]. The reported Co_3O_4@MnO_2 core-shell NS array electrode has shown a synergetic effect between Co_3O_4 triangle NSs and MnO_2 NSs and has shown remarkable electrochemical properties along with excellent cycle performance. The Co_3O_4@MnO_2 core-shell NSs are potential candidates for supercapacitors due to their high electrochemical performance. Using simple calcination of nickel-based MOFs, they created extremely porous nanostructured NiO/C yolk-shell nanocomposites. By integrating sodium and hierarchical MoS_2 nanostructures using a simple hydrother-mal process, the porosity and electrochemical properties were improved even more. The as-synthesized Na-doped MoS_2@NiO/C nanocomposites with hierarchical poros-ity were shown to have improved electrochemical performance for acting as super-capacitors [48]. The new graphitic carbon nitride (g-C_3N_4) nanocomposites coated with oxygen vacancies-rich ZnO (OZCN) were created by direct thermal breakdown of melamine in air from zeolitic imidazolate framework precursor. Because of the synergetic impact of g-C_3N_4 and oxygen vacancies-rich ZnO, the as-prepared OZCN nanocomposites demonstrated strong capacitive performance (3000 F g^{-1} at 3 A g^{-1}) and outstanding cycle stability [49]. Due to improved charge transfer between differ-ent metal ions, bimetallic MOFs may induce rich redox reactions, thereby enhancing the performance of supercapacitors even further. The "one-for-all" method is used in this study to create both positive and negative electrodes for hybrid asymmetric SCs (ASCs) from a single bimetallic MOF. A simple approach was used to create the bimetallic Zn/Co-MOF with cuboid-like features. After post-heating the as-syn-thesized Zn/Co-MOF and washing with HCl, the MOF-derived nanoporous carbons (NPC) were formed, and bimetallic oxides ($ZnCo_2O_4$) were obtained by sintering the Zn/Co-MOF in the air [50]. In another study, a unique two-step approach for the fab-rication of NiO/Ni architecture is encapsulated in N-doped carbon nanotubes (NiO/Ni/NCNTs). A solution reaction was used to create 3D columnar nickel-based MOFs (Ni-MOFs). The Ni^{2+} in Ni-MOFs was then partially reduced to Ni to catalyze the synthesis of NCNTs via inert atmosphere calcination. These nanocomposites of NiO/Ni/NCNTs can provide not only additional reactive sites for electrochemical reactions but also conductive pathways for electron transport [51].

12.3.3 MXene/MOF Composites

MXene/MOF composites have emerged as promising materials for energy storage applications. Metal-organic frameworks (MOFs) and MXenes, a class of two-dimen-sional transition metal nitrides and carbides, possess unique properties that make them suitable for energy storage and conversion. MOF-based materials, including pristine MOFs and MOF composites, have attracted significant interest in this field. MOF composites, such as Pt nanoparticles@MOF and MOF/graphene composites, have been investigated for their strength in energy conversion and storage. These composites combine the tunable porosity with the large surface area of MOFs with the excellent electrical conductivity and mechanical properties of MXenes. The inte-gration of MXenes into MOF composites can enhance the electrochemical perfor-mance and stability of the materials, making them suitable for applications such as supercapacitors and lithium-ion batteries. The low-temperature reduction strategy has been employed to synthesize Si/Ti_3C_2 MXene/MOF composites, exhibit improved electrochemical performance and cycling stability. The combination of MXenes and MOFs in these composites offers synergistic effects, leading to enhanced energy storage capabilities. Further research is needed to optimize the composition and structure of MXene/MOF composites and explore their full potential in energy storage applications. MOF nanocomposites have been extensively studied for their potential in supercapacitor applications. The fusion of MOFs with other materials, such as graphene, has been shown to enhance the performance of supercapacitors. Thin film coatings of MOFs supported on solid substrates or as free-standing membranes have also been explored for functionalizing surfaces and improving supercapacitor performance. In a study, nanocrystalline MOFs (nMOFs) doped with graphene were successfully incorporated into supercapacitor devices, demonstrating high capacitance and long life cycle behaviour. The researchers examined a series of 23 different nMOFs with varying structures, pore sizes, and metal ions, and found that some of them exhibited exceptional capacitance. The charge/discharge profiles, cyclic voltammetry curves, and cycling performance of these nMOFs followed the general behaviour observed in other supercapacitors. Another study focused on the synthesis of La and Ce mixed MnO_2 nanostructure/rGO nanocomposites for supercapacitor applications The nanocomposite electrodes exhibited high specific capacitance and energy density, making them suitable for supercapacitor electrode fabrication. The electrochemical performance of the nanocomposites was examined using electrochemical impedance spectroscopy, charge-discharge measurements, and cyclic voltammetry. Overall, the research on MOF nanocomposites for supercapacitors shows promising results. The combination of MOFs with other materials, such as graphene or MnO_2, can enhance the capacitance and energy density of supercapacitors. Further research and development in this field can lead to the design of MOF-based nanocomposites with upgraded capability and stability for energy storage applications.

MXene (2D titanium carbide) has been extensively researched and evaluated for energy storage reasons in recent years. It possesses great features like as hydrophilicity, metallic conductivity, and, most importantly, high surface redox reactivity, which is critical for energy storage applications. In three electrode configurations, the as-prepared FeCu MOF/MXene electrode demonstrated a high specific capacity of 440 mA h g^{-1}. After 10,000 alternate GCD cycles, the as-prepared MXene-FeCuMOF//AC ASC demonstrated good cyclic stability of 89% [52]. $NiCo_2S_4$ @ Co_3S_4 nanocages with superior yolk-shell structure are introduced to anchor on the surface of $Ti_3C_2T_x$ via ion exchange reaction and electrostatic attraction, forming the distinctive hierarchical structure of $NiCo_2S_4$ @Co_3S_4/Ti_3C_2Tx composed of $NiCo_2S_4$ @Co_3S_4 YSNs adsorbed on 2D $Ti_3C_2T_x$ nanosheets [53]. $ZnCo_2O_4$ (ZCO) particles produced from ZnCo-MOF (ZCM) adsorb on (MX) nanosheets and form a mesoporous structure that may allow flexible ion transport paths. The ZCO particles not only serve as activation sites for free charge mobility, but they also aid in the avoidance of MX nanosheet agglomeration. MX@ZCO, because of its innovative composite structure, has a high specific capacity of 260 mAh g^{-1} at a current density of 1 mA g^{-1} [54]. Composites produced from MXene (Ti3C2Tx) and cobalt-MOF (Co-NC/Ti_3C_2-T). Interestingly, the optimized composite Co-NC/Ti_3C_2_800 exhibits improved electrocatalytic oxygen reduction reaction (ORR) activity with a 1.04 V onset potential (Eonset) versus reversible hydrogen electrode (RHE), 4.8 mA/cm2 current density (JL), and 0.93 V half-wave potential (E1/2) versus RHE [55]. In 1 M LiCl electrolyte, ideal Co-Fe oxide/Ti_3C_2TX composite paper exhibits a high volumetric capacitance of 2467.6 F cm^3. When combined into a flexible symmetrical supercapacitor, the specific areal capacitance of 356.4 mF cm^2 is excellent [56].

12.3.4 REDOX POLYMER NANOCOMPOSITES

Redox polymer nanocomposites are a class of materials that combine redox-active polymers with other components to create unique properties and functionalities. These nanocomposites have attracted significant attention due to their potential applications in various fields, including biosensing, catalysis, energy storage, and electronics. The enhanced properties of redox polymer nanocomposites can be attributed to the synergistic effect between the components. For example, the PANI/Cu9S5 hybrid nanofibers exhibited superior catalytic performance due to the strong interactions between PANI and Cu9S5 components [57]. Similarly, the MoS_2/PPy nanocomposite showed enhanced catalytic properties, with the MoS_2 nanosheets coated with PPy forming a unique algae-like structure [57]. Conducting polymer-based nanocomposites, including redox polymer nanocomposites, have been widely studied for their enhanced properties and applications. These nanocomposites combine the advantages of conducting polymers with other components, leading to improved or new distinct properties. For instance, hollow graphene/conducting polymer composite fibre electrodes have shown outstanding electrochemical performance and high flexibility, making them promising for wearable electronics. The synthesis and characterization of redox polymer nanocomposites involve various techniques. Fourier transform infrared spectroscopy (FTIR) and transmission electron microscopy (TEM) are commonly used to investigate the functionality and morphology of nanocomposites. Cyclic voltammetry (CV) is employed to study the electrochemical properties and optimize the synthetic conditions. Other techniques, such as X-ray diffraction and UV-vis spectroscopy, are also used to analyze the structural and optical properties of the nanocomposites. Redox polymer nanocomposites have shown great prospects for applications in supercapacitors. By pulse electrodepositing poly(3,4-ethylenedioxythiophene, or PEDOT), a conducting polymer, onto thin films of the cerium-based metal-organic framework (Ce-MOF-808), it is possible to create nanocomposites. A pseudocapacitance is produced by the highly porous Ce-MOF-808's reversible electrochemical reactivity, and an exceptional double-layer capacitance can be produced by the electronically conducting PEDOT while also facilitating electronic conduction between the MOF's redox-active cerium sites. The composite can therefore outperform both pristine electrodeposited PEDOT and pristine MOF as the active components for supercapacitors [58]. To create PANI/MIL-101 nanocomposites, Wang et al. [59] created coordinated PANI with the unsaturated metal sites in MIL-101 via its electron-rich imine function to make a flexible supercapacitor. A solid-state asymmetric flexible supercapacitor device based on Co-BTC coated on a nanowire microsphere was created by Lin et al. [60] utilizing a straightforward non-calcined method. The mechanical flexibility and stability of the entire system have been improved by the micro-spherical shape, while the nanowire architecture enhanced the electronic conductivity of the MOFs. The constructed device's energy density was discovered to be 34.4 Wh kg^{-1}, and its maximum power density was 375 W kg^{-1}. According to Jafari et al. [61], PANI and rod-like HKUST-1 (MOF-199) were both made utilizing a hydrothermal technique with ammonium persulfate (APS) as an oxidizing agent. A straightforward mixing of $Cu_3(BTC)_2$ (HKUST-1) and PANI with a mass ratio of 25:75, known as HP, was used to create the composite. To increase the energy density of the composites compared to the traditional method, in which active inks are prepared from PANI/MOF composites, polymer binder, and conducting materials, followed by drop-casting or spin coating onto the current collector, Cheng et al. [62] proposed to grow Ni-MOF nanosheet onto PANI decorated NF. By releasing electroactive sites and lowering the internal resistance, this novel method might address issues with the conventional method. In a nutshell, in situ, oxidative polymerization in an aniline solution was used to deposit PANI onto NF.

One notable application is the use of redox polymer nanocomposites in the development of stable polyaniline-based supercapacitors. By introducing a second redox system, such as quinones, into the supercapacitor, a tunable redox shuttle is created that controls electron transfer processes at the polyaniline-modified electrodes. Quinones, with their small size and excellent electrochemical reversibility at low pH, act as mediating agents and provide superior stability to the supercapacitor. This stability is crucial in preventing the conversion of porous polyaniline to a highly reactive state, ensuring the long-term performance and durability of the supercapacitor. The use of redox polymer nanocomposites in supercapacitors holds promise for the development of high-performance energy storage devices with enhanced stability and longevity.

12.4 MOF FOR THE CIRCULAR ECONOMY (TRASH TO TREASURE)

The circular economy is an economic model that aims to maximize resource efficiency, minimize waste generation, and promote sustainable consumption and production. It is a departure from the traditional linear economy, which follows a "take-make-dispose" approach. In a circular economy, resources are kept in use for as long as possible through strategies such as recycling, reusing, and remanufacturing. This approach reduces the need for extracting new raw materials and minimizes the environmental impact of resource extraction and waste disposal (Figure 12.3). By closing the loop and creating a circular flow of materials, the circular economy promotes sustainable development and addresses the challenges of resource scarcity and environmental degradation. The recent advancement in the creation of MOFs for CH_4 storage for automotive applications and selective CO_2 capture from post-combustion flue gas. Several effective techniques are being utilized to enhance CO_2 adsorption absorption at low pressures. The authors emphasize the flexible and stiff MOFs with CH_4 storage capacities that are near the target set by the US Department of Energy while also discussing both traditional and new MOF regeneration procedures. They conclude by discussing the challenges of using MOFs for CH_4 storage Fe-BDC(W) has been synthesized by using waste products [63]. Fe-based metalorganic framework material. The Fe salt is produced by recycling rust, while the benzene dicarboxylic acid (BDC) linker is made from recycled polyethene terephthalate

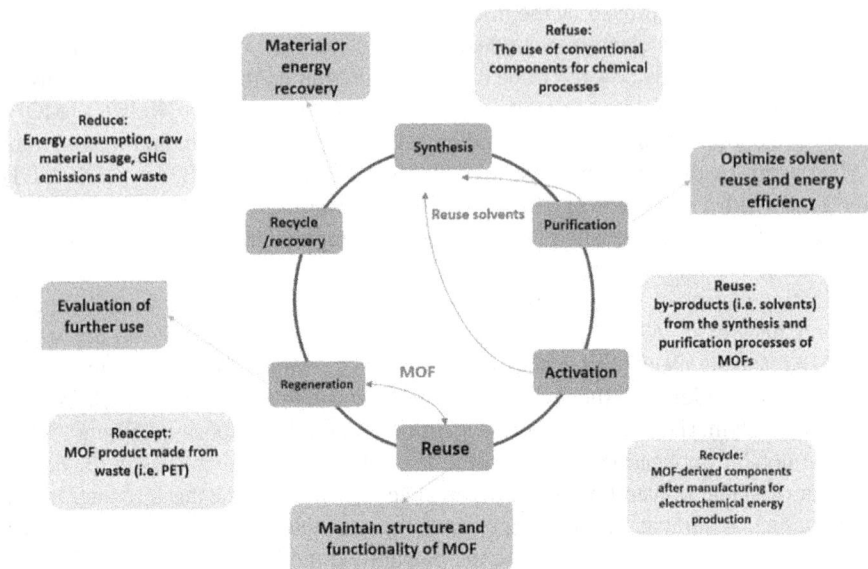

FIGURE 12.3 MOF in the circular economy.

Source: Reproduced with permission from [71]. Copyright 2022 Elsevier.

(PET) bottles. Sustainable energy storage from waste materials aims to develop energy storage systems that are both ecologically friendly and commercially successful. A supercapacitor uses the produced MOF as its active component. MOFs have been incorporated into circularity through a variety of industries. In contrast, the recycling of trash (such as LiB) can be utilized to create various MOF composites (such as Al-MOF, Ni-Mn MOF), according to Lagae-Capelle et al. [64]. MOFs have been generated from depolymerized PET for high-purity organic ligands for building MOFs. In addition, PET can be used for the environmentally friendly green synthesis of MOFs for hydrogen storage from PET bottles [65] and the sustainable manufacture of MOF catalysts (MOF methane dry reforming catalysts) [66]. Additionally, Jamil et al. [67] studied the application of MOF catalysts for the creation of biodiesel from used cooking oil by incorporating the essential elements that make up the circular economy. MOFs can be recycled without losing their capacity to adsorb CO_2 [68]. The use of MOFs for the elimination of harmful compounds from aquatic environments is a major area of interest. By using MOF in wastewater streams, aqueous environments might be purified without using an endless amount of energy [69,70].

12.4.1 Purification and Activation of MOF in the Interpretation of Circular Economy

MOFs are versatile materials with a wide range of applications, involving gas storage, gas separation, and catalysis. However, before MOFs can be used effectively, they often require purification and activation techniques to optimize their performance.

Purification involves removing any impurities or residual solvents from the MOF structure, ensuring its purity and stability. Activation, on the other hand, refers to the process of preparing the MOF for its intended application by creating an open and accessible pore structure. Several techniques are commonly employed for MOF purification and activation. Thermal treatment is a widely used method for activation, involving heating the MOF to remove any adsorbed or trapped molecules. Solvent exchange is another technique, where the MOF is immersed in a different solvent to replace the original solvent and remove any impurities. Freeze-drying and super-critical CO_2 techniques are also employed for activation, allowing for the removal of solvents while preserving the MOF structure [72]. In addition, to these techniques, there are also advanced characterization methods used to assess the structural properties and stability of MOFs during purification and activation. High-throughput computational analysis can provide insights into the activation status and stability of MOFs, helping to identify potential collapse or amorphization issues. Mechanical properties, such as bulk and shear moduli, can serve as predictors for collapse events and guide the optimization of activation techniques. Experimental techniques like in situ two-dimensional (2D)-powder X-ray diffraction (PXRD) can provide valuable observations about structural changes at the microscopic level during the activation process. Overall, the purification and activation of MOFs are crucial steps in their utilization for various applications. These techniques ensure the purity, stability, and accessibility of the MOF structure, optimizing its performance in gas separation, storage, and catalysis. Advanced characterization methods play a vital role in understanding the structural properties and stability of MOFs during these processes, guiding the development of effective purification and activation techniques.

12.5 CONCLUSIONS

MOFs have become attractive materials for a variety of circular economy applications. Their high porosity, chemical tunability, and functionalization capabilities make them suitable for energy storage, gas separation, catalysis, and other applications. By optimizing their performance in CO_2 capture and developing functional MOF materials, they can contribute to reducing emissions, promoting resource efficiency, and enabling the circular use of materials. In summary, the structure of MOFs is characterized by the formation of robust bonds between organic and inorganic units, resulting in crystalline materials with high thermal and chemical stability and ultrahigh porosity. The ability to chemically modify MOFs without changing their underlying topology offers great flexibility in tailoring their properties. The porous properties of MOFs allow for various functionalizations and applications, making them a promising class of materials in fields such as gas separation, gas storage, and catalysis. Further research is focused on expanding the metrics of MOFs, characterizing their porous properties, and exploring dynamic crystal transformations for the development of next-generation porous functions.

The limits of individual pure MOFs can be eliminated by using MOF nanocomposites, which have the necessary properties to create high-performance supercapacitor electrodes. It has long been known that MOF nanocomposites with optimally characterized chemical and thermal stabilities, electrical conductivity, flexibility surface areas, and so on, would result in extremely versatile energy storage devices. Several synthetic approaches for creating MOF-based nanocomposites have been outlined and covered in this chapter. Remarkably, for MOF-based nanocomposites in supercapacitor applications, thermolysis under various atmospheric conditions continues to be the most often used technique. Building MOF-based nanocomposites can also increase the number of flexible supercapacitors in the family. Several important points about these nanocomposites might be summed up as follows: to create unique MOF nanocomposites, a variety of MOFs, including MOF/conductive polymer, MOF/metal oxide/metal, and MOF/redox polymer, can serve as sacrificial templates or precursors. Various MOF precursors and functional species result in unique nanocomposites with unique compositions, morphologies, and architectures. Therefore, for high-performance flexible supercapacitor applications, great consideration must be given to the choice of parent MOFs and composite materials. By utilizing the synergistic effects of the individual components, a composite of MOF-derived nanoparticles with a variety of functions, such as carbon, conductive polymers, metal oxides, can successfully improve the cycle life, flexibility, and energy/power density of flexible devices.

Furthermore, there are not many documented synthesis techniques for flexible supercapacitors based on MOF nanocomposites, and their uses are still mostly confined to the lab. However, to create MOF-based nanocomposites with fascinating features for high-performing flexible supercapacitor devices for industrial-scale applications, future research efforts should make use of many types of advanced materials. Redox polymer nanocomposites are a promising class of materials with unique properties and applications. These nanocomposites combine redox-active polymers with other components, such as conductive materials, to enhance their performance in various fields. The synthesis and characterization of redox polymer nanocomposites involve a range of techniques to investigate their functionality, morphology, and electrochemical properties.

Furthermore, supercapacitors are promising energy storage devices with high power density and long cycle life. The choice of electrode material and the development of advanced materials and fabrication techniques are key factors in improving the performance of supercapacitors. Further research is needed to optimize the energy density and explore new applications for these devices.

MOF-based nanocomposites have gained significant attention in the field of energy storage and supercapacitor applications. These nanocomposites consist of MOFs combined with other materials to enhance their energy storage capabilities. One study investigated the role of polymer shells in the interfacial regions of ferroelectric polymer nanocomposites filled with core-shell structured polymer@BaTiO$_3$. The researchers found that the electrical properties of the interfacial regions play a critical role in determining the energy storage density of the nanocomposites. Researchers also demonstrated that a strong interaction between the polymer shells and the matrix can suppress dielectric loss and enhance energy storage capability. Additionally, the leakage currents in the nanocomposites were found to affect the dielectric loss and energy storage efficiency. The discharged energy density, which characterizes the effective energy storage capability, varied among the different nanocomposites. Another study focused on enhancing the energy storage capability of polymer nanocomposites by combining bio-inspired fluoro-polydopamine with barium titanate nanowires. The researchers aimed to overcome the trade-off between high dielectric constant and breakdown strength, which limits the increase in energy storage capability. They reported that the new strategy of combining fluoro-polydopamine with barium titanate nanowires resulted in polymer nanocomposites with superior energy storage capability, rivalling or exceeding some advanced nanoceramics-based materials. The discharged energy densities of these nanocomposites were also compared to a commercial biaxially oriented polypropylene film. In addition to MOF-based nanocomposites, other types of nanocomposites have also been explored for energy storage applications. Overall, MOF-based nanocomposites and other types of nanocomposites show great potential for energy storage and supercapacitor applications. The combination of different materials allows for the enhancement of energy storage capabilities, making these nanocomposites promising candidates for future energy storage devices. Further research and development in this field can lead to the advancement of innovative materials for diverse applications. Further research and development in MOF synthesis, characterization, and application will continue to advance their potential in the circular economy.

REFERENCES

[1] Wang, D. G., Liang, Z., Gao, S., Qu, C., & Zou, R. Metal-Organic Framework-Based Materials for Hybrid Supercapacitor Application. *Coord. Chem. Rev.*, *404*, 213093, 2020.

[2] Xu, Y., Li, Q., & Pang, H. Recent Advances in Metal Organic Frameworks and Their Composites for Batteries. *Nano Futures*, *4*(3), 032007, 2020.

[3] Xiao, X., Zou, L., Pang, H., & Xu, Q. Synthesis of Micro/Nanoscaled Metal—Organic Frameworks and Their Direct Electrochemical Applications. *Chem. Soc. Rev.*, *49*(1), 301, 2020.

[4] Yan, J., Liu, T., Liu, X., Yan, Y., & Huang, Y. Metal-Organic Framework-Based Materials for Flexible Supercapacitor Application. *Coord. Chem. Rev.*, *452*, 214300, 2022.

[5] Chmiola, J., Largeot, C., Taberna, P. L., Simon, P., & Gogotsi, Y. Desolvation of Ions in Subnanometer Pores and Its Effect on Capacitance and Double-layer Theory. *Angew. Chem.*, *120*(18), 3440, 2008.

[6] Zhai, Y., Dou, Y., Zhao, D., Fulvio, P. F., Mayes, R. T., & Dai, S. Carbon Materials for Chemical Capacitive Energy Storage. *Adv. Mater.*, *23*(42), 4828, 2011.

[7] Xie, J., Gu, P., & Zhang, Q. Nanostructured Conjugated Polymers: Toward High-Performance Organic Electrodes for Rechargeable Batteries. *ACS Energy Lett.*, *2*(9), 1985, 2017.

[8] Xie, J., Zhao, C. E., Lin, Z. Q., Gu, P. Y., & Zhang, Q. Nanostructured Conjugated Polymers for Energy-Related Applications Beyond Solar Cells. *Chem. Asian J.*, *11*(10), 1489, 2016.

[9] Li, H. Y., Zhao, S. N., Zang, S. Q., & Li, J. Functional Metal—Organic Frameworks as Effective Sensors of Gases and Volatile Compounds. *Chem. Soc. Rev.*, *49*(17), 6364, 2020.

[10] Koo, W. T., Jang, J. S., & Kim, I. D., Metal-Organic Frameworks for Chemiresistive Sensors. *Chem*, *5*(8), 1938, 2019.

[11] Fang, X., Zong, B., & Mao, S., Metal—Organic Framework-Based Sensors for Environmental Contaminant Sensing. *Nano-Micro Lett.*, *10*(4), 64, 2018.

[12] Wei, Y. S., Zhang, M., Zou, R., & Xu, Q., Metal—Organic Framework-Based Catalysts with Single Metal Sites. *Chem. Rev.*, *120*(21), 12089, 2020.

[13] Alhumaimess, M. S., Metal—Organic Frameworks and Their Catalytic Applications. *J. Saudi Chem. Soc.*, *24*(6), 461, 2020.

[14] Pascanu, V., González Miera, G., Inge, A. K., & Martín-Matute, B., Metal—Organic Frameworks as Catalysts for Organic Synthesis: A Critical Perspective. *J. Am. Chem. Soc.*, *141*(18), 7223, 2019.

[15] Li, D., Xu, H. Q., Jiao, L., & Jiang, H. L., Metal-Organic Frameworks for Catalysis: State of the Art, Challenges, and Opportunities. *Energy Chem.*, *1*(1), 100005, 2019.

[16] Farha, O. K., & Hupp, J. T., Rational Design, Synthesis, Purification, and Activation of Metal—Organic Framework Materials. *Acc. Chem. Res.*, *43*(8), 1166, 2010.

[17] Li, J. R., Ma, Y., McCarthy, M. C., Sculley, J., Yu, J., Jeong, H. K., Balbuena, P. B., & Zhou, H. C., Carbon Dioxide Capture-Related Gas Adsorption and Separation in Metal-Organic Frameworks. *Coord. Chem. Rev.*, *255*(15–16), 1791, 2011.

[18] Xu, W., & Yaghi, O. M., Metal—Organic Frameworks for Water Harvesting from Air, Anywhere, Anytime. *ACS Cent. Sci.*, *6*(8), 1348, 2020.

[19] Hanikel, N., Prévot, M. S., & Yaghi, O. M., MOF Water Harvesters. *Nat. Nanotechnol.*, *15*(5), 348, 2020.

[20] Nemiwal, M., & Kumar, D., Metal Organic Frameworks as Water Harvester from Air: Hydrolytic Stability and Adsorption Isotherms. *Inorg. Chem. Commun.*, *122*, 108279, 2020

[21] Kalmutzki, M. J., Diercks, C. S., & Yaghi, O. M., Metal-Organic Frameworks for Water Harvesting from Air. *Adv. Mater.*, *30*(37), 1704304, 2018.

[22] Jayaramulu, K., Geyer, F., Schneemann, A., Kment, T., Otyepka, M., Zboril, R., Vollmer, D., & Fischer, R. A., Hydrophobic Metal—Organic Frameworks. *Adv. Mater.*, *32*, 1900820, 2019.

[23] Mehek, R., Iqbal, N., Noor, T., Amjad, M. Z. B., Ali, G., Vignarooban, K., & Khan, M. A., Metal—Organic Framework Based Electrode Materials for Lithium-ion Batteries: A Review. *RSC Adv.*, *11*(47), 29247, 2021.

[24] Ye, Z., Jiang, Y., Li, L., Wu, F., & Chen, R., Rational Design of MOF-Based Materials for Next-Generation Rechargeable Batteries. *Nano-Micro Lett.*, *13*(1), 2021.

[25] Wang, Z., Tao, H., & Yue, Y., Metal-Organic-Framework-Based Cathodes for Enhancing the Electrochemical Performances of Batteries: A Review. *Chem. Electrochem.*, *6*(21), 5358, 2019.

[26] Li, Y., Zhang, J., & Chen, M., MOF-Derived Carbon and Composites as Advanced Anode Materials for Potassium Ion Batteries: A Review. *SM&T*, *26*, e00217, 2020.

[27] Liang, Z., Qu, C., Guo, W., Zou, R., & Xu, Q., Pristine Metal-Organic Frameworks and Their Composites for Energy Storage and Conversion. *Adv. Mater.*, *30*(37), 1702891, 2017.

[28] Li, C., Liu, L., Kang, J., Xiao, Y., Feng, Y., Cao, F. F., & Zhang, H., Pristine MOF and COF Materials for Advanced Batteries. *Energy Stor. Mater.*, *31*, 115, 2020.

[29] Xu, B., Zhang, H., Mei, H., & Sun, D., Recent Progress in Metal-Organic Framework-Based Supercapacitor Electrode Materials. *Coord. Chem. Rev.*, *420*, 213438, 2020.

[30] Gao, H., Shen, H., Wu, H., Jing, H., Sun, Y., Liu, B., Chen, Z., Song, J., Lu, L., Wu, Z., & Hao, Q., Review of Pristine Metal—Organic Frameworks for Supercapacitors: Recent Progress and Perspectives. *Energy Fuel.*, *35*(16), 12884, 2021.

[31] Baumann, A. E., Burns, D. A., Liu, B., & Thoi, V. S., Metal-Organic Framework Functionalization and Design Strategies for Advanced Electrochemical Energy Storage Devices. *Commun. Chem.*, *2*(1), 2019.

[32] Cherusseri, J., Pandey, D., Sambath Kumar, K., Thomas, J., & Zhai, L., Flexible Supercapacitor Electrodes Using Metal—Organic Frameworks. *Nanoscale*, *12*(34), 17649, 2020.

[33] Mohanty, A., Jaihindh, D., Fu, Y. P., Senanayak, S. P., Mende, L. S., & Ramadoss, A., An Extensive Review on Three Dimension Architectural Metal-Organic Frameworks Towards Supercapacitor Application. *J. Power Sources*, *488*, 229444, 2021.

[34] Hu, A., Pang, Q., Tang, C., Bao, J., Liu, H., Ba, K., Xie, S., Chen, J., Chen, J., Yue, Y., Tang, Y., Li, Q., & Sun, Z., Epitaxial Growth and Integration of Insulating Metal—Organic Frameworks in Electrochemistry. *J. Am. Chem. Soc.*, *141*(28), 11322, 2019.

[35] Sun, L., Campbell, M. G., & Dincă, M., Electrically Conductive Porous Metal-Organic Frameworks. *Angew. Chem. Int. Ed. Engl.*, *55*(11), 3566, 2016.

[36] Sun, J., Cui, B., Chu, F., Yun, C., He, M., Li, L., & Song, Y., Printable Nanomaterials for the Fabrication of High-Performance Supercapacitors. *Nanomater*, *8*(7), 528, 2018.

[37] Chen, T., & Dai, L., Carbon Nanomaterials for High-Performance Supercapacitors. *Mater Today*, *16*(7–8), 272, 2013.

[38] Miller, J. R., & Burke, A., Electrochemical Capacitors: Challenges and Opportunities for Real-World Applications. *Electrochem. Soc. Interface*, *17*(1), 53, 2008.

[39] Tian, X., Xiao, B., Xu, X., Xu, L., Liu, Z., Wang, Z., Yan, M., Wei, Q., & Mai, L., Vertically Stacked Holey Graphene/Polyaniline Heterostructures with Enhanced Energy Storage for On-Chip Micro-Supercapacitors. *Nano Res.*, *9*(4), 1012, 2016.

[40] Perry IV, J. J., Perman, J. A., & Zaworotko, M. J., Design and Synthesis of Metal—Organic Frameworks Using Metal—Organic Polyhedra as Supermolecular Building Blocks. *Chem. Soc. Rev.*, *38*(5), 1400, 2009.

[41] Guillerm, V., Kim, D., Eubank, J. F., Luebke, R., Liu, X., Adil, K., Lah, M. S., & Eddaoudi, M., A Supermolecular Building Approach for the Design and Construction of Metal—Organic Frameworks. *Chem. Soc. Rev.*, *43*(16), 6141, 2014.

[42] Mahmood, A., Guo, W., Tabassum, H., & Zou, R., Metal-Organic Framework-Based Nanomaterials for Electrocatalysis. *Adv. Energy Mater.*, *6*(17), 1600423, 2016.

[43] Jiao, L., Seow, J. Y. R., Skinner, W. S., Wang, Z. U., & Jiang, H. L., Metal—Organic Frameworks: Structures and Functional Applications. *Mater. Today*, *27*, 43, 2019.

[44] Furukawa, H., Cordova, K. E., O'Keeffe, M., & Yaghi, O. M., The Chemistry and Applications of Metal-Organic Frameworks. *Science*, *341*(6149), 2013.

[45] Khokhar, S., Anand, H., & Chand, P., Current Advances of Nickel Based Metal Organic Framework and Their Nanocomposites for High Performance Supercapacitor Applications: A Critical Review. *J. Energy Storage*, *56*, 105897, 2022.

[46] Wang, B. R., Hu, Y., Pan, Z., & Wang, J., MOF-Derived Manganese Oxide/Carbon Nanocomposites with Raised Capacitance for Stable Asymmetric Supercapacitor. *RSC Adv.*, *10*(57), 34403, 2020.

[47] Yin, X., Liu, H., Cheng, C., Li, K., & Lu, J., MnO2 Nanosheets Decorated MOF-Derived Co3O4 Triangle Nanosheet Arrays for High-Performance Supercapacitors. *Mater. Technol.*, *37*(12), 2188, 2021.

[48] Zheng, S. Q., Lim, S. S., Foo, C. Y., Haw, C. Y., Chiu, W. S., Chia, C. H., & Khiew, P. S., Fabrication of Sodium and MoS2 Incorporated NiO and Carbon Nanostructures for Advanced Supercapacitor Application. *J. Energy Storage*, *63*, 106980, 2023.

[49] Shen, J., Wang, P., Jiang, H., Wang, H., Pollet, B. G., Wang, R., & Ji, S., MOF Derived Graphitic Carbon Nitride/Oxygen Vacancies-Rich Zinc Oxide Nanocomposites with Enhanced Supercapacitive Performance. *Ionics*, *26*(10), 5155, 2020.

[50] He, D., Gao, Y., Yao, Y., Wu, L., Zhang, J., Huang, Z. H., & Wang, M. X., Asymmetric Supercapacitors Based on Hierarchically Nanoporous Carbon and ZnCo2O4 from a Single Biometallic Metal-Organic Frameworks (Zn/Co-MOF). *Front. Chem.*, *8*, 2020.

[51] Wang, L., Jiao, Y., Yao, S., Li, P., Wang, R., & Chen, G., MOF-derived NiO/Ni Architecture Encapsulated into N-doped Carbon Nanotubes for Advanced Asymmetric Supercapacitors. *Inorg. Chem. Front.*, *6*(6), 1553, 2019.

[52] Adil, M., Olabi, A. G., Abdelkareem, M. A., Alawadhi, H., Bahaa, A., ElSaid, K., & Rodriguez, C., In-Situ Grown Bimetallic FeCu MOF-MXene Composite for Solid-State Asymmetric Supercapacitors. *J. Energy Storage*, *68*, 107817, 2023.

[53] Wu, W., Liu, T., Diwu, J., Li, C., & Zhu, J., Metal-Organic Framework—Derived NiCo2S4@Co3S4 Yolk-Shell Nanocages/Ti3C2Tx MXene for High-Performance Asymmetric Supercapacitors. *J. Alloys Compd.*, *954*, 170213, 2023.

[54] Kitchamsetti, N., & Kim, D., A Facile Method for Synthesizing MOF Derived ZnCo2O4 Particles on MXene Nanosheets as a Novel Anode Material for High Performance Hybrid Supercapacitors. *Electrochim. Acta*, *441*, 141824, 2023.

[55] Parse, H. B., Patil, I., Kakade, B., & Swami, A., Cobalt Nanoparticles Encapsulated in N-Doped Carbon on the Surface of MXene (Ti3C2) Play a Key Role for Electroreduction of Oxygen. *Energy Fuel.*, *35*(21), 17909, 2021.

[56] Xie, W., Wang, Y., Zhou, J., Zhang, M., Yu, J., Zhu, C., & Xu, J., MOF-Derived CoFe2O4 Nanorods Anchored in MXene Nanosheets for All Pseudocapacitive Flexible Supercapacitors with Superior Energy Storage. *Appl. Surf. Sci.*, *534*, 147584, 2020.

[57] Yang, Z., Wang, C., & Lu, X., Conducting Polymer-Based Peroxidase Mimics: Synthesis, Synergistic Enhanced Properties and Applications. *Sci. China Mater.*, *61*(5), 653, 2018.

[58] Chang, Y. L., Tsai, M. D., Shen, C. H., Huang, C. W., Wang, Y. C., & Kung, C. W., Cerium-Based Metal—Organic Framework-Conducting Polymer Nanocomposites for Supercapacitors. *Mater. Today Sustain.*, *23*, 100449, 2023.

[59] Wang, Q., Shao, L., Ma, Z., Xu, J., Li, Y., & Wang, C., Hierarchical Porous PANI/MIL-101 Nanocomposites Based Solid-State Flexible Supercapacitor. *Electrochim. Acta*, *281*, 582, 2018.

[60] Zhang, H., Wang, J., Sun, Y., Zhang, X., Yang, H., & Lin, B., Wire Spherical-Shaped Co-MOF Electrode Materials for High-Performance All-Solid-State Flexible Asymmetric Supercapacitor Device. *J. Alloys Compd.*, *879*, 160423, 2021.

[61] Jafari, E. A., Moradi, M., Borhani, S., Bigdeli, H., & Hajati, S., Fabrication of Hybrid Supercapacitor Based on Rod-Like HKUST-1@polyaniline as Cathode and Reduced Graphene Oxide as Anode. *Phys. E: Low-Dimens. Syst. Nanostructures.*, *99*, 16, 2018.

[62] Cheng, Q., Tao, K., Han, X., Yang, Y., Yang, Z., Ma, Q., & Han, L., Ultrathin Ni-MOF Nanosheet Arrays Grown on Polyaniline Decorated Ni Foam as an Advanced Electrode for Asymmetric Supercapacitors with High Energy Density. *Dalton Trans.*, *48*(13), 4119, 2019.

[63] Deka, R., Mal, D. D., & Mobin, S. M., Upcycling Rust and Plastic Waste into an Fe MOF for Effective Energy Storage Applications: Transformation of Trash to Treasure. *Dalton Trans.*, *52*(24), 8204, 2023.

[64] Lagae-Capelle, E., Cognet, M., Madhavi, S., Carboni, M., & Meyer, D., Combining Organic and Inorganic Wastes to Form Metal—Organic Frameworks. *Mater.* *13*(2), 441, 2020.

[65] Ren, J., Dyosiba, X., Musyoka, N. M., Langmi, H. W., North, B. C., Mathe, M., & Onyango, M. S., Green Synthesis of Chromium-Based Metal-Organic Framework (Cr-MOF) from Waste Polyethylene Terephthalate (PET) Bottles for Hydrogen Storage Applications. *Int. J. Hydrog. Energy*, *41*(40), 18141, 2016.

[66] Karam, L., Miglio, A., Specchia, S., El Hassan, N., Massiani, P., & Reboul, J. PET Waste as Organic Linker Source for the Sustainable Preparation of MOF-Derived Methane Dry Reforming Catalysts. *Mater. Adv.*, *2*(8), 2750, 2021.

[67] Jamil, U., Husain Khoja, A., Liaquat, R., Raza Naqvi, S., Nor Nadyaini Wan Omar, W., & Aishah Saidina Amin, N., Copper and Calcium-Based Metal Organic Framework (MOF) Catalyst for Biodiesel Production from Waste Cooking Oil: A Process Optimization Study. *Energy Convers. Manag.*, *215*, 112934, 2020.

[68] Li, B., Dong, X., Wang, H., Ma, D., Tan, K., Jensen, S., Deibert, B. J., Butler, J., Cure, J., Shi, Z., Thonhauser, T., Chabal, Y. J., Han, Y., & Li, J., Capture of Organic Iodides from Nuclear Waste by Metal-Organic Framework-Based Molecular Traps. *Nature Commun.*, *8*(1), 2017.

[69] Abdollahi, N., Moussavi, G., & Giannakis, S., A Review of Heavy Metals' Removal from Aqueous Matrices by Metal-Organic Frameworks (MOFs): State-of-the Art and Recent Advances. *J. Environ. Chem. Eng.*, *10*(3), 107394, 2022.

[70] Pandis, P. K., Kalogirou, C., Kanellou, E., Vaitsis, C., Savvidou, M. G., Sourkouni, G., Zorpas, A. A., & Argirusis, C. Key Points of Advanced Oxidation Processes (AOPs) for Wastewater, Organic Pollutants and Pharmaceutical Waste Treatment: A Mini Review. *Chem. Eng.*, *6*(1), 8, 2022.

[71] Vaitsis, C., Kanellou, E., Pandis, P. K., Papamichael, I., Sourkouni, G., Zorpas, A. A., & Argirusis, C., Sonochemical Synthesis of Zinc Adipate Metal-Organic Framework (MOF) for the Electrochemical Reduction of CO2: MOF and Circular Economy Potential. *Sustain. Chem. Pharm.*, *29*, 100786, 2022.

[72] Mohamed, S. A., Kim, Y., Lee, J., Choe, W., & Kim, J., Understanding the Structural Collapse during Activation of Metal—Organic Frameworks with Copper Paddlewheels. *Inorg. Chem.*, *61*(25), 9702, 2022.

13 A Circular Economy Approach to Solar Energy

*Ashok Prabhakar**

13.1 INTRODUCTION

The utilization of solar photovoltaics (PV) has shown to be an important factor in the switch to renewable energy needed to reduce climate change during the past several years, and it will continue in a cost-effective manner [1]. The SDGs goal of the United Nations may be attained but, largely it depends on solar panel usage [2]. The US Department of Energy predicts that 30 years from now PV cell and CSP may be utilized to increase the power generation up to 1.6 terawatts (TW) [3].

13.1.1 SOLAR ENERGY DEMAND, PRODUCTION, AND WASTE GENERATION

In the last few years, the installed solar PV capacity has grown quickly throughout the world. Over 126 GW of solar PV capacity was added worldwide in 2020, according to IRENA [4]. Since 2011, when the capacity was at 73 GW, more than 713 GW of solar power has been built, multiplying about ten times over the last ten years, expanding at an average of 70 GW each year. Beginning in 2026, it is anticipated that approximately 70 GW of solar capacity will need to be retired each year due to the solar modules' approximate 25-year life span, after which they become unprofitable. This ends with 280 million panels or around 4 million panels per GW as waste [ARENA]. Solar PV systems are among those posing some of the biggest difficulties.

By the early 2030s, it is expected that there will be significant annual waste (1.7–8 million tons) generation due to PV market and the volume of retired PV panels. This number will approach 60–78 million tons of waste by 2050 [5,6]. Much of the anticipated garbage will come from G20 member countries. Solar energy might make sense economically, but environmental performance has shown it to be unfavorable. Higher levels of circularity result from keeping products and components in use in addition to enhanced recycling possibilities. Reuse and repair techniques can extend the life span of PV panels. The waste material generation from PV panel is forecasted up to 2050 and is depicted in Figure 13.1. It is predicted that 4% of the installed PV in the 2030s will be retired each year. Due to the continually rising demand, waste levels will also rise; by 2050 [5,7].

In the United States alone, this amount is predicted to be between 7.5 and 10 million tonnes. According to estimates made by IRENA [4] and the IEA-PVPS (2016) [6],

* ashok04che@gmail.com

DOI: 10.1201/9781003269779-13

Predicted Waste Generation

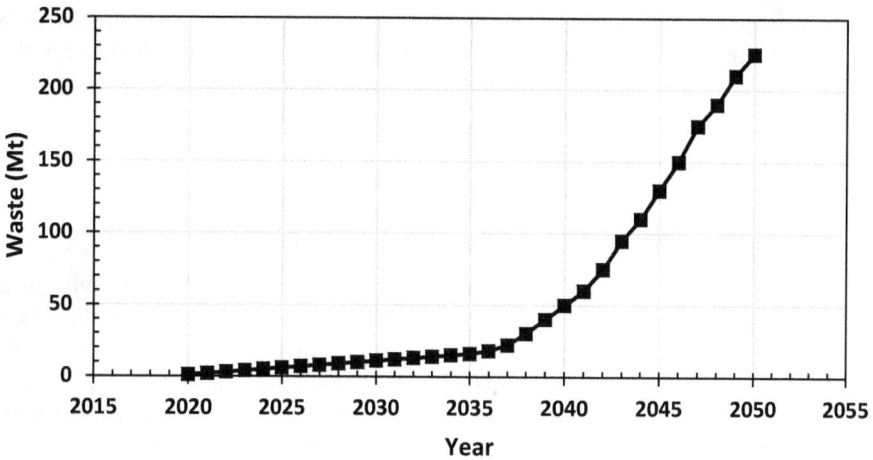

FIGURE 13.1 Cumulative waste projection from solar PV panels [4].

TABLE 13.1
Crystalline Silicon Photovoltaic Panel: Major Components [8]

Components	% wt.
Glass	70 (Sb 0.1%–1% of glass)
Photovoltaic frame	18 (Al)
Encapsulation layer	5.1
Silicon	3.65
Back-sheet layer	1.5 (polyvinyl fluoride)
Cables	1 (Cu and polymers)
Internal conductor	0.53 (Al)
Internal conductor	0.11 (C)
Ag	0.053
Other metals	0.053 (Sn, Pb)

if these end-of-life (EOL) PVs were to be recovered, they could provide enough raw material to produce new PV panels with a capacity equal to 630 GW. The major components and composition of a type of c-Si PV technology are provided in Table 13.1.

13.1.2 SOLAR WASTE HANDLING

The sincere and systematic management of PV panels presents enticing chances to recycle important materials and explore new business prospects. Using the concepts of the circular economy (CE), used solar PV panels can be repaired, reused, and recycled. By 2050, nearly $8.8 billion value addition will be received from the recycling of the discarded PV panels. A comprehensive and robust regulatory framework must be in place now to address the solar waste issue and this can be done via various stakeholders (i.e., government, industry, public-private joint efforts) to utilize the highest potential of the CE. Moreover, apart from policy framework the technical aspect of dealing with solar waste, also need the detailed investigation. Reuse and repair techniques can extend the life span of PV panels [9] as they could promote resource-efficient product design, product reuse [10] and value addition from byproducts or "waste" [11].

13.1.2.1 Business Model for Circular Economy

According to [12] circular business models seek to maximize the longevity of a product's and material's value while minimizing their negative effects on the environment. The product-service system (PSS) business model is one of the more well-known circular business models. They reflect a range of value propositions that businesses can provide to their clients, constantly combining aspects from both products and services.

According to Tukker (2015) [13], PSS are able to separate revenue generating from material and product use as well as waste production. This is because most PSS models still hold the manufacturer accountable for the product's use and EOL phase, add additional value to extend product life, and make decisions to opt for new design that support to successfully implement CE strategies like redoing of waste uses via its repairing, remanufacturing, or recycling. However, because PSS models do not necessarily imply the CE or sustainability results, they must be properly created [14]. The challenge of increasing the uptake of solar PV as well as the waste issue associated with solar PV investments may be addressed by CE strategies and business models in order to meet Sustainable Development Goal (SDG) 12. Furthermore, according to recent research, organizations that adopted circular strategies were able to maintain their resilience during the COVID-19 outbreak [15]. Finally, with the capital investment of durable goods, organizational market segments can struggle with a divided incentive problem. Split incentive issues arise when property owners are unable to fully appreciate the advantages of their investments, leading to less than ideal investment choices for both owners and users. They are especially prevalent when decisions about energy use and investments have unrelated financial effects, which is frequently the case in market segments of big organizations [16].

The majority of research focuses on the PV recycling technique to successfully implement the CE approach [17,18].

13.2 CIRCULAR ECONOMY STRATEGY

13.2.1 VARIOUS PERSPECTIVES OF THE CIRCULAR ECONOMY

Since PV panels include essential raw materials, dangerous substances, and high-value elements like glass, silver, silicon, and lead, recycling has historically been the standard procedure for decommissioned PV. The PV panel laminated structure creates difficult to separate components, and is a significant burden to date [18]. The CE perspective also includes the repairing and reuse of those PV panels which are yet to reach their end of technical life. Based on the idea of maintaining items at their original and optimum value addition, for longest feasible time, as CE emphasizes the utilization of whole technical life time, rather than recycling [10]. According to life cycle assessment (LCA) results, the 30-year technical life of PV panels through reuse or life extension is also advantageous from a more general sustainability standpoint. Repowering the installation has a greater environmental impact per kWh of electricity produced because it requires replacing and recycling panels more frequently (every 10 or 15 years). Even when component repairs are necessary or when panels must be carried over long distances for reuse, the benefits of recycling or the higher efficiency of the new panels do not fully offset this [19]. According to estimates from IRENA and IEA-PVPS (2016) [20], up to 80% of the PV waste stream may contain goods that were flawed or failed during manufacture, shipping, or the first few years of use. About 45%–65% of these panels can be fixed or renovated, according to partners in the Horizon 2020 CIRCUSOL project and consultants' estimates [10]. However, there are still a number of underutilized value creation opportunities that can be realized through PV reuse in circular business models [21,22]. Latunussa et al., summarized the list of recovered materials mentioned in Table 13.2 per ton of solar modules [8].

For the implementation of the circular business model, there are several factors starting from supply chain and sourcing management to challenges associated with regulations along with restrictions in financial access and less acceptance in the market [23,24]. Additionally, there are the usual obstacles to implementing circularity throughout supply chains. Investment hazards, lack of waste handling, low quality resource, a lack of market demand, less acceptance, and a lack of understanding of data technology prospects are a few of these [25].

Big data, artificial intelligence, the internet of things, data analytics, blockchain, and smart products are just a few examples of the technologies that make it possible

TABLE 13.2
Various Components Extracted from 1 Ton of PV Panels (EOL) [8]

Materials	Unit
Al (Resource)	0.1826 ton
Glass (Resource)	0.686 ton
Cu (Resource)	0.00438 ton
Si (Resource)	0.03468 v
Ag (Resource)	0.0005 ton
Fly ash (Resource)	0.002 ton
Liquid waste ($H_2O + 2Ca(OH)_2 + HNO_3$ (Resource)	0.30613 ton
Electricity (Resource)	248840 KJ
Heat Energy (Resource)	502840 KJ
Sludge (Waste)	0.05025 ton
Dirty glass (Waste)	0.014 ton

for businesses to optimize their processes and systems [26], reduce waste, encourage the extension of product life through improved performance and early maintenance of products, and lower transaction costs [27].

13.3 PATHWAYS TO CIRCULAR ECONOMY VIA PV MAKING/RECYCLING

To enhance the economic and to make environmentally friendly PV systems, a wide variety of CE techniques can be implemented into the PV production process.

13.3.1 MATERIAL STRENGTH AND RECYCLING

Prior to the official development and integration of CE practices, efforts to lower the material [28,29] and energy [30,31] intensity of PV manufacture over the past two decades have produced positive economic and environmental outcomes [31,32]. A sizeable portion (40%) of the mass of high-demand silicon (solar-grade) is lost as cutting loss during the wafer sawing from the c-Si ingot (which accounts for 25% of the total PV cell cost), according to the ITRPV52. Research and development (R&D) has focused on kerf loss reduction by switching to better sawing methods and kerf-free process of wafering [33,34]. Along with R&D, the reutilization of silicon with improved standards and guidelines recovered from PV panel production waste will speed up CE practices. Impurities that could weaken the PV cell are a major apprehension in the reutilization of silicon recovered through kerf losses and ingot cuts performance [35]. The verification of purity and characteristics of recovered silicon need to be rechecked with the virgin silicon to ensure the degree and type of contaminants from kerf losses [36].

Through R&D, which aimed at improving the recovery process to reduce impurities and analysing the balance among cell performance, material costings, and effects on environmental for virgin silicon (solar-grade) to secondary silicon [37] considering range of silicon manufacturing conditions for circular economy. Additionally, using kerf loss as feedstock for other applications (such as hydrogen production [38] or Li-ion batteries) [39] may be more advantageous from an economic and environmental standpoint than landfilling. Materials extracted from a PV at EOL are used again in the production of PVs in closed-loop recycling. Apart from silicon, other material recovery (such as glass and silver) can also be extracted and reused for PV making which will benefit the energy return and ultimately environmental protection by reducing greenhouse gases (GHG), but several studies need to be done before its final implementation.

13.3.2 REUTILIZATION OF MATERIALS FROM OTHER INDUSTRIES VIA RECYCLING

The recycling technique provides huge potential for the reutilization of materials extracted from non-PV systems in PV manufacture, which can be a more advantageous way to get raw materials from an economic and environmental standpoint. Post-consumer plastic trash, for instance, can be recycled in the manufacturing of PV module encapsulants [40].

13.3.3 USE OF ELECTRICITY FROM RENEWABLE ENERGY SOURCES

Energy is one of the highly impactful part for the success of solar circular economy. More than 50% of the energy footprint along with climate effect of c-Si PV panel is due to energy consumed in the extraction and getting silicon as pure material [41]. The climate footprint of PV systems can be greatly increased by the CO_2 released during the early phases of the PV life cycle [37,42]. The climatic footprint of PV modules can be greatly reduced by using electricity from renewal source which will help boost CE strategy [37]. Industry is looking into opportunities to reduce its carbon footprint as a result of the potential to use renewable electricity, the manufacturing process can be made carbon free [43]. A promising area of research is creating market-recognized labels and valuations for the advantages of lower GHG emissions along with energy, which should be based on new, reliable quantitative indicators. Collaboration with stakeholders from business and non-governmental organizations is the greatest way to achieve direction.

13.3.4 DESIGN OF PV MODULE FOR CIRCULAR ECONOMY

In order to make PV cost-competitive with other electricity sources, the traditional method of developing PV modules must be replaced by the latest techniques to lower the manufacturing price, to increase durability and dependability so as to enhance the efficiency of the PV module [44]. As global PV installations have reached several terawatts, the key parameter of PV system design must be redefined to address the sustainable challenges. This is necessary because the stock of raw materials needed and waste production increased exponentially [45]. PV waste recycling [46–49] can be hampered by factors like dangerous components in c-Si PV modules (such as lead), fluorine, and difficulties removing the ethylene vinyl acetate (EVA) laminate.

These new sustainability problems can be addressed by innovative PV module designs and materials to improve recyclability, switch from materials with limited supply to materials with abundant supply [50], and lessen hazards to environment and human well-being. The choice of nonhazardous and ecologically friendly materials can be guided by machine learning (ML) and artificial intelligence (AI) techniques during the construction of PV modules [51]. Utilizing recyclable components in the PV module can improve recycling at EOL and reduce landfilling [52].

PV systems can be produced [53]. for less money by substituting plentiful materials for limited ones. The apprehensions for plant life and human being throughout the usage and EOL stages can be reduced by replacing hazardous chemicals in the PV module. For c-Si modules, switching to a fluorine-free design instead of fluorinated backsheets can reduce the dangers to human wellbeing and enable recycling methods at high-temperature for quicker and more effective recycling of the used PV module [48–49]. Eliminating lead solder can avoid possible lead emissions during thermal recycling and possibly avoid the classification of c-Si PV modules as hazardous waste [54]. Frameless modules make recycling easier, have less of an environmental and energy impact, and contain less aluminum. They also require less transportation and don't need to be deframed before recycling [55] an unlaminated construction [56] or by removing the requirement for the thermal, chemical, or mechanical operations necessary to remove the EVA during PV recycling, edge sealants instead of EVA save the time and energy needed for recycling [54]. The examples concentrate on crystalline silicon, the most popular module technology now. If newer technologies, such as perovskite, get a large market share, other difficulties may arise or cease to be pertinent. Design for circularity techniques might need trade-offs in other PV system life cycle stages. Copper metallization has the potential to reduce PV module durability and cell performance [57]. A laminate-free design [58] affects the PV module's longevity and energy generation profile, which affects the system's performance overall. Lead-free substitutes may be more expensive and raise soldering temperatures, which could lead to thermal-mechanical stress and breaking of the silicon wafer during manufacture [59]. According to preliminary field testing [60], modules with fluorine-free backsheets are less durable as compared to fluorinated one.

A comprehensive evaluation of the trade-offs that material and design decisions impose on the module's technical performance (such as its ability to generate power) and the life cycle impact [61,62] will aid in choosing the circularity alternative's most environmentally friendly design. This will give the design for circularity techniques top priority since they produce the greatest overall net economic and environmental benefit for the PV system.

13.3.5 SUSTAINABLE ROLE OF PURCHASE FOR CIRCULAR ECONOMY

To increase sustainability of PV modules during the journey of PV supply chain will need to be environmentally friendly and PV suppliers should incorporate sustainable procurement requirements as outlined in the regulations [63,64].

13.3.5.1 Low-Carbon Solar to Mitigate GHG Emissions

The circular economy strategy will work based on two precautions: (1) reduction in the utilization of CO_2–based fuel and (2) Green electricity (less carbon) in PV making. Studies have shown that placing PV manufacturing in less CO_2-intensive regions considerably reduces energy and GHG payback times [37,41,65]. Based on these findings, certain PV manufacturers and supply chain participants are currently implementing CE measures to reduce the embodied carbon of PV [43]. Recovery + reutilization of scrap and EOL materials can both offer similar advantages, but they both need to be quantified and promoted to enhance awareness and uptake.

13.3.5.2 Modules (PV) Free from Fluorine and Lead

R&D has concentrated on lowering the hazardous elements in PV panel to bad impact on environment and human health concerns during usage and recycling [66]. Prediction indicates that the level of lead and fluorine will decline as the focus of industry is to produce PV panels free of these elements [55]. This will lead to the satisfaction and motivation of the customer that the EOL of such modules are non-hazardous and will result into reduced cost and lesser impact on environment. Suppliers will also be motivated to be part of the lead and fluorine–free module business, which will help them to get a high score of sustainability, among all others [67,68].

13.3.5.3 Environmental and Social Benefit

PV manufacturing CE techniques have a great potential to boost social benefits and environmental justice outcomes. Utilizing renewable energy sources lessens dependency on fossil fuels, hence reducing climate change and its repercussions, including fatalities due to the release of air pollutants from the burning of fossil fuels, which disproportionately affect low-income and minority groups as well as the developing globe, aggravating socioeconomic inequalities [69,70]. CE strategies in PV manufacturing can significantly reduce the likelihood of environmental and health hazards, which have in the past had an impact on communities near PV manufacturing facilities, by adhering to closed-loop recycling and putting more emphasis on substituting hazardous materials with environmentally benign materials in the supply chain.

The rankings of socially responsible PV suppliers published by scorecards [68] encourage supply chain transparency to avoid the sourcing of conflict minerals and prevent violations of worker rights [71] and health and safety regulations. Additionally, the creation of industrial alliances that prioritize CE techniques throughout the PV supply chain might boost the competitiveness of PV manufacturing and, consequently, the potential for job growth in the US PV sector [72]. All CE techniques that lower material needs also lower the costs felt in frontline communities close to the resource extraction businesses.

13.4 KEY ASPECTS AT END-OF-LIFE USES

Crystalline silicon is anticipated to make up 85% of overall installations, based on historical data of the US market; projected PV EOL materials by scenario through 2050 are shown in Figure 13.2. For utility scale installations, PV module life spans

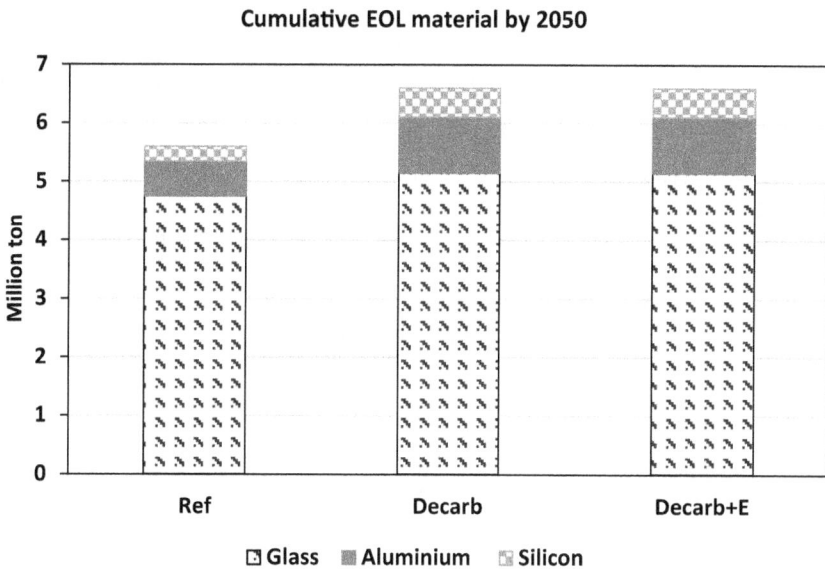

FIGURE: 13.2 PV EOL material mass by Solar Futures Study scenario through 2050 [74].

Cumulative Cumulative manufacturing scrap by 2050

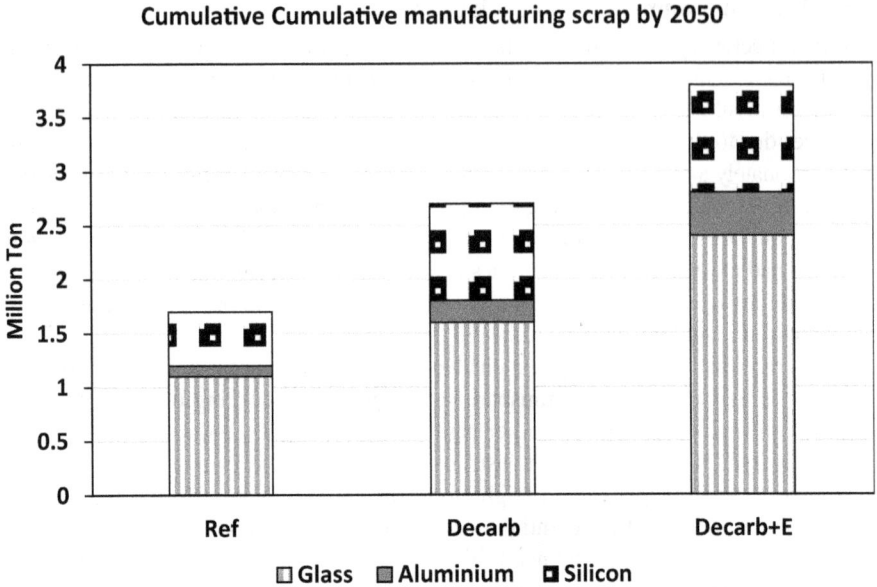

FIGURE 13.3   Mass of c-Si PV manufacturing scrap by Solar Futures Study [74].

have reached 32 years or more currently [73]. This indicates that a module placed by the year 2028 will remain valid till 2060. As a result, fewer modules used in these scenarios will become obsolete by 2050. The pre-2050 EOL modules in figure 13.2, are a result of older installations and early failures [74]. Both the Decarb and Decarb+E scenarios from the Solar Futures Study produce roughly 6.43 million metric tonnes of total EOL material in 2050. Glass brings up a sizable portion of the total EOL materials in all scenarios since it makes up a sizable portion of PV module weight.

PViCE (PV in the Circular Economy model) has the distinction of accounting for manufacturing scrap separately and in addition to EOL materials. The scrap from manufacturing industry is half of the waste obtained from EOL but is more closely timed to meet the demand for virgin materials, highlighting the significance of effective manufacturing and recycling of scrap to reduce the need for virgin materials [75]. Due to manufacturing inefficiencies, such as ingot wafering, silicon makes up a sizable amount of manufacturing trash, whereas glass manufacture is a more resource-efficient manufacturing method. The cumulative amount of silicon scrap generated during production is actually more than the amount created by EOL modules, which shows that the majority of modules used in study scenarios of solar future and will still be in use by 2050. The greatest manufacturing scrap both annually and overall, as was predicted, is shown in Figure 13.3 [74].

PViCE's ability to track a deployed cohort by location is another innovative feature. The Southeast is the region with the highest variation between scenarios in terms of relative magnitude differences. Such geographical outcomes could help stakeholders plan ahead for EOL materials on a regional level, which could result in a more effective use of funding for infrastructure for recycling and other EOL management processes. The development of an effective infrastructure can quicken trends towards a sustainable CE. For instance, tighter circular loops could prolong the field life of PV modules and generate higher-skilled local jobs in the sustainability industry. Industries (like glass manufacture) might be able to exploit a EOL supply of local materials from PV panel if circular loops are extended [75].

It is possible to identify potential routes towards circular, symbiotic, cross-sector economic opportunities by conducting further studies of recycling, PV module life extensions, and the accompanying social, environmental, and economic implications of each. This will help communities handle the difficulties they will face in the future decades. Finally, replacing EOL materials with virgin materials is a significant prospective method to reduce material demands in a developing PV market while enhancing supply chains' robustness. A policy like this would also lessen the costs mining has on the environment and social justice, provide markets for recycling facilities wanting to sell recycled products, lessen the demand for essential commodities, and so on. Decommissioning will surpass installations in 2032 for reference scenario and by 2038 for Decarb+E. Thus, a portion of the required PV installations could be offset by module reuse, refurbishment, and recycling, especially after 2030.

As reported the silver, aluminum, and silicon material requirements for the Decarb+E scenario can be partially satisfied with EOL materials. Less than 20% of the material requirements prior to 2040 can be met by EOL material at the high deployment pace. When deployment slows down after 2040, EOL material can meet 25% to 30% of these material demands. The cohort structure of PViCE considers the material composition of the generations of PV modules that are being retired for determining EOL material. The annual deployment rate looks to be very changeable as a result of ReEDS simulation and optimization techniques [75]. The abrupt drop in the deployment rate in 2040 is not industry feasible and needs to be viewed as an anomaly.

13.4.1 Addressing EOL Issues for Circular Economy

13.4.1.1 Recycling

The PV CE approach that has been used and studied the most is recycling. A thorough examination of the research on recycling of crystalline silicon PV modules demonstrates that R&D has focused heavily on recovering materials [76]. In recycling systems, materials extracted from PV employed in non-PV applications, according to the literature. The recovered materials from crystalline silicon PV modules can be used for various other industries such as battery, cement, ceramic, paper, tiles, clay bricks, and medicine [77–80]. On the other hand, there is potential for reusing components from non-PV items in PV systems. The recent research findings identified problems for c-Si PV recycling by thoroughly evaluating current R&D initiatives and business operations, which are mentioned below [81,82].

Some trace materials which are present in a c-Si module (tin, Pb, Cu, and Ag), at much lower quantities than glass and silicon, have received less attention from research and development. The lack of emphasis on R&D may be linked to mass-based recovery standards stipulated in PV recycling rules, which require that at least a part will be recycled. For instance, a WEEE directive stipulates that 85% and 80% of the mass of a PV module, respectively, must be recovered or recycled [10]. Because bulk materials make up around 90% of the module weight, they are prime candidates for recycling [83]. However, the exponential rise in PV waste could spur a broadening of the standards' application to embrace both bulk and trace material recovery. This will assist in the environmentally appropriate management of hazardous compounds, such as lead, and reduce toxicity and environmental concerns in the event of wrong management at EOL [84]. The trace components are recovered in this investigation as part of bigger aggregates that also contain other bulk minerals, which makes them unsuitable for direct reuse and necessitates additional downstream processing [85]. Additionally, only 33.2% of the module's silver mass is retrieved, which could result in a loss of money.

The deficiencies highlight the need for a low-cost, process recycling process to recover elements from c-Si PV modules with high efficiencies [81]. There aren't many thorough evaluations of the PV recycling industry's economic feasibility that are publicly accessible. Such evaluations could be used to improve the policies and incentives for moving to a CE for PV systems. The income generated by the recovered material from recycling process is not self-sufficient to run this, hence additional fees are required to sustain this process. Estimations reveal significant heterogeneity in recycling costs according on the recycling technology [86,87] even if commercial operations already exist [82,88].

A repository of the various PV module designs will help to continuously update and could be useful for addressing recycling difficulties caused by the variety in module design and material content. The repository might make it easier for PV manufacturers and recyclers to work together and share information, address manufacturer concerns about data privacy, increase transparency about the design and material composition of old and new PV modules, guarantee repeatability of results and compliance with toxicity tests (like TCLP259), and help better tailor recycling strategies. Through ecolabels, material passports, radio-frequency identification, and bill of materials, manufacturers can communicate with recyclers [89,90].

In order to meet the increasing demand for PV capacity additions worldwide, waste modules will need to be collected and transported from various sites to centralized recycling facilities. Transporting PV modules produces carbon dioxide, which can be reduced by the module recycling at the installed location. Recycling in smaller-scale facilities, however, will forgo the chance to benefit from economies of scale in a centralized recycling plant. As a result, when switching to a CE for PV systems, there is a possibility to situate recycling facilities in the best locations to reduce the financial and environmental costs of PV recycling [91–94].

13.4.1.2 Repairs

Without processing the module destructively, repair can fix flaws in components including the junction box, backsheet, bypass diodes, encapsulant, broken glass, and connections [10]. The module degradation, improved power generation, improved strength, and reliability can be well handled by the repairing process. External components repairing, such as backsheets and junction boxes, is important as it extends the life span of PV panels, which is a crucial CE approach, and avoids the financial and environmental constraints [95]. The potential of PV system repair estimations indicating that the market for PV repair and maintenance is anticipated to be worth $9 billion by 2025 serve as evidence [96]. Additionally, the cost barrier to purchasing PV systems in price-sensitive regions is lowered by the availability of repaired PV modules on the used market at lower prices than new PV modules [97].

Despite the potential of repairing and reusing EOL PV modules, there is a lack of data in public to understand the reliability and various standards to ensure the performance and quality. This will help to address mechanism to price these modules and boost market confidence in them. Additionally, it is necessary to thoroughly evaluate [98] and contrast the environmental and financial trade-offs between module repair and reuse and alternative CE measures.

13.4.2 ENVIRONMENTAL AND SOCIAL BENEFIT

As one of the job categories with the quickest rate of growth in the US economy, the additional attributes like recycling of waste material, repair of PV system, and maintenance may result in more employment possibilities [72]. To assess the trade-offs associated with higher employment prospects in repair, which may reduce the amounts of PV waste and, consequently, the creation recycling job or landfilling, more research is needed. Additionally, there is a need to improve planning and cooperation between different stakeholders to identify pertinent skill sets and develop training programmers. The potential harm to human health is reduced and the consequences for environmental justice are enhanced by handling hazardous materials. In the past, minority and low-income populations have disproportionately been affected by the detrimental effects of waste management activities on their health [99].

The development of tools to investigate stakeholder behavioural responses to encouragements (e.g., recycling versus landfilling PV waste), policy inducements, market signals and determine the social outcomes of CE strategies (e.g., reduction in energy poverty) is critical if policies are to be designed that maximize potential social benefits and improve environmental justice outcomes. To ascertain the social effects of CE at PV EOL, frameworks like social LCA, agent-based modelling, discrete event simulation, and social LCA, as well as techniques like ML and AI, are interesting possibilities [100].

13.5 INDIAN PERSPECTIVE OF CIRCULAR ECONOMY IN SOLAR ENERGY

13.5.1 INSTALLED PV CAPACITY IN INDIA

In India, policymakers started talking about the advancement and adoption of solar energy in third term of the Five Year Plan (1961–1966) [101]. Its introduction into the Indian power sector was not a normal progression because it was the latest technological development at that time. The implementation of solar energy as an electricity source took place after many debates and a Commission for Additional Sources of Energy (CASE) was established in the sixth term of the Five Year Plan (i.e., 20 years later, in 1980–1985). Soon after, the NASPAD was put into place, and it allowed the

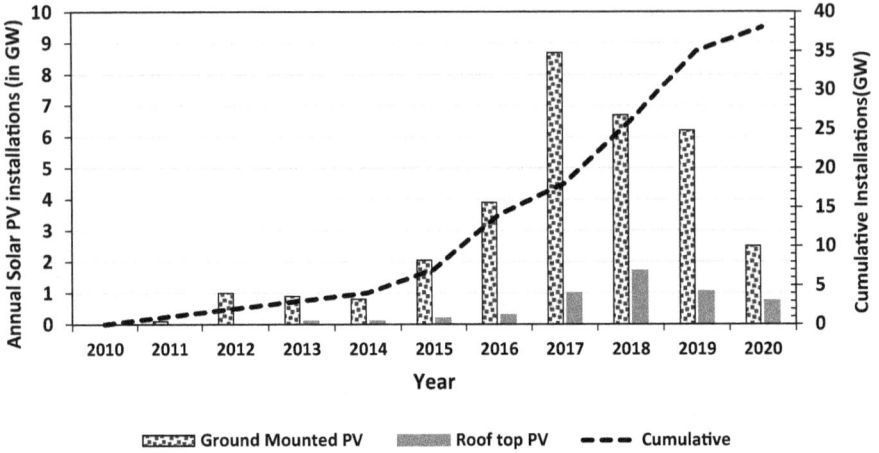

FIGURE 13.4 Installations of rooftop and ground-mounted PV systems 2009–2020 (in GW) [103].

manufacturing of solar PV cells for the first time and modules with capacities of 10.35 kW, 21.07 kW, and 31.75 kW.

As part of the National Action Plan on Climate Change (NAPCC-2008), the government announced the National Solar Mission (JNNSM) in 2010 in the name of Jawaharlal Nehru. By 2022, 22 GW of grid-connected power generation plants were scheduled to be operational [101]. This was the very first solar energy installation plan nationwide by government, albeit having a conservative aim. The goal was reset to 100 GW installation of solar energy by 2022, and this was done by the new government in 2015. Only 3.2 GW of total solar capacity might be installed nationwide by 2014 [102]. However, solar installations experienced exponential growth once the national RE expansion programme was announced in 2015, as shown in Figure 13.4.

13.5.2 INSTALLED PV CAPACITY PROJECTION FOR TARGETED EXPANSION OF RE

With 95% of the global market currently made up of c-Si PV technology, thin amorphous technology still has a very small share of the solar PV market and little is known about its LCA [104]. Therefore, only c-Si PV technology is used in this study to analyse the proportion of the anticipated capacity development. The study employs the global market share as a stand-in for the Indian solar market because market share data for solar PV technologies in the Indian context is lacking. As a result, the capacity expansion for c-Si PV modules is projected to reach 95 GW by 2022, or 95% of the targeted 100 GW solar capacity expansion.

The E3-India model, an effect assessment tool created from the widely used E3ME global model framework, is used to project installed capacity. The model is employed to simulate the consequences of India's energy and economic policies. The model, which is based on the Keynes-Leontief-Klein framework, is made to offer policymakers a multidimensional policy effect analysis to evaluate a programme's

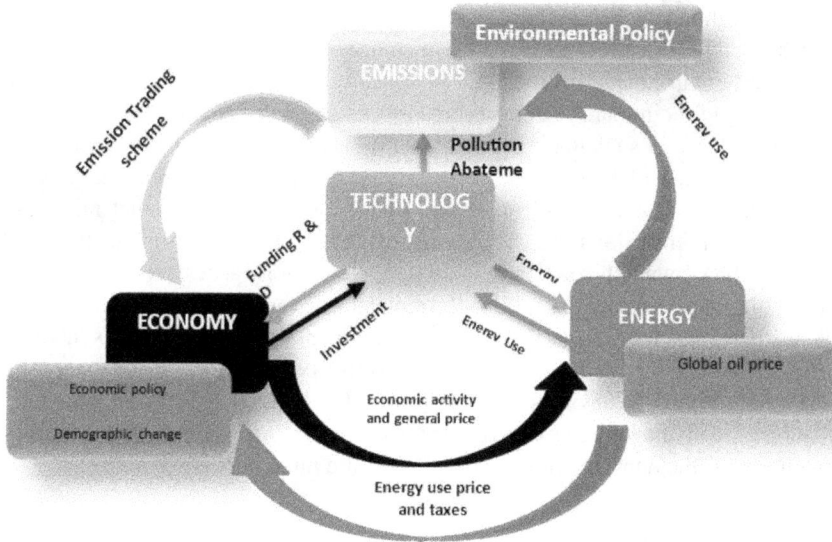

FIGURE 13.5 E3-India model structure [105].

merits from the perspectives of the economy, energy, and environment. Figure 13.5 depicts the feedback mechanism between these three dimensions.

13.6 CONCLUSION

The Solar Futures Study describes how solar technology can reduce GHG emissions from the power industry [1]. Additionally, solar energy reduces air pollutants like NOx and SO_2, which combine with one another to create ozone and fine particulate matter ($PM_{2.5}$) in the environment. According to preliminary analysis, reducing these emissions in accordance with the Solar Futures Study Decarb scenario (reductions in the power sector) and Decarb+E scenario (reductions in the power and transportation sectors) could result in air quality and health benefits between 2021 and 2050 that are worth between $300 billion and $400 billion. Both possibilities drastically reduce water withdrawals from the electricity industry. Regarding these conventional environmental measures, solar deployment under the Solar Futures Study scenarios makes significant contributions to addressing crucial national and worldwide issues like climate change, respiratory health consequences (including premature death), and future water availability under a changing climate.

Based on deployment forecasts made in 2014, a benchmark global projection of PV EOL materials anticipated that 80 million metric tonnes will be produced globally overall by 2050. We discover that combining more precise and up-to-date data on PV reliability and performance, the Solar Futures Study scenarios, and EOL materials from PV modules in the United States are anticipated to account for less than 10% of this global total, or 6.5 million metric tonnes. In a benchmark report, the materials value for all EOL PV modules was estimated as $15 billion, and if necessary, 2 billion new PV modules, with sufficient raw materials our findings were recovered in closed-loop recycling, which is comparable to 630 GW of capacity.

Social and environmental justice can be benefited via solar CE techniques. For instance, design for circularity strategies that reduce the usage of hazardous materials can stop the release of such materials from having harmful health effects. Incentives such as scorecards, ecolabels, and preferred purchasing programmes encourage supply chain and manufacturing transparency, stop the sourcing of conflict minerals, and enhance worker health and safety. Reuse and PSS lower initial investment costs for household PV ownership and may increase PV electricity access. Recycling lessens the social, environmental, and health implications that PV manufacturing supply chains have upstream, boosts domestic employment prospects, creates synergies with related businesses, and guards against harmful health effects from hazardous material releases into the environment. Recycling also stops the unequal responsibilities that historically have been placed on underprivileged communities when landfills and facilities for the handling of hazardous waste are built.

REFERENCES

[1] IEA. *World Energy Outlook 2021*. IEA, Paris, 2021.
[2] Global Solar Council. Solar Power Lights the Way Towards the SDGs with Broad Benefits for Green Recovery Plans [WWW Document]. *PV Magazine*, 2020. www.pv-magazine.com/press-releases/solar-power-lights-the-waytowards-the-sdgs-with-broad-benefits-for-green-recovery-plans/ (accessed 8.31.22).
[3] DOE. *Solar Futures Study*, GO-102021–5621. U.S. Department of Energy, Washington, DC, 2021.
[4] IRENA. www.irena.org/Energy-Transition/Policy/Circular-economy
[5] Gautam, A.; Shankar, R.; Vrat, P. End-of-Life Solar Photovoltaic E-Waste Assessment in India: A Step Towards a Circular Economy. *Sustainable Production and Consumption* 2021, 26, 65–77. https://doi.org/10.1016/j.spc.2020.09.011.
[6] IRENA, IEA-PVPS. *End-of-Life Management: Solar Photovoltaic Panels*. International Renewable Energy Agency and International Energy Agency Photovoltaic Power Systems, 2016. https://iea-pvps.org/key-topics/irena-iea-pvps-end-of-life-solar-pv-panels-2016/
[7] EPRI. *Improving PV Sampling Methods for End-of-Life Leach Testing*, 2016. https://www.epri.com/research/products/000000003002014825
[8] Latunussa, C. E. L.; Ardente, F.; Blengini, G. A.; Mancini, L. Life Cycle Assessment of an Innovative Recycling Process for Crystalline Silicon Photovoltaic Panels. *Solar Energy Materials and Solar Cells* 2016a, 156, 101–111. https://doi.org/10.1016/j.solmat.2016.03.020.
[9] Radavičius, A.; van Heide, W.; Palitzsch, T.; Rommens, J.; Denafas, M. Tvaronavičienė Circular Solar Industry Supply Chain Through Product Technological Design Changes. *IRD* 2021, 3, 10–30. https://doi.org/10.9770/IRD.2021.3.3(1)
[10] Tsanakas, J. A.; Heide, A. van der; Radavičius, T.; Denafas, J.; Lemaire, E.; Wang, K.; Poortmans, J.; Voroshazi, E. Towards a Circular Supply Chain for PV Modules: Review of Today's Challenges in PV Recycling, Refurbishment and Re-Certification. *Progress in Photovoltaics: Research and Applications* 2020, 28 (6), 454–464. https://doi.org/10.1002/pip.3193.
[11] Bocken, N. M. P., et al. Product Design and Business Model Strategies for a Circular Economy. *Journal of Industrial and Production Engineering* 2016, 33, 308–320. https://doi.org/10.1080/21681015.2016.1172124.

[12] Bocken, N.; Antikainen, M. Circular Business Model Experimentation: Concept and Approaches. In Dao, D.; Howlett, R. J.; Setchi, R.; Vlacic, L. (Eds.), *Sustainable Design and Manufacturing 2018, Smart Innovation, Systems and Technologies*, Springer International Publishing, Cham, 2019, pp. 239–250. https://doi.org/10.1007/978-3-030-04290-5_25.

[13] Tukker, A. Product Services for a Resource-Efficient and Circular Economy—A Review. *Journal of Cleaner Production* 2015, 97, 76–91. https://doi.org/10.1016/j.jclepro.2013.11.049; Special Volume: Why have 'Sustainable Product-Service Systems' not been widely implemented?

[14] Moro, S. R.; Cauchick-Miguel, P. A.; de Sousa Mendes, G. H. Adding Sustainable Value in Product-Service Systems Business Models Design: A Conceptual Review Towards a Framework Proposal. *Sustainable Production and Consumption* 2022, 32, 492–504.

[15] Lize, B.; Brusselaers, J.; Vrancken, K. C.; Deckmyn, S.; Marynissen, P. Toward Resilient Organizations after COVID-19: An Analysis of Circular and Less Circular Companies. *Resources, Conservation and Recycling* 2023, 188, 106681.

[16] Stephen, B.; Diana, H. Policy Options for the Split Incentive: Increasing Energy Efficiency for Low-Income Renters. *Energy Policy* 2012, 48, 506–514.

[17] Contreras-Lisperguer, R.; Muñoz-Cerón, E.; Aguilera, J.; de la Casa, J. A Set of Principles for Applying Circular Economy to the PV Industry: Modeling a Closed-Loop Material Cycle System for Crystalline Photovoltaic Panels. *Sustainable Production and Consumption* 2021, 28, 164–179.

[18] Radavičius, T.; van der Heide, A.; Palitzsch, W.; Rommens, T.; Denafas, J.; Tvaronavičienė, M. Circular Solar Industry Supply Chain Through Product Technological Design Changes. *Insights into Regional Development* 2021, 3 (3), 10–30. https://doi.org/10.9770/IRD.2021.3.3(1)

[19] PVPS, IEA. *Preliminary Environmental & Financial Viability Analysis of Circular Economy Scenarios for Satisfying PV System Service Lifetime.* Report IEA-PVPST12–21, 2021. https://iea-pvps.org/wp-content/uploads/2021/11/IEA_PVPS_T12_Preliminary-EnvEcon-Analysis-of-module-reuse_2021_report.pdf

[20] IRENA, IEA-PVPS. *End-of-Life Management: Solar Photovoltaic Panels.* International Renewable Energy Agency and International Energy Agency Photovoltaic Power Systems, 2016. https://www.irena.org/-/media/Files/IRENA/Agency/Publication/2016/IRENA_IEAPVPS_End-of-Life_Solar_PV_Panels_2016.pdf

[21] Lundqvist, H. K. T. *Circular Economy Among Swedish Solar PV Firms*, 2020. https://lup.lub.lu.se/luur/download?func=downloadFile&recordOId=9018867&fileOId=9018878

[22] Rabaia, M. K. H.; Semeraro, C.; Olabi, A-G. Recent Progress Towards Photovoltaics' Circular Economy. *Journal of Cleaner Production* 2022, 373, 133864.

[23] CEPS Barriers and Enablers for Implementing Circular Economy Business Models RR2021–01, 2021, p. 58. https://circulareconomy.europa.eu/platform/sites/default/files/rr2021-01_barriers-and-enablers-for-implementing-circular-economy-business-models.pdf

[24] Van Opstal, W.; Smeets, A.; Duhoux, T.; Le Blévennec, K.; Gillabel, J. Mid-Term Report on Co-Created Business Models VITO. *Mol* 2021. https://scholar.google.com/scholar_lookup?title=Mid-term%20Report%20on%20Co-created%20Business%20Models&publication_year=2021&author=W.%20Van%20Opstal&author=A.%20Smeets&author=T.%20Duhoux&author=K.%20Le%20Bl%C3%A9vennec&author=J.%20Gillabel

[25] Kumar, P.; Singh, R. K.; Kumar, V. Managing Supply Chains for Sustainable Operations in the Era of Industry 4.0 and Circular Economy: Analysis of Barriers. *Resources, Conservation and Recycling* 2021, 164, Article 105215. https://doi.org/10.1016/j.resconrec.2020.105215.

[26] Stankovic, M., et al. Industry 4.0: Opportunities Behind the Challenge. *Background Paper for UNIDO General Conference* 2017, 17.

[27] MacArthur, E. *Foundation Intelligent Assets: Unlocking the Circular Economy Potential*, 2016, p. 74. https://www.ellenmacarthurfoundation.org/intelligent-assets-unlocking-the-circular-economy-potential

[28] International Technology Roadmap for Photovoltaic (ITRPV). *International Technology Roadmap for Photovoltaic (ITRPV) Results 2018*, 2019. https://pv-manufacturing. org/wp-content/uploads/2019/03/ITRPV-2019.pdf

[29] International Technology Roadmap for Photovoltaic (ITRPV). *International Technology Roadmap for Photovoltaic (ITRPV) 2019 Results*, 2020. https://resources.solarbusinesshub.com/ solar-industry-reports/item/international-technology-roadmap-for-photovoltaic-itrpv-2019

[30] Dwarakanath, T. R.; Wender, B. A.; Seager, T.; Fraser, M. P. Towards Anticipatory Life Cycle Assessment of Photovoltaics. In *2013 IEEE 39th Photovoltaic Specialists Conference (PVSC)*. IEEE, 2013, pp. 2392–2393. Tampa, FL, United States Duration: Jun 16 2013– Jun 21 2013.

[31] Bhandari, K. P.; Collier, J. M.; Ellingson, R. J.; Apul, D. S. Energy Payback Time (EPBT) and Energy Return on Energy Invested (EROI) of Solar Photovoltaic Systems: A Systematic Review and Meta-analysis. *Renewable and Sustainable Energy Reviews* 2015, 47, 133–141. https://doi.org/10.1016/j.rser.2015.02.057.

[32] Woodhouse, M.; Smith, B.; Ramdas, A.; Margolis, R. *Crystalline Silicon Photovoltaic Module Manufacturing Costs and Sustainable Pricing: 1H 2018 Benchmark and Cost Reduction Road Map*, National Renewable Energy Laboratory, 2019. Denver West Parkway Golden, CO 80401 303-275-3000, https://www.nrel.gov/docs/fy19osti/72134.pdf

[33] Kumar, A.; Melkote, S. N. Diamond Wire Sawing of Solar Silicon Wafers: A Sustainable Manufacturing Alternative to Loose Abrasive Slurry Sawing. *Procedia Manufacturing* 2018, 21, 549–566.

[34] Li, J.; Lin, Y.; Wang, F.; Shi, J.; Sun, J.; Ban, B.; Liu, G.; Chen, J. Progress in Recovery and Recycling of Kerf Loss Silicon Waste in Photovoltaic Industry. *Separation and Purification Technology* 2021, 254. https://doi.org/10.1016/j.seppur.2020.117581.

[35] Davis, J. R.; Rohatgi, A.; Hopkins, R. H.; Blais, P. D.; Rai-Choudhury, P.; McCormick, J. R.; Mollenkopf, H. C. Impurities in Silicon Solar Cells. *IEEE Transactions on Electron Devices* 1980, 27 (4), 677–687.

[36] SEMI PV17—Specification for Virgin Silicon Feedstock Materials for Photovoltaic Applications, 2021. https://store-us.semi.org/products/pv01700-semi-pv17-specification -for-virgin-silicon-feedstock-materials-for-photovoltaic-applications#:~:text=This%20 Specification%20covers%20virgin%20silicon,distilled%20silane%20or% 20halosilane%20compounds.

[37] Ravikumar, D.; Wender, B.; Seager, T. P.; Fraser, M. P.; Tao, M. A Climate Rationale for Research and Development on Photovoltaics Manufacture. *Applied Energy* 2017, 189, 245–256. https://doi.org/10.1016/j.apenergy.2016.12.050.

[38] Kao, T.-L.; Huang, W.-H.; Tuan, H.-Y. Kerf Loss Silicon as a Cost-Effective, High Efficiency, and Convenient Energy Carrier: Additive-Mediated Rapid Hydrogen Production and Integrated Systems for Electricity Generation and Hydrogen Storage. *Journal of Materials Chemistry A* 2016, 4 (33), 12921–12928. https://doi.org/10.1039/c6ta03657k.

[39] Kim, J.; Kim, S. Y.; Yang, C. M.; Lee, G. W. Possibility of Recycling SiOx Particles Collected at Silicon Ingot Production Process as an Anode Material for Lithium Ion Batteries. *Scientific Reports* 2019, 9 (1), 13313.

[40] DuPont Teijin Films. *Mylar UVPHET—Sustainability Without Compromise*, 2020. https://www.intersolar.de/news/sustainability-without-compromise-intersolar-award-winner-dupont-teijin-films

[41] Yue, D.; You, F.; Darling, S. B. Domestic and Overseas Manufacturing Scenarios of Silicon-Based Photovoltaics: Life Cycle Energy and Environmental Comparative Analysis. *Solar Energy* 2014, 105, 669–678. https://doi.org/10.1016/j.solener.2014.04.008.

[42] Ravikumar, D.; Seager, T. P.; Chester, M. V.; Fraser, M. P. Intertemporal Cumulative Radiative Forcing Effects of Photovoltaic Deployments. *Environmental Science & Technology* 2014, 48 (17), 10010–10018.

[43] Ultra Low Carbon Solar Alliance. *The Ultra Low-Carbon Solar Era Is Here.* https://ultralowcarbonsolar.org/ultra-low-carbon-solar/

[44] Lazard. *Levelized Cost of Energy and Levelized Cost of Storage,* 2020. https://www.lazard.com/research-insights/levelized-cost-of-energy-levelized-cost-of-storage-and-levelized-cost-of-hydrogen-2020/

[45] Weckend, S.; Wade, A.; Heath, G. A. *End of Life Management: Solar Photovoltaic Panels*; NREL/TP-6A20–73852, 1561525; IRENA, 2016; p NREL/TP-6A20–73852, 1561525. https://doi.org/10.2172/1561525.

[46] Ravikumar, D.; Seager, T.; Sinha, P.; Fraser, M. P.; Reed, S.; Harmon, E.; Power, A. Environmentally Improved CdTe Photovoltaic Recycling through Novel Technologies and Facility Location Strategies. *Progress in Photovoltaics: Research and Applications* 2020, 28 (9), 887–898.

[47] Ardente, F.; Latunussa, C. E. L.; Blengini, G. A. Resource Efficient Recovery of Critical and Precious Metals from Waste Silicon PV Panel Recycling. *Waste Management* 2019, 91, 156–167. https://doi.org/10.1016/j.wasman.2019.04.059.

[48] Aryan, V.; Font-Brucart, M.; Maga, D. A Comparative Life Cycle Assessment of End-of Life Treatment Pathways for Photovoltaic Backsheets. *Progress in Photovoltaics: Research and Applications* 2018, 26 (7), 443–459. https://doi.org/10.1002/pip.3003.

[49] Fraunhofer UMSICHT. *End-of-Life Pathways for Photovoltaic Backsheets,* 2017. https://www.coveme.com/files/immagini/green-solutions/Final_Report_End-of-Life_Pathways_For_Photovoltaic_Backsheets.pdf

[50] Lennon, A.; Yao, Y.; Wenham, S. Evolution of Metal Plating for Silicon Solar Cell Metallisation. *Progress in Photovoltaics: Research and Applications* 2013, 21 (7), 1454–1468.

[51] PV Magazine. *Safely Meeting Demand for Renewable Energy with Innovative Material Design for Health and Sustainability,* 2020. https://www.pv-magazine.com/2020/10/05/dont-publish-safely-meeting-demand-for-renewable-energy-with-innovative-material-design-for-health-and-sustainability/

[52] DSM. *The Recyclable Backsheet.* www.dsm.com/dsm-insolar/en_US/technologies/pv-backsheets/the-recyclable-backsheet.html (accessed 1.1.2021).

[53] Karas, J.; Michaelson, L.; Munoz, K.; Jobayer Hossain, M.; Schneller, E.; Davis, K. O.; Bowden, S.; Augusto, A. Degradation of Copper-plated Silicon Solar Cells with Damp Heat Stress. *Progress in Photovoltaics: Research and Applications* 2020, 28 (11), 1175–1186. https://doi.org/10.1002/pip.3331.

[54] Goris, M. J. A. A.; Rosca, V.; Geerligs, L. J.; de Gier, B. Production of Recyclable Crystalline Si PV Modules 2015, 1925–1929. DOI 10.4229/EUPVSEC20152015-5EO.1.2.

[55] Norgren, A.; Carpenter, A.; Heath, G. *Design for Recycling Principles Applicable to Selected Clean Energy Technologies- Crystalline-Silicon Photovoltaic Modules, Electric Vehicle Batteries, and Wind Turbine Blades,* 2020. https://link.springer.com/article/10.1007/s40831-020-00313-3

[56] Einhaus, R.; Madon, F.; Degoulange, J.; Wambach, K.; Denafas, J.; Lorenzo, F. R.; Abalde, S. C.; Garcia, T. D.; Bollar, A. Recycling and Reuse Potential of NICE PV Modules. In *Proceedings of the 2018 IEEE 7th World Conference on Photovoltaic Energy Conversion (WCPEC)* (A Joint Conference of 45th IEEE PVSC, 28th PVSEC & 34th EU PVSEC), Waikoloa, HI, USA, 10–15 June 2018. [Google Scholar]

[57] Phua, B.; Shen, X.; Hsiao, P.-C.; Kong, C.; Stokes, A.; Lennon, A. Degradation of Plated Silicon Solar Module Due to Copper Diffusion: The Role of Capping Layer Formation and Contact Adhesion. *Solar Energy Materials and Solar Cells* 2020, 215. https://doi.org/10.1016/j.solmat.2020.110638.

[58] Couderc, R.; Amara, M.; Degoulange, J.; Madon, F.; Einhaus, R. Encapsulant for Glass PV Modules for Minimum Optical Losses: Gas or EVA? *Energy Procedia* 2017, 124, 470–477.

[59] Song, H.-J.; Yoon, H. S.; Ju, Y.; Kim, S. M.; Shin, W. G.; Lim, J.; Ko, S.; Hwang, H. M.; Kang, G. H. Conductive Paste Assisted Interconnection for Environmentally Benign Lead-Free Ribbons in c-Si PV Modules. *Solar Energy* 2019, 184, 273–280. https://doi.org/10.1016/j.solener.2019.04.011.

[60] DuPont. *DuPont Global Field Reliability Report*, 2020. https://skypower.com/wp-content/uploads/2020/08/Global-Field-Reliability-Report-2020.pdf

[61] Ravikumar, D.; Sinha, P.; Seager, T. P.; Fraser, M. P. An Anticipatory Approach to Quantify Energetics of Recycling CdTe Photovoltaic Systems. *Progress in Photovoltaics: Research and Applications* 2016, 24 (5), 735–746. https://doi.org/10.1002/pip.2711.

[62] Ravikumar, D.; Seager, T. P.; Cucurachi, S.; Prado, V.; Mutel, C. Novel Method of Sensitivity Analysis Improves the Prioritization of Research in Anticipatory Life Cycle Assessment of Emerging Technologies. *Environmental Science & Technology* 2018, 52 (11), 6534–6543.

[63] Hemlock Semiconductor. *Making Solar Energy Even Cleaner*, 2021. https://www.hscpoly.com/solar-is-one-of-the-cleanest-power-sources-weve-got-but-it-could-be-even-greener/

[64] PV Magazine. French Regulator Proposes Tightening up Controversial Carbon Footprint Rules. *PV Magazine*, 2019. https://www.pv-magazine.com/2019/03/11/french-regulator-proposes-tightening-up-controversial-carbon-footprint-rules/

[65] Grant, C. A.; Hicks, A. L. Effect of Manufacturing and Installation Location on Environmental Impact Payback Time of Solar Power. *Clean Technologies and Environmental Policy* 2019, 22 (1), 187–196. https://doi.org/10.1007/s10098-019-01776-z.

[66] Dziedzic, A.; Graczyk, I. *Lead-Free Solders and Isotropically Conductive Adhesives in Assembling of Silicon Solar Cells-Preliminary Results*. IEEE, 2003, pp. 127–132. DOI:10.1109/ISSE.2003.1260499, Information Security Solutions Europe http://www.isse.eu.com/

[67] NSF International Standard/American National Standard. *NSF/ANSI 457-2019-Sustainability Leadership Standard for Photovoltaic Modules and Photovoltaic Inverters*, 2019. https://globalelectronicscouncil.org/wp-content/uploads/NSF-457-2019-1.pdf

[68] The Silicon Valley Toxics Coalition. *Solar Scorecard Guidelines*. www.solarscorecard.com/2016-17/score-guidelines.php.

[69] Vohra, K.; Vodonos, A.; Schwartz, J.; Marais, E. A.; Sulprizio, M. P.; Mickley, L. J. Global Mortality from Outdoor Fine Particle Pollution Generated by Fossil Fuel Combustion: Results from GEOS-Chem. *Environmental Research* 2021. https://doi.org/10.1016/j.envres.2021.110754.

[70] Diffenbaugh, N. S.; Burke, M. Global Warming Has Increased Global Economic Inequality. *Proceedings of the National Academy of Sciences of the United States of America* 2019, 116 (20), 9808–9813.

[71] Politico. *Fears Over China's Muslim Forced Labor Loom Over EU Solar Power*. www.politico.eu/article/xinjiang-china-polysilicon-solar-energy-europe/ (accessed 1.1.2021).

[72] U.S. Bureau of Labor Statistics. *Solar Photovoltaic Installers*, 2021. https://www.bls.gov/oes/2021/may/oes472231.htm

[73] Wiser, R.; Bolinger, M.; Seel, J. *Benchmark Utility-Scale PV Operational Expenses and Project Lifetimes; LBNL Technical Brief.* Lawrence Berkeley National Laboratory, Berkeley, CA, 2020.

[74] Curtis, T. L.; Buchanan, H., Smith, L.; Heath, G. *A Circular Economy for Solar Photovoltaic System Materials: Drivers, Barriers, Enablers, and U.S. Policy Considerations*. National Renewable Energy Laboratory, Golden, CO, 2021. NREL/TP-6A20-74550. www.nrel.gov.docs/fy21osti/74550.

[75] Heath, G.; Ravikumar, D.; Ovaitt, S.; Walston, L.; Curtis, T.; Millstein, D.; Mirletz, H.; Hartmann, H.; McCall, J. *Environmental and Circular Economy Implications of Solar Energy in a Decarbonized U.S. Grid*. National Renewable Energy Laboratory, Golden, CO, 2022. NREL/TP-6A20-80818. www.nrel.gov/docs/fy22osti/80818.pdf.

[76] International Technology Roadmap for Photovoltaic (ITRPV). *International Technology Roadmap for Photovoltaic (ITRPV) 2013 Results*, 2014. https://www.semi.org/sites/semi.org/files/docs/ITRPV_2014_Roadmap_Revision1_140324.pdf

[77] Eshraghi, N.; Berardo, L.; Schrijnemakers, A.; Delaval, V.; Shaibani, M.; Majumder, M.; Cloots, R.; Vertruyen, B.; Boschini, F.; Mahmoud, A. Recovery of Nano-Structured Silicon from End-of-Life Photovoltaic Wafers with Value-Added Applications in Lithium-Ion Battery. *ACS Sustainable Chemistry & Engineering* 2020, 8 (15), 5868–5879. https://doi.org/10.1021/acssuschemeng.9b07434.

[78] Lin, K.-L.; Lee, T.-C.; Hwang, C.-L. Effects of Sintering Temperature on the Characteristics of Solar Panel Waste Glass in the Production of Ceramic Tiles. *Journal of Material Cycles and Waste Management* 2014, 17, 194–200. https://doi.org/10.1007/s10163-014-0240-3.

[79] Guojian, C.; Chao, S.-J.; Cheng, A.; Hsu, H.-M.; Chang, J.-R.; Teng, L.-W.; Chen, S.-C.; Muhammad, Y. Durability Quality Research of Concrete Containing Solar PV Cells. *MATEC Web of Conferences* 2015, 27. https://doi.org/10.1051/matecconf/20152701004.

[80] Qin, B.; Lin, M.; Zhang, X.; Xu, Z.; Ruan, J. Recovering Polyethylene Glycol Terephthalate and Ethylene-Vinyl Acetate Copolymer in Waste Solar Cells via a Novel Vacuum-Gasification-Condensation Process. *ACS ES&T Engineering* 2020. https://doi.org/10.1021/acsestengg.0c00091.

[81] Heath, G. A.; Silverman, T. J.; Kempe, M.; Deceglie, M.; Ravikumar, D.; Remo, T.; Cui, H.; Sinha, P.; Libby, C.; Shaw, S.; Komoto, K.; Wambach, K.; Butler, E.; Barnes, T.; Wade, A. Research and Development Priorities for Silicon Photovoltaic Module Recycling to Support a Circular Economy. *Nature Energy* 2020, 5 (7), 502–510. https://doi.org/10.1038/s41560-020-0645-2.

[82] Deng, R.; Chang, N. L.; Ouyang, Z.; Chong, C. M. A Techno-Economic Review of Silicon Photovoltaic Module Recycling. *Renewable and Sustainable Energy Reviews* 2019, 109, 532–550.

[83] Farrell, C. C.; Osman, A. I.; Doherty, R.; Saad, M.; Zhang, X.; Murphy, A.; Harrison, J.; Vennard, A. S. M.; Kumaravel, V.; Al-Muhtaseb, A. H.; Rooney, D. W. Technical Challenges and Opportunities in Realising a Circular Economy for Waste Photovoltaic Modules. *Renewable and Sustainable Energy Reviews* 2020, 128, 109911. https://doi.org/10.1016/j.rser.2020.109911.

[84] Motta, C. M.; Cerciello, R.; De Bonis, S.; Mazzella, V.; Cirino, P.; Panzuto, R.; Ciaravolo, M.; Simoniello, P.; Toscanesi, M.; Trifuoggi, M.; Avallone, B. Potential Toxicity of Improperly Discarded Exhausted Photovoltaic Cells. *Environ Pollut* 2016, 216, 786–792. https://doi.org/10.1016/j.envpol.2016.06.048.

[85] Akimoto, Y.; Iizuka, A.; Shibata, E. High-Voltage Pulse Crushing and Physical Separation of Polycrystalline Silicon Photovoltaic Panels. *Minerals Engineering* 2018, 125, 1–9. https://doi.org/10.1016/j.mineng.2018.05.015.

[86] Veolia. *Veolia Opens the First European Plant Entirely Dedicated to Recycling Photovoltaic Panels*, 2018. https://www.reuters.com/article/us-solar-recycling/europes-first-solar-panel-recycling-plant-opens-in-france-idUSKBN1JL28Z/

[87] Renew Economy. *Australia's First Solar Panel Recycling Facility to Be Established in Adelaide*, 2021. https://reneweconomy.com.au/australias-first-solar-panel-recycling-facility-to-be-established-in-adelaide/

[88] Tao, M.; Fthenakis, V.; Ebin, B.; Steenari, B.-M.; Butler, E.; Sinha, P.; Corkish, R.; Wambach, K.; Simon, E. S. Major Challenges and Opportunities in Silicon Solar Module Recycling. *Progress in Photovoltaics: Research and Applications* 2020, 28 (10), 1077–1088. https://doi.org/10.1002/pip.3316.

[89] Arup. *Circular Photovoltaics—Circular Business Models for Australia's Solar Photovoltaics Industry*, 2020. https://www.arup.com/perspectives/publications/promotional-materials/section/circular-business-models-for-australia-solar-photovoltaics

[90] Chowdhury, B.; Chowdhury, M. U. *RFID-Based Real-Time Smart Waste Management System* 2007, pp. 175–180. DOI:10.1109/ATNAC.2007.4665232

[91] Choi, J.-K.; Fthenakis, V. Crystalline Silicon Photovoltaic Recycling Planning: Macro and Micro Perspectives. *Journal of Cleaner Production* 2014, 66, 443–449. https://doi.org/10.1016/j.jclepro.2013.11.022.

[92] Choi, J.-K.; Fthenakis, V. Design and Optimization of Photovoltaics Recycling Infrastructure. *Environmental Science & Technology* 2010, 44 (22), 8678–8683.

[93] Goe, M.; Gaustad, G.; Tomaszewski, B. System Tradeoffs in Siting a Solar Photovoltaic Material Recovery Infrastructure. *Journal of Environmental Management* 2015, 160, 154–166. https://doi.org/10.1016/j.jenvman.2015.05.038.

[94] Guo, Q.; Guo, H. A Framework for End-of-Life Photovoltaics Distribution Routing Optimization. *Sustainable Environment Research* 2019, 29 (1).

[95] PV Europe. *Repairing Solar Modules: Sometimes Easier Than Buying New Ones.* www.pveurope.eu/solar-generator/repairing-solar-modules-sometimes-easierbuying-new-ones.

[96] Mackenzie, W. *Annual Solar Repairs and Maintenance Spend to Grow to $9 Billion by 2025.* www.woodmac.com/press-releases/annual-solar-repairs-and-maintenancespend-to-grow-to-$9-billion-by-2025/ (accessed 1.1.2020).

[97] ASES (American Solar Energy Society). *PV Recycling Webinar*, 2020. Boulder, Colorado.

[98] Wade, A.; Sinha, P.; Drozdiak, K.; Brutsch, E. *Beyond Waste—the Fate of End-of Life Photovoltaic Panels from Large Scale PV Installations in the EU the Socio-Economic Benefits of High Value Recycling Compared to Re-Use* 2017, pp. 25–29. Conference: 33rd EU PVSECAt: Amsterdam, The Netherlands.

[99] Kramar, D. E.; Anderson, A.; Hilfer, H.; Branden, K.; Gutrich, J. J. A Spatially Informed Analysis of Environmental Justice: Analyzing the Effects of Gerrymandering and the Proximity of Minority Populations to U.S. Superfund Sites. *Environmental Justice* 2018, 11 (1), 29–39. https://doi.org/10.1089/env.2017.0031.

[100] Ghoreishi, M.; Happonen, A. Key Enablers for Deploying Artificial Intelligence for Circular Economy Embracing Sustainable Product Design: Three Case Studies. *AIP Conference Proceedings* 2020, 2233, 050008. https://doi.org/10.1063/5.0001339.

[101] Kapoor, K.; Pandey, K. K.; Jain, A. K.; Nandan, A. Evolution of Solar Energy in India: A Review. *Renewable and Sustainable Energy Reviews* 2014, 40, 475–487. https://doi.org/10.1016/j.rser.2014.07.118.

[102] MERCOM. *India Solar Market—December 2018.* Mercom India Research, 2018. https://mercomindia.com/wp-content/uploads/2018/12/Mercom-Intersolar-India-Solar-Market-Update-Whitepaper.pdf

[103] Prabhu, V. S.; Shrivastava, S.; Mukhopadhyay, K. Life Cycle Assessment of Solar Photovoltaic in India: A Circular Economy Approach. *Circular Economy and Sustainability* 2022, 2, 507–534. https://doi.org/10.1007/s43615-021-00101-5.

[104] Fraunhofer ISE. *Photovoltaics Report. Fraunhofer Institute of Solar Energy Systems.* ISE, 2020. www.ise.fraunhofer.de/content/dam/ise/de/documents/publications/studies/Photovoltaics-Report.pdf

[105] Pollitt, H. *Introduction to the E3-India Macroeconomic Model.* IORA Conference, 2017. www.e3indiamodel.com/wpcontent/uploads/2018/04/pollitt-intro-e3-india-macroeconomic-model-IORA.pdf

14 Recycling within the Chemical Industry

The Circular Economy Approach—Industrial Perspectives

Brijnandan Singh Dehiya*

14.1 INTRODUCTION

The prevailing linear economic model of taking resources, manufacturing chemicals and plastics, and disposing of waste has posed tremendous sustainability challenges for the chemical industry, including significant greenhouse gas emissions, pollution, and waste generation [1]. However, growing recognition of planetary boundaries and resource constraints has led to the rising adoption of circular economy thinking aimed at enhanced recycling, reuse, and recovery of materials across chemical life cycles [2]. This chapter provides an overview of how the principles of the circular economy can potentially transform chemical production, design, waste management, and industrial symbiosis to boost sustainability.

The traditional linear model of resource throughput has resulted in substantial environmental impacts from the global chemicals sector [9]. The resultant products, such as plastics and chemical pollution, affect all ecosystems and are increasingly hazardous to human health [4]. At the same time, emissions from production and waste contribute to global climate change [27]. The industry's dependence on non-renewable feedstock like petroleum and natural gas also raises supply concerns [5]. In contrast, the circular economy offers a holistic systems solution, emphasizing closed-loop material flows, waste reduction, renewable inputs, and extending resource lifetime through superior design [6, 7]. This effect goes beyond incremental efficiency gains, requiring a fundamental rethinking of technologies, business models, product design, and consumption patterns [30, 54, 56, 64, 65].

The circular economy aims to shift away from the current linear "take-make-dispose" approach by decoupling economic activity from finite resource consumption [9]. Its focus on cycling materials through multiple product life cycles provides a regenerative model for the sustainable use of resources [10]. As depicted in

* brijnandan.che@dcrustm.org

DOI: 10.1201/9781003269779-14

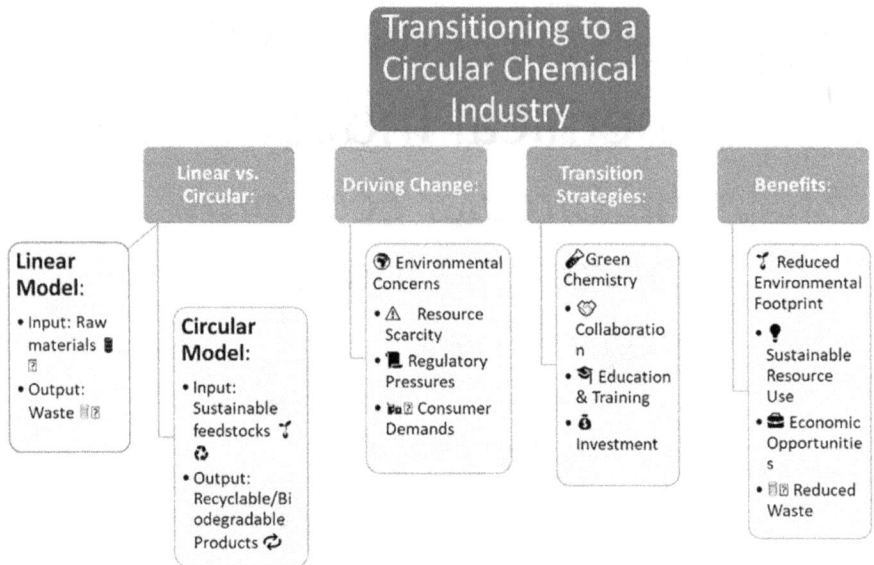

FIGURE 14.1   Shifting to a circular chemical industry.

Figure 14.1, a transition towards circularity across chemical value chains promises significant environmental and economic benefits [11, 12].

For the chemicals sector, the critical circular economy strategies can include [2, 50, 61]:

1. Process optimization and renewable energy use (e.g., hydrogen) in manufacturing.
2. Designing for recyclability, biodegradability, and using bio-based feedstocks.
3. Enhanced collection systems and chemical recycling.
4. Industrial symbiosis networking to reuse wastes and byproducts.

Circular production emphasizes enhancing resource efficiency in manufacturing through process optimization, waste minimization, recycling, and renewable energy [14]. For example, using best practices, steam cracking plants can improve energy efficiency by over 10% [15]. Preventing chemical leaks and enhancing real-time monitoring further minimize emissions [16]. Product design also plays a crucial role through strategies like recyclable polymer development, compostable chemicals, and bio-based materials [17]. For instance, Braskem produces "I'm Green" polyethylene from sugarcane [18]. Closed-loop supply chains that facilitate collection and take-back after use also boost circularity [19].

In waste management, improving collection infrastructure and advancing recycling processes like solvent distillation, plastic pyrolysis, and chemical separation allows enhanced recovery of valuable chemicals, fuels, and metals from discards [20, 21]. This approach preserves embedded energy and reduces demands for virgin inputs while minimizing landfill disposal [22]. Industrial symbiosis networks further integrate circular flows by fostering byproducts, energy, water, and resource

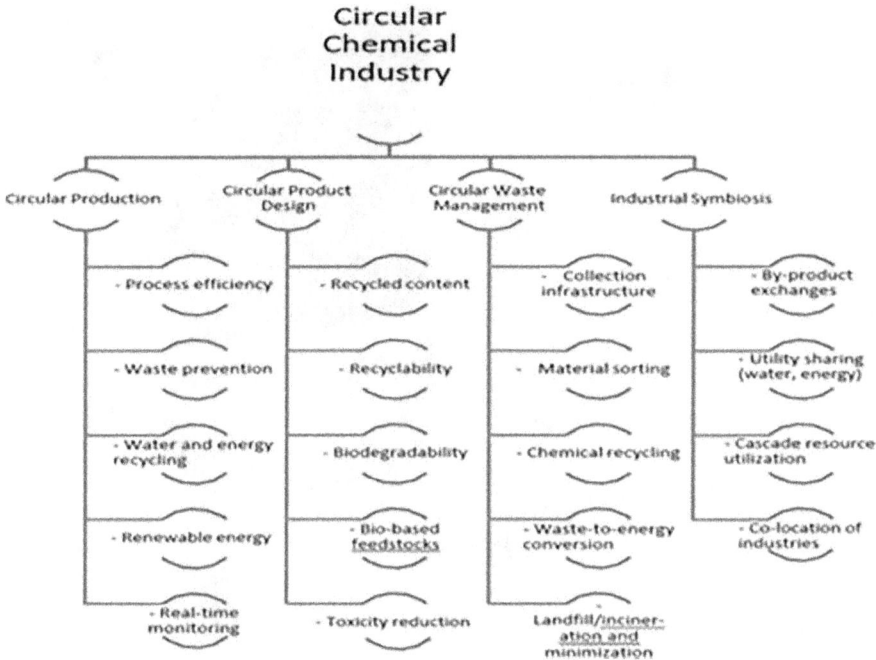

FIGURE 14.2   Key aspects of a circular chemical industry.

exchanges between manufacturers [23]. The Kalundborg (Denmark) Eco-Industrial Park annually trades approximately 3 million tons of residual resources [24]. Such collaboration closes material loops regionally. Overall, this multi-pronged approach can drive the transition towards a sustainable circular chemical industry, as outlined in Figure 14.2.

However, transitioning from the linear status quo faces numerous challenges, including high upfront costs for new infrastructure, policy and regulatory gaps, technological limitations around chemical recycling, lack of consumer awareness, and organizational inertia among firms wedded to existing linear business models [26, 27, 28]. Overcoming these barriers will require solutions across innovation, collaboration, investment, and governance [29]. Advances in separation techniques, biorefineries, bio-based chemicals, and recycling technology can expand possibilities [30]. Collaboration throughout value chains and the formation of industry symbiosis networks will enable necessary systemic shifts [31].

Policy reforms like extended producer responsibility, recycled content mandates, landfill taxes, and zero waste targets are crucial to reshaping market incentives around circularity [32]. Moreover, access to finance must rapidly scale to support the transition, which is estimated to require upwards of $4 trillion in cumulative investment globally [33].

Rising resource constraints worldwide reinforce the urgency of enhancing circular resource flows. As indicated in Figure 14.3, global consumption of crucial chemical feedstock and fertilizers has surged over 50% since 2000 [34]. Reserves

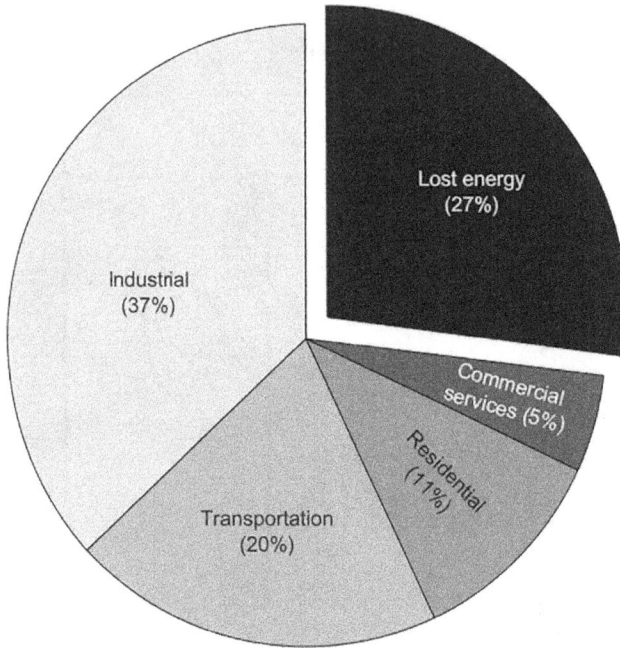

FIGURE 14.3   Percentage share of sectors on energy consumption [81].

Source: Copyright 2018. Reproduced with permission from Elsevier.

face depletion while extraction costs increase for remaining stocks [35]. This supply outlook makes a compelling case for maximizing recovery, recycling, and reuse.

In summary, adopting circular economy principles promises to enable sustainable growth in the chemical industry through significantly improved resource efficiency, waste and emission reductions, renewable energy systems, product and process redesign, recycling, industrial symbiosis, and cradle-to-cradle material flows. The chapters in this book detail strategies, developments, and evidence driving the transition toward circularity across chemical production, design, waste management, and inter-industry collaboration. While the shift poses challenges, circularity has become imperative for the future due to rising resource constraints, environmental awareness, and a supportive policy landscape. The chemical industry can lead the transition by embracing innovation, collaboration, systems thinking, and sustainable economics. This book provides an invaluable reference for experts and practitioners seeking insights on propelling the industry toward an inclusive, just, and circular future.

The circular economy is an economic system aimed at eliminating waste and promoting the continual use of resources [37]. The chemical industry must transition to a circular economy model for sustainable development [38]. The key objectives are to reduce reliance on virgin materials, implement closed-loop production processes, and improve recycling and recovery of waste [39]. Circular economy principles can significantly enhance resource efficiency and reduce environmental impacts in the chemical industry.

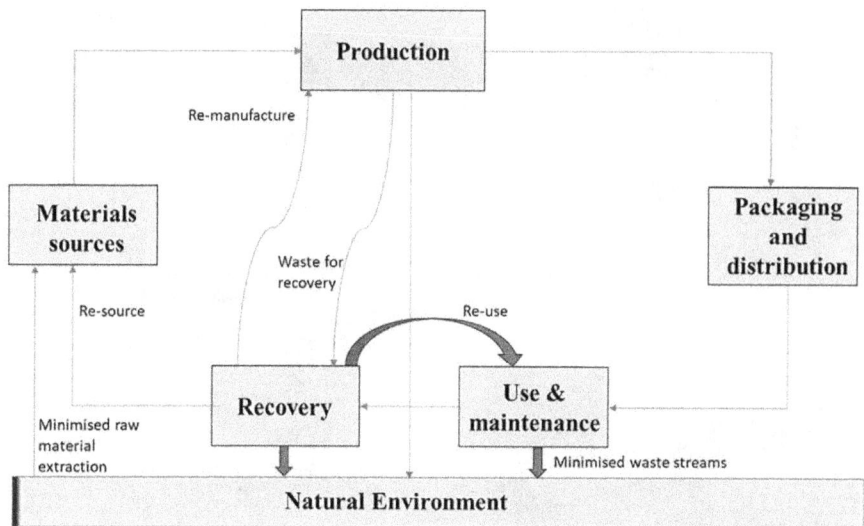

FIGURE 14.4 Flow diagram: closed-loop chemical industry.

Chemical recycling allows the reclamation of economic value from waste by producing oils, fuels, solvents, and monomers [47]. New separation methods also enable the recovery of high-value chemicals from residues [46]. This approach also delivers sustainability while maintaining competitiveness. However, barriers like costs, policy gaps, and reluctance to change linear business models hinder progress [48]. Solutions across innovation, collaboration, investment, and governance are vital for the industry to adopt circularity. Figure 14.4 depicts the flow diagram of closed-loop chemical industry.

14.1.1 KEY PRINCIPLES OF CIRCULAR ECONOMY IN CHEMICAL INDUSTRIES

The circular economy entails several core principles for the chemical industry:

1. Reduce waste and pollution through improved production efficiencies and better product designs.
2. Reuse materials and resources to maximize utilization—for example, reusing solvents or iron scrap.
3. Recycle and recover valuable materials from waste streams. This method allows continued capture of economic value.
4. Renewable energy, such as solar or wind, can reduce fossil fuel dependence.

Figure 14.5 illustrates the chemical industry manufacturing schematic. Several core principles underpin a transition towards a circular economy in the chemical sector, enabling enhanced sustainability through recycling, waste reduction, and resource efficiency [42].

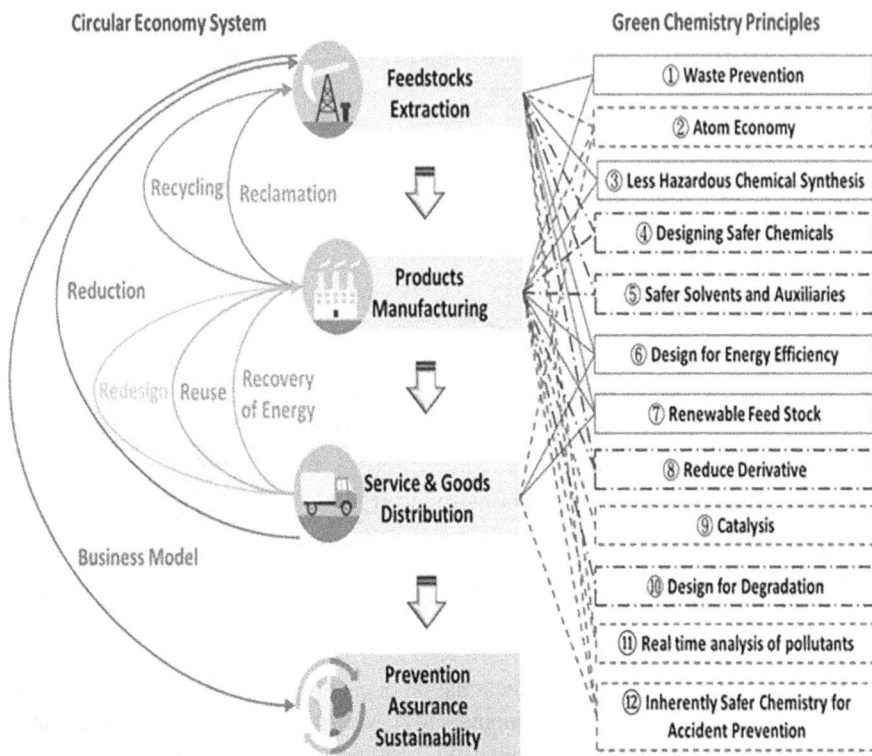

FIGURE 14.5 Circular economy and green chemistry principles [82].

Source: Copyright 2020. Reproduced with permission from Elsevier.

14.1.1.1 Reduce Waste and Pollution

Minimizing hazardous emissions and waste is critical. Strategies include improving production efficiency, modifying plant operations, investing in pollution control systems, and training employees in prevention [53]. For example, through process optimizations, European ethylene producers reduced carbon emissions per ton of production by over 20% from 2008 to 2018 [52]. Real-time monitoring, leak detection, and maintenance help minimize chemical releases [39]. Overall, circular production redesigns processes to avoid pollution rather than relying solely on end-of-pipe treatments.

14.1.1.2 Reuse Materials and Resources

Extensive reuse of chemicals across multiple production cycles reduces virgin input needs [44]. Scrap metals, solvents, and polymers can often be deployed numerous times before replacement. Developing industrial symbiosis networks allows firms to exchange byproducts and wastes, keeping materials circulating in the economy. For instance, the Kalundborg (Denmark) industrial park annually exchanges 2.9 million tons of residual resources [43].

TABLE 14.1
Estimated Global Savings from Improved Resource Efficiency [45]

S. No	Recovery Source	Estimated Value (in $ billions)
1	Plastics	80–120
2	Metals	300–600
3	Fertilizers	110–185
4	Building Materials	15–65
5	Critical Minerals	75–100
6	Food and Organics	155–400

14.1.1.3 Recycle and Recover Valuables from Waste

Circular systems recapture and purify materials from residuals and effluents for reuse via separation techniques, chemical recycling, and other waste valorization processes [44]. A US study found over $60 billion of recoverable resources across major waste streams [45]. Recycling preserves raw materials, captures embedded economic value, and reduces demands on finite resources [46].

14.1.1.4 Utilize Renewable Energy

Adopting regionally available renewable energy like solar, wind, or bio-based fuels aligns chemical production with circular economy goals [47]. Waste-to-energy conversion of biomass can also assist in meeting process energy demands [48]. This results in lower CO_2 emissions than fossil fuel reliance.

In summary, these interlinked principles aim to transition chemical manufacturing into a more restorative, closed-loop system focused on resource optimization and the circular flow of materials. Table 14.1 shows the estimated global savings from improved resource efficiency.

14.2 CIRCULAR ECONOMY IN CHEMICAL PRODUCTION

Key strategies here include the following:

1. Reusing raw materials and byproducts across production processes reduces the need for virgin inputs.
2. Closed-loop production systems allow continued recycling of solvents, acids, metals, and other chemicals.
3. Process optimization and heat integration improve energy efficiency in chemical production.
4. Improved waste management and treatment technology reduce emissions and waste generation.

Transitioning chemical production to a more circular model is critical for reducing waste, pollution, and resource consumption. Multiple strategies can be adopted across manufacturing processes, operations, and supply chains to transition from the traditional linear take-make-dispose model.

To maximize utilization, a core principle is reusing raw materials and byproducts across as many production cycles as possible. Solvents, acids, scrap metals, and plastics can often be reused across multiple batches before needing replacement [63]. This process reduces reliance on continual inputs of virgin feedstocks, enabling significant material savings. In the EU, chemical byproduct reuse could deliver over €1 billion in annual cost savings by 2030 while avoiding 3 million tons of waste (McKinsey, 2021). Additionally, establishing closed-loop production systems allows used chemicals, solvents, and water to be continually recycled within the manufacturing plant [53]. This approach may require separation and purification technologies to recover materials from process waste streams. Modeling indicates that the global adoption of closed loops could reduce greenhouse gas emissions from chemical manufacture by 10%–15% by 2050 (IEA, 2020 [49]).

Optimizing energy use through heat integration, cogeneration, and other process efficiency measures is also crucial for operational circularity, reducing energy waste [73, 80]. Renewable energy adoption, e.g., wind, solar, or biofuels, aligns chemical production with circular economy principles by utilizing sustainable energy flows [64, 70, 78]. Modifying manufacturing equipment, such as improved valves, pumps, and reactors, can prevent chemical leaks, minimizing worker waste and exposure risks (OECD, 2018 [37]). Real-time process monitoring and preventive maintenance can further support process optimization and safety. Redesigning production lines to meet circular economy criteria could create 80,000 new jobs in the EU chemical sector by 2030 [23, 38].

Incorporating advanced treatment technologies also enables improved gaseous, liquid, and solid waste management. Solutions range from membrane filtration to allow water recycling to scrubber systems for air pollutant removal before emission (Rainey et al., 2022 [74]). Such technologies are critical for preventing environmental contamination and allowing resource recovery.

Additionally, supply chain management plays an important role. Initiatives like chemical leasing shift away from purely transactional interactions between chemical producers and users towards service-based models focused on optimizing chemical application and recovery [66, 69, 70]. Overall, a combination of process innovation, new technologies, and supply chain collaboration is positioning chemical production as a leader of the global circular economy.

However, significant investment is needed to overcome technological and infrastructure barriers [51]. One estimate suggests a cumulative spend of over $3 trillion is required globally by 2050 to realize a circular chemical industry (Accenture, 2021 [62, 63]). Policy measures like extended producer responsibility and material taxes could help drive this investment by incentivizing circular design (Geissdoerfer et al., 2017 [61]). Additionally, improving consumer awareness around circular chemicals can accelerate market uptake.

Integrating reuse, recycling, renewable energy, waste minimization, and supply chain collaboration throughout chemical manufacturing is critical to enabling circular production systems [55]. The transition promises significant sustainability benefits through reduced resource demand, lower carbon emissions, and minimized environmental health impacts from chemical pollution. Expert leadership from industry, governments, academia, and civil society is essential to guide the chemical sector through this system-level transformation toward circularity.

14.3   CIRCULAR ECONOMY IN CHEMICAL PRODUCT DESIGN

Sustainable chemical product design is crucial involving:

1. Designing products for durability, reuse, and recyclability.
2. Developing biodegradable and compostable chemicals and polymers.
3. Use of renewable raw materials like bio-based feedstock.
4. Closed-loop supply chains that recover products after use for recycling.

The reuse, recover, and recycle loops for a circular chemical industry are shown in Figure 14.6. For circular economy principles to be applied throughout the product life cycle, chemical products must be redesigned. Plastics, polymers, and chemicals can be designed to for circular economy using several innovative strategies. The goal is to improve product durability, reusability, and recycling potential. The use of products with a long life span and multiple cycles of use minimizes the generation of waste [9, 19, 54]. The circularity of industrial chemicals can be enhanced by designing them in robust, reusable containers rather than single-use drums. In addition, recycling feedstocks allows them to displace some virgin inputs. Among the

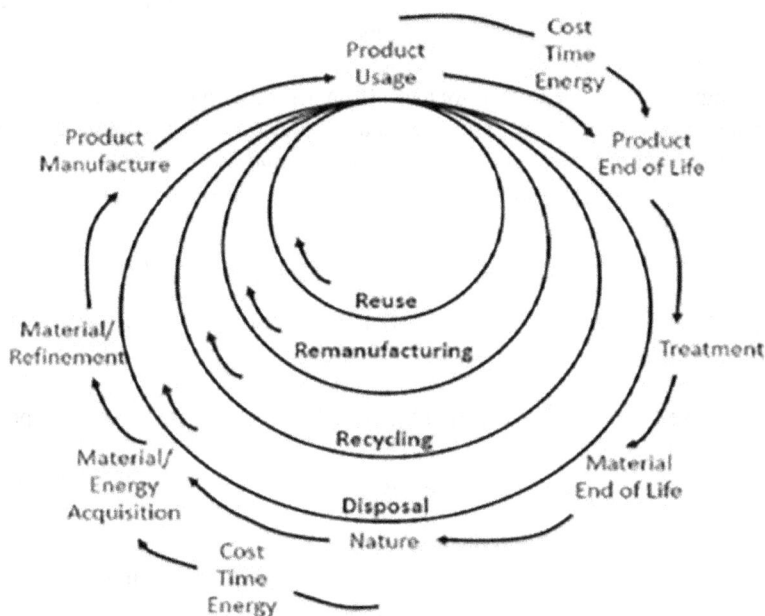

FIGURE 14.6   The reuse, recover, recycle loops for a circular chemical industry.

Source: Mihelcic et al. (2003) [53].

materials that could be used in this approach are recycled plastic, industrial scrap metal, and waste oils [3, 41, 44, 47]. Material recovery rates can also be increased by designing products to be easily recycled at the end of their lives.

Plastic pollution impacts humans and ecosystems negatively when biodegradable and compostable chemicals and polymers are used [36,76]. Bio-based, non-toxic inputs can synthesize chemicals using green chemistry principles. With consumer demand growing, major chemical companies have begun offering biodegradable polymer products.

Separation and purification of chemicals can also be achieved through chemical recycling after use [69, 70]. Thus, molecules can retain their embedded energy while avoiding waste. Chemicals can also be tagged digitally to facilitate recycling (Accenture [20, 41]). Materials can also be designed to be remanufactured using chemicals, allowing them to last longer before recycling. Among the strategies are the development of reusable resins for 3D printing and modular chemical inputs to simplify disassembly (EMF, 2021 [19]). Introducing new business models like chemical leasing also promotes more efficient use of chemicals by users and take-back systems for unused chemicals [6, 13]. This can reduce waste by promoting mindful usage. It is difficult, however, to develop viable biodegradable or recyclable alternatives due to various factors, including consumer behavior, costs, and technological limitations (Hartley et al., 2020 [67]). Upstream chemicals must be redesigned, and downstream systems must be aligned to handle new recycling processes and products.

Some estimates estimate that the transition to circular chemicals will require $500 billion in R&D and infrastructure [3, 50, 63, 79]. The development of sustainable options can be accelerated through incentives like green public procurement and product standards. By leveraging green chemistry principles and circular economy principles in chemical product design, it is possible to reduce environmental impacts significantly. We must move away from resource-intensive, waste-producing chemicals in our products of tomorrow to a circular economy across supply chains and product lifetimes. A more sustainable and circular industry requires leadership from chemical manufacturers.

14.4   CIRCULAR ECONOMY IN CHEMICAL WASTE MANAGEMENT

Effective management of chemical waste streams enables recovery of value, such as [11]:

1. Improved collection and sorting facilities for post-consumer chemical products and hazardous wastes.
2. Enhancing recycling and recovering valuable chemicals, solvents, and metals from waste.
3. Reducing landfilling and incineration, leading to resource loss.
4. Adopting new waste-to-chemical, waste-to-energy, and separation technologies.

Resource recovery and environmental impact can be maximized by implementing circular economy principles in managing chemical wastes. Key strategies must be implemented across waste collection, treatment, and disposal processes to improve recycling and divert materials from landfills. As a first step, improved collection systems and sorting infrastructure promote the segregation and circular processing of post-consumer chemical products, hazardous wastes, and industrial byproducts (Velenturf & Purnell, 2021 [77]). A standardized labeling system also simplifies automation management. Industrial symbiosis can also be achieved by exchanging wastes and byproducts (Patwa et al., 2021 [71]). Many technologies are available to recycle and purify chemicals, solvents, fuels, and metals [52, 58]. Chemical compounds contain high embodied energy, which can be preserved through recycling, and recycling can reduce virgin inputs.

Plastic waste can be recycled chemically into oils, chemicals, or monomers for use in manufacturing [21, 73, 80]. Reusing solvents after they have been purified and distilled is also possible. The development of new polymers is also being done to simplify recycling. It is estimated that the global potential for recycling plastic waste will reach 55% by 2030, resulting in a reduction of over $30 billion in annual resource consumption (WEF, 2020 [50]). Through gasification, pyrolysis, and chemical catalysis, waste-to-chemical techniques can also produce sustainable biofuels, synthetic gases, and other products from biomass sources [70]. Anaerobic digestion or incineration of waste can produce renewable energy. The emissions from such processes, however, must be avoided.

To avoid pollution and waste, it is crucial to minimize landfilling and incineration of chemical waste (Worrell & Vesilind, 2012 [80]). When disposal is necessary, improved incinerator design and operation can reduce health and environmental impacts. Additionally, landfills are expensive, preventing waste from being diverted. To scale up circular chemical waste management, significant investment in collection, sorting, and recycling infrastructure is essential [60, 62, 64, 72]. Manufacturers can also support collecting and recycling after consumer use through more robust policy measures, such as extended producer responsibility.

For chemical waste management to transition from a linear to a circular model, multiple approaches are needed, including technological innovation, infrastructure development, policy reform, and system-wide coordination. The key to a circular economy is to capture the residual value of waste chemicals. Collaboration between industry, government, researchers, and civil society will be essential to accelerating chemical recycling adoption worldwide.

14.5 CHALLENGES AND OPPORTUNITIES

Some key barriers exist to a circular chemical recycling economy [8, 59, 60]:

1. Resistance to changing traditional linear business models reliant on virgin material inputs [58].
2. Limited investment capability and lack of technological expertise.
3. Lack of regulations and incentives to promote circular economy practices.

However, there are opportunities for innovation, competitive advantage, and cost savings [12, 25].

Chemical industries have traditionally consumed and generated waste linearly, posing tremendous challenges to sustainability (Geissdoerfer et al., 2017 [10]). Recycling, reuse, and resource recovery are reshaping the sector, however, as a circular economy approach becomes more recognized. There are several barriers, but there are also several opportunities [40].

Business models that rely heavily on virgin input materials impede the change of entrenched linear business models (Kirchherr et al., 2018 [62]). The "take-make-dispose" model has been highly profitable for the petrochemical industry, disincentivizing investments in circular innovations. Without a convincing business case, many companies hesitate to embrace circularity (Ritzén & Sandström, 2017 [75]). Organizational inertia results from today's linear model's known benefits.

Chemical recycling, purification, and industrial symbiosis are also restricted by a lack of investment capability and technical expertise (Geissdoerfer et al., 2017 [61]). There are substantial capital costs associated with modifying processes, purchasing new equipment, and constructing waste collection infrastructure, among others (Iles & Martin, 2013). In addition, firms often require knowledge about life cycle analysis, circular design, and sustainability-oriented engineering (Kirchherr et al., 2018 [62]).

Moreover, current policy landscapes do not adequately promote circular practices (Ritzén & Sandström, 2017 [75]). Laws extending producer responsibility and landfill taxes are limited, and recycled content mandates do not exist. To motivate change, stronger policy signals are needed (Geissdoerfer et al., 2017 [61]). It is insufficient to rely on voluntary initiatives alone.

The circular business model, however, offers opportunities for growth, cost savings, and first-mover advantage (Lacy & Rutqvist, 2016 [69]). Recycling and circular material flows can reduce reliance on virgin inputs. It is also possible to save money by avoiding waste disposal costs. The circular economy can also foster better collaboration across supply chains and the development of innovative products and processes (Ritzén & Sandström, 2017 [75]).

Several innovative circular strategies have emerged, such as chemical leasing, product-as-a-service, and sharing platforms, which decouple revenues from material volumes (EMF, 2021 [29]). In addition to predictive maintenance, real-time optimization, and other efficiency gains enabled by digitalization, analytics support circular design [19]. There is likely to be an increase in sustainability demands from downstream users, favoring circular suppliers.

It faces obstacles to achieving more remarkable recycling, reuse, and resource recovery, but there is a clear rationale for change. The chemical industry can overcome barriers by establishing forward-looking policies, collaborating across value chains, and investing in R&D (Kirchherr et al., 2018 [26]). The transition to a circular economy is inevitable due to resource constraints and environmental awareness. The first movers in the chemical value chain will reap significant benefits as they lead the way to circularity, becoming the new norm.

As a result of its traditionally linear resource consumption and waste production model, the chemical industry faces tremendous sustainability challenges (Geissdoerfer et al., 2017 [10]). Recycling, reuse, and resource recovery are at the forefront of reshaping the sector with a circular economy approach. Inherent linear business models reliant heavily on virgin inputs are one of the primary impediments

(Kirchherr et al., 2018 [62]). Petrochemical companies have been highly profitable on a "take-make-dispose" model, which disincentivizes investment in circular innovations. Many firms are reluctant to introduce circularity without a compelling business case (Ritzén & Sandström, 2017 [75]). As a result of the linear model's well-known benefits, organizations tend to become inert. Chemical recycling, purification, and industrial symbiosis are also restricted by a lack of investment capability and technical expertise (Geissdoerfer et al., 2017 [61]). Modifying processes, purchasing new equipment, or constructing waste collection infrastructure is costly (Iles & Martin, 2013). Circular product design and engineering require specialized expertise (Kirchherr et al., 2018 [62]).

A lack of regulations and incentives further hinders the implementation of circular practices in the policy landscape (Ritzén & Sandström, 2017 [75]). Some examples are a lack of extended producer responsibility laws, low landfill taxes, and a lack of recycled content mandates. Change must be motivated by stronger policy signals (Geissdoerfer et al., 2017 [61]). By implementing circular business models, first-mover advantage, growth, and cost savings can be gained (Lacy & Rutqvist, 2016 [69]).

Several innovative circular strategies have emerged, such as chemical leasing, product-as-a-service, and sharing platforms, which decouple revenues from material volumes (EMF, 2021 [29]). In addition to predictive maintenance, real-time optimization, and other efficiency gains enabled by digitalization, analytics support circular design [19]. The demand for sustainability will likely grow among downstream users, making circular suppliers more attractive.

14.6 CONCLUSIONS

In summary, circular economies can enable sustainable growth in the chemical industry by reducing waste, reusing energy, and recycling resources. New collaborations, policy changes, and regulations are needed to support its continued adoption. As the chemical industry moves towards the circular economy, it looks promising.

To achieve global sustainability objectives, chemical industries must transition to a circular economy model emphasizing enhanced materials recycling, recovery, and reuse. While the rationale and urgency for change are compelling, given rising resource constraints and environmental impacts associated with current chemical production paradigms, adoption is constrained by entrenched linear business models, infrastructure costs, and policy gaps.

With the help of a circular economy, chemical manufacturing, product design, waste management, and industrial symbiosis can be improved significantly, emissions of carbon and chemicals can be reduced, waste can be minimized, green chemicals can be developed, and economic growth can be decoupled from feedstock availability.

Manufacturing processes, supply chains, new collaboration models, recycling technologies, biodegradable materials, energy systems, and infrastructure all fall under the scope of circular innovation. Despite initial costs, circularity offers long-term savings from enhanced resource productivity and avoided disposal costs. Competitive advantages can be gained by first movers who integrate sustainability into their business models at an early stage.

The transition can be managed in the context of large-scale industrial facilities optimized for linear throughput, a policy environment that has allowed externalization of pollution costs, and consumer expectations shaped by decades of petrochemical-derived products. Globally, a shift towards circularity appears inevitable through collaboration, governance, innovation, and creative financing. In an era of diminishing resources and heightened environmental awareness, chemical companies must act urgently and aggressively to adopt circular designs and find new value in waste.

REFERENCES

[1] Silvério, A.C., Ferreira, J., Fernandes, P.O. and Dabić, M. (2023) 'How does circular economy work in industry? Strategies, opportunities, and trends in scholarly literature', *Journal of Cleaner Production*, 412, p. 137312. https://doi.org/10.1016/j.jclepro.2023.137312

[2] Londoño, N.A.C. and Cabezas, H. (2021) 'Perspectives on circular economy in the context of chemical engineering and sustainable development', *Current Opinion in Chemical Engineering*, 34, p. 100738.

[3] World Resources Institute (2021) *The Plastics Waste Problem*. Available at: https://research.wri.org/gfr/plastics-waste-management.

[4] Center for International Environmental Law (2019) *Plastic & Climate: The Hidden Costs of a Plastic Planet*. Available at: www.ciel.org/wp-content/uploads/2019/05/Plastic-and-Climate-FINAL-2019.pdf.

[5] International Energy Agency (2018) *The Future of Petrochemicals: Towards More Sustainable Plastics and Fertilisers*. Available at: https://iea.blob.core.windows.net/assets/bee4ef3a-8876-4566-98cf-7a130c013805/The_Future_of_Petrochemicals.pdf.

[6] Blomsma, F. and Brennan, G. (2017) 'The emergence of circular economy: A new framing around prolonging resource productivity', *Journal of Industrial Ecology*, 21(3), pp. 603–614. Available at: https://onlinelibrary.wiley.com/doi/abs/10.1111/jiec.12603.

[7] Kirchherr, J., Reike, D. and Hekkert, M. (2017) 'Conceptualizing the circular economy: An analysis of 114 definitions', *Resources, Conservation & Recycling*, 127, pp. 221–232. Available at: www.sciencedirect.com/science/article/pii/S0921344917302835.

[8] Ritzén, S. and Sandström, G.Ö. (2017) 'Barriers to the circular economy—integration of perspectives and domains', *Procedia CIRP*, 64, pp. 7–12. Available at: www.sciencedirect.com/science/article/pii/S2212827116307385.

[9] Ellen MacArthur Foundation (2015) *Towards a Circular Economy: Business Rationale for Accelerated Transition*. Available at: https://archive.ellenmacarthurfoundation.org/assets/downloads/publications/TCE_Ellen-MacArthur-Foundation_26-Nov-2015.pdf.

[10] Geissdoerfer, M., Savaget, P., Bocken, N.M. and Hultink, E.J. (2017) 'The circular economy—a new sustainability paradigm?', *Journal of Cleaner Production*, 143, pp. 757–768. Available at: www.sciencedirect.com/science/article/pii/S0959652616321023.

[11] Londoño, N.A.C. and Cabezas, H. (2021) 'Perspectives on circular economy in the context of chemical engineering and sustainable development', *Current Opinion in Chemical Engineering*, 34, p. 100738. Available at: www.sciencedirect.com/science/article/pii/S2211339821000382.

[12] Accenture (2021) Available at: www.accenture.com/content/dam/accenture/final/accenture-com/document/Accenture-Chemicals-Circular-Economy-Growth.pdf

[13] Nasr, N., Konash, A. and Müller, J. (2022) 'Reusable production materials in circular economy', *Resources, Conservation & Recycling*, 179, p. 106178. Available at: www.sciencedirect.com/science/article/pii/S0921344921005442.

[14] Organisation for Economic Cooperation and Development (2018) *Improving Resource Efficiency and the Circularity of Economies for a Greener World.* Available at: www. oecd.org/publications/improving-resource-efficiency-and-the-circularity-of-econo-mies-for-a-greener-world-1b38a38f-en.htm.

[15] Ren, J., Manan, Z.A. and Klemeš, J.J. (2022) 'Progress in energy saving and consumption reduction in petrochemical industry towards circular economy and sustainability', *Processes*, 10(3), p. 472. Available at: www.mdpi.com/2227-9717/10/3/472.

[16] Paramati, S.R., Utture, S.C., Banerjee, R., Das, P.K. and Gupta, A.K. (2021) 'Towards circular economy in petrochemical industry: A framework and review', *Journal of Cleaner Production*, 294, p. 126294. Available at: www.sciencedirect.com/science/article/pii/S0959652621010841.

[17] Schneiderman, D.K. and Hillmyer, M.A. (2021) '50th-anniversary perspective: There is a great future in sustainable polymers', *Macromolecules*, 54(13), pp. 5295–5310. Available at: https://pubs.acs.org/doi/10.1021/acs.macromol.1c00293.

[18] Braskem (2022) *I'm Green Polyethylene.* Available at: www.braskem.com.br/im-green-plastic.

[19] Ellen MacArthur Foundation (2021) *The Vision of a Circular Economy for Plastics.* Available at: https://emf.thirdlight.com/link/rgzteqdzzyeq-kz@xyp1/@/preview/1.

[20] Al-Salem, S.M. (2019) 'The role of recycling and incineration in the context of plastic waste management', in: Letcher, T.M. and Vallero, D. (eds.) *Plastic Waste Management.* Boca Raton, FL: CRC Press.

[21] Huang, J., Veksha, A., Chan, W.P., Giannis, A. and Lisak, G. (2022) 'Chemical recycling of plastic waste for sustainable material management: A prospective review on catalysts and processes', *Renewable and Sustainable Energy Reviews*, 154, p. 111866. Available at: www.sciencedirect.com/science/article/pii/S1364032121007770.

[22] Hopewell, J., Dvorak, R. and Kosior, E. (2009) 'Plastics recycling: Challenges and opportunities', *Philosophical Transactions of the Royal Society B: Biological Sciences*, 364(1526), pp. 2115–2126. Available at: https://royalsocietypublishing.org/doi/10.1098/rstb.2008.0311.

[23] Martin, A.N. (2015) 'Industrial ecology and regional competitiveness', in: Carayannis, E.G., Varblane, U. and Roolaht, T. (eds.) *Handbook of Regions and Competitiveness.* Northampton, MA: Edward Elgar Publishing, pp. 205–220. Available at: www. elgaronline.com/display/edcoll/9781783475001/9781783475001.00020.xml.

[24] Kalundborg Symbiosis (2022) *Sustainable Production in Symbiosis.* Available at: www.symbiosis.dk/en/.

[25] IEA (2018) *The Future of Petrochemicals: Towards More Sustainable Plastics and Fertilisers.* Paris: IEA. https://doi.org/10.1787/9789264307414-en.

[26] Kirchherr, J., et al. (2018) 'Barriers to the circular economy: Evidence from the European Union (EU)', *Ecological Economics*, 150, pp. 264–272. Available at: www. sciencedirect.com/science/article/pii/S0921800918304494.

[27] Ritzén, S. and Sandström, G.Ö. (2017) 'Barriers to the circular economy—integration of perspectives and domains', *Procedia CIRP*, 64, pp. 7–12. Available at: www. sciencedirect.com/science/article/pii/S2212827116307385.

[28] World Economic Forum (2020) *The Future of Petrochemicals: Growth Surges Ahead.* Available at: http://www3.weforum.org/docs/WEF_Future_of_Petrochemicals_2020.pdf.

[29] Ellen MacArthur Foundation (2021) *Universal Circularity Indicators: Navigating the Circular Economy Transition.* Available at: https://ellenmacarthurfoundation.org/universal-policy-goals/overview.

[30] Geissdoerfer, M., Pieroni, M.P.P., Pigosso, D.C.A. and Soufani, K. (2020) 'Circular business models: A review', *Journal of Cleaner Production*, 277, p. 123741. Available at: www.sciencedirect.com/science/article/pii/S0959652620337025.

[31] Ashton, W.S., Chertow, M.R. and Althaf, S. (2022) 'Industrial symbiosis: Novel supply networks for the circular economy', in: Bals, L., Tate, W.L. and Ellram, L.M. (eds.) *Circular Economy Supply Chains: From Chains to Systems*. Bingley: Emerald Publishing Limited, pp. 29–48.

[32] Kirchherr, J., et al. (2018) 'Barriers to the circular economy: Evidence from the European Union (EU)', *Ecological Economics*, 150, pp. 264–272. Available at: www.sciencedirect.com/science/article/pii/S0921800918304494.

[33] Accenture (2021) *Achieving a Circular Economy for Chemicals*. Available at: www.accenture.com/us-en/insights/chemicals/circularity-in-chemicals.

[34] Index Mundi (2022) *Commodity Data*. Available at: www.indexmundi.com/commodities/.

[35] European Commission (2020) *Study on the EU's List of Critical Raw Materials*. Available at: https://op.europa.eu/en/publication-detail/-/publication/ff34ea21-ee55-11ea-991b-01aa75ed71a1/language-en.

[36] Thompson R. C., Moore C. J., Vom Saal F. S., Swan S. H. Plastics, the environment and human health: current consensus and future trends. Philos Trans R Soc Lond B Biol Sci. 2009 Jul 27;364(1526):2153–66. doi: 10.1098/rstb.2009.0053. PMID: 19528062; PMCID: PMC2873021. doi: https://doi.org/10.1098%2Frstb.2009.0053

[37] Organisation for Economic Cooperation and Development (2018) *Improving Resource Efficiency and the Circularity of Economies for a Greener World*. Available at: www.oecd.org/publications/improving-resource-efficiency-and-the-circularity-of-economies-for-a-greener-world-1b38a38f-en.htm.

[38] European Chemical Industry Council (2019) *Landscape of the European Chemical Industry 2018*. Available at: https://cefic.org/a-pillar-of-the-european-economy/landscape-of-the-european-chemical-industry/.

[39] Cusack, C. and Mihelcic, J.R. (2022) *Engineering, Design, and Technology for Circular Economy*. New York, NY: Elsevier.

[40] Fabian Takacs, Dunia Brunner, Karolin Frankenberger, Barriers to a circular economy in small- and medium-sized enterprises and their integration in a sustainable strategic management framework, *Journal of Cleaner Production*, Volume 362, 2022, 132227, ISSN 0959-6526, https://doi.org/10.1016/j.jclepro.2022.132227.

[41] Nasr, N., Konash, A. and Müller, J. (2022) 'Reusable production materials in circular economy', *Resources, Conservation & Recycling*, 179, p. 106178.

[42] https://sustainability-innovation.asu.edu/kaiteki/research/circular-economy-in-the-chemical-industry/

[43] Kalundborg Symbiosis (2022) *Sustainable Production in Symbiosis*. Available at: www.symbiosis.dk/en/.

[44] Ragaert, K., Delva, L. and Van Geem, K. (2017) 'Mechanical and chemical recycling of solid plastic waste', *Waste Management*, 69, pp. 24–58.

[45] U.S. National Academy of Sciences (2022) *Accelerating Tech Discovery & Deployment for a Circular Economy*. Available at: www.nist.gov/circular-economy.

[46] Worrell, E. and Reuter, M.A. (eds.) (2014) 'Recycling: A key factor for resource efficiency', in: Worrell, E. and Reuter, M.A. (eds.) *Handbook of Recycling*. Boston, MA: Elsevier, pp. 3–8.

[47] Carmona, M., et al. (2022) 'Achieving a circular economy of PET derived from renewable energy and biomass', *Nature Reviews Materials*, 7(2), pp. 117–137.

[48] Luo, G., Jahan, M.S., Wang, Z., Lyu, Z., Yang, T. and Nasr, Y. (2019) 'State of the art of waste-to-energy technologies towards enhanced energy recovery and circular economy', *Renewable and Sustainable Energy Reviews*, 101, pp. 334–346.

[49] International Renewable Energy Agency (2022) *Renewable Energy in Industry: Part 2- Pathways*. Available at: www.irena.org/-/media/Files/IRENA/Agency/Publication/2020/Dec/IRENA_Renewable_energy_in_industry_2020_part_2.pdf.

[50] World Economic Forum (2020) *The Future of Petrochemicals: Growth Surges Ahead*. https://www.iea.org/reports/the-future-of-petrochemicals

[51] Kirchherr, J., et al. (2018) 'Barriers to the circular economy: Evidence from the European Union (EU)', *Ecological Economics*, 150, pp. 264–272.

[52] Nasr, N. and Konash, A. (2022) 'The circular economy and resource use reduction: A case study of long-term resource efficiency measures in a medium manufacturing company', *Cleaner Production Letters*, 3, p. 100025.

[53] Mihelcic, J.R., Crittenden, J.C., Small, M.J., Shonnard, D.R., Hokanson, D.R., Zhang, Q. and Schnoor, J.L. (2003) 'Sustainability science and engineering: The emergence of a new metadiscipline', *Environmental Science & Technology*, 37(23), 5314–5324.

[54] Ellen MacArthur Foundation (2013) *Towards the Circular Economy: Opportunities for Business*. Available at: https://archive.ellenmacarthurfoundation.org/publications/towards-the-circular-economy-vol-1-an-economic-and-business-rationale-for-an-accelerated-transition.

[55] https://www.plantengineering.com/articles/the-circular-economys-impact-on-the-oil-and-gas-industry/

[56] Kirchherr, J., Reike, D. and Hekkert, M. (2017) 'Conceptualizing the circular economy: An analysis of 114 definitions', *Resources, Conservation & Recycling*, 127, pp. 221–232.

[57] General Insights on Circular Economy and Its Application in Industries, Ellen MacArthur Foundation. (2013) *Towards the Circular Economy: Economic and Business Rationale for an Accelerated Transition*. https://www.ellenmacarthurfoundation.org/towards-a-circular-economy-business-rationale-for-an-accelerated-transition

[58] Atanda, L., et al. (2022) 'Innovative Strategies for Effective Resource Recovery from Industrial Wastes', *ACS Sustainable Chemistry & Engineering*. https://doi.org/10.1021/acssuschemeng.1c06187

[59] Lobo, A., et al. (2022) 'Barriers to transitioning towards smart circular economy: A systematic literature review', in: Dey, N., et al. (eds.) *IoT and Smart Cities*. Singapore: Springer.

[60] Kazancoglu, I., et al. (2020) 'A conceptual framework for barriers of circular supply chains for sustainability in the textile industry', *Sustainable Development*, 28(5), pp. 1477–1492.

[61] Geissdoerfer, M., Savaget, P., Bocken, N.M. and Hultink, E.J. (2017) 'The circular economy—A new sustainability paradigm?', *Journal of Cleaner Production*, 143, pp. 757–768.

[62] Kirchherr, J., Reike, D. and Hekkert, M. (2018) 'Barriers to the circular economy: Evidence from the European Union (EU)', *Ecological Economics*, 150, pp. 264–272.

[63] Accenture (2021) *Achieving a Circular Economy for Chemicals*. Available at: www.accenture.com/_acnmedia/PDF-171/Accenture-Chemicals-Circular-Economy-POV.pdf.

[64] Boston Consulting Group (2018) *Rethinking Recycling in the Circular Economy*. Available at: www.bcg.com/en-us/publications/2018/rethinking-recycling-circular-economy.

[65] Ellen MacArthur Foundation and McKinsey (2022) *Circular Economy in the Chemical Industry: Mobilizing Value Through Data and Digitalization*. Available at: https://emf.thirdlight.com/link/ojx13zj7t1yd-n15zs/@/preview/1.

[66] Geissdoerfer, M., Savaget, P., Bocken, N.M.P. and Hultink, E.J. (2017) 'The circular economy—a new sustainability paradigm?', *Journal of Cleaner Production*, 143, pp. 757–768.

[67] Hartley, K., Van Santen, R. and Kirchherr, J. (2020) 'Policies for transitioning towards a circular economy: Expectations from the European Union (EU)', *Resources, Conservation and Recycling*, 155, p. 104634.

[68] Iles, A. and Martin, A.N. (2013) 'Expanding bioplastics production: Sustainable business innovation in the chemical industry', *Journal of Cleaner Production*, 45, pp. 38–49.

[69] Lacy, P. and Rutqvist, J. (2016) *Waste to Wealth: The Circular Economy Advantage*. London: Palgrave Macmillan.

[70] Luo, G., et al. (2019) 'State of the art of waste-to-energy technologies towards enhanced energy recovery and circular economy', *Renewable and Sustainable Energy Reviews*, 101, pp. 334–346.

[71] Patwa, N., et al. (2021) 'Towards a circular economy: An emerging economies context', *Journal of Business Ethics*, 168(1), pp. 71–91.

[72] Ritzén, S. and Sandström, G.Ö. (2017) 'Barriers to the circular economy—integration of perspectives and domains', *Procedia CIRP*, 64, pp. 7–12.

[73] Ragaert, K., Delva, L. and Van Geem, K. (2017) 'Mechanical and chemical recycling of solid plastic waste', *Waste Management*, 69, pp. 24–58.

[74] Rainey, T.J., et al. (2022) 'A review of water recovery technologies for reusing wastewater in the circular economy', *Journal of Environmental Management*, 304, p. 114080.

[75] Ritzén, S. and Sandström, G.Ö. (2017) 'Barriers to the circular economy—integration of perspectives and domains', *Procedia CIRP*, 64, pp. 7–12.

[76] Schneiderman, D.K. and Hillmyer, M.A. (2021) '50th-anniversary perspective: There is a great future in sustainable polymers', *Macromolecules*, 54(13), pp. 5295–5310.

[77] Velenturf, A.P.M. and Purnell, P. (2021) 'Principles for a sustainable circular economy: Moving from theory to practice', *Sustainability Science*, 16(6), pp. 1695–1708.

[78] Wang, M., et al. (2011) 'Post-combustion CO_2 capture with chemical absorption: A state-of-the-art review', *Chemical Engineering Research and Design*, 89(9), pp. 1609–1624.

[79] World Economic Forum (2020) *The Future of Petrochemicals: Growth Surges Ahead*. Available at: www3.weforum.org/docs/WEF_Future_of_Petrochemicals_2020.pdf (accessed: 26 February 2023)

[80] Worrell, W.A. and Vesilind, P.A. (2001) *Solid Waste Engineering*. BROOKS/COLE, Pacific Grove, CA, 2002 ISBN: 9780534378141, 0534378145

[81] Khaligh, A. and Onar, O.C. (2018) *Power Electronics Handbook* (Fourth Edition), pp. 725–765. Elsevier, https://www.sciencedirect.com/science/article/abs/pii/B9780128114070000258, ISBN 978-0-12-811407-0.

[82] Chen, T.L., Kim, H., Pan, S.Y., Tseng, P.C., Lin, Y.P. and Chiang, P.C. (2020) 'Implementation of green chemistry principles in circular economy system towards sustainable development goals: Challenges and perspectives', *Science of the Total Environment*, 716, 136998.

15 The Future Perspectives of Global Circular Economy Transition and Comprehensive Cost Analysis

Surinder Singh*, Himanshi Bansal, Surendra Kumar Sharma, Suresh Sundaramurthy, and Sushil Kumar Kansal

15.1 CIRCULAR ECONOMY INTRODUCTION

The circular economy (CE) paradigm is the concurrent accounting of economic and environmental factors aimed at extending the useful life span of goods, feedstocks and materials in the economy, recycling resources, and reducing waste production. Its goal is to replace the outdated linear economic model (taking material, goods manufacture, utilization and disposition) with a circular loop framework that alters the conventional product life span by focusing on the utility of industrial feedstocks and materials (e.g., metals, water, energy) and generating a closed loop for economically improved production and waste utilization. The industrial wastes and its management or optimized re-channelling and recycling is a major challenge to sustainable economy. The circular economy encourages moving away from the linear economy model and marching towards increasing resource potential and waste valorization. It is thought to be a critical paradigm shift in the global industrial system as it focuses on optimum use of waste to useful quality resources, instead of simple recycling of materials which can be sometimes misleading in waste management policy [1]. Researchers, managers, industry personnel and policymakers have all been influenced by circular economy paradigm [2–4]. CE is acknowledged as the most critical and valuable framework to reconcile economic growth, environmental protection, resource recovery and industrial development. Although the term "circular economy" appears to be novel, it is actually just a return to the conventional organizational structure of the economy and the way that nature is (already) self-organized [5]. Despite the fact that available literature in CE area has been continuously growing, still it is limited in the manner; how CE analyses economic and competitive opportunities

* ssbdcet@gmail.com

DOI: 10.1201/9781003269779-15

in the face of various industrial management regimes. Despite its enormous scope, circular economy (CE) is sometimes criticized for not being sustainable from an economic, ecological and social context, or simultaneously not in these three domains [6]. Therefore, it is crucial for a circular economy transition in industries around the globe for initiating sustainable economy and ecosystem conservation, that environmentally benign and productive strategies having less focus on natural resources be developed, and that the resources are optimally recovered from waste streams and reprocessed. As a result, the CE strategy has grown in popularity in the industrial, professional, education and policy spheres all over the globe when it comes to finding solutions for sustainability issues through variable circular economy tools and tactics.

No other economic framework related to sustainability has garnered as much attention till CE has gained in the last decade itself. In addition to its catchy name, the success of CE can be credited to the systems integrated approach, technological, sustainable and product-level industrial production and consumption with resource recycling. Additionally, CE implementation can result in significant social, technological, institutional, and economic change that touches on a number of sustainability transitional elements, including business growth models, innovative and resource recovering tactics, institutional wisdom and preparedness aspects, governance principles and sustainability tools. However there are certain issues in a smooth and worldwide transition from linear to circular economy including lack of quantitative indicators and value retention mechanisms and specific industry-oriented monitoring frameworks [7]. Also, there aren't many real-world instances of groundbreaking CE solutions that actually transform how things are done [1,8].

Mass-based indicators such as recycling rates are not true indicators of circular economy or sustainability; instead what is required are the ecological impact–based indicators that can assess the value of positive environmental impact by applying circular economy principles. Setting targets and monitoring environmental effects and material usage are essential for CE to be implemented successfully [8]. Three categories of CE indicators (i.e., resources efficacy metric, feedstocks and material flows metric, and the product-centric metric) were put forth by Parchomenko et al. (2019) [8]. Only a few CE metrics evaluated value maintenance, value change, and long life span among all of these categories. Commonly analysed CE elements were efficient disposal of waste, optimum resource usage, resources re-looping efficiency and trash management. Financial and economic instability affects both individual businesses and industrial economies more frequently as a result of economic issues such supply-demand shrinkage, poor ownership system, deregulated market scenario and inappropriate incentive frameworks [9]. Many researchers have discussed their own research frameworks in order to quantify circular economy approach and underlying principles, CE indicators and documented analysis for specific industries/sectors with respect to circular economy transition. For example, Munaro et al., 2023 have reported elaborative study on construction sector from circular economy perspective (i.e. resource reuse, waste generation and GHG emissions) [10]. They have discussed various definitions of CE and also elaborated the barriers and drivers for circular economy transition. They concluded that lack of CE awareness and communication is the major bottleneck in implementation of circular economy in construction sector. Sayyedi et al., 2023 have investigated about the implications of global plastic pollution in marine ecosystems in their study. They have used artificial intelligence (AI)-based and waste reduction models (WARM) to relate energy recovery from ocean plastic waste [11]. Rodrigues et al., 2022 have proposed a CMCC model for CE transition by involving industry experts' opinions and enumerating the cost of indicators using the model, so as to justify the measures adopted [12]. Ahmed et al., 2022 have suggested a CE framework which enable holistic assessment using both qualitative and quantitative indicators using a systematic selection methodology for CE indicators covering micro, meso and macro levels of CE approach [13]. Devi et al., 2022 have investigated techno-economic analysis approach and suggested circular bio-economy levels in their study on biomethane production using CE approach [14].

Andooz et al., 2023 have described the CE approach in their work related to the pyrolysis process and have discussed its socio-economic and ecological aspects and suggested future strategies on pyrolysis [15]. Aarikka-Stenroos et al., 2023 have reported the different drivers, indicators and barriers in nutrient recycling in their research using CE framework. The CE ecosystem actors were carefully identified and boundaries were specified in their study and drivers and barriers interrelation and interaction was discussed; the study was able to present driving variables in many socio-economic situations and circumstances, as well as using quantitative research methodologies in CE paradigm [16]. Similar such CE frameworks and analysis were reported on e-business applications by Fatima et al., 2023 [17]; circular supply chains by Carissimi et al., 2023 [18]; on rural hybrid renewable energy systems by Li et al., 2023 [19]; on drivers and barriers in CE transition by Neves et al., 2022 [20]; for bio-ethanol production in Ghana using CE approach by Tulashie et al., 2023 [21] and EMAS environmental framework for CE transition by Dorado et al., 2022 [22].

15.1.1 PRINCIPLES OF CIRCULAR ECONOMY

CE happens to be sustainable production and waste optimization framework having co-opted school of thought (concepts) from fields such as industrial symbiosis, life cycle assessment, product-service metrics, cradle-to-cradle framework, industrial ecology and bio-mimicry [23]. There are nine "R" principles or strategies that are critical to the CE transition. Starting with the "product stage," foremost resource usage can be made efficient ("Reduce") and the product is not discarded after intended usage but put into the CE loop as "Reuse," "Recycle," "Remanufacture," and "Repurpose." Interestingly, it is possible to re-extend the product's life through "Repair" and "Refurbish" mechanisms. Further, the production process must be improved or redesigned to make it sustainable using "Rethink" and "Recover" terminology [24–26]. The 10th "R" that can be added to circular framework is "Responsible," both in using goods and production of materials and services. Uvarova et al., 2023 have provided a roadmap consisting of 60 "R" CE principles divided into "reduce, reuse, recycle and reverse" strategies and generated a framework to access CE and structure tasks for CE implementation [27]. They studied extensive literature (about 148 articles), formulated 60 "R" CE principles and had discussions with industry people and practitioners. Their coding of CE principles and classification of CE principles are shown in Figure 15.1.

RQ1: What are the CE principles that can be adopted by companies and how can these principles be classified?			RQ2: What are the perspectives of the CE principles and how are they related to the strategy and business models of companies?
Collecting & reformulating CE principles	**Adoption of the coding system for CE principles**	**Grouping (classifying) CE principles**	**Perspectives of CE principles, and their alignment with business models and strategies**
• Identification of unique 60 CE principles during the literature review • Upon necessity reformulation of principles with the first letter "R"	• #R1 "reuse" - yellow • #R2 "reduce"- green • #R3 "recycle - blue • #R4 "reverse logistics" – pink • Cross-cutting principles - grey	• 16 principles as #R1 • 29 principles as #R2 • 8 principles as #R3 • 7 principles as #R4 • 3 cross-cutting principles	• The perspectives of CE principles by intervention to the business model and a strategy • The intervention of CE principles with Porter's generic strategies
Contribution from the 1st focus group discussion with experts			Contribution from the 2nd focus group discussion with experts

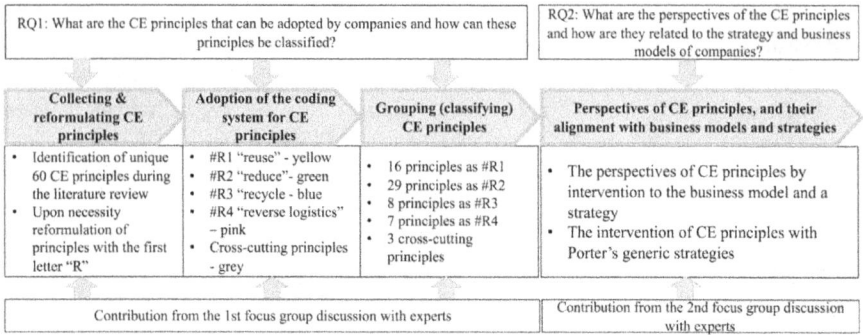

FIGURE 15.1   The process of definition and classification of CE principles [27].

Source: Copyright 2023. Reproduced with permission from Elsevier.

Lehmusto et al., 2022 have presented a mathematical framework in their research for assessing life cycle cost for electrical energy storage using lithium ion batteries (LIB) and application of CE to energy storage in the marine environment [28]. They comprehended that LIB, after completion of its life, still bears some remaining value and capacity that can be employed in other kinds of applications with less stringent performance criteria, providing a sizeable profit margin. The continuous rapid expansion of LIB uses in electric vehicles for land transportation, stationary systems, marine systems and consumer electronics suggested that the industries for battery production as well as material re-circulation and re-use will experience rapid growth in the ensuing decades.

Bressanelli et al., 2021 had devised a framework in their study for CE indicators for the electric and electronics domain as shown in Figure 15.2 [23].

The industry 4.0 framework is a key component of the industrial revolution, and over the past few years, it has helped to create new, inventive and more restorative business models. The CE is developing into a well-known economic model that offers potential advantages to business, society, and the environment. Fatima et al., 2023 have described the CE e-business models (4.0 and 5.0) in their research work as shown in Figure 15.3 [17].

15.2   DRIVERS OF CIRCULAR ECONOMY TRANSITION

Inefficient processes for manufacturing goods and uncontrolled consumption patterns have led to the depletion of resources and an increase in environmental catastrophes; these issues have grown more troubling in the current time for various countries. The introduction of "sustainable development goals" and "climate action plans" are only two of the targeted strategic initiatives that nations have taken to address these problems [17]. These tactics recognize that ecological, societal and government-based (ESG) measurements and indicators are becoming more important in promoting a firm's growth. ESG also aids stakeholders in comprehending the organization's commitment to managing risks and opportunities associated with current sustainable and environmental challenges. Much effort has been displayed by

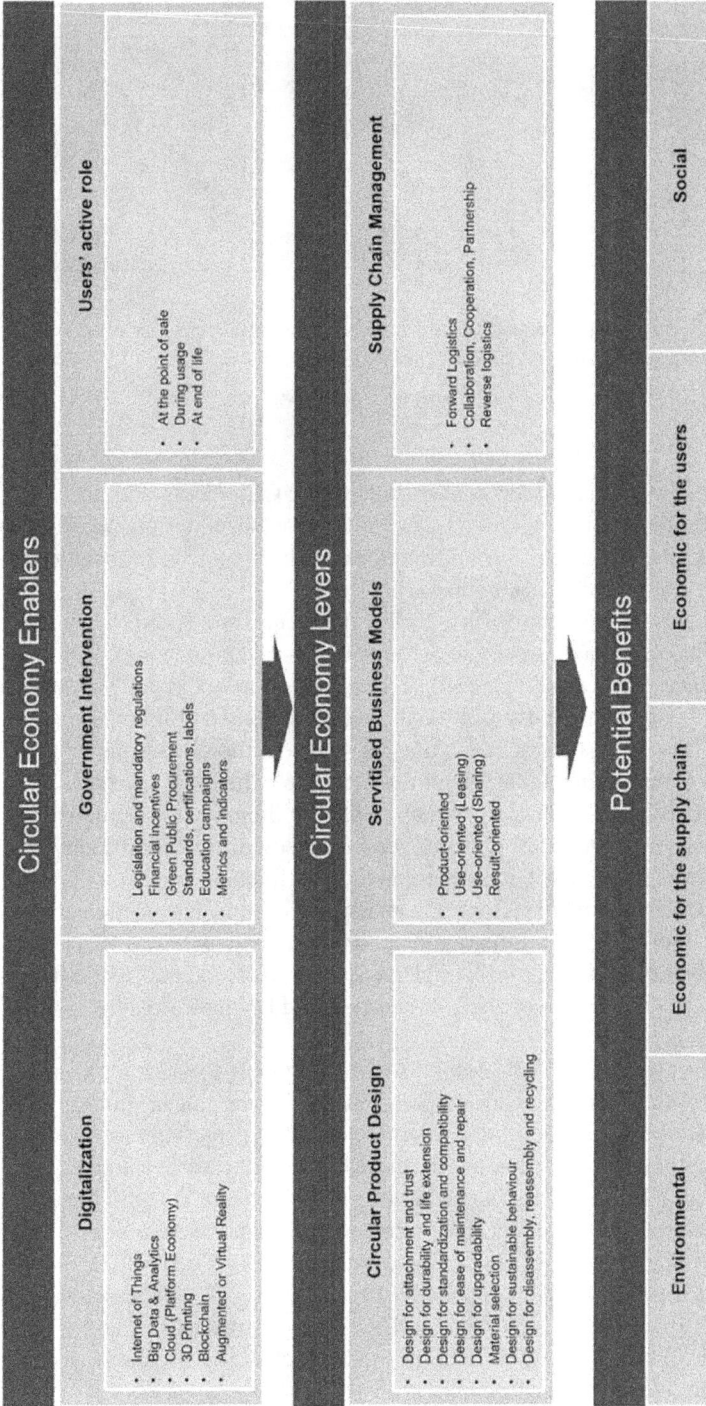

FIGURE 15.2 Research framework for electrical and electronics industry [23].

Source: Copyright 2021. Reproduced with permission from Elsevier.

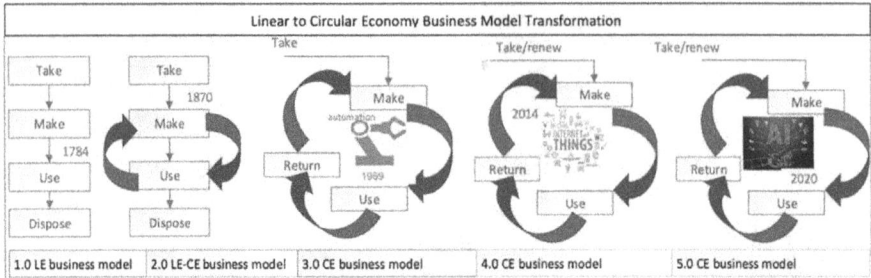

FIGURE 15.3 Transforming a traditional linear economy into a circular economy e-business models (4.0 and 5.0 circular economy business models) [17].

Source: Copyright 2023. Reproduced with permission from Elsevier.

the industry personnel and policymakers along with CE practitioners for transition towards resource optimization–based CE, although no specific framework exists as such. There exist a plethora of varied frameworks which emphasize specific aspects and which are not related to each other [7].

As CE is coherent to sustainability, certain policy measures need to be integrated to social, technological and economic drivers of industrial production. As per factual data, it is predicted that worldwide feedstock consumption would grow to 190 billion tonnes in 2060, up from the estimated material consumption of 65 billion tonnes in 2010 [29]. When a consumer's needs are not satisfied, these materials will in fact be thrown or dumped in sanitary landfills after the exhaustion of their life span. The CE provides opportunities to save up to $380 billion by converting end-of-life (EOL) materials and goods to reformed, recycled, reused, refurbished and remanufactured products, with predictions that this figure will climb to nearly $639 billion in European Union company sectors. The major barriers in CE transition are knowledge, financial, organizational, regulation and control and technical barriers which need to be overcome to successfully implement CE framework in different industries (e.g. oil and gas) [30]. Barros et al., 2023 has reported key indicators for business, as shown in Figure 15.4 [31].

As per Hartley et al. (2023), despite so much support to CE and efforts being undertaken by practitioners, industry personnel and government, global industrial production primarily remained linear till date [32]. This is partially attributed to the barriers which prevent transition to CE and which need to be overcome, including market scarcity for circularity, cultural barriers, individual and institutional issues, technical constraints and bottlenecks, and regularity issues such as non-support for CE which has made it time-consuming and expensive. The circularity at worldwide level has declined from 9.1% in the year 2018 to 7.2% in 2023 [32] due to enhanced raw material extraction and major emphasis on virgin feedstocks instead of recycled materials. This means only 7.2% materials are recycled globally out of 100 billion tonnes extracted from the earth.

This emphasises the need for shift or transition towards CE in a holistic, sustainably sound, gradually achievable, industrially profitable and regulatory supportive

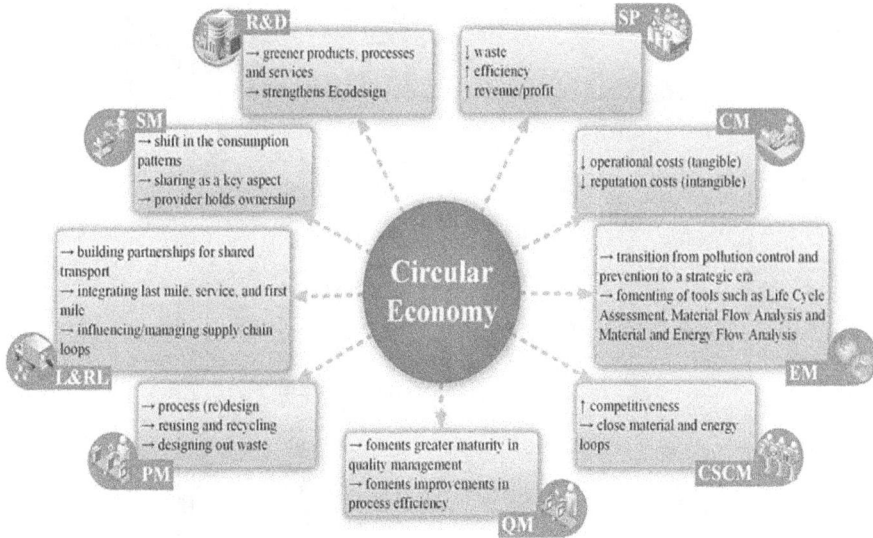

FIGURE 15.4 Key-impact map: contributions of circular economy to sustainable business management [31].

Source: Copyright 2021. Reproduced with permission from Elsevier.

manner. Hence main drivers for CE transition are environmental concerns, global factors (institutional), socio-economic factors, public support, individual and government support and initiatives such as advocacy of UN Sustainable Development Goals 7, 9, and 12 related to CE framework and principles. According to Neves et al. (2022), a country's demographic age distribution can be a good indicator of whether or not it will exhibit or propel CE. Younger citizens are more likely to abandon the so-called linear model of extractive industrial paradigm than older people. It is therefore necessary to implement policies aimed at educating seniors about the advantages and significance of a CE. The tendency to purchase things made from recycled materials declines as per capita income rises as more wealth is also available for expenditure. Policymakers should pay close attention to this finding. Income disparity, in turn, makes the transition to a CE more challenging. Middle-class people are more likely to practice green behaviour, suggesting that they are more conscious of environmental challenges [20]. Baldassarre et al., 2023 in their study have pinpointed and detailed the critical drivers and barriers to plastic waste and its recycling using a review of the literature and in-depth interviews with important parties such as automakers, suppliers, recyclers, specialists and trade groups [33]. They have identified four major barriers and drivers using literature and stakeholder's opinions and exhibited them in a plastic value chain using coloured legends as shown in Figure 15.5.

Ali et al., 2022 in their study have proposed a tool set in the form of behavioural indicators pertinent to the banking sector and ethical business models for enacting a strategic organizational framework. Their study's conclusions showcased that the moderators for the association between environmental aspects and CE transition

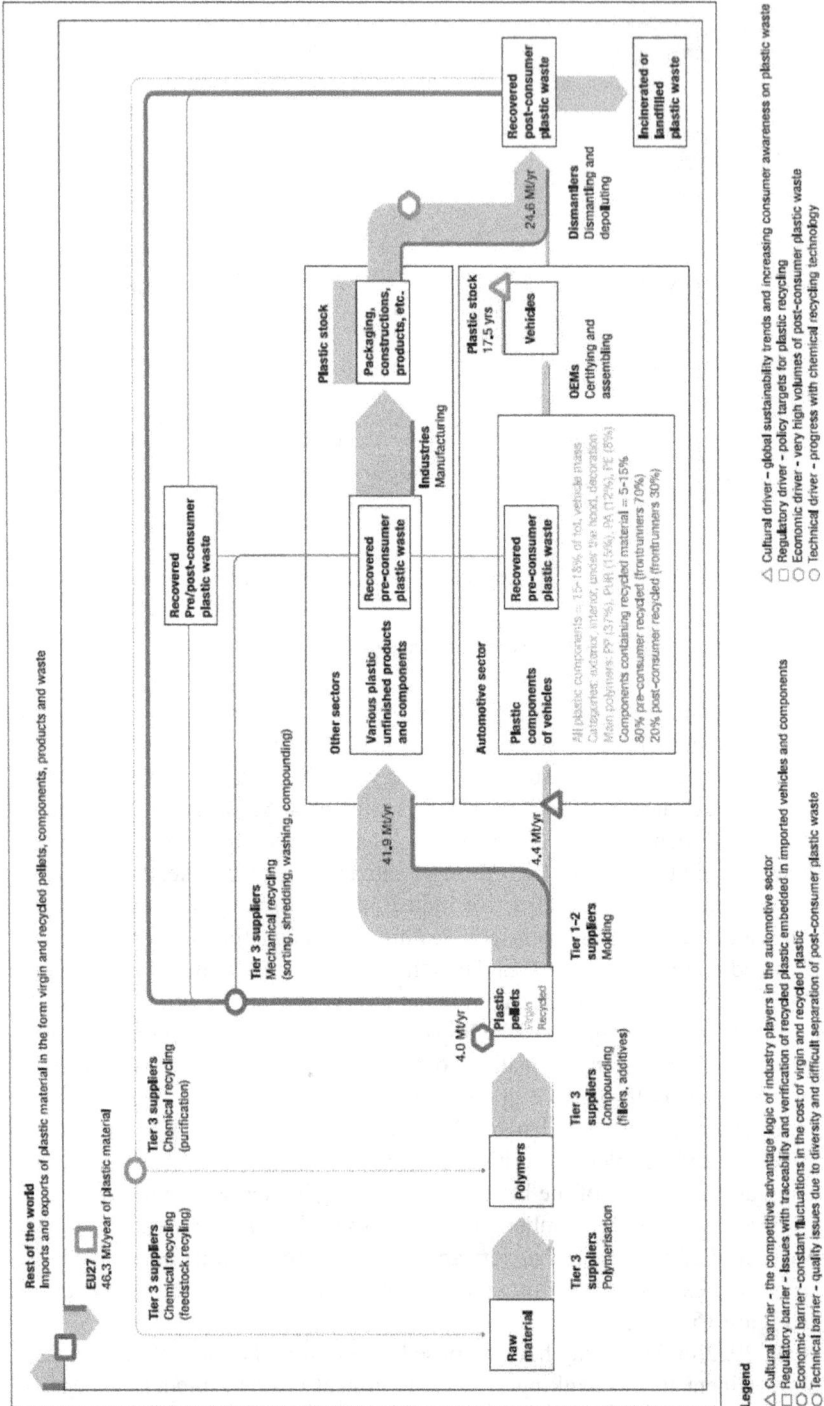

FIGURE 15.5 Value chain of virgin and recycled plastic in the EU automotive sector. Based on literature and stakeholder inputs [33].

Source: Copyright 2022. Reproduced with permission from Elsevier.

include gender, age, and awareness about environmental degradation [34]. Gue et al., 2022 have investigated in their theoretical framework about four scenarios on growing product demand, establishing circular business models, applying government laws and developing service sectors by simulating these factors in an environmental-extended input-output model. Each scenario's GDP growth, material intensity, and regulatory restrictions were assessed for a CE adoption and its propagation [35]. Sharma et al., 2023 have documented barriers in CE transition in the oil and gas industry [30]. Silverio et al., 2022 in their research have examined the barriers, indicators and drivers for CE transition using five clusters based literature mining [36]. They have put forth a framework to assist managers in setting priorities in order to avoid minimal gains connected to circular and sustainable strategies. Their framework will also assist practitioners in evaluating CE from novel approach and designing aggressive strategies without compromising the economic development and sustainability.

15.3 LONG-TERM VALUE MAXIMIZATION USING CIRCULAR ECONOMY

In pursuit to shift to a CE transition, while maximizing long-term value, it is necessary to weigh the financial consequences of circular practices and make long-term strategic decisions that will both improve sustainability and generate economic gains using holistic cost assessment strategies. A guiding philosophy for organizations is long-term value maximization, which emphasizes long-term value (wealth) creation for all stakeholders rather than just short-term financial advantages. It acknowledges that placing a priority on long-term success can result in more robust and stable firms, improved stakeholder relations, and beneficial societal contributions. Companies committed to the CE place a high priority on product design that extends product life cycles and makes it simple to repair, refurbish, and reuse products. This framework promotes the use of modular components that can be modified or replaced while creating goods. Choosing materials that are recyclable, renewable, and have minimal environmental impact is a part of adopting a circular strategy. Through effective material procurement and utilization, organizations aim to reduce the use of virgin resources and waste generation.

In order to maximize long-term value, costs must be thoroughly evaluated using approaches other than traditional accounting. This entails taking into account both the direct and indirect costs of adopting circular business practices, such as those related to product development, sourcing raw materials, manufacturing, distribution and end-of-life management, as well as any potential advantages like longer product life cycles and lower waste disposal costs.

Eco-innovation has been recognized a novel method for boosting productivity and competitiveness while also having favourable effects on the environment and society. To change the prevailing linear economy model and to construct a socio-economic system based on the CE approach using circular cost analysis, Eco-innovation strategy can be employed as a transformative tool [37]. Chishti et al., 2023 in their study have investigated drivers that influence the production of electricity generation and had identified environmental policy, energy transition, CE and geopolitical risk as novice determinants for their econometric models. For its empirical analysis, their study used a variety of cutting-edge techniques (e.g. quantile VAR, quantile slope estimate, and wavelet-based correlation methods). The beneficial effects across quantiles provided the evidence that CE is vital in supporting the global electricity generation process, according to their results [38].

15.4 COMPREHENSIVE COST ANALYSIS IN CIRCULAR ECONOMY

CE is a crucial strategy in the shift towards a more sustainable economic paradigm. Further, CE is regarded as a feasible socio-technical method for achieving economic and ecological sustainability rather than mere a material reuse-recycling concept [39]. In CE, used materials and waste are considered to be fresh raw materials for the economy. It is attributed as a framework suitable for demands of businesses and countries to lower input costs as well as aspirations to function in a more sustainable way [37]. In the context of the CE, comprehensive cost analysis goes beyond the conventional accounting of direct monetary expenses. It includes a wider range of financial and nonfinancial expenses and gains related to the complete life cycle of a good or process. The objective is to record, not only the immediate costs (capital, operational and maintenance) but also the long-term social and environmental effects. In a circular environment, this method offers a more accurate depiction of the real cost and value of goods and processes from ecological and economic viewpoint. In this process it is imperative to consult and take the viewpoint of major stakeholders and CE experts while framing a circular design–based product or service [40].

There is a stark difference in conventional and circular cost analysis. The traditional system focuses on the production, distribution, and disposal expenses that are directly related to wealth. It frequently disregards external factors like social costs or long-term environmental implications. Conventional cost systems tend to promote cost-cutting strategies that might not be viable in the long-term and have a limited opportunity to assess the total value that a product generates. Whereas circular cost analysis adopts a comprehensive approach to benefits and costs across a product's life cycle, includes external factors like resource use, pollution and societal advantages. It also promotes sustainable behaviours that consider the full impact of decisions into account and reveals a true contribution of a product's impact on sustainability. The world's top nations are currently attempting to eliminate the barriers prohibiting the establishment of a CE model and are attempting to make decisions that will strengthen the transition to CE as much as possible at a fast rate [39]. There are many things to take into account when performing an extensive cost analysis within the framework of the CE. This kind of study includes a wider variety of costs and benefits, both tangible and intangible, compared to conventional cost analyses. Some of the important factors include direct and indirect costs, environmental and social costs and benefits, long-term considerations, research and innovation costs, supply-chain measures, external costs, risks assessment and regulatory implications, end-of-life costs and operational efficiency. There are some intangible factors, too, which need to be included such as strengthened corporate social responsibility, enhanced company culture and increased stakeholder trust. The detailed circular cost analysis is shown in Table 15.1. Farrukh et al., 2023 have also emphasized that a green Lean

TABLE 15.1
Cost-Benefit Factors in Comprehensive (Circular) Cost Analysis

S. No	Cost-Benefit Factors	Details of Cost-Benefits Included
1	Direct Monetary Costs	Costs associated with labor, materials, production, distribution, etc.
2	Indirect Monetary Costs	Costs related to regulatory compliance, pollution control, waste management, etc.
3	Environmental Costs-Benefits	1. Costs related to extraction of resources, use of energy, emissions, waste production, etc. 2. Potential environmental advantages of reduced resource use and emissions.
4	Social Costs-Benefits	1. Costs incurred on employee well-being, community involvement, health and safety. 2. Social benefits like increased quality of life, local economic growth and the creation of jobs.
5	Innovation and Research Costs	1. Research and development expenditures for environmentally friendly products and processes. 2. Potential advantages of market differentiation and technical progress/advancement.
6	End of Life Costs and Benefits	1. Expenses related to recycling, reuse or ethical and safe disposal. 2. Benefits from minimal waste going to sanitary landfills and lower environmental impacts.
7	Long-Term Issues	1. Impact of pollution and resource depletion on future generations, costs associated with product disposal. 2. Possibility of increased product longevity and less waste production.
8	External Costs	Incorporation of external costs and benefits that are not usually accounted for in traditional analyses (e.g., cost of hiring experts, Environment Impact Assessment (EIA) costs, data collection and analysis costs, cost due to biodiversity loss, cost of educating/awareness among customers, corporations and policymakers about the wider effects of their decisions).
9	Risk and Sensitivity Analysis	1. Time and money to create various scenarios, perform computations, and assess the outcomes while conducting a sensitivity and risk assessment study. 2. Costs related to analytical tools and software. 3. Although sensitivity analysis improves decision-making, it also adds a level of complexity that could cause decisions to be delayed, particularly if several scenarios need to be considered.
10	Policy and Regulatory Compliance Costs	Expenses related to changing corporate procedures to comply with new circular economy requirements. Costs related to adopt environmental laws and regulations, as well as any internationalization-oriented initiatives.

Six Sigma approach is also a circular manufacturing framework and by consultation with senior industry mangers one can formulate the correct CE strategy for the particular industry or process (e.g. flexible packaging industry) [41]. This process industry was chosen in particular because it deals with a number of barriers related to the CE, such as the depletion of natural resources, lack of recycling, production of solid waste, air pollution, energy footprints and marine contamination.

As each process/industry/unit could have different factors to take into account, it is crucial to customize the CE analysis for a particular project, thing or activity which needs to be undertaken. The objective is to present a thorough analysis of the advantages and disadvantages of CE methods, taking into account monetary, environmental, social and ethical considerations.

15.5   COST-BENEFIT ANALYSIS OF CIRCULAR ECONOMY TRANSITION

The cost-benefit analysis of a shift to a CE entails evaluating the short- and long-term effects, tangible and non-tangible effects (e.g. economic, environmental, social, and corporate environment) in adopting circular practices as opposed to conventional linear economy ones. The factors which need to be considered in a circular or comprehensive cost analysis are enumerated in Table 15.1.

Hence cost benefit analysis in CE evaluates the direct costs, indirect costs, environmental advantages and the social benefits like jobs to the youth. The strategy emphasizes trade-offs, future sustainability and provides advice for making wise decisions. Sensitivity analysis takes into account how important factors and their impacts influence the industry production and the environment.

15.6   FUTURE TRENDS AND PROSPECTS

The forthcoming trends in CE ideas include technology progress, resource optimization, digitalization, technological application of the internet of things (IOT), industrial internet of things (IIOT), AI and data analytics in CE, sharing economy models and bio-mimicry. CE transition is affected by changes in corporate structures, political support, market availability for recycled products, resource (waste) valorisation, individual (e.g. start-ups) and public initiatives (plastic recycling and segregation), and the effect of consumer behaviour on circular practices. Along with difficulties like technology adoption and unexpected repercussions, cooperation, economic resilience, and global adoption are underlined.

Discussed in the following sections are some examples of CE applications in underdeveloped countries by local people with creativeness in mind [43].

15.6.1   ColdHubs: Solar-Powered, Cooling-as-a-Service Solution

In Nigeria, food waste is a thing of the past, thanks to a post-harvest, solar-powered ColdHubs technology, which also reduces emissions. Nnaemeka Ikegwuonu, a businesswoman, founded ColdHubs, about one square-meter storage facilities that can keep food fresh for up to 21 days. It is estimated that 42,000 tonnes of food/

food products, equivalent to more than 1 million kilograms of CO_2, were prevented from going to waste in 2020 attributed to ColdHubs' 54 units installed in Nigeria by Nnaemeka.

15.6.2 BANGLADESH'S AWARD-WINNING HOSPITAL

A remote hospital in Bangladesh that has been dubbed "The World's Best New Building" is making headlines. The Shyamnagar village-based Friendship Hospital was constructed with equal access to healthcare for underprivileged communities in mind utilizing regional, sustainable materials and local craftsmen. The hospital's grounds are divided into inpatient and outpatient areas by a canal that was purposely built by the architects to blend with the nearby riverine landscape while providing natural cooling. Additionally, rainwater collected on-site is stored in two water tanks at either end of the canal [40].

15.6.3 BRAZILIAN COSMETICS BRAND NATURA FLIPS
THE SCRIPT ON DEFORESTATION

Brazil has witnessed 33.12% of the deforestation in the tropics, primarily as a result of industrial cattle raising. Natura, a cosmetics company, has successfully changed this narrative by using natural ingredients extracted from the Amazon rainforest to make its cosmetic products. The "standing forest" theory, which contends that a tree has considerably more value standing up than being cut down, is the foundation of Natura's unconventional economic model. This way of thinking has helped Natura maintain roughly 2 million hectares of rainforest while also distributing the company's earnings to nearby people.

15.6.4 NETHERLANDS BECAME THE GLOBAL BICYCLE CAPITAL

Massive demonstrations over rising traffic fatalities and the destruction of "historic neighbourhoods" to make way for highways took place in the Netherlands during the 1970s. This combined with the continuous energy crisis created a perfect storm, forcing the Netherlands government to give safe and clean mobility as its first priority. Cycling became the new standard in Netherlands when 20,000 kilometers of bike lanes were built. With 17 million citizens, the Netherlands already has 23 million bicycles and more than 25% of all trips are performed by bicycles.

The preceding examples justify the idea for CE transition by community and individuals in our neighbourhood, cities and countries with an aim to make our surroundings and planet clean and sustainable with ample natural resources and a better place to live. The transition to CE involves creating more fairness and social cohesion and to place the economy at the service of human needs rather than humans at the service of economic growth, hence the CE aspires to restructure the economic system and environmental restoration.

In summary it can be concluded that CE transition at present is evolving by leaps and bounds. There is a great scope and importance of CE transition in the coming decades. At present, the world needs the utmost full resource/feedstocks optimization, waste minimization, channeling the product and goods manufacturing through the circular loops, sensitize environmental concern and ensure sustainable production and development for its present and future generations. The key points that will affect the CE transition in future are:

1. **Innovation and Technology:** The way resources are utilized, recycled, and repurposed will change as a result of advancements in technologies like AI, IOT, IIOT, machine learning and biotechnology.
2. **Data Analytics and Digital Platforms:** These will optimize material flows, enabling more effective circular systems as a result of the digital transformation.
3. **New Business Models:** Product-as-a-service, sharing platforms and circular supply chains will promote economic development while cutting down on waste generation.
4. **Government Support:** Throughout the world governments will implement regulations that support circular economies, increase producer accountability and reward sustainability. Regulatory provisions will strengthen the aim to have global CE transition.
5. **Consumer Influence:** Consumers who are informed and concerned about the environment will advocate for sustainable products and promote circular economies.
6. **Cross-industry Cooperation:** Businesses will work together to create circular supply chains that will cut waste and improve resource sharing.
7. **Sustainable Materials:** Biomimicry and biodegradable materials will produce goods and packaging that are more environmentally friendly.
8. **Circular Mindset:** A move from linear to circular thinking will become crucial to company strategy and decision-making and will advocate environmental safety and sustainability.
9. **Economic Resilience:** By minimizing reliance on scarce resources, circular business practices will improve resource security.
10. **Challenges and Solutions:** Through constant innovation and group efforts, barriers to CE adoption, behavior change and unintended consequences will be addressed.
11. **Global Adoption:** CE ideas will spread, paving the way for a future that is more robust and sustainable.
12. **Potential for Transformation:** The CE has the potential to alter industries, economies and lifestyles while promoting sustainable growth.

REFERENCES

[1] Yang, C. K., Ma, H. W., Liu, K. H., et al. Measuring circular economy transition potential for industrial wastes. *Sustainable Production and Consumption*, *40*, 376–388. 2023
[2] Caferra, R., Tsironis, G., Morone, A., et al. Is the circular economy proposed as sustainability in firm mission statements? A semantic analysis. *Environmental Technology & Innovation*, *32*, 103304. 2023

[3] Kirchherr, J., Yang, N. H. N., Schulze-Spüntrup, F., et al. Conceptualizing the circular economy (revisited): An analysis of 221 definitions. *Resources, Conservation and Recycling, 194*, 107001. 2023

[4] Ghosh, A., Bhattacharjee, D., Bhola, P., et al. Exploring the practicality of circular economy through its associates: A case analysis-based approach. *Journal of Cleaner Production*, 138457. 2023

[5] Arauzo-Carod, J. M., Kostakis, I., Tsagarakis, K. P. Policies for supporting the regional circular economy and sustainability. *The Annals of Regional Science, 68*(2), 255–262. 2022

[6] Kulakovskaya, A., Wiprächtiger, M., Knoeri, C., et al. Integrated environmental-economic circular economy assessment: Application to the case of expanded polystyrene. *Resources, Conservation and Recycling, 197*, 107069. 2023

[7] Parchomenko, A., Nelen, D., Gillabel, J., Rechberger, H. Measuring the circular economy—A multiple correspondence analysis of 63 metrics. *Journal of Cleaner Production, 210*, 200–216. 2019

[8] Haupt, M., Hellweg, S. Measuring the environmental sustainability of a circular economy. *Environmental and Sustainability Indicators, 1*, 100005. 2019

[9] Geissdoerfer, M., Savaget, P., Bocken, N. M. The circular economy—A new sustainability paradigm? *Journal of Cleaner Production, 143*, 757–768. 2017

[10] Munaro, M. R., Tavares, S. F. A review on barriers, drivers, and stakeholders towards the circular economy: The construction sector perspective. *Cleaner and Responsible Consumption*, 100107. 2023

[11] Kowsari, E., Ramakrishna, S., Gheibi, M. Marine plastics, circular economy, and artificial intelligence: A comprehensive review of challenges, solutions, and policies. *Journal of Environmental Management, 345*, 118591. 2023

[12] Rodríguez, R. M., Labella, Á., Nuñez-Cacho, P. A comprehensive minimum cost consensus model for large scale group decision making for circular economy measurement. *Technological Forecasting and Social Change, 175*. 2022

[13] Ahmed, A. A., Nazzal, M. A., Darras, B. M. A comprehensive multi-level circular economy assessment framework. *Sustainable Production and Consumption, 32*, 700–717. 2022

[14] Devi, M. K., Manikandan, S., Kumar, P. S. A comprehensive review on current trends and development of biomethane production from food waste: Circular economy and techno economic analysis. *Fuel, 351*, 128963. 2023

[15] Andooz, A., Eqbalpour, M., Kowsari, E., et al. A comprehensive review on pyrolysis from the circular economy point of view and its environmental and social effects. *Journal of Cleaner Production*, 136021. 2023

[16] Aarikka-Stenroos, L., Kokko, M., Pohls, E. L. Catalyzing the circular economy of critical resources in a national system: Case study on drivers, barriers, and actors in nutrient recycling. *Journal of Cleaner Production, 397*, 136380. 2023

[17] Fatimah, Y. A., Kannan, D., Govindan, K., et al. Circular economy e-business model portfolio development for e-business applications: Impacts on ESG and sustainability performance. *Journal of Cleaner Production*, 137528. 2023

[18] Carissimi, M. C., Creazza, A., Pisa, M. F., et al. Circular economy practices enabling circular supply chains: An empirical analysis of 100 SMEs in Italy. *Resources, Conservation and Recycling, 198*, 107126. 2023

[19] Li, S., Zhang, L., Liu, X., et al. Collaborative operation optimization and benefit-sharing strategy of rural hybrid renewable energy systems based on a circular economy: A Nash bargaining model. *Energy Conversion and Management, 283*, 116918. 2023

[20] Neves, S. A., Marques, A. C. Drivers and barriers in the transition from a linear economy to a circular economy. *Journal of Cleaner Production, 341*, 130865. 2022

[21] Tulashie, S. K., Dodoo, D., Ketu, E., Adiku, S. G. K., et al. Environmental and socio-economic benefits of a circular economy for bioethanol production in the northern part of Ghana. *Journal of Cleaner Production, 390*, 136131. 2023

[22] Dorado, A. B., Leal, G. G., de Castro Vila, R. EMAS environmental statements as a measuring tool in the transition of industry towards a circular economy. *Journal of Cleaner Production, 369*, 133213. 2022.

[23] Bressanelli, G., Pigosso, D. C., Saccani, N., Perona, M. Enablers, levers and benefits of circular economy in the electrical and electronic equipment supply chain: A literature review. *Journal of Cleaner Production, 298*, 126819. 2021

[24] Morseletto, P. Targets for a circular economy. *Resources, Conservation and Recycling, 153*, 104553. 2020

[25] Courtens, F. M., Haezendonck, E., Dooms, M. Accelerating the circular economy transition process for gateway ports: The case of the Port of Zeebrugge. *Maritime Transport Research, 4*, 100088. 2023

[26] Blomsma, F., Brennan, G. The emergence of circular economy: A new framing around prolonging resource productivity. *Journal of Industrial Ecology, 21*(3), 603–614. 2017

[27] Uvarova, I., Atstaja, D., Volkova, T., et al. The typology of 60R circular economy principles and strategic orientation of their application in business. *Journal of Cleaner Production, 409*, 137189. 2023

[28] Lehmusto, M., Santasalo-Aarnio, A. Mathematical framework for total cost of ownership analysis of marine electrical energy storage inspired by circular economy. *Journal of Power Sources, 528*, 231164. 2022

[29] International Resource Panel. *Global Resources Outlook 2019: Summary for Policymakers, United Nations Environ.* Program, 2019: 1–23. www.resourcepanel.org/reports/global-resources-outlook-2019.

[30] Sharma, M., Joshi, S., Prasad, M., Bartwal, S. Overcoming barriers to circular economy implementation in the oil & gas industry: Environmental and social implications. *Journal of Cleaner Production, 391*, 136133. 2023

[31] Barros, M. V., Salvador, R., do Prado, G. F., et al. Circular economy as a driver to sustainable businesses. *Cleaner Environmental Systems, 2*, 100006. 2021

[32] Hartley, K., Schülzchen, S., Bakker, C. A., Kirchherr, J. A policy framework for the circular economy: Lessons from the EU. *Journal of Cleaner Production, 412*, 137176. 2023

[33] Baldassarre, B., Maury, T., Mathieux, F. Drivers and barriers to the circular economy transition: The case of recycled plastics in the automotive sector in the European Union. *Procedia CIRP, 105*, 37–42. 2022

[34] Ali, Q., Parveen, S., Yaacob, H., et al. Environmental beliefs and the adoption of circular economy among bank managers: Do gender, age and knowledge act as the moderators?. *Journal of Cleaner Production, 361*, 132276. 2022

[35] Gue, I. H. V., Tan, R. R., Chiu, A. S., et al. Environmentally-extended input-output analysis of circular economy scenarios in the Philippines. *Journal of Cleaner Production, 377*, 134360. 2022

[36] Silvério, A. C., Ferreira, J., Fernandes, P. O., et al. How does circular economy work in industry? Strategies, opportunities, and trends in scholarly literature. *Journal of Cleaner Production*, 137312. 2023

[37] De Jesus, A., Mendonça, S. Lost in transition? Drivers and barriers in the eco-innovation road to the circular economy. *Ecological Economics, 145*, 75–89. 2018

[38] Chishti, M. Z., Dogan, E., Zaman, U. Full-length effects of the circular economy, environmental policy, energy transition, and geopolitical risk on sustainable electricity generation. *Utilities Policy, 82*, 101585. 2023

[39] Kuzior, A., Arefiev, S., Poberezhna, Z. Informatization of innovative technologies for ensuring macroeconomic trends in the conditions of a circular economy. *Journal of Open Innovation: Technology, Market, and Complexity, 9*(1), 10–20. 2023

[40] Ho, O. T. K., Gajanayake, A., Iyer-Raniga, U. Transitioning to a state-wide circular economy: Major stakeholder interviews. *Resources, Conservation & Recycling Advances,* 200163. 2023

[41] Farrukh, A., Mathrani, S., Sajjad, A. Green-lean-six sigma practices and supporting factors for transitioning towards circular economy: A natural resource and intellectual capital-based view. *Resources Policy, 84,* 103789. 2023

[42] www.circle-economy.com/blog/9-examples-of-the-circular-economy-in-action. Accessed on 20th August 2023

Index

2,5-furandicarboxylic acid, 111

A

acrylic monomers, 115
activated carbon (AC), 137
aerogel, 277–278
ammonia fiber or freeze explosion, 185
ash and minerals, 106
asymmetric supercapacitors (ASSC), 136

B

beneficiary relation, 31
betalains, 279
bimetallic catalysts, 256
biodegradable plastics and composting, 260
biomass-derived materials, 139
biomass gasification, 258
biomass policies project, 182
biomethane, 199, 349
bionic engineering, 90
business model for circular economy, 310

C

calcium hydroxide, 186
carbon-based material, 139, 152, 163, 290
carbon dioxide explosion pretreatment, 187
carbon footprints, 213, 229, 245, 289
carbon nanotubes (CNTS), 168
carbon quantum dots (CQDS), 166
carotenoids, 279, 283
cellulose, 100
cellulose microfibrils, 99
char (pyrolysis solid), 254
charge carrier separation and transfer, 78
charging-discharging, 136
chemical upcycling, 261
circular economy and decentralized energy
     systems, 18
circular economy in fossil fuel industries, 16
circular economy in various industries, 10
circular economy processes or strategies, 8
circular supply chain management (CSCM), 40
classical packaging, 273
closed-loop recycling, 252
coal-fired electricity, 55
coal gasification, 77, 93
Coldhubs: solar-powered, cooling-as-a-service
     solution, 358
combustion, 188

composting or co-composting techniques, 193
comprehensive cost analysis in circular
     economy, 356
conditions of supply, 210–211
controlled size and shape synthesis, 89
control methods of microgrid structures, 207
conventional gasification, 257
conventional pyrolysis, 254
cooperative change, 37
cradle-to-cradle design, 3
cradle-to-grave, 54, 56

D

decentralization and local resilience, 203
defect engineering, 89
design for recycling, 252
design of PV module for circular economy, 313
dichloromethane, 105
dilute acid pretreatment, 184
dioxins, 244
d-lactic acid, 112–113
doping of metals and non-metals, 87
drivers of circular economy transition, 350
droop control, 207
dual-cell hybrid photocatalytic system, 87
dye-sensitized solar cells (DSSCS), 163
dye-sensitized tandem electrolysis cells, 81

E

eco-innovation, 355
electrical and electronic equipment (EEE), 10
electric double-layer capacitors (EDLC), 290
electrochemical cells, 57
electrode material, 138
emulsions as templates for energy material
     synthesis, 276
emulsion-templated porous materials, 276
energy material efficiency and optimization, 15
energy storage and circular solutions, 16
environmental management (EM), 41
enzymatic recycling, 260
equity between generations, 35
execution is context dependent, 38
extended producer responsibility (EPR), 235
extractives, 105

F

faradaic redox process, 290
feeder connection, 210

365

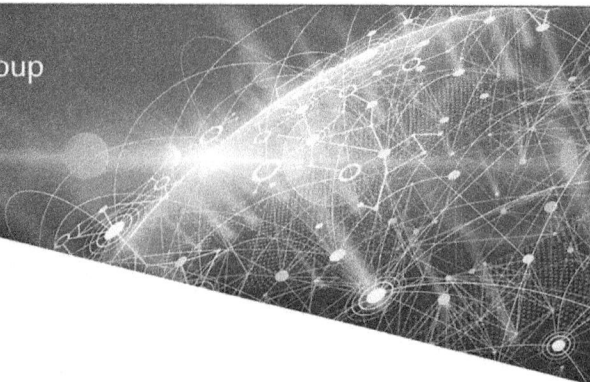

For Product Safety Concerns and Information please contact our EU
representative GPSR@taylorandfrancis.com
Taylor & Francis Verlag GmbH, Kaufingerstraße 24, 80331 München, Germany